Alloy Steels

Special Issue Editor
Robert Tuttle

MDPI • Basel • Beijing • Wuhan • Barcelona • Belgrade

MDPI

Special Issue Editor
Robert Tuttle
Saginaw Valley State University
USA

Editorial Office
MDPI AG
St. Alban-Anlage 66
Basel, Switzerland

This edition is a reprint of the Special Issue published online in the open access journal *Metals* (ISSN 2072-6651) from 2016–2017 (available at: http://www.mdpi.com/journal/metals/special_issues/alloy_steels).

For citation purposes, cite each article independently as indicated on the article page online and as indicated below:

Lastname, F.M.; Lastname, F.M. Article title. *Journal Name* **Year**, *Article number*, page range.

First Editon 2018

Cover photo courtesy of Robert Tuttle.

ISBN 978-3-03842-883-1 (Pbk)
ISBN 978-3-03842-884-8 (PDF)

Table of Contents

About the Special Issue Editor

Robert Tuttle, Professor of Mechanical Engineering, earned his Ph.D. in Metallurgical Engineering from the University of Missouri-Rolla in 2004. He has been a faculty member at Saginaw Valley State University (SVSU) since 2004 where he teaches materials science, manufacturing processes, metalcasting, and electron microscopy. His teaching has been at the undergraduate and graduate levels. Dr. Tuttle has been heavily involved in the American Foundry Society where he is currently serving his third term as the Steel Division Chair. He also advises the Student Chapter of the American Foundry Society at SVSU. Dr. Tuttle has published over forty papers related to metalcasting and steels. His current work has focused on understanding the nucleation of steels during solidification.

Preface to "Alloy Steels"

The subject of alloy steels covers a broad range of steels and end use markets. Alloy steels are used in the automotive, aerospace, industrial machinery, and the energy sectors. When this Special Issue was started, the goal was to provide readers with a collection of papers from around the world that reflect the field. Within this book are twenty-three papers that cover the microstructure, manufacturing, corrosion, and application of alloy steels. These papers cover a broad range of alloys and research focuses that provide the cutting-edge answers needed for today's markets. The review by Dr. Mohammed provides an excellent review about welding austenitic and duplex stainless steels.

As editor, my job is much simpler than those of the authors or reviewers. They are the ones who have struggled in the lab conducting experiments and analyzing the data. Their considerable effort in providing high-quality papers should never be underestimated or unappreciated. Reviewers also deserve appreciation for the time in reading the submitted papers and providing valuable feedback to improve them. The countless staff at MDPI also contributed to this by helping everyone involved with the day-to-day functions required in publishing this issue. I want to thank all of these individuals for helping make this Special Issue the second largest ever published by MDPI.

Robert Tuttle
Special Issue Editor

metals

MDPI

Editorial

Alloy Steels

Robert Tuttle

Department of Mechanical Engineering, Saginaw Valley State University, 209 Pioneer Hall, 7400 Bay Rd, University Center, MI 48710, USA; rtuttle@svsu.edu; Tel.: +1-989-964-4676

Received: 15 January 2018; Accepted: 7 February 2018; Published: 8 February 2018

Since their invention in 1865, alloy steels have found broad application in multiple industries; the automotive, aerospace, heavy equipment, and pipeline industries to name a few. Alloy steels include a tremendous variation in alloying content. They range from the 1–2 wt. % Cr or Ni in some low alloy steels to the 15–18 wt. % Cr content of many stainless steels. The topic of alloy steels contains both the common 4140 and 316 alloys to more exotic alloys such as the Hadfield steels. These steels can form a wide variety of microstructures such as pearlite, bainite, or martensite, which result in an equally broad range of properties. It is this range that has made them useful to so many industries. In some cases, these are the only steel alloys that can provide the required combination of properties. Their use in the automotive industry has been key to the development of safer vehicles and improved fuel efficiency. Our modern world would not be possible without the advanced alloy steels employed to safely transport oil through pipelines. Therefore, continued development is necessary to expand markets, improve products, and enhance the human condition. It is this importance that has lead us at *Metals* to create the special issue on alloy steels that you are reading. What follows are 23 papers from a wide range of authors and nationalities which represents the current state of the art in alloy steel research.

This issue, like alloy steels themselves, covers diverse set of articles. There are articles on manufacturing, microstructure, heat treatment, corrosion, and service conditions. This expansive range reflects the multifaceted nature of alloy steels. Even the individual areas are extensively represented. In manufacturing, there are articles on the effects of welding [1–4], electroslag remelting [5,6], and rolling [7–9]. As is typical in any discussion of steels, many papers focus on microstructure–property relations [10–14]. These form the basis for improving the alloys and their processing. Another large section of work focuses on topics of more interest to those who are the final customers of the steel industry. Many alloy steels are heat treated and understanding the effects of heat treatment and heat treating parameter selection ensure the correct microstructure and properties are attained. Readers will find these topics addressed by several authors in our pages [10,15–18]. Those readers interested in improving the corrosion resistance of alloy steels will find several pieces on this topic [2,19,20]. An aspect often ignored by many journals is performance under actual service conditions in the final product, there are two articles on wear resistance and pipeline life that remind us of the importance of understanding the final product [21–23]. Data from the final products created from these steels always provide a powerful insight which the best labs can never replicate, and their inclusion in this special issue is a significant contribution to this special issue.

Of particular interest to the readers of *Metals* should be the excellent review article by Mohammed et al. that presents a wide review of welded austenitic and duplex stainless steels [3]. Their work covers the effects of heat input on the microstructure, corrosion resistance, and mechanical properties of this diverse class of steels. Mohammaed et al. also review the current state of stress corrosion cracking work. They distill the results of over 140 papers in the field of stainless steels and deliver an accurate view of our current understanding of these alloys. Readers will find this an invaluable asset in building their understanding of these technical issues. The bibliography alone is worth reading this article.

As can be seen from the topics in this special issue of *Metals*, the broad world of alloy steels remains a current area of research and innovation. My goal as editor at the outset of this issue was to include papers covering the large variety of research and industrial work being done to address the challenges facing alloy steels. While it is impossible to cover everything in a single issue, this issue provides the reader with an excellent understanding of the modern problems needing solutions to create the alloy steels of the future. Hopefully, the readers will find it as enlightening as everyone here at *Metals* believes.

Being my first time as an editor, I would like to thank the staff at MDPI and *Metals* for their help. In particular, I appreciate the fine editorial assistants and the managing editor for assisting me in this. They guided me through the process and helped me understand my options with reviews. I hope I correctly balanced the request of our reviewers and the needs of our authors in creating as fair a peer review process as possible. I want to thank all our reviewers for the tireless efforts in examining and commenting on our papers. This effort is always key to a successful journal. All of the authors also deserve recognition for their contributions and tireless work to promote alloy steels. Finally, I would like to thank you the reader since without you there would be no need to write papers or publish regular issues, let alone special issues.

Conflicts of Interest: The author declares no conflict of interest other than his own passion for alloy steels and their development.

References

1. Tutar, M.; Aydin, H.; Bayram, A. Effect of Weld Current on the Microstructure and Mechanical Properties of a Resistance Spot-Welded TWIP Steel Sheet. *Metals* **2017**, *7*, 519. [CrossRef]
2. Li, J.; Liu, X.; Li, G.; Han, P.; Liang, W. Characterization of the Microstructure, Mechanical Properties, and Corrosion Resistance of a Friction-Stir-Welded Joint of Hyper Duplex Stainless Steel. *Metals* **2017**, *7*, 138. [CrossRef]
3. Mohammed, G.; Ishak, M.; Aqida, S.; Abdulhadi, H. Effects of Heat Input on Microstructure, Corrosion and Mechanical Characteristics of Welded Austenitic and Duplex Stainless Steels: A Review. *Metals* **2017**, *7*, 39. [CrossRef]
4. Corpace, F.; Monnier, A.; Grall, J.; Manaud, J.-P.; Lahaye, M.; Poulon-Quintin, A. Resistance Upset Welding of ODS Steel Fuel Claddings—Evaluation of a Process Parameter Range Based on Metallurgical Observations. *Metals* **2017**, *7*, 333. [CrossRef]
5. Liu, Y.; Zhang, Z.; Li, G.; Wang, Q.; Wang, L.; Li, B. Effect of Current on Structure and Macrosegregation in Dual Alloy Ingot Processed by Electroslag Remelting. *Metals* **2017**, *7*, 185. [CrossRef]
6. Liu, Y.; Zhang, Z.; Li, G.; Wang, Q.; Wang, L.; Li, B. The Structural Evolution and Segregation in a Dual Alloy Ingot Processed by Electroslag Remelting. *Metals* **2016**, *6*, 325. [CrossRef]
7. Huang, Y.; Wang, S.; Xiao, Z.; Liu, H. Critical Condition of Dynamic Recrystallization in 35CrMo Steel. *Metals* **2017**, *7*, 161. [CrossRef]
8. Calvillo, N.; Soria, M.; Salinas, A.; Gutiérrez, E.; Reyes, I.; Carrillo, F. Influence of Thickness and Chemical Composition of Hot-Rolled Bands on the Final Microstructure and Magnetic Properties of Non-Oriented Electrical Steel Sheets Subjected to Two Different Decarburizing Atmospheres. *Metals* **2017**, *7*, 229. [CrossRef]
9. Karavaeva, M.; Abramova, M.; Enikeev, N.; Raab, G.; Valiev, R. Superior Strength of Austenitic Steel Produced by Combined Processing, including Equal-Channel Angular Pressing and Rolling. *Metals* **2016**, *6*, 310. [CrossRef]
10. Zhou, M.; Xu, G.; Tian, J.; Hu, H.; Yuan, Q. Bainitic Transformation and Properties of Low Carbon Carbide-Free Bainitic Steels with Cr Addition. *Metals* **2017**, *7*, 263. [CrossRef]
11. Liu, H.; Fu, P.; Liu, H.; Sun, C.; Gao, J.; Li, D. Carbides Evolution and Tensile Property of 4Cr5MoSiV1 Die Steel with Rare Earth Addition. *Metals* **2017**, *7*, 436. [CrossRef]
12. Gong, N.; Wu, H.-B.; Yu, Z.-C.; Niu, G.; Zhang, D. Studying Mechanical Properties and Micro Deformation of Ultrafine-Grained Structures in Austenitic Stainless Steel. *Metals* **2017**, *7*, 188. [CrossRef]
13. Tian, J.; Xu, G.; Zhou, M.; Hu, H.; Wan, X. The Effects of Cr and Al Addition on Transformation and Properties in Low-Carbon Bainitic Steels. *Metals* **2017**, *7*, 40. [CrossRef]

14. Białobrzeska, B.; Konat, Ł.; Jasiński, R. The Influence of Austenite Grain Size on the Mechanical Properties of Low-Alloy Steel with Boron. *Metals* **2017**, *7*, 26. [CrossRef]

15. Luo, Y.; Guo, H.; Sun, X.; Mao, M.; Guo, J. Effects of Austenitizing Conditions on the Microstructure of AISI M42 High-Speed Steel. *Metals* **2017**, *7*, 27. [CrossRef]

16. Zhang, Y.; Li, J.; Shi, C.-B.; Qi, Y.-F.; Zhu, Q.-T. Effect of Heat Treatment on the Microstructure and Mechanical Properties of Nitrogen-Alloyed High-Mn Austenitic Hot Work Die Steel. *Metals* **2017**, *7*, 94. [CrossRef]

17. Ning, A.; Mao, W.; Chen, X.; Guo, H.; Guo, J. Precipitation Behavior of Carbides in H13 Hot Work Die Steel and Its Strengthening during Tempering. *Metals* **2017**, *7*, 70. [CrossRef]

18. Liu, C.; Liu, X.; Yang, S.; Li, J.; Ni, H.; Ye, F. The Effect of Niobium on the Changing Behavior of Non-Metallic Inclusions in Solid Alloys Deoxidized with Mn and Si during Heat Treatment at 1473 K. *Metals* **2017**, *7*, 223. [CrossRef]

19. Kim, M.; Abro, M.; Lee, D. Corrosion of Fe-(9~37) wt. %Cr Alloys at 700–800 °C in (N_2, H_2O, H_2S)-Mixed Gas. *Metals* **2016**, *6*, 291. [CrossRef]

20. Okonkwo, P.; Shakoor, R.; Benamor, A.; Amer Mohamed, A.; Al-Marri, M. Corrosion Behavior of API X100 Steel Material in a Hydrogen Sulfide Environment. *Metals* **2017**, *7*, 109. [CrossRef]

21. Stawicki, T.; Białobrzeska, B.; Kostencki, P. Tribological Properties of Plough Shares Made of Pearlitic and Martensitic Steels. *Metals* **2017**, *7*, 139. [CrossRef]

22. Sroka, M.; Zieliński, A.; Dziuba-Kałuża, M.; Kremzer, M.; Macek, M.; Jasiński, A. Assessment of the Residual Life of Steam Pipeline Material beyond the Computational Working Time. *Metals* **2017**, *7*, 82. [CrossRef]

23. Yu, C.; Shiue, R.-K.; Chen, C.; Tsay, L.-W. Effect of Low-Temperature Sensitization on Hydrogen Embrittlement of 301 Stainless Steel. *Metals* **2017**, *7*, 58. [CrossRef]

Review

Effects of Heat Input on Microstructure, Corrosion and Mechanical Characteristics of Welded Austenitic and Duplex Stainless Steels: A Review

Ghusoon Ridha Mohammed [1,2,*], Mahadzir Ishak [1], Syarifah N. Aqida [1] and Hassan A. Abdulhadi [1,2]

[1] Faculty of Mechanical Engineering, University Malaysia Pahang, 26600 Pekan, Pahang, Malaysia; mahadzir@ump.edu.my (M.I.); aqida@ump.edu.my (S.N.A.); h.shamary@gmail.com (H.A.A.)

[2] Institute of technology, Middle Technical University, Foundation of Technical Education, Baghdad-Alzafaranya 10074, Iraq

* Correspondence: Ghusoon_ridha@yahoo.com & PMM_14004@stdmail.ump.edu.my; Tel.: +60-129-457-480

Academic Editor: Robert Tuttle
Received: 19 November 2016; Accepted: 18 January 2017; Published: 30 January 2017

Abstract: The effects of input heat of different welding processes on the microstructure, corrosion, and mechanical characteristics of welded duplex stainless steel (DSS) are reviewed. Austenitic stainless steel (ASS) is welded using low-heat inputs. However, owing to differences in the physical metallurgy between ASS and DSS, low-heat inputs should be avoided for DSS. This review highlights the differences in solidification mode and transformation characteristics between ASS and DSS with regard to the heat input in welding processes. Specifically, many studies about the effects of heat energy input in welding process on the pitting corrosion, intergranular stress, stress-corrosion cracking, and mechanical properties of weldments of DSS are reviewed.

Keywords: duplex stainless steel; heat input; pitting corrosion; intergranular stress; stress corrosion cracking; mechanical properties

1. Introduction

Welding is a fabrication process of generating a perpetual joint result from the melting of the surface of the parts to be joined together, with or without the utilization of pressure and a filler material. In welding, applying heat source defined as the energy input and studying it is essential to study the welding process. Energy input is described as the amount of energy entered per unit length of weld from a moving heat source. The energy input (heat input) is formulated in joules per meter or millimeter. It can be determined as the ratio of total input power in Watts to welding speed. The heat input per unit length (H) is calculated according to the following equation [1]:

$$H = fEI/V \tag{1}$$

where f = Efficiency of heat transfer; E = Volts; I = Amperes; and V = Velocity of heat source (mm/s). Heat is the controlling factor accountable for thermochemical responses that happen in weld pools, and led upon cooling to modify the chemistry of weld metal. This is confirmed by Sun and Wu [2] when they stated that welding heat input is the key factor impacting heat and mass transfer, liquid flow, and the thermal cycle in the pool.

Stainless steel is recognized as the steel that resists corrosion. This resistance to corrosion is consequent of chromium oxide film created by chromium on the surface of the metal forming a passive layer that isolates and preserves the surface. Austenitic stainless steel (ASS) is widely employed in caustic environments [3–7]. A main downside of ASS is the sensitivity to chloride-induced stress-corrosion cracking (SCC) [8,9]. Although ferritic stainless steel is more resistant to such corrosion, it is inferior to ASS in terms of ductility and weldability. Given an appropriate composition and thermo-mechanical processing, duplex stainless steel (DSS), which exhibits an austenite–ferrite dual-phase structure, can be obtained. DSS presents many benefits upon single-phase grades, such as increased yield strength and resistance against SCC [10–13].

DSS is a popular constitutional material in the oil, gas, and manufacturing sectors. In particular, DSS is employed in chemical, wastewater, and marine engineering fields, as well as in desalination industries and marine constructions [14–19]. Given the high corrosion resistance of DSS, this material is favorable for shipbuilding, petrochemical, paper, and nuclear industries and can gradually substitute the expensive 300 ASS [20]. DSS is also the preferred material for petroleum and refining industries [21]. The important mechanical properties of DSS help reduce thickness and are especially required in transportation to address the demand of the industry sector [17,22–27].

For the commonly used ASS, the material should be held in the solution-annealed condition for optimal corrosion resistance. As such, weld thermal cycles are selected to ensure rapid cooling of the weld metal and the adjacent Heat-Affected Zone (HAZ), thereby preventing the formation of deleterious phases, which can adversely affect the corrosion properties [28,29]. Consumables should be selected to ensure the proportion of 5% to 10% delta ferrite in the welded microstructure, which is primordial to inhibit solidification cracking. The heat input and temperature are limited to a maximum of 1.5 KJ/mm and 100 °C, respectively, to avoid extensive precipitation of brittle phases in weld metal when very slow cooling is applied.

Directions for welding technologies indicate that excessive dilution with the base metal should be avoided [30]. Basically, instructions to weld DSS are recommended on the expertise with ASS. Thus, when high heat inputs and slow cooling rates are applied (Figure 1), the properties of DSS weldment can accurately advance [31].

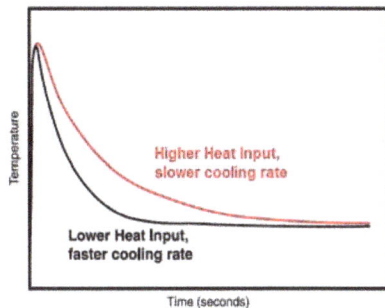

Figure 1. Heat input influences cooling rate [32].

This review does not aim to specify the welding conditions for DSS, but rather to draw attention to the difference in transforming behavior between ASS and DSS and consequently determine how this difference was affected by heat input. Therefore, this review highlights the differences in the physical metallurgy between ASS and DSS and discusses the existing works about the weldment properties of DSS.

2. Metallurgy

2.1. Duplex Structure

DSS consists of ferrite and austenite compounds. Table 1 presents the typical compositions of the common DSS. Many other grades of DSS exist, but most of them would fall in the same compositional range (Table 1). Clearly, a wide range of composition exists, extending from 2101 LDX to the super DSS 2507 [33]. A pseudo-binary phase diagram for Cr and Ni with 70% Fe is illustrated in Figure 2.

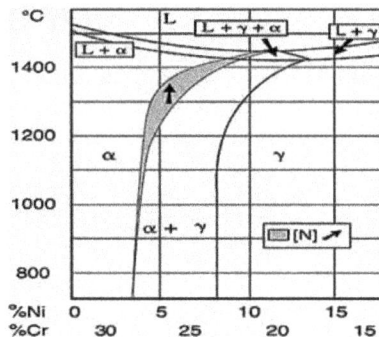

Figure 2. Pseudo-binary Fe-Cr-Ni phase at 70% Fe section illustrating areas of detrimental phase formation [34].

The composition of a representative DSS falls in the $(\alpha + \gamma)$ phase field. For many stainless steel compositions, the austenite phase is expanded; hence, the ferrite phase is separated into high and low-temperature ferrites, which are denoted as delta ferrite and alpha ferrite (formed by the transformation of austenite), correspondingly. Ferrite exists continuously from solidification to room temperature for DSS, and such ferrite is denoted as alpha ferrite. Given that $\alpha/(\alpha + \gamma)$ and $(\alpha + \gamma)/\gamma$ phase boundaries are not vertical, the ferrite-to-austenite ratio in a particular grade would depend on the exact composition, as well as the thermo-mechanical processing [35–37].

Table 1. Typical compositions (in % atom fractions) of some commonly used duplex stainless steels [38]. LDX (Lean Duplex): DX (Duplex): EN (European standard): No. (Number): UNS (Unified Numbering System for Metals and Alloys).

Grade	EN No./UNS	Type	Approx. Composition					
			C	Cr	Ni	Mo	N	Mn
2101 LDX	1.4162/S32101	Lean	0.04	21.0–22.0	1.35–1.70	0.3–0.8	0.2–0.25	4–6
DX 2202	1.4062/S32202	Lean	0.03	21.5–24.0	1.0–2.8	0.45	0.18–0.26	2.0
2304	1.4362/S32304	Lean	0.03	21.5–24.5	3–5.5	0.05–0.6	0.05–0.2	2.5
2205	1.4462/S32205	Standard	0.03	22–23	4.5–6.5	3.0–3.5	0.14–0.2	2.0
2507	1.4410/S32750	Super	0.03	24–26	6–8	3–5	0.24–0.32	1.2

Most modern DSS are designed with a similar austenite-to-ferrite ratio. As a result of the duplex microstructure, partition of alloying components occurs among the phases. Notably, the solidification method of DSS is definite from that of ASS with remaining delta ferrite. Generally, the microstructure of (American Iron and Steel Institute) AISI 304 ASS is completely austenitic at room temperature. In the process of regular welding, the cooling rate is accelerated; thus, the ferrite–austenite alteration cannot be completed, and some δ-ferrite is kept at room temperature after solidification. High input heat leads to a low cooling rate, which further contributes to the change from delta-ferrite phase to austenite phase in weld metal with stainless steel [39–41]. Consequently, the duplex $\delta + \gamma$ structure will be the definitive composition of the weld metal of ASS [42–45].

In the phase equilibrium, the influence of various alloying elements can be quantified by employing equivalent amounts of Ni and Cr. For a very long time, many equations have been used to calculate these two equivalent amounts [46–48]. The Schaeffler diagram [46] and the DeLong diagram [47] shown in Figure 3 are possibly the most commonly used approaches to calculate the ferrite contents in stainless steel weld metals. The equations reported by Datta [49] are listed below:

$$Cr_{eq} = 1.5(Mo) + (Cr) + 5.5(Al) + 5(V) + 2(Si) + 1.75(Nb) + 1.5(Ti) + 0.75(W)$$ (2)

$$Ni_{eq} = (Co) + 0.3(Cu) + 0.5(Mn) + (Ni) + 30(C) + 25(N)$$ (3)

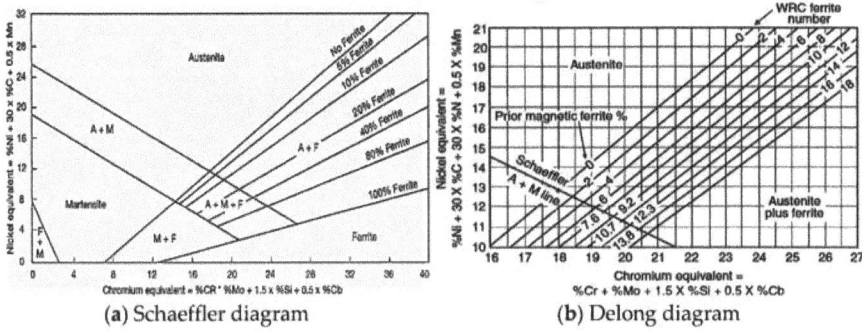

(a) Schaeffler diagram

(b) Delong diagram

Figure 3. Constitution diagrams to predict phase content in stainless-steel weld metal from chemical composition: (**a**) Schaeffler diagram [46]; and (**b**) Delong diagram [47].

These equations have also been previously discussed in detail for stainless steel welding [50,51]. The literature indicated that a Schaeffler alloy with a Cr_{eq}/Ni_{eq} ratio under 1.48 would stabilize essentially as austenite; moreover, delta ferrite is formed from the chromium- and molybdenum-enriched residual melts between the austenite cells or dendrites. As expected, this delta ferrite would be enriched in Cr and Mo.

When the alloy presents a Cr_{eq}/Ni_{eq} ratio between 1.48 and 1.95, primary solidification occurs in the ferrite–austenite mode. This process also results in the segregation of alloying elements upon solidification, with the gamma and alpha phases enriched in austenite- and ferrite-forming elements, respectively. For steels of Cr_{eq}/Ni_{eq} ratio, the remaining ferrite can also undergo solid-state transformation to austenite. Weldments of DSS with a Cr_{eq}/Ni_{eq} ratio above 1.95 stiffen as a single-phase ferrite.

With the high diffusivity of Cr and Mo in ferrite, the ferrite solidification is not accompanied by significant segregation. The austenite is formed from the solid-state ferrite via a Widmanstätten mechanism [51]. Segregation of ferrite stabilizers to alpha, and of austenite stabilizers to gamma, occurs during the solid-state transformation but not during solidification when the Cr_{eq}/Ni_{eq} ratio is less than 1.95. Given this solid-state segregation, the phase balance and amount of segregation will be controlled by factors including the cooling rates for castings and weldments, whereas the thermo-mechanical processing conditions and annealing treatment are very important for wrought products. Additionally, the solidification of duplex weldments occurring in the single-phase ferritic mode largely influences the precipitation of other phases [52–54].

2.2. Precipitation of Other Phases

As the outcome of DSS and super DSS (SDSS) thermal history, one of the significant concerns to gain the required mechanical performance and resistance to corrosion is to understand their microstructural evolution. Derivation of the time–temperature history of the occurrence in welding processes or treatments of technical heat might lead to the deposition of various compounds (e.g., Cr_2N) and a few other intermetallic phases (e.g., σ phase) as illustrated in Figure 4a,b. Such microstructural characteristics are often related to harmful effects on the corrosion resistance, as well as the mechanical behavior of the steel [55]. In fact, σ period deposition can devalue the corrosion features, in which Mo and Cr are retained in this phase. Moreover, such deposition can reduce the toughness of DSS and SDSS [56–60].

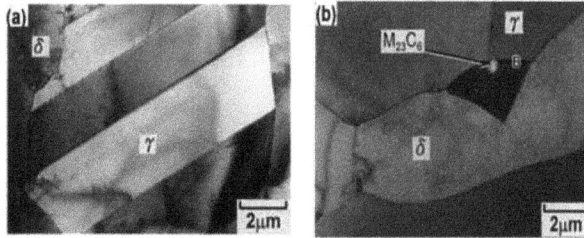

Figure 4. Transmission electron micrograph of duplex stainless steel: (**a**) as received; and (**b**) image of σ phase contained with a $M_{23}C_6$ carbide particle after solution treated at 1080 °C [58].

High toughness can be obtained by implementing suitable solution temperatures and cooling rates [61–63]. This finding is consistent with a previous report [64] indicating that the optimal solution temperature and cooling rate result in high toughness. In addition to austenite and ferrite, other phases may occur depending on the thermal history of the steel, when DSS is exposed to a temperature range of 300 to 1000 °C [65–68]. Examples of such phases are chromium nitrides, carbides, or carbonitride, as well as gamma-phase, chi-phase, R-phase, alpha-prime precipitation, [69–72], alpha precipitation, copper precipitates, and martensite in gamma phase. Generally, the formation of secondary phases influences the corrosion resistance and mechanical properties [73]. The absence of these phases is a result of the rapid cooling in the weld zone and prompt growth and nucleation compared with the fusion process [74]. The temperature–time precipitation curves for various phases observed in a 2205 type alloy are shown in Figure 5.

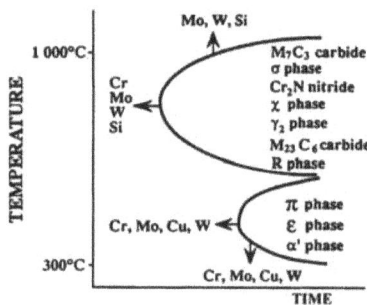

Figure 5. Isothermal cooling curve for the ternary Fe-Cr-Ni system showing the effect of alloying addition on the precipitation of the secondary phases that can form upon cooling [75].

Most of the modern DSS are rich in nitrogen (0.1% to 0.2%). Nitrogen, an austenite stabilizer, is added as both a solid-solution hardener and resistance promoter against pitting corrosion in chloride-containing media. As previously reported, chromium nitrides precipitate in the ferritic phase when a DSS is quenched from a very high annealing temperature [76–78]. Chromium nitrides precipitated when rapid cooling occurred in weld metal or the HAZ of welded DSS [78–81]. Furthermore, fine Cr_2N precipitates were observed when worked on the rod of UNS S31803 DSS treated in two stages [82], that is, after treatment at 1050 °C for 1 h and then at 800 °C from 100 to 31,622 min, with water as a quenching medium for the two stages.

Nevertheless, whether the regulating action is the nucleation rate remains incompletely understood [28,58,83], although the growth kinetics of this response have been widely investigated [84,85]. The amount of Cr_2N precipitation increases with the rise of annealing temperature. As the annealing temperature increases, the volume fraction of austenite decreases, and the ferrite must take up more nitrogen in the solid solution. Although these levels of nitrogen are soluble in ferrite at high temperatures, Cr_2N precipitates in the ferrite upon rapid cooling because nitrogen is relatively insoluble in ferrite at low temperatures. For weldments, the same mechanism is operative, and the HAZ regions were heated to 1300 °C and higher, at which the steel subsisted wholly ferritic with nitrogen in the solid solution. The degree of sensitivity is associated with the great supplement of Cr and Mo drain region outcome from intermetallic phases.

Under certain welding conditions, this region can experience a very rapid cooling cycle, and severe Cr_2N precipitation occurs in an almost completely ferritic HAZ. If the HAZ experiences slower cooling, austenite is formed and nitrogen is dissolved in the austenite, thus reducing the amount of Cr_2N precipitation. Additionally, the cooling rate has been proven to decline as the interpass temperature increases [86].

For ASS, the heat input should be reduced to accelerate the cooling time in the weld metal and the adjoining HAZ, thereby ensuring that deleterious phases, such as $M_{23}C_6$, do not precipitate during the welding process [87]. By contrast, very low heat inputs can disastrously affect the renitence of DSS welding against pitting corrosion [88].

3. Properties of Duplex Stainless Steel Weldments

3.1. Pitting Corrosion

El-Batahgy et al. [89] studied the influence of laser welding parameters (simultaneous heat inputs and shielding gas) on the corrosion resistance of autogenously bead-on plates; their results affirmed that the corrosion average of plates combined contracts with the rise in welding speed, indicating the drop in heat input. This effect of heat input can be justified by the cooling time [21,90]. When the welding process occurs with less heat input, the rapid cooling time leads to a scant measure of austenite and major chromium nitride precipitation.

The above phenomenon can drive consumption of Cr, N_2 all over the residue, which causes a harmful result on the pitting impedance. The use of N_2 as a shielding gas to substitute argon, below similarly stream situation, has also resulted in an exceptional reduction in corrosion speed of connecting pieces. This finding is supported by Srinivasan [91] and Lothongkum [92] and advances the corrosion characteristics of laser joint parts manufactured with nitrogen as a shielding gas relevant to improve the ferrite–austenite ratio in weld metal WM and HAZ. The corrosion morphology of grains around precipitates and the precipitate-free zone were left intact and clarified on the substructure of the chromium exhaustion around the Cr_2N precipitates in ferrite.

Slow cooling rates permit enough time for restoration (Cr redistribution) of the depleted zones around the Cr_2N precipitates. This phenomenon is considered a beneficial effect from high heat inputs. Yasuda et al. [93] studied the influence of heat input on the resistance of a 2205 alloy against pitting corrosion. HAZs were then simulated by heating the samples of 2205 alloy to varying temperatures, quenching in water and in compressed air, and then by air cooling to yield cooling rates of 300 °C/s, 40 °C/s, or 20 °C/s.

Cr_2N was identified in the ferrite phase after cooling at fast and slow rates. Therefore, the beneficial effect of slower cooling rates on resistance against pitting corrosion was explained by the restoration of Cr-depleted areas nearby residues. Higher austenite volume fractions taking more nitrogen into solid solution with a subsequent decrease in the amount of Cr_2N precipitated in ferrite may also contribute to the beneficial effect of slow cooling rates. In terms of weld heat input, the heat inputs greater than 1 kJ/mm can improve the pitting resistance.

Plasma arc welded 2304 DSS was investigated in terms of pitting corrosion resistance by using Cr_{eq}/Ni_{eq} ratio; the results revealed that the microstructure is more balanced with austenite phase after thermal cycles with low Cr_{eq}/Ni_{eq}, and the pitting corrosion resistance decreased with the increase of Cr_{eq}/Ni_{eq} [94]. Moreover, high heat input and satisfactory time intended for ferrite–austenite transformation produce superior austenite content, and heat input plays the main role in the corrosion resistance and microstructure of DSS joints. This finding clearly explains that heat input is directly related to the strength and metallurgical aspects of DSS [31,95,96].

In addition, Lundquist et al. [97] studied the effects of welding conditions on the resistance of 2205 [85,98] and 2304 DSS against pitting corrosion. The influence of heat input from 0.5 to 3.0 kJ/mm was examined for Tungsten Inert Gas (TIG) welded bead-on-tube welds with and without the addition of a filler metal. The pitting resistance was remarkably improved with increasing heat input (Table 2). The only weld beads made with the highest heat input of 3.0 kJ/mm passed the $FeCl_3$ test for duplicate specimens. At 25 °C, welds made by heat input of 2.0 kJ/mm by using filler metal and at 2.5 kJ/mm were autogenously resistant to pitting. When a filler metal is used, lower heat inputs can be tolerated without affecting the resistance against pitting corrosion. The reason for the detrimental effect of low heat inputs was also attributed to an appreciable amount of Cr_2N precipitation in ferrite grains. The amount of precipitation diminished for higher heat inputs was virtually absent at 3.0 kJ/mm. This finding can be explained by the austenite reformation at the expense of nitride precipitates.

In addition to TIG welding, Lundqvist et al. [97] also performed Shielding Metal Arc (SMA) butt welding on 20 mm-thick 2205 plate by using heat inputs from 2.0 to 6.0 kJ/mm. Although the entire top surfaces of the weld metal passed the pitting test in 10% $FeCl_3$ 6HP at 30 °C irrespective of heat input, the weld metal on the root side, which was the first to be deposited, failed. To further investigate this phenomenon, tests on critical pitting temperature were conducted under 3% NaCl.

Table 2. Pitting tests on TIG-welded bead-on-tube welds of 2205 X/2 = Specimens attacked/specimens tested solution: 10% $FeCl_3 \cdot 6H_2O$.

Filler Metal	Temperature (°C)	Heat Input, kJ/mm					
		0.5	1.0	1.5	2.0	2.5	3.0
Sandvik	25	-	2/2	2/2	0/2	0/2	0/2
22.8.3.L	30	-	2/2	2/2	2/2	1/2	0/2
None	25	2/2	2/2	2/2	2/2	0/2	0/2
	30	2/2	2/2	2/2	2/2	1/2	0/2

Critical pitting temperatures of 48, 43 and 40 °C were obtained at heat inputs of 2.0, 4.0 and 6.0 kJ/mm, correspondingly. Microstructural evaluation revealed that extremely fine austenite precipitated in the first- and second-weld beads. A higher heat input during subsequent weld passes led to a reformation of more austenite. In addition to nitrides, fine precipitates of austenite, which were presumed to be reformed at temperatures as low as 800 °C, also negatively affected the pitting resistance. However, the reformed austenite is less detrimental to pitting resistance than Cr_2N precipitates. Lundquist et al. also investigated the use of nitrogen as a shielding gas. Although nitrogen provides adequate protection against oxidation, nitrogen also diffuses into the weld metal and HAZ. Such diffusion increases the amount of austenite in the root run and effectively suppresses the amount of chromium nitride precipitated, which can possibly enhance the resistance to pitting corrosion.

It is generally agreed that very low heat inputs should be avoided on DSS and that much higher heat input can be tolerated for austenitic. This affects the costs of weld fabrication since joints can be made in fewer passes [99]. It should be noted that heat input alone does not determine the cooling rate, but that thickness of the parent metal and interpass temperatures also should be considered. The heat inputs for SAF 2304 and SAF 2205 are specified to be 0.5 to 2.5 kJ/mm, respectively, with the upper limit not considered [97,100]. The choice of heat input pertinent with the material thickness, as the reform of sufficient austenite, needs heat inputs in the upper part of the thick material.

Jang et al. [101] refer to the influence of shielding gas component using N_2 on the impedance to pitting corrosion at the Hyper Duplex Stainless Steel HDSS within quiet massive chloride environment. After welding, the resistance of the HDSS tube to pitting corrosion was produced using N_2 with Ar shielding gas and was increased due to the decrease of α-phase in the HAZ and weld metal. Furthermore, the improved corrosion resistance is assigned to the reduction of Pitting Resistance Equivalent Number (PREN) variation between the α-phase and γ-phase in the weld metal [16,102].

Furthermore, Wang et al. [103] used dissimilar metals to investigate the consequence of welding conditions in terms of welding process type. Gas Tungsten Arc Welding (GTAW) joint A and Shielded Metal Arc Welding (SMAW) joint B were applied sequentially, and ER2209 welding wire was used to join two different materials: 2205 DSS and 16MnR low alloy high-strength steel. To assess the corrosion resistance of weld metal, the plates were partly tightened with A/B glue and a corrosion mixture of 3.5% NaCl. Figure 6 and Table 3 present the results of electrochemical corrosion experiments with 2205 DSS base metal and weld metal, correspondingly. However, their corrosion possibilities are relatively uneven: joint B < joint A < 2205 DSS. The corrosion possibility is a constant index of electrochemical corrosion renitency and exhibits the sensitivity to corrosion of the material [77,91,104].

Table 3. Electrochemical parameters of DSS BM and weld metals [103]. DSS: duplex stainless steel. BM: Base Metal.

Samples	Joint A	Joint B	DSS BM
Corrosion potential/E_{corr} (V)	−0.394	−0.463	−0.251
Corrosion current/I_{corr} (A)	0.2932	0.3041	0.2862

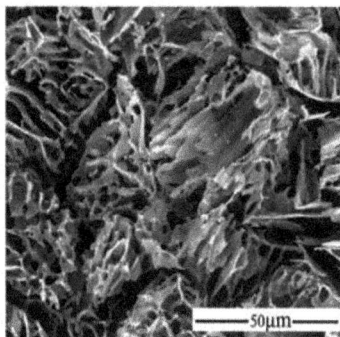

Figure 6. Scanning electron microscopy (SEM) micrograph of pitting corrosion.

The weld metal of DSS is more resistant to pitting corrosion than the base metal. Furthermore, the enhanced response of weld metal to pitting resistance is associated with the addition of Cr and Ni elements [105]. This finding is supported by Olsson [106], who reported that the stability of passive films gained from Cr and Ni would reduce the comprehensive decay of Fe and Cr. Overall, the weld microstructure is influenced by the differences in the heat input between joint A and joint B and leads to change in the configuration state of the element surface unfavorable layer.

For the common DSS 2205, its composition range was considered as UNS S31803. However, the N content associated with UNS S31803 may be reduced to 0.08%, and a plane is assured to be minimized for reliable HAZ and autogenous fusion zone properties in the same status of the welding process. In addition to the increased requirement of minimum nitrogen for S32205 versus

S31803, the minimum Mo and Cr contents are enhanced. During welding, nitrogen controls the ferrite/austenite phase equality [107].

Susceptibility to solidification cracking is one of the problems coupled with fusion welding of these materials, which is comparatively higher than that of 304L ASS [108]. Nitride precipitation was noted in HAZ. The presence of δ-ferrite with higher Cr content may deleteriously affect the corrosion resistance because of the potential difference between the δ-ferrite and austenite phases [109], but this factor may exert less influence on the corrosion resistance than the others. The corrosion properties of the welds also include grain boundary effects [110,111]. Experimentally, Shamanian and Yousefieh [112] studied the differences in heat input result on a DSS, a microstructure of UNS S32760 in seawater. Remarkably, the heat input at about 0.95 kJ/mm presents the largest component of corrosion, that is, the lack of harmful phases resulting in sigma and Cr_2N, as well as the attribution of equal ferrite–austenite.

The GTAW process pulsed current has benefited the representative GTA process, that is, the ferrite–austenite ratio within the base metal and weld metal relies on the welding energy input. After welding in DSS pipes of four distinct sizes, metallography tests completed by scanning electron microscopy (SEM) and energy-dispersive X-ray spectroscopy (EDX) were conducted to show the phases of chemical compositions presented in the microstructure, and polarization curves of various specimens were analyzed. As a result, the formation of sigma and Cr_2N phases decreased the potential formation of corrosion for a fine and coarse structure in the weld metal. The undercooling level will be prospected within a welding process, given that in the cap region, the weld metal will chill down more rapidly because of lower ambient temperatures at the weld surface and adjacent parent material at the beginning. No evidence shows the existence of secondary austenite (γ′), which is present in all of the examined weld metal conditions. Furthermore, areas containing the intermetallic phases of any of the generally prospected carbides were not observed [113].

The detrimental effects of very high heat inputs or multi-pass welding are the formation of secondary austenite and Cr_2N in the HAZ. Interestingly, the number of thermal cycles is one of the most related criteria to estimate the deleterious effect of sigma phase development, as previously reported [99,114,115].

3.2. Intergranular Corrosion

At grain boundaries, the precipitation of the chromium carbides causes susceptibility to intergranular corrosion, which can be a serious problem in ASS if these materials are held for prolonged periods in the sensitization region (about 500 to 750 °C) [116–118]. This kind of corrosion occurs from the Cr deficiency nearby $M_{23}C_6$ precipitates. As for combating intergranular corrosion in stainless steels, one way is to reduce carbon content below 0.3% to delay the deposition of $M_{23}C_6$, which helps to avoid any damaging consequence.

One of the exceptional features of DSS is the resistance against sensitization. The ferrite phase provided the largest amount of Cr in carbides as a result of high diffusivity of Cr and the high Cr content in the ferrite phase. Accordingly, an extremely broad Cr-consuming region dwells at the ferrite part of the interface. The austenite phase offered a really scanty Cr volume, leading to a very thin but penetrating Cr-consuming region on the austenite side of the interface [82]. Intergranular corrosion in HAZ decreased with the increasing number of weld passes. According to the results, as the number of passes increased, the corrosion attack of grain boundaries decreased. This behavior is ascribed to the prolonged retention time above the sensitization temperature range, which promotes the solubilization of chromium carbides in the grain matrix, thereby avoiding sensitization [119].

The elegant resistance of duplex alloys against sensitization is achieved by recharging the small region with diffusing Cr from the inside of the austenite grain. Despite this inherent resistance, many commercial DSS contain less than 0.03% carbon, thus further reducing the risk of sensitization. In addition to the effect of heat input on the pitting corrosion of DSS weldments as previously detailed,

Sridhar [78] and Lundquist [97] studied the effect on resistance to intergranular corrosion. In both cases, excellent resistance was reported, even in the case of high heat inputs.

For the resistance against intergranular corrosion and SCC, no detrimental effect from high heat inputs was found. The tensile elongation and toughness of welds on DSS are affirmed to increase when the heat input is raised.

3.3. Stress-Corrosion Cracking

DSS demonstrated desirable impedance to SCC. DSS is not as resistant as ferritic SS but more resistant than austenitic against SCC [15]. The susceptibility of DSS to SCC depends on many factors, including alloying elements [120,121], microstructures [122,123], applied stresses [124,125], and environment [126–129].

Slow-strain rate tests (2.2×10^{-6} s^{-1}) at 35% boiling water MgCl$_2$ solution and 125 °C were performed to determine the sensitivity to SCC [98]. Additional tests in an inoperative environment (glycerin) were conducted. All specimens with welded joints tested in glycerin at 125 °C showed sufficient plasticity, and their fracture surfaces were entirely ductile. Detailed examinations revealed that these samples broke in parent material 316L steel close to HAZ of the weld. Samples tested in MgCl$_2$ solution broke in a brittle manner. The welded specimens were brittle or mixed, and the ductile-brittle shape of fracture surfaces. Thus, the consequence of SCC is this succession of plasticity [130].

The ways of crack diffusion ordinarily continue together with state limits or transverse ferrite grains. Cracks were frequently prevented on extended, upright austenite grains, or overpass through [131]. The heat input did not influence the susceptibility to SCC.

3.4. Mechanical Properties

The ferrite content of DSS can considerably influence the yield strength, tensile ductility, and toughness of the alloy [89,97,112,132]. As such, it further assumed that the heat input will manipulate these properties in a welded part because the austenite-to-ferrite stability is related to the cooling rate [133]. The toughness was reportedly improved as the heat input was top-up on SAF 2205 [97]. This finding was further correlated to the reduction of ferrite content. They also point out that sigma phase and 475 °C embrittlement are not troubled when welding modern DSS, owing to well-balanced chemical compositions and favorable austenite-to-ferrite ratios. From a strength and toughness viewpoint, the arc powers do not need to be maximized for 2205 [134]. The findings also proved that even 1% of sigma phase formation in DSS is sufficient to cause embrittlement. Furthermore, the coarse-grained ferrite structure formed near the fusion line is responsible for the reduction in impact toughness of the DSS weld [135–137].

For austenite antirust steels, the important relationship between the changes, as well as heat input in microstructural development, is highly related to the relevant mechanical performance in the weldments of AISI 304L stainless steel by GTAW [138]. Moreover, little heat input joints show high ductility and tensile strength, but great heat input joints demonstrate small tensile ductility and strength. The HAZ photomicrographs of these weldments show that grain coarsening influence is inducted, which is inclined to recede the joint part and may thus influence the functional performance of the weld joint in practice. With high heat input, the grain-coarsening degree within this zone is comparatively higher, but a small heat input helps to inhibit the grain growth by subjecting the zone to abrupt heat gradients [139]. Hardness is relatively small along the large weld zone of heat input, which includes long dendrites with larger interdendritic spacing, but hardness is improved along with the small weld zone of heat input, which contains comparatively few dendrites along with smaller interdendritic spacing. All HAZs in distinct weldments undergo grain coarsening, of which the extent is increasingly enhanced with the improvement of welding heat input.

Mourad et al. [140] studied the effects of gas tungsten arc as well as laser beam welding on 2205 properties of DSS. The effects of gas tungsten arc welding and beam welding of carbon dioxide laser on the microstructure and size of fusion zone, mechanical as well as corrosion properties of DSS grade 6.4 mm-thick 2205 plates were comparatively studied. The ferrite–austenite balance of both HAZ and weld metal are affected by heat input, which is a welding process function.

The impacts of heat input on the mechanical properties of UNS S31803 DSS plates from friction stir welding were studied recently [31]. The fracture morphology, residence of elements, and their distribution on joint zone were comprehensively examined through SEM associated with energy dispersive spectroscopy (EDS). The fracture occurred in the base metal and partially penetrated the weld zone. The fracture occurred in the weld region, and the strength of friction welded joints was reduced than the parent material [86]. The fracture did not expand extensively in the base metal and partly entered the weld zone. The fracture occurred in the weld region, and the strength of the friction-welded joints decreased more than the parent material. No intermetallic phases were detected by X-ray diffraction. Outcomes of tensile strength confirmed that the joint strength can be kept when high heat input was applied. However, at room temperature, toughness declined as the heat input grew. Microhardness was improved with the rise of heat input owing to the grain perfection.

After a detailed literature review, most of the investigations are found to have focused on the effects of heat transfer/heat input on DSS by using various welding techniques. The characteristics of various materials processed after different welding process were also analyzed. Clearly, the irregular heat input variations of various fusion welding on DSS cause drastic changes in phase balance, which needs to be balanced. Knowledge about microstructural evolution of such kind of steels, as well as the result of thermal history, is fundamental to achieve the expected mechanical behavior and corrosion resistance. The time–temperature history, derived from industrial heat treatments or welding processes [141–143], may lead to precipitation of various compounds (e.g., chromium carbides and nitrides) and some other intermetallic phases (e.g., σ phase). The formation of such compounds leads to losses in both corrosion resistance and fracture toughness [144]. However, extensive research is needed to deal with the effects of heat input on DSS welds for in-depth understanding of this phenomenon. Few reports exist on different welding processes of DSS grades. Thus, the full-depth knowledge about heat input throughout the welding process of DSS is notably essential for controlling and improving the welding quality. As such, the current research investigated the effects of heat input on mechanical measures such as tensile strength, Charpy impact toughness, microhardness, metallurgical specifics, and corrosion of DSS welded joints.

4. Conclusions

The amount of heat input in the welding process affects the properties of both DSS and ASS. The final properties depend on the variety of solidification modes and the transformation characteristics of these two alloys, especially the corrosion properties.

The following conclusions can be drawn from this review:

(1) DSS solidifies in the single-phase ferritic mode, whereas ASS solidifies in the austenitic or austenitic-ferritic mode.
(2) Austenite phase in weldments of DSS is formed by solid-state transformation, which is strongly affected by the cooling rate.
(3) The resultant ferrite-to-austenite ratio is dependent on the energy input during the welding process.
(4) In duplex weldments, low heat inputs result in high volume fractions of ferrite and severe precipitation of chromium nitrides, which adversely affect mechanical and corrosion properties.
(5) High heat inputs requisite sufficient time in the DSS welding for austenite reformation at high temperature.
(6) The number of thermal cycles is significantly associated criteria to evaluate the deleterious impact of sigma phase extension.
(7) Susceptibility to SCC and the resistance against intergranular corrosion did not affect from high heat inputs.

Acknowledgments: The authors are grateful to the Universiti Malaysia Pahang (UMP), Pekan, Malaysia, and Automotive Engineering Centre (AEC), Pekan, Malaysia, for financial support given under GRS 1503107 (GRS/8/2015/UMP).

Author Contributions: Ghusoon and Ishak conceived and designed the Review; Hassan and Aqida analyzed the data; Ishak and Ghusoon contributed reagents/materials/analysis tools; Ghusoon wrote the paper

Conflicts of Interest: The authors declare no conflict of interest. And "The founding sponsors had no role in the design of the study; in the collection, analyses, or interpretation of data; in the writing of the manuscript, and in the decision to publish the results".

References

1. Jarvis, B.L.; Tanaka, M. 3—Gas tungsten arc welding. In *New Developments in Advanced Welding*; Woodhead Publishing: Witney, UK, 2005; pp. 40–80.
2. Sun, J.; Chuansong, W. The effect of welding heat input on the weldpool behavior in MIG welding. *Sci. China Ser. E Technol. Sci.* **2002**, *45*, 291–299. [CrossRef]
3. Rondelli, G.; Vicentini, B. Susceptibility of highly alloyed austenitic stainless steels to caustic stress corrosion cracking. *Mater. Corros.* **2002**, *53*, 813–819. [CrossRef]
4. Rondelli, G.; Vicentini, B.; Sivieri, E. Stress corrosion cracking of stainless steels in high temperature caustic solutions. *Corros. Sci.* **1997**, *39*, 1037–1049. [CrossRef]
5. Parnian, N. Failure analysis of austenitic stainless steel tubes in a gas fired steam heater. *Mater. Des.* **2012**, *36*, 788–795. [CrossRef]
6. Betova, I.; Bojinov, M.; Hyökyvirta, O.; Saario, T. Effect of sulphide on the corrosion behaviour of AISI 316L stainless steel and its constituent elements in simulated kraft digester conditions. *Corros. Sci.* **2010**, *52*, 1499–1507. [CrossRef]
7. Chasse, K.; Raji, S.; Singh, P. Effect of chloride ions on corrosion and stress corrosion cracking of duplex stainless steels in hot alkaline-sulfide solutions. *Corrosion* **2012**, *68*, 932–949. [CrossRef]
8. Elsariti, S.M. Behaviour of stress corrosion cracking of austenitic stainless steels in sodium chloride solutions. *Procedia Eng.* **2013**, *53*, 650–654. [CrossRef]
9. Alyousif, O.M.; Nishimura, R. Stress corrosion cracking and hydrogen embrittlement of sensitized austenitic stainless steels in boiling saturated magnesium chloride solutions. *Corros. Sci.* **2008**, *50*, 2353–2359. [CrossRef]
10. Lo, K.H.; Shek, C.H.; Lai, J. Recent developments in stainless steels. *Mater. Sci. Eng. R Rep.* **2009**, *65*, 39–104. [CrossRef]
11. Bonollo, F.; Tiziani, A.; Ferro, P. Welding processes, microstructural evolution and final properties of duplex and superduplex stainless steels. In *Duplex Stainless Steels*; John Wiley & Sons, Inc.: Hoboken, NJ, USA, 2013; pp. 141–159.
12. Kisasoz, A.; Gurel, S.; Karaaslan, A. Effect of annealing time and cooling rate on precipitation processes in a duplex corrosion-resistant steel. *Metal Sci. Heat Treat.* **2016**, *57*, 544–547. [CrossRef]
13. Srikanth, S.; Saravanan, P.; Govindarajan, P.; Sisodia, S.; Ravi, K. Development of Lean Duplex Stainless Steels (LDSS) with Superior Mechanical and Corrosion Properties on Laboratory Scale. *Adv. Mater. Res.* **2013**, *794*, 714–730. [CrossRef]
14. Alvarez-Armas, I.; Degallaix-Moreuil, S. *Duplex Stainless Steels*; John Wiley & Sons: Hoboken, NJ, USA, 2013.
15. Bhattacharya, A.; Singh, P.M. Electrochemical behaviour of duplex stainless steels in caustic environment. *Corros. Sci.* **2011**, *53*, 71–81. [CrossRef]
16. Lai, R.; Cai, Y.; Wu, Y.; Li, F.; Hua, X. Influence of absorbed nitrogen on microstructure and corrosion resistance of 2205 duplex stainless steel joint processed by fiber laser welding. *J. Mater. Process. Technol.* **2016**, *231*, 397–405. [CrossRef]
17. Baddoo, N.R. Stainless steel in construction: A review of research, applications, challenges and opportunities. *J. Construct. Steel Res.* **2008**, *64*, 1199–1206. [CrossRef]
18. Olsson, J.; Snis, M. Duplex—A new generation of stainless steels for desalination plants. *Desalination* **2007**, *205*, 104–113. [CrossRef]
19. Chandler, K.A. 3—Marine environments. In *Marine and Offshore Corrosion*; Butterworth-Heinemann: Amsterdam, The Netherlands, 1985; pp. 38–50.
20. Sarlak, H.; Atapour, M.; Esmailzadeh, M. Corrosion behavior of friction stir welded lean duplex stainless steel. *Mater. Des.* **2015**, *66*, 209–216. [CrossRef]

21. Pekkarinen, J.; Kujanpää, V. The effects of laser welding parameters on the microstructure of ferritic and duplex stainless steels welds. *Phys. Procedia* **2010**, *5*, 517–523. [CrossRef]
22. El Bartali, A.; Evrard, P.; Aubin, V.; Herenú, S.; Alvarez-Armas, I.; Armas, A.; Degallaix-Moreuil, S. Strain heterogeneities between phases in a duplex stainless steel. Comparison between measures and simulation. *Procedia Eng.* **2010**, *2*, 2229–2237. [CrossRef]
23. Saha Podder, A.; Bhanja, A. Applications of Stainless Steel in Automobile Industry. *Adv. Mater. Res.* **2013**, *794*, 731–740. [CrossRef]
24. Cunat, P.J. *Stainless Steel in Structural Automotive Applications*; SAE International: Paris, France, 2002.
25. Hariharan, K.; Balachandran, G.; Prasad, M.S. Application of cost-effective stainless steel for automotive components. *Mater. Manuf. Process.* **2009**, *24*, 1442–1452. [CrossRef]
26. Pouranvari, M.; Alizadeh-Sh, M.; Marashi, S. Welding metallurgy of stainless steels during resistance spot welding part I: Fusion zone. *Sci. Technol. Weld. Join.* **2015**, *20*, 502–511. [CrossRef]
27. Dur, E.; Cora, Ö.N.; Koç, M. Effect of manufacturing conditions on the corrosion resistance behavior of metallic bipolar plates in proton exchange membrane fuel cells. *J. Power Sources* **2011**, *196*, 1235–1241. [CrossRef]
28. Urena, A.; Otero, E.; Utrilla, M.; Munez, C. Weldability of a 2205 duplex stainless steel using plasma arc welding. *J. Mater. Process. Technol.* **2007**, *182*, 624–631. [CrossRef]
29. Zhong, Y.; Zhou, C.; Chen, S.; Wang, R. Effects of temperature and pressure on stress corrosion cracking behavior of 310s stainless steel in chloride solution. *Chin. J. Mech. Eng.* **2016**, *29*, 1–7. [CrossRef]
30. Rahmani, M.; Eghlimi, A.; Shamanian, M. Evaluation of microstructure and mechanical properties in dissimilar austenitic/super duplex stainless steel joint. *J. Mater. Eng. Perform.* **2014**, *23*, 3745–3753. [CrossRef]
31. Mohammed, A.M.; Shrikrishna, K.A.; Sathiya, P.; Goel, S. The impact of heat input on the strength, toughness, microhardness, microstructure and corrosion aspects of friction welded duplex stainless steel joints. *J. Manuf. Process.* **2015**, *18*, 92–106.
32. Funderburk, R.S. A look at input. *Weld. Innov.* **1999**, *16*, 1–4.
33. Lai, J.K.L.; Shek, C.H.; Lo, K.H. *Stainless Steels: An Introduction and Their Recent Developments*; Bentham Science Publishers: Beijing, China, 2012.
34. Cieslak, M.; Savage, W. Weldability and solidification phenomena of cast stainless steel. *Weld. J.* **1980**, *5*, 136s–146s.
35. Tseng, K.H. Development and application of oxide-based flux powder for tungsten inert gas welding of austenitic stainless steels. *Powder Technol.* **2013**, *233*, 72–79. [CrossRef]
36. Kolenič, F.; Kovac, L.; Drimal, D. Effect of laser welding conditions on austenite/ferrite ratio in duplex stainless steel 2507 welds. *Weld. World* **2011**, *55*, 19–25. [CrossRef]
37. Medina, E.; Medina, J.M.; Cobo, A.; Bastidas, D.M. Evaluation of mechanical and structural behavior of austenitic and duplex stainless steel reinforcements. *Construct. Build. Mater.* **2015**, *78*, 1–7. [CrossRef]
38. Alvarez-Armas, I. Duplex stainless steels: Brief history and some recent alloys. *Recent Patents Mech. Eng.* **2008**, *1*, 51–57. [CrossRef]
39. Nowacki, J.; Łukojć, A. Structure and properties of the heat-affected zone of duplex steels welded joints. *J. Mater. Process. Technol.* **2005**, *164*, 1074–1081. [CrossRef]
40. Hwang, S.W.; Ji, J.H.; Lee, E.G.; Park, K.-T. Tensile deformation of a duplex Fe–20Mn–9Al–0.6C steel having the reduced specific weight. *Mater. Sci. Eng. A* **2011**, *528*, 5196–5203. [CrossRef]
41. Betini, E.G.; Cione, F.C.; Mucsi, C.S.; Colosio, M.A.; Rossi, J.L.; Orlando, M.T.D.A. Experimental Study of the Temperature Distribution in Welded Thin Plates of Duplex Stainless Steel for Automotive Exhaust Systems. *SAE Int.* **2016**. [CrossRef]
42. Hunter, A.; Ferry, M. Phase formation during solidification of AISI 304 austenitic stainless steel. *Scr. Mater.* **2002**, *46*, 253–258. [CrossRef]
43. Yan, J.; Gao, M.; Zeng, X. Study on microstructure and mechanical properties of 304 stainless steel joints by TIG, laser and laser-TIG hybrid welding. *Opt. Lasers Eng.* **2010**, *48*, 512–517. [CrossRef]
44. Fu, J.; Yang, Y.; Guo, J.; Ma, J.; Tong, W. Formation of two-phase coupled microstructure in AISI 304 stainless steel during directional solidification. *J. Mater. Res.* **2009**, *24*, 2385–2390. [CrossRef]
45. Eghlimi, A.; Shamanian, M.; Eskandarian, M.; Zabolian, A.; Szpunar, J.A. Characterization of microstructure and texture across dissimilar super duplex/austenitic stainless steel weldment joint by austenitic filler metal. *Mater. Charact.* **2015**, *106*, 208–216. [CrossRef]

46. Schaeffler, A. Constitution diagram for stainless steel weld metal. *Met. Prog.* **1949**, *56*, 680–680B.
47. Long, C.; DeLong, W. Ferrite content of austenitic stainless steel weld metal. *Weld. J.* **1973**, *52*, 281.
48. Wegrzyn, T. Delta ferrite in stainless steel weld metals. *Weld. Int.* **1992**, *6*, 690–694. [CrossRef]
49. Datta, P.; Upadhyaya, G. Sintered duplex stainless steels from premixes of 316L and 434L powders. *Mater. Chem. Phys.* **2001**, *67*, 234–242. [CrossRef]
50. Suutala, N. Effect of solidification conditions on the solidification mode in austenitic stainless steels. *Metall. Trans. A* **1983**, *14*, 191–197. [CrossRef]
51. Suutala, N.; Takalo, T.; Moisio, T. Ferritic-austenitic solidification mode in austenitic stainless steel welds. *Metall. Trans. A* **1980**, *11*, 717–725. [CrossRef]
52. Baeslack, W.A.; Duquette, D.J.; Savage, W.F. The effect of ferrite content on stress corrosion cracking in duplex stainless steel weld metals at room temperature. *Corrosion* **1979**, *35*, 45–54. [CrossRef]
53. Menendez, H.; Devine, T. The influence of microstructure on the sensitization behavior of duplex stainless steel welds. *Corrosion* **1990**, *46*, 410–418. [CrossRef]
54. Ogawa, T.; Koseki, T. Effect of composition profiles on metallurgy and corrosion behavior of duplex stainless steel weld metals. *Weld. J.* **1989**, *68*, 181.
55. Utu, I.D.; Mitelea, I.; Urlan, S.; Crăciunescu, C. Transformation and precipitation reactions by metal active gas pulsed welded joints from X2CrNiMoN22-5-3 duplex stainless steels. *Materials* **2016**, *9*, 606. [CrossRef]
56. El Koussy, M.; El Mahallawi, I.; Khalifa, W.; Al Dawood, M.; Bueckins, M. Effects of thermal aging on microstructure and mechanical properties of duplex stainless steel weldments. *Mater. Sci. Technol.* **2004**, *20*, 375–381. [CrossRef]
57. Ahn, Y.; Kang, J. Effect of aging treatments on microstructure and impact properties of tungsten substituted 2205 duplex stainless steel. *Mater. Sci. Technol.* **2000**, *16*, 382–388. [CrossRef]
58. Chen, T.; Yang, J. Effects of solution treatment and continuous cooling on σ-phase precipitation in a 2205 duplex stainless steel. *Mater. Sci. Eng. A* **2001**, *311*, 28–41. [CrossRef]
59. Chen, T.; Weng, K.; Yang, J. The effect of high-temperature exposure on the microstructural stability and toughness property in a 2205 duplex stainless steel. *Mater. Sci. Eng. A* **2002**, *338*, 259–270. [CrossRef]
60. Wessman, S.; Pettersson, R.; Hertzman, S. On phase equilibria in duplex stainless steels. *Steel Res. Int.* **2010**, *81*, 337–346. [CrossRef]
61. Sieurin, H.; Sandström, R. Austenite reformation in the heat-affected zone of duplex stainless steel 2205. *Mater. Sci. Eng. A* **2006**, *418*, 250–256. [CrossRef]
62. Kim, S.K.; Kang, K.Y.; Kim, M.-S.; Lee, J.M. Low-temperature mechanical behavior of super duplex stainless steel with sigma precipitation. *Metals* **2015**, *5*, 1732–1745. [CrossRef]
63. Bouyne, E.; Joly, P.; Houssin, B.; Wiesner, C.; Pineau, A. Mechanical and microstructural investigations into the crack arrest behaviour of a modern $2\frac{1}{4}$Cr-1 Mo pressure vessel steel. *Fatigue Fract. Eng. Mater. Struct.* **2001**, *24*, 105–116. [CrossRef]
64. Santos, T.F.; Marinho, R.R.; Paes, M.T.; Ramirez, A.J. Microstructure evaluation of UNS S32205 duplex stainless steel friction stir welds. *Rem Rev. Esc. Minas* **2013**, *66*, 187–191. [CrossRef]
65. Kasper, J. The ordering of atoms in the chi-phase of the iron-chromium-molybdenum system. *Acta Metall.* **1954**, *2*, 456–461. [CrossRef]
66. Byun, S.H.; Kang, N.; Lee, T.H.; Ahn, S.K.; Lee, H.W.; Chang, W.S.; Cho, K.M. Kinetics of Cr/Mo-rich precipitates formation for 25Cr-6.9Ni-3.8Mo-0.3N super duplex stainless steel. *Met. Mater. Int.* **2012**, *18*, 201–207. [CrossRef]
67. Kim, S.M.; Kim, J.S.; Kim, K.T.; Park, K.T.; Lee, C.S. Effect of ce addition on secondary phase transformation and mechanical properties of 27Cr–7Ni hyper duplex stainless steels. *Mater. Sci. Eng. A* **2013**, *573*, 27–36. [CrossRef]
68. Hsieh, C.C.; Wu, W. Overview of intermetallic sigma (σ) phase precipitation in stainless steels. *ISRN Metall.* **2012**, *2012*, 16. [CrossRef]
69. Vinoth Jebaraj, A.; Ajaykumar, L. Influence of microstructural changes on impact toughness of weldment and base metal of duplex stainless steel AISI 2205 for low temperature applications. *Procedia Eng.* **2013**, *64*, 456–466. [CrossRef]
70. Wu, H.; Tsay, L.; Chen, C. Laser beam welding of 2205 duplex stainless steel with metal powder additions. *ISIJ Int.* **2004**, *44*, 1720–1726. [CrossRef]

71. Kingklang, S.; Uthaisangsuk, V. Investigation of hot deformation behavior of duplex stainless steel grade 2507. *Metall. Mater. Trans. A* **2016**, *48*, 95–108. [CrossRef]

72. Deng, B.; Jiang, Y.; Xu, J.; Sun, T.; Gao, J.; Zhang, L.; Zhang, W.; Li, J. Application of the modified electrochemical potentiodynamic reactivation method to detect susceptibility to intergranular corrosion of a newly developed lean duplex stainless steel LDX2101. *Corros. Sci.* **2010**, *52*, 969–977. [CrossRef]

73. Chan, K.W.; Tjong, S.C. Effect of secondary phase precipitation on the corrosion behavior of duplex stainless steels. *Materials* **2014**, *7*, 5268–5304. [CrossRef]

74. Ajith, P.M.; Sathiya, P.; Aravindan, S. Characterization of microstructure, toughness, and chemical composition of friction-welded joints of UNS S32205 duplex stainless steel. *Friction* **2014**, *2*, 82–91. [CrossRef]

75. Jinlong, L.; Tongxiang, L.; Limin, D.; Chen, W. Influence of sensitization on microstructure and passive property of AISI 2205 duplex stainless steel. *Corros. Sci.* **2016**, *104*, 144–151. [CrossRef]

76. Chasse, K.R.; Singh, P.M. Hydrogen embrittlement of a duplex stainless steel in alkaline sulfide solution. *Corrosion* **2011**, *67*, 015002-1–015002-12. [CrossRef]

77. Guo, Y.; Hu, J.; Li, J.; Jiang, L.; Liu, T.; Wu, Y. Effect of annealing temperature on the mechanical and corrosion behavior of a newly developed novel lean duplex stainless steel. *Materials* **2014**, *7*, 6604–6619. [CrossRef]

78. Sridhar, N.; Tormoen, G.; Hackney, S.; Anderko, A. Effect of aging treatments on the repassivation potential of duplex stainless steel S32205. *Corrosion* **2009**, *65*, 650–662. [CrossRef]

79. Geng, S.; Sun, J.; Guo, L.; Wang, H. Evolution of microstructure and corrosion behavior in 2205 duplex stainless steel GTA-welding joint. *J. Manuf. Process.* **2015**, *19*, 32–37. [CrossRef]

80. Schmidt-Rieder, E.; Tong, X.; Farr, J.; Aindow, M. In situ electrochemical scanning probe microscopy corrosion studies on duplex stainless steel in aqueous NaCl solutions. *Br. Corros. J.* **2013**, *31*, 139–146. [CrossRef]

81. Ramkumar, K.D.; Thiruvengatam, G.; Sudharsan, S.; Mishra, D.; Arivazhagan, N.; Sridhar, R. Characterization of weld strength and impact toughness in the multi-pass welding of super-duplex stainless steel UNS 32750. *Mater. Des.* **2014**, *60*, 125–135. [CrossRef]

82. Arıkan, M.E.; Arıkan, R.; Doruk, M. Determination of susceptibility to intergranular corrosion of UNS 31803 type duplex stainless steel by electrochemical reactivation method: A comparative study. *Int. J. Corros.* **2012**, *2012*, 1–14. [CrossRef]

83. Zhan, X.; Dong, Z.; Wei, Y.; Ma, R. Simulation of grain morphologies and competitive growth in weld pool of Ni–Cr alloy. *J. Cryst. Growth* **2009**, *311*, 4778–4783. [CrossRef]

84. Łabanowski, J.; Świerczyńska, A.; Topolska, S. Effect of microstructure on mechanical properties and corrosion resistance of 2205 duplex stainless steel. *Pol. Marit. Res.* **2014**, *21*, 108–112. [CrossRef]

85. Bermejo, M.V.; Karlsson, L.; Svensson, L.E.; Hurtig, K.; Rasmuson, H.; Frodigh, M.; Bengtsson, P. Effect of shielding gas on welding performance and properties of duplex and superduplex stainless steel welds. *Weld. World* **2015**, *59*, 239–249. [CrossRef]

86. Hazra, M.; Rao, K.S.; Reddy, G.M. Friction welding of a nickel free high nitrogen steel: Influence of forge force on microstructure, mechanical properties and pitting corrosion resistance. *J. Mater. Res. Technol.* **2014**, *3*, 90–100. [CrossRef]

87. Cárcel-Carrasco, F.J.; Pascual-Guillamón, M.; Pérez-Puig, M.A. Effects of X-rays radiation on AISI 304 stainless steel weldings with AISI 316L filler material: A study of resistance and pitting corrosion behavior. *Metals* **2016**, *6*, 102. [CrossRef]

88. Neissi, R.; Shamanian, M.; Hajihashemi, M. The effect of constant and pulsed current gas tungsten arc welding on joint properties of 2205 duplex stainless steel to 316L austenitic stainless steel. *J. Mater. Eng. Perform.* **2016**, *25*, 2017–2028. [CrossRef]

89. El-Batahgy, A.M.; Khourshid, A.F.; Sharef, T. Effect of laser beam welding parameters on microstructure and properties of duplex stainless steel. *Mater. Sci. Appl.* **2011**, *2*, 1443–1451. [CrossRef]

90. Mohammed, G.R.; Ishak, M.; Aqida, S.N.; Abdulhadi, H.A. The effect of fiber laser parameters on microhardness and microstructure of duplex stainless steel. *MATEC Web Conf.* **2017**, *90*, 01024. [CrossRef]

91. Srinivasan, P.B.; Muthupandi, V.; Dietzel, W.; Sivan, V. An assessment of impact strength and corrosion behaviour of shielded metal arc welded dissimilar weldments between UNS 31803 and IS 2062 steels. *Mater. Des.* **2006**, *27*, 182–191. [CrossRef]

92. Lothongkum, G.; Wongpanya, P.; Morito, S.; Furuhara, T.; Maki, T. Effect of nitrogen on corrosion behavior of 28Cr–7Ni duplex and microduplex stainless steels in air-saturated 3.5 wt% NaCl solution. *Corros. Sci.* **2006**, *48*, 137–153. [CrossRef]

93. Yasuda, K.; Kimura, M.; Kawasaki, H.; Works, C.; Uegaki, T. Optimizing welding condition for excellent corrosion resistance in duplex stainless steel linepipe. *Kawasaki Steel Giho* **1988**, *20*, 197–202.
94. Jiang, Y.; Tan, H.; Wang, Z.; Hong, J.; Jiang, L.; Li, J. Influence of Cr_{eq}/Ni_{eq} on pitting corrosion resistance and mechanical properties of UNS S32304 duplex stainless steel welded joints. *Corros. Sci.* **2013**, *70*, 252–259. [CrossRef]
95. Moura, V.S.; Lima, L.D.; Pardal, J.M.; Kina, A.Y.; Corte, R.R.A.; Tavares, S.S.M. Influence of microstructure on the corrosion resistance of the duplex stainless steel UNS S31803. *Mater. Charact.* **2008**, *59*, 1127–1132. [CrossRef]
96. Sadeghian, M.; Shamanian, M.; Shafyei, A. Effect of heat input on microstructure and mechanical properties of dissimilar joints between super duplex stainless steel and high strength low alloy steel. *Mater. Des.* **2014**, *60*, 678–684. [CrossRef]
97. Lundquist, B.; Norberg, P.; Olsson, K. Influence of different welding conditions on mechanical properties and corrosion resistance of sandvik SAF 2205 (UNS S31803). In Proceedings of the Conference Duplex Stainless Steels, the Hague, Netherlands, October 1986; pp. 16–29.
98. Łabanowski, J. Mechanical properties and corrosion resistance of dissimilar stainless steel welds. *Arch. Mater. Sci. Eng.* **2007**, *28*, 27–33.
99. Hosseini, V.A.; Bermejo, M.A.V.; Gårdstam, J.; Hurtig, K.; Karlsson, L. Influence of multiple thermal cycles on microstructure of heat-affected zone in TIG-welded super duplex stainless steel. *Weld. World* **2016**, *60*, 233–245. [CrossRef]
100. Busschaert, F.; Cassagne, T.; Pedersen, A.; Johnsen, S. New challenges for the use of duplex stainless steels at low temperatures. *Rev. Métall.* **2013**, *110*, 185–197. [CrossRef]
101. Jang, S.H.; Kim, S.T.; Lee, I.S.; Park, Y.S. Effect of shielding gas composition on phase transformation and mechanism of pitting corrosion of hyper duplex stainless steel welds. *Mater. Trans.* **2011**, *52*, 1228–1236. [CrossRef]
102. Hosseini, V.A.; Wessman, S.; Hurtig, K.; Karlsson, L. Nitrogen loss and effects on microstructure in multipass TIG welding of a super duplex stainless steel. *Mater. Des.* **2016**, *98*, 88–97. [CrossRef]
103. Wang, S.; Ma, Q.; Li, Y. Characterization of microstructure, mechanical properties and corrosion resistance of dissimilar welded joint between 2205 duplex stainless steel and 16MnR. *Mater. Des.* **2011**, *32*, 831–837. [CrossRef]
104. Yang, L.; Zhang, Z.Z. Study on Weldablity of Dissimilar Steel between 16MnR and S31803. *Adv. Mater. Res.* **2012**, *391–392*, 768–772. [CrossRef]
105. Tavares, S.; Pardal, J.; Lima, L.; Bastos, I.; Nascimento, A.; de Souza, J. Characterization of microstructure, chemical composition, corrosion resistance and toughness of a multipass weld joint of superduplex stainless steel UNS S32750. *Mater. Charact.* **2007**, *58*, 610–616. [CrossRef]
106. Olsson, C.O.A.; Landolt, D. Passive films on stainless steels-chemistry, structure and growth. *Electrochim. Acta* **2003**, *48*, 1093–1104. [CrossRef]
107. Kotecki, D.J. Some pitfalls in welding of duplex stainless steels. *Soldag. Insp.* **2010**, *15*, 336–343. [CrossRef]
108. Dong, W.; Kokawa, H.; Sato, Y.S.; Tsukamoto, S. Nitrogen desorption by high-nitrogen steel weld metal during CO_2 laser welding. *Metall. Mater. Trans. B* **2005**, *36*, 677–681. [CrossRef]
109. Lin, C.-M.; Tsai, H.-L.; Cheng, C.-D.; Yang, C. Effect of repeated weld-repairs on microstructure, texture, impact properties and corrosion properties of AISI 304L stainless steel. *Eng. Fail. Anal.* **2012**, *21*, 9–20. [CrossRef]
110. Lu, Z.; Shoji, T.; Meng, F.; Xue, H.; Qiu, Y.; Takeda, Y.; Negishi, K. Characterization of microstructure and local deformation in 316NG weld heat-affected zone and stress corrosion cracking in high temperature water. *Corros. Sci.* **2011**, *53*, 1916–1932. [CrossRef]
111. Lu, Z.; Shoji, T.; Xue, H.; Meng, F.; Fu, C.; Takeda, Y.; Negishi, K. Synergistic effects of local strain-hardening and dissolved oxygen on stress corrosion cracking of 316NG weld heat-affected zones in simulated BWR environments. *J. Nuclear Mater.* **2012**, *423*, 28–39. [CrossRef]
112. Yousefieh, M.; Shamanian, M.; Saatchi, A. Influence of heat input in pulsed current GTAW process on microstructure and corrosion resistance of duplex stainless steel welds. *J. Iron Steel Res. Int.* **2011**, *18*, 65–69. [CrossRef]
113. Gideon, B.; Ward, L.; Biddle, G. Duplex stainless steel welds and their susceptibility to intergranular corrosion. *J. Miner. Mater. Charact. Eng.* **2008**, *7*, 247–263. [CrossRef]

114. Tan, H.; Wang, Z.; Jiang, Y.; Yang, Y.; Deng, B.; Song, H.; Li, J. Influence of welding thermal cycles on microstructure and pitting corrosion resistance of 2304 duplex stainless steels. *Corros. Sci.* **2012**, *55*, 368–377. [CrossRef]

115. Silva, E.; Marinho, L.; Filho, P.; Leite, J.; Leite, J.; Fialho, W.; de Albuquerque, V.; Tavares, J. Classification of induced magnetic field signals for the microstructural characterization of sigma phase in duplex stainless steels. *Metals* **2016**, *6*, 164. [CrossRef]

116. Shimada, M.; Kokawa, H.; Wang, Z.; Sato, Y.; Karibe, I. Optimization of grain boundary character distribution for intergranular corrosion resistant 304 stainless steel by twin-induced grain boundary engineering. *Acta Mater.* **2002**, *50*, 2331–2341. [CrossRef]

117. Lin, S.X.; Bao, W.K.; Gao, J.; Wang, J.B. Intergranular Corrosion of Austenitic Stainless Steel. *Appl. Mech. Mater.* **2012**, *229–231*, 14–17. [CrossRef]

118. Vlčková, I.; Jonšta, P.; Jonšta, Z.; Váňová, P.; Kulová, T. Corrosion fatigue of austenitic stainless steels for nuclear power engineering. *Metals* **2016**, *6*, 319. [CrossRef]

119. Mirshekari, G.R.; Tavakoli, E.; Atapour, M.; Sadeghian, B. Microstructure and corrosion behavior of multipass gas tungsten arc welded 304L stainless steel. *Mater. Des.* **2014**, *55*, 905–911. [CrossRef]

120. Tsai, W.T.; Chen, M.S. Stress corrosion cracking behavior of 2205 duplex stainless steel in concentrated nacl solution. *Corros. Sci.* **2000**, *42*, 545–559. [CrossRef]

121. Kim, K.; Zhang, P.; Ha, T.; Lee, Y. Electrochemical and stress corrosion properties of duplex stainless steels modified with tungsten addition. *Corrosion* **1998**, *54*, 910–921. [CrossRef]

122. Nilsson, J.O.; Kangas, P.; Wilson, A.; Karlsson, T. Mechanical properties, microstructural stability and kinetics of σ-phase formation in 29Cr-6Ni-2Mo-0.38N superduplex stainless steel. *Metall. Mater. Trans. A* **2000**, *31*, 35–45. [CrossRef]

123. Strubbia, R.; Hereñú, S.; Marinelli, M.; Alvarez-Armas, I. Short crack nucleation and growth in lean duplex stainless steels fatigued at room temperature. *Int. J. Fatigue* **2012**, *41*, 90–94. [CrossRef]

124. Brandolt, C.; Rosa, M.; Ramos, L.; Schroeder, R.; Malfatti, C.; Müller, I. Temperature influence on SCC behaviour of duplex stainless steel. *Mater. Sci. Technol.* **2016**, 1–7. [CrossRef]

125. Li, H.; Jiao, W.; Feng, H.; Li, X.; Jiang, Z.; Li, G.; Wang, L.; Fan, G.; Han, P. Deformation characteristic and constitutive modeling of 2707 hyper duplex stainless steel under hot compression. *Metals* **2016**, *6*, 223. [CrossRef]

126. Bellezze, T.; Giuliani, G.; Roventi, G.; Fratesi, R.; Andreatta, F.; Fedrizzi, L. Corrosion behaviour of austenitic and duplex stainless steels in an industrial strongly acidic solution. *Mater. Corros.* **2016**, *67*, 831–838. [CrossRef]

127. Tseng, C.M.; Tsai, W.T. Environmentally assisted cracking behavior of single and dual phase stainless steels in hot chloride solutions. *Mater. Chem. Phys.* **2004**, *84*, 162–170. [CrossRef]

128. El-Yazgi, A.; Hardie, D. Stress corrosion cracking of duplex and super duplex stainless steels in sour environments. *Corros. Sci.* **1998**, *40*, 909–930. [CrossRef]

129. Örnek, C.; Idris, S.A.; Reccagni, P.; Engelberg, D.L. Atmospheric-induced stress corrosion cracking of grade 2205 duplex stainless steel—Effects of 475 °C embrittlement and process orientation. *Metals* **2016**, *6*, 167. [CrossRef]

130. Chattoraj, I. Stress Corrosion Cracking of Duplex Stainless Steels. *Adv. Mater. Res.* **2013**, *794*, 552–563. [CrossRef]

131. Bhattacharya, A.; Singh, P.M. Effect of heat treatment on corrosion and stress corrosion cracking of S32205 duplex stainless steel in caustic solution. *Metall. Mater. Trans. A* **2009**, *40*, 1388–1399. [CrossRef]

132. Muthupandi, V.; Bala Srinivasan, P.; Seshadri, S.K.; Sundaresan, S. Effect of weld metal chemistry and heat input on the structure and properties of duplex stainless steel welds. *Mater. Sci. Eng. A* **2003**, *358*, 9–16. [CrossRef]

133. Esmailzadeh, M.; Shamanian, M.; Kermanpur, A.; Saeid, T. Microstructure and mechanical properties of friction stir welded lean duplex stainless steel. *Mater. Sci. Eng. A* **2013**, *561*, 486–491. [CrossRef]

134. Reddy, G.M.; Rao, K.S. Microstructure and mechanical properties of similar and dissimilar stainless steel electron beam and friction welds. *Int. J. Adv. Manuf. Technol.* **2009**, *45*, 875–888. [CrossRef]

135. Kordatos, J.; Fourlaris, G.; Papadimitriou, G. The effect of cooling rate on the mechanical and corrosion properties of SAF 2205 (UNS 31803) duplex stainless steel welds. *Scr. Mater.* **2001**, *44*, 401–408. [CrossRef]

136. Calliari, I.; Straffelini, G.; Ramous, E. Investigation of secondary phase effect on 2205 DSS fracture toughness. *Mater. Sci. Technol.* **2010**, *26*, 81–86. [CrossRef]
137. Ibrahim, O.; Ibrahim, I.; Khalifa, T. Impact behavior of different stainless steel weldments at low temperatures. *Eng. Fail. Anal.* **2010**, *17*, 1069–1076. [CrossRef]
138. Kumar, S.; Shahi, A.S. On the influence of welding stainless steel on microstructural development and mechanical performance. *Mater. Manuf. Process.* **2014**, *29*, 894–902. [CrossRef]
139. Song, T.; Jiang, X.; Shao, Z.; Mo, D.; Zhu, D.; Zhu, M. The interfacial microstructure and mechanical properties of diffusion-bonded joints of 316L stainless steel and the 4J29 kovar alloy using nickel as an interlayer. *Metals* **2016**, *6*, 263. [CrossRef]
140. Mourad, A.H.I.; Khourshid, A.; Sharef, T. Gas tungsten arc and laser beam welding processes effects on duplex stainless steel 2205 properties. *Mater. Sci. Eng. A* **2012**, *549*, 105–113. [CrossRef]
141. Shrikrishna, K.A.; Sathiya, P. Effects of post weld heat treatment on friction welded duplex stainless steel joints. *J. Manuf. Process.* **2015**, *21*, 196–200.
142. Atapour, M.; Sarlak, H.; Esmailzadeh, M. Pitting corrosion susceptibility of friction stir welded lean duplex stainless steel joints. *Int. J. Adv. Manuf. Technol.* **2015**, *83*, 721–728. [CrossRef]
143. Abdulhadi, H.A.; Aqida, S.N.; Ishak, M.; Mohammed, G.R. Thermal fatigue of die-casting dies: An overview. *MATEC Web Conf.* **2016**, *74*, 00032. [CrossRef]
144. Chagas de Souza, G.; da Silva, A.L.; Tavares, S.S.M.; Pardal, J.M.; Ferreira, M.L.R.; Filho, I.C. Mechanical properties and corrosion resistance evaluation of superduplex stainless steel UNS S32760 repaired by gtaw process. *Weld. Int.* **2016**, *30*, 432–442. [CrossRef]

metals

MDPI

Article

Corrosion of Fe-(9~37) wt. %Cr Alloys at 700–800 °C in (N₂, H₂O, H₂S)-Mixed Gas

Min Jung Kim, Muhammad Ali Abro and Dong Bok Lee *

School of Advanced Materials Science and Engineering, Sungkyunkwan University, Suwon 16419, Korea;
abc1219@skku.edu (M.J.K.); abromdali@gmail.com (M.A.A.)
* Correspondence: dlee@skku.ac.kr; Tel.: +82-31-290-7371

Academic Editor: Robert Tuttle
Received: 18 October 2016; Accepted: 15 November 2016; Published: 23 November 2016

Abstract: Fe-(9, 19, 28, 37) wt. %Cr alloys were corroded at 700 and 800 °C for 70 h under 1 atm of N_2, 1 atm of $N_2/3.2\%H_2O$ mixed gas, and 1 atm of $N_2/3.1\%H_2O/2.42\%H_2S$ mixed gas. In this gas composition order, the corrosion rate of Fe-9Cr alloy rapidly increased. Fe-9Cr alloy was always non-protective. In contrast, Fe-(19, 28, 37) wt. %Cr alloys were protective in N_2 and $N_2/3.2\%H_2O$ mixed gas because of the formation of the Cr_2O_3 layer. They, however, became nonprotective in $N_2/3.1\%H_2O/2.42\%H_2S$ mixed gas because sulfidation dominated to form the outer FeS layer and the inner Cr_2S_3 layer containing some $FeCr_2S_4$.

Keywords: Fe-Cr alloy; oxidation; sulfidation; H_2S corrosion

---

1. Introduction

Fe-Cr alloys are widely used as high-temperature structural materials. They oxidize when exposed to air or oxygen at high temperatures. When the Cr content in iron was ~5 wt. %, triple oxide layers such as $Fe_2O_3/Fe_3O_4/FeO$ formed, and the oxidation rate was mainly controlled by the growth rate of FeO that formed on the alloy side [1]. The non-stoichiometric wustite grows much faster than the nearly stoichiometric Fe_3O_4 and Fe_2O_3. With an increase in the Cr content to ~10 wt. %, dispersed particles of the $FeCr_2O_4$ spinel formed more inside the FeO layer, and $FeCr_2O_4$ particles blocked the diffusion of Fe^{2+} ions to make the FeO layer thinner. With the further increase in the Cr content to ~15 wt. %, a mixed spinel, $Fe(Fe,Cr)_2O_4$, formed, which decreased the oxidation rate significantly. When the Cr content exceeded ~20 wt. %, the oxidation rate dropped sharply, forming a thin, continuous Cr_2O_3 layer containing a small amount of dissolved Fe ions [1–3]. When Fe-Cr alloys were exposed to S_2 gas at 1 atm, an FeS layer formed below 1.86 wt. %Cr, an outer $Fe_{1-x}S$ layer and an inner (FeS, $FeCr_2S_4$) mixed layer formed in the range of 1.86–38.3 wt. %Cr, and a solid solution of $FeS-Cr_2S_3$ formed above 38.3 wt. %Cr [4]. Although Cr decreased the sulfidation rate, even Fe-Cr alloys with high Cr contents displayed insufficient corrosion resistance. This is attributed to the fact that sulfidation rates of common metals are 10–100 times faster than oxidation rates because the sulfides have much larger defect concentrations and lower melting points than the corresponding oxides [5]. The sulfidation of Fe-(20, 25, 30) wt. %Cr steels in $94Ar/5H_2/1H_2S$ mixed gas at 600 °C for 718 h resulted in the formation of the outer FeS layer and the inner $FeCr_2S_4$ layer [6]. On the other hand, the corrosion of conventional oxidation-resistant alloys by water vapor and H_2S gas has been a serious problem [3]. Water vapor and H_2S gas release hydrogen atoms, which ingress in the metals interstitially, form hydrogen clusters, and cause hydrogen embrittlement. Water vapor that is present in many industrial gases can form metal hydrides, and change not only the reaction at the scale/metal interface but also the mass transfer in scales, accelerating the corrosion rate [3,7]. In this study, Fe-Cr alloys were corroded at 700 and 800 °C in (N₂, H₂O, H₂S) mixed gas in order to understand their corrosion behavior in hostile (H₂O, H₂S)-containing environments for practical applications. The aim of this study is to examine the influence of the Cr content and the (N_2, H_2O, H_2S)-containing gas on the high-temperature corrosion of Fe-Cr alloys, which has not been adequately investigated before.

2. Experimental Procedures

Four kinds of hot-rolled ferritic Fe-Cr alloy sheets, viz., Fe-(8.5, 18.5, 28.3, 36.9) wt. %Cr, were prepared. They are termed as Fe-(9, 19, 28, 37)Cr, respectively, in this study. They were homogenized at 900 °C for 1 h under vacuum, cut into a size of 2 mm × 10 mm × 15 mm, ground up to a 1000-grit finish with SiC papers, ultrasonically cleaned in acetone, and corroded at 700 and 800 °C for 70 h under 1 atm of total pressure. Each test coupon was suspended by a Pt wire in a quartz reaction tube within the hot zone of an electrical furnace (Ajeon, Seoul, Korea), as shown in Figure 1. Three kinds of corrosion atmospheres were employed, viz. 1 atm of N_2, (0.968 atm of N_2 plus 0.032 atm of H_2O) that was achieved by bubbling the N_2 gas through the water bath kept at 25 °C, and (0.9448 atm of N_2 plus 0.031 atm of H_2O plus 0.0242 atm of H_2S) that was achieved by bubbling N_2 gas through the water bath kept at 25 °C and simultaneously flowing the N_2-5%H_2S gas into the quartz reaction tube. The N_2 gas was 99.999% pure, and H_2S gas was 99.5% pure. Nitrogen gas was blown into the reaction tube during heating and cooling stages. After finishing the corrosion test in N_2, N_2/3.2%H_2O, and N_2/3.1%H_2O/2.42%H_2S gas, the test coupons were furnace-cooled, and characterized by a scanning electron microscope (SEM, Jeol JSM-6390A, Tokyo, Japan), a high-power X-ray diffractometer (XRD, Mac Science M18XHF-SRA, Yokohama, Japan) with Cu-Kα radiation operating at 40 kV and 300 mA, and an electron probe microanalyzer (EPMA, Shimadzu, EPMA 1600, Kyoto, Japan).

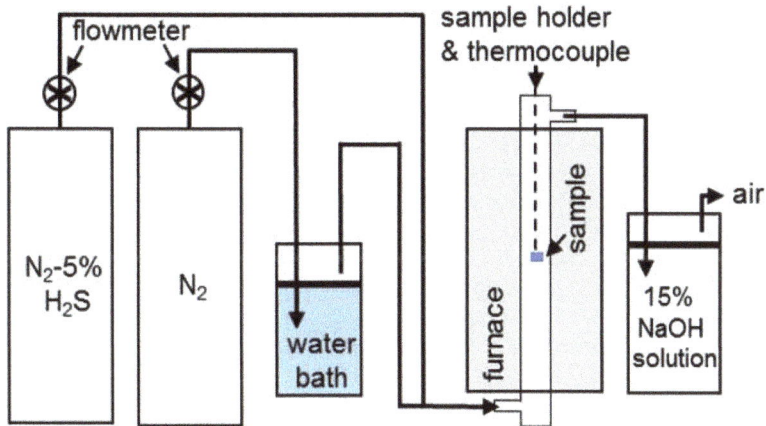

Figure 1. Corrosion testing apparatus.

3. Results and Discussion

Table 1 lists the weight gains of Fe-(9, 19, 28, 37)Cr alloys due to corrosion at 700 and 800 °C for 70 h, which were measured using a microbalance before and after corrosion. Fe-9Cr always displayed the worst corrosion resistance, gaining excessive weight. For example, Fe-9Cr oxidized fast even in the N_2 gas through the reaction with impurities such as 3 ppm H_2O and 2 ppm O_2 in the N_2 gas (99.999% pure). Fe-9Cr oxidized faster in the N_2/H_2O gas than in the N_2 gas because of water vapor [7]. Water vapor dissociates into oxygen and hydrogen, oxidizes the metal, and forms voids within the oxide scale according to the equation [1,3],

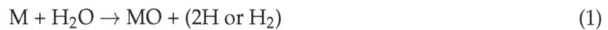

$$M + H_2O \rightarrow MO + (2H \text{ or } H_2) \tag{1}$$

Table 1. Weight gain of Fe-(9, 19, 28, 37)Cr alloys measured after corrosion at 700 and 800 °C for 70 h under 1 atm of N_2, $N_2/3.2\%H_2O$, and $N_2/3.1\%H_2O/2.42\%H_2S$ gas.

Temp.	Gas	Weight Gain (mg/cm^2)			
		9Cr	19Cr	28Cr	37Cr
700 °C	N_2	195	1–2	1–2	1–2
	N_2/H_2O	220	1–2	1–2	1–2
	$N_2/H_2O/H_2S$	2050	470	200	70
800 °C	N_2	235	1–2	1–2	1–2
	N_2/H_2O	400	1–2	1–2	1–2
	$N_2/H_2O/H_2S$	massive spalling	1530	690	550

Fe-9Cr corroded the most seriously in $N_2/H_2O/H_2S$ gas, because H_2S was much more harmful than H_2O. H_2S dissociates into hydrogen and sulfur. Sulfur forms non-protective metal sulfides according to the following equation:

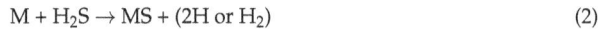

$$M + H_2S \rightarrow MS + (2H \text{ or } H_2) \tag{2}$$

Hydrogen, which is released from H_2S and H_2O, dissolves and ingresses into the alloy and the scale interstitially, generates lattice point defects, forms hydrogen clusters and voids, causes hydrogen embrittlement, produces volatile hydrated species, and accelerates cracking, spallation and fracture of the scale. Hence, no metals are resistant to H_2O/H_2S corrosion. As listed in Table 1, Fe-(19, 29, 37)Cr displayed much better corrosion resistance in N_2 and $N_2/3.2\%H_2O$ with weight gains of 1–2 mg/cm^2 than Fe-9Cr. Fe-(19, 29, 37)Cr formed 0.3- to 1.3-μm-thick, adherent oxide scales. However, even Fe-(19, 29, 37)Cr failed in $N_2/3.1\%H_2O/2.42\%H_2S$ with large weight gains, forming non-adherent, fragile sulfide scales as thick as 35–750 μm. This scale failure made the weight gains measured in $N_2/H_2O/H_2S$ gas inaccurate. In $N_2/H_2O/H_2S$ gas, the amount of local cracking, spallation and void formation in the scale varied for each test run. Although the accurate measurement of weight gains in $N_2/H_2O/H_2S$ gas was impossible, it was clear that weight gains due to scaling decreased sharply with the addition of Cr.

Figure 2 shows the XRD patterns of scales formed after corrosion at 800 °C for 70 h. The corrosion of Fe-9Cr in N_2 and N_2/H_2O resulted in the formation of Fe_2O_3 and Fe_3O_4, as shown in Figure 2a,b. Oxide scales formed on Fe-9Cr in N_2 and N_2/H_2O were 90 and 100 μm thick, respectively. Since X-rays could not penetrate such thick oxide scales, FeO and Cr-oxides such as $FeCr_2O_4$, which might form next to the alloy [1], were absent in Figure 2a,b. In contrast, Fe-(19, 28, 37)Cr alloys oxidized at much slower rates in N_2 and N_2/H_2O than Fe-9Cr alloy, as listed in Table 1. Fe-(19, 28, 37)Cr alloys formed the protective Cr_2O_3 scale, as typically shown in Figure 2c,d. Here, the Fe-Cr peaks were strong owing to the thinness of the oxide scales. In Fe-(19, 28, 37)Cr alloys, Cr was dissolved in the α-Fe matrix.

Figure 3 shows the EPMA analytical results on the scales formed on Fe-9Cr after corrosion at 700 °C for 70 h. The oxide scales that formed after corrosion in N_2 and N_2/H_2O were about 90 and 140 μm thick, respectively. The scale morphology and elemental distribution in N_2 gas were similar to those in N_2/H_2O gas, as shown in Figure 3, indicating that the same oxidation mechanism operated in N_2 and N_2/H_2O gas. Voids were sporadically scattered in both oxide scales, below which the oxygen-affected zone (OAZ) existed. Voids formed owing to the volume expansion during scaling, hydrogen released from the water vapor, and the Kirkendall effect arose due to the outward diffusion of cations during scaling. In both oxide scales, the outer layer consisted of iron oxides, while the inner layer consisted of (Fe,Cr) mixed oxides. This indicated that Fe^{2+} and Fe^{3+} ions were more mobile than Cr^{3+} ions. The oxidation in N_2 and N_2/H_2O gas was mainly controlled by the outward diffusion of iron ions through the inner (Fe,Cr) mixed oxide layer. Iron oxidized preferentially in N_2 and N_2/H_2O gas because iron is the base element and its oxide, FeO, is a non-stoichiometric compound with a relatively fast growth rate.

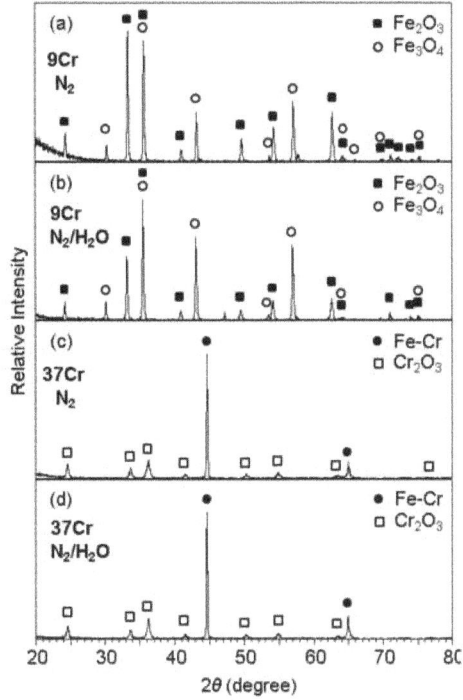

Figure 2. XRD patterns taken after corrosion testing at 800 °C for 70 h. (**a**) Fe-9Cr in N$_2$; (**b**) Fe-9Cr in N$_2$/3.2%H$_2$O; (**c**) Fe-37Cr in N$_2$; (**d**) Fe-37Cr in N$_2$/3.2%H$_2$O.

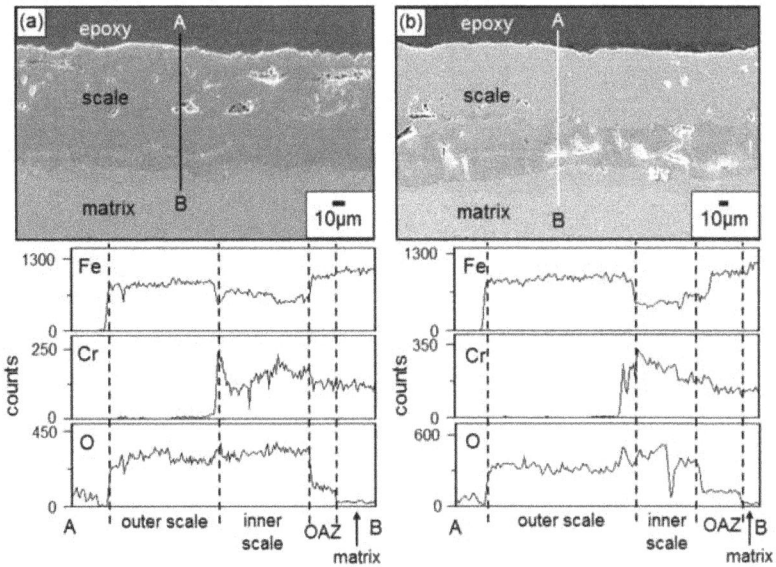

Figure 3. EPMA cross-section and line profiles of Fe-9Cr after corrosion at 700 °C for 70 h in (**a**) N$_2$; and (**b**) N$_2$/3.2%H$_2$O.

Figure 4 shows the EPMA analytical results on the scales formed on Fe-37Cr after corrosion at 700 °C for 70 h. The oxide scales that formed after corrosion in N_2 and N_2/H_2O were about 0.6 and 1.1 μm thick, respectively. In N_2 and N_2/H_2O gas, the Cr_2O_3 scale formed (Figure 2c,d), in which Fe was dissolved (Figure 4). The complete dissolution of Fe_2O_3 in Cr_2O_3 is possible, because Cr_2O_3 and Fe_2O_3 have the same rhombohedral structure [8]. Like Fe-37Cr, Fe-(19, 28)Cr also formed a thin Cr_2O_3 scale containing some Fe when they corroded in N_2 and N_2/H_2O gas. Once the thin but protective Cr_2O_3 scale formed, the outward diffusion of iron ions was suppressed so that good corrosion resistance was achieved.

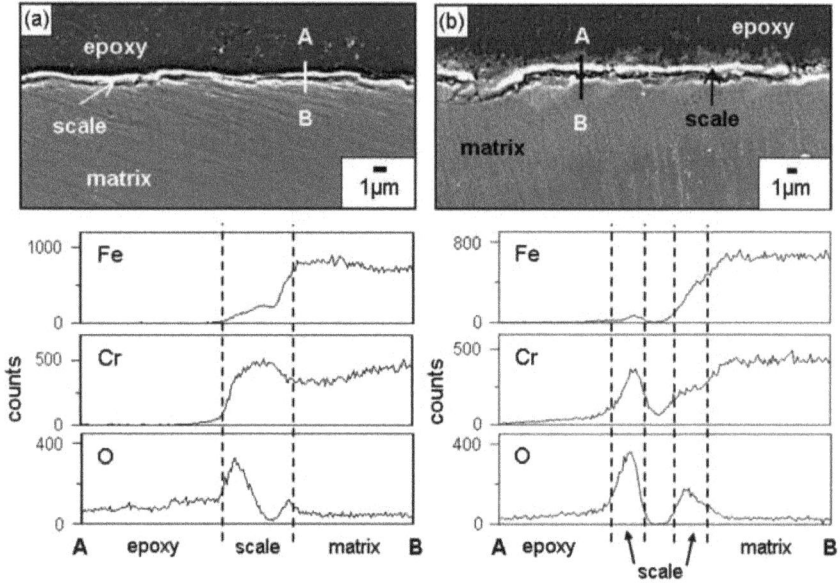

Figure 4. EPMA cross-section and line profiles of Fe-37Cr after corrosion at 700 °C for 70 h in (**a**) N_2; and (**b**) $N_2/3.2\%H_2O$.

In $N_2/3.1\%H_2O/2.42\%H_2S$ gas, Fe-(9-37)Cr alloys could not form Cr_2O_3, and corroded fast, as typically shown in Figure 5. The scales formed on Fe-(9, 19, 28, 37)Cr alloys consisted primarily of the outer FeS layer (Figure 5a), and the inner Cr_2S_3 layer containing some $FeCr_2S_4$ (Figure 5b). Since FeS grows fast owing to its high non-stoichiometry, outer FeS grains were coarser than the inner (Cr_2S_3, $FeCr_2S_4$) mixed grains. In Figure 5c, cracks propagated inter- and trans-granularly due mainly to the excessive growth stress generated in the thick outer scale. The scale shown in Figure 5d was about 100 μm thick, and had cracks and voids. A small amount of Cr was dissolved in the outer FeS layer (Figure 5e). The preferential sulfidation of iron in the outer FeS layer decreased the sulfur potential underneath, and thereby increased the oxygen potential in the inner Cr_2S_3-rich layer, leading to the incorporation of oxygen in the inner Cr_2S_3-rich layer. FeS is a p-type metal-deficit compound, which grows fast by the outward diffusion of Fe^{2+} ions [5,9,10]. Its defect chemical equation is as follows.

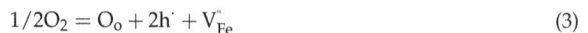

$$1/2O_2 = O_o + 2h^{\cdot} + V_{Fe}^{''} \tag{3}$$

here, O_o, $h^{\cdot}$ and $V_{Fe}^{''}$ mean the O atom on the O site, the electron hole in the valence band with a + 1 charge, and the iron vacancy with a − 2 charge. The defect chemical reaction for the dissolution of Cr_2S_3 in FeS is as follows.

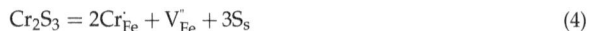

$$Cr_2S_3 = 2Cr_{Fe}^{\cdot} + V_{Fe}^{''} + 3S_s \tag{4}$$

Hence, the doping of Cr^{3+} ions would increase the concentration of iron vacancies, leading to the enhancement of the FeS growth. Oxygen was incorporated in the inner Cr_2S_3-rich layer (Figure 5e). However, no oxides were detected in Figure 5b, because their amount was small or oxygen was dissolved in the sulfide scales. Grains in the inner layer were fine owing to the nucleation and growth of Cr_2S_3, together with some $FeCr_2S_4$ and probably some oxides. In $N_2/3.1\%H_2O/2.42\%H_2S$ gas, Fe-(9, 19, 28, 37)Cr alloys sulfidized preferentially owing to the high sulfur potential in the test gas.

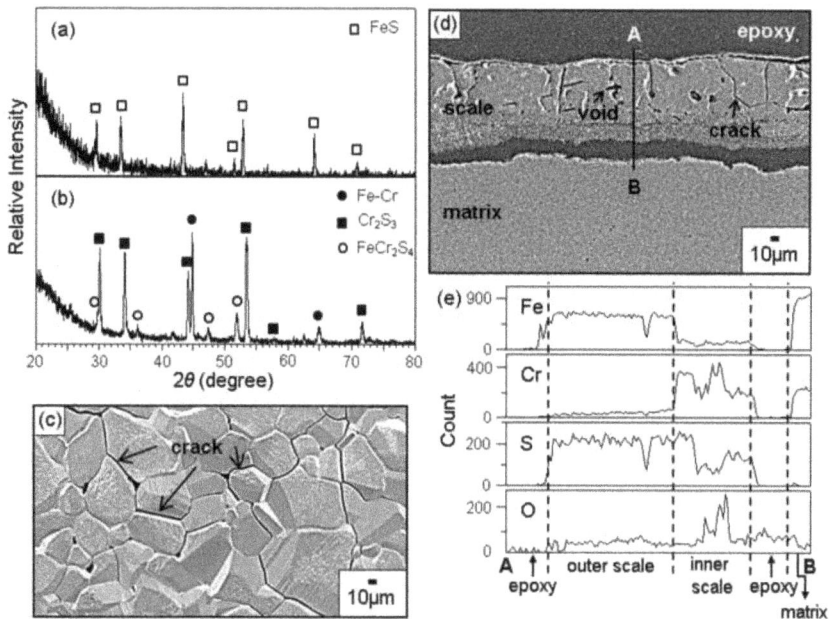

Figure 5. Fe-19Cr after corrosion at 700 °C for 40 h in $N_2/3.1\%H_2O/2.42\%H_2S$. (**a**) XRD pattern after corrosion; (**b**) XRD pattern taken after grinding off the outer scale; (**c**) SEM top view; (**d**) EPMA cross-section; (**e**) EPMA line profiles of along A–B denoted in (**d**).

4. Conclusions

When Fe-9Cr alloy corroded at 700 and 800 °C in N_2 and $N_2/3.2\%H_2O$ gas, thick, porous oxide scales formed, which consisted of the outer iron oxide layer and the inner (Fe,Cr) mixed oxide layer. Under the same corrosion condition, Fe-(19, 28, 37)Cr alloys formed thin, dense, protective Cr_2O_3 oxide layers, in which iron was dissolved to a certain extent. In $N_2/3.1\%H_2O/2.42\%H_2S$ gas, Fe-(9, 19, 28, 37)Cr alloys corroded fast, forming thick, non-adherent, fragile scales, which consisted of the outer FeS layer and the inner Cr_2S_3 layer containing some $FeCr_2S_4$. The preferential sulfidation of Fe-(9, 19, 28, 37)Cr alloys in the H_2S-containing gas was responsible for the poor corrosion resistance of Fe-(9, 19, 28, 37)Cr alloys.

Acknowledgments: This research was supported by Basic Science Research Program through the National Research Foundation of Korea (NRF) funded by the Ministry of Education (2016R1A2B1013169), Korea.

Author Contributions: Min Jung Kim conceived and designed the experimental procedure and drafted the paper. Muhammad Ali Abro prepared the samples, conducted the experiments and analyzed the data. All the results were discussed with Dong Bok Lee who supervised the experimental work and finalized the paper.

Conflicts of Interest: The authors declare no conflict of interest.

References

1. Birks, N.; Meier, G.H.; Pettit, F.S. *Introduction to High Temperature Oxidation of Metals*, 2nd ed.; Cambridge University: New York, NY, USA, 2006.
2. Khanna, A.S. *Introduction to High Temperature Oxidation and Corrosion*; ASM Int.: Metals Park, OH, USA, 2002; p. 135.
3. Young, D. *High Temperature Oxidation and Corrosion of Metals*; Elsevier: Oxford, UK, 2008.
4. Mrowec, S.; Walec, T.; Werber, T. High-temperature sulfur corrosion of iron-chromium alloys. *Oxid. Met.* **1969**, *1*, 93.
5. Mrowec, S.; Przybylski, K. Transport properties of sulfide scales and sulfidation of metals and alloys. *Oxid. Met.* **1985**, *23*, 107. [CrossRef]
6. Schulte, M.; Rahmel, A.; Schutze, M. The Sulfidation Behavior of Several Commercial Ferritic and Austenitic Steels. *Oxid. Met.* **1998**, *49*, 33. [CrossRef]
7. Saunders, S.R.J.; McCartney, L.N. Current Understanding of Steam Oxidation—Power Plant and Laboratory Experience. *Mater. Sci. Forum* **2006**, *119*, 522. [CrossRef]
8. Asteman, H.; Norling, R.; Svensson, J.-E.; Nylund, A.; Nyborg, L. Quantitative AES depth profiling of iron and chromium oxides in solid solution, $(Cr_{1-x}Fe_x)_2O_3$. *Surf. Interface Anal.* **2002**, *34*, 234. [CrossRef]
9. Mrowec, S.; Wedrychowska, M. Kinetics and mechanism of high-temperature sulfur corrosion of Fe-Cr-Al alloys. *Oxid. Met.* **1979**, *13*, 481. [CrossRef]
10. Danielewski, M.; Mrowec, S.; Stolosa, A. Sulfidation of iron at high temperatures and diffusion kinetics in ferrous sulfide. *Oxid. Met.* **1982**, *17*, 77. [CrossRef]

metals

MDPI

Article

Superior Strength of Austenitic Steel Produced by Combined Processing, including Equal-Channel Angular Pressing and Rolling

Marina V. Karavaeva *, Marina M. Abramova, Nariman A. Enikeev, Georgy I. Raab and Ruslan Z. Valiev

Institute of Physics of Advanced Materials, Ufa State Aviation Technical University, 12 K. Marx str., Ufa 450008, Russia; abramovamm@yandex.ru (M.M.A.); nariman.enikeev@ugatu.su (N.A.E.); giraab@mail.ru (G.I.R.); rzvaliev@yahoo.com (R.Z.V.)
* Correspondence: karma11@mail.ru; Tel.: +7-917-781-7784

Academic Editor: Hugo F. Lopez
Received: 27 October 2016; Accepted: 29 November 2016; Published: 8 December 2016

Abstract: Enhancement in the strength of austenitic steels with a small content of carbon can be achieved by a limited number of methods, among which is ultrafine-grained (UFG) structure formation. This method is especially efficient with the use of severe plastic deformation (SPD) processing, which significantly increases the contribution of grain-boundary strengthening, and also involves a combination of the other strengthening factors (work hardening, twins, etc.). In this paper, we demonstrate that the use of SPD processing combined with conventional methods of deformation treatment of metals, such as rolling, may lead to additional strengthening of UFG steel. In the presented paper we analyze the microstructure and mechanical properties of the Cr–Ni stainless austenitic steel after a combined deformation. We report on substantial increases in the strength properties of this steel, resulting from a consecutive application of SPD processing via equal-channel angular pressing and rolling at a temperature of 400 °C. This combined loading yields a strength more than 1.5 times higher than those produced by either of these two techniques used separately.

Keywords: stainless steel; severe plastic deformation; strength; ultrafine-grained materials

---

1. Introduction

Severe plastic deformation (SPD) processing significantly improves the mechanical properties of a broad range of metallic materials due to the formation of an ultrafine-grained (UFG) structure, ensuring the concurrent action of several mechanisms of strengthening thanks to the hardening contributions of solid solution, precipitations and particles, defect structures and, primarily, grain refinement [1–7]. A high-strength state is provided by controlling the microstructural parameters that are sensitive to SPD processing regimes. SPD parameters that have the greatest effect on the microstructure are strain, temperature and loading route. The latter has an effect on both the kinetics of microstructural evolution and the homogeneity of the produced microstructure. For example, when studying the microstructure transformation of Ti alloys with a change of the deformation path, it was shown that the substitution of a monotonic loading with an essentially non-monotonic one enabled activation of new slip systems and thus intensified the process of microstructural refinement [8,9]. With respect to SPD processing, it was demonstrated that the so-called route of equal-channel angular pressing (ECAP) has a great effect on structural evolution [10,11]. The best results, in terms of microstructure refinement and enhancement of mechanical properties, were obtained when using routes B and Bc, in which the billet is rotated by 90° around its axis between ECAP passes. Such a turn changes the schemes of the principal stresses and strains in a material, and as a result, the deformation process becomes non-monotonic. A similar result was obtained for the cyclic HPT when sufficient grain refinement in Ni and Fe was reported to be achieved at a smaller deformation level than for the one-direction HPT [12]. A vivid example of non-monotonic loading is the SPD technique of multiple forging, in which the change of the scheme of principal stresses is achieved as a result of a consecutive rotation of the billet around three axes [13,14].

It is possible to realize the non-monotonic loading process through a consecutive processing of billets by different methods. This procedure has already been tested successfully for Ti-based [15], Cu-based [16] and Al-based alloys [17,18]. At the first stage of processing, SPD by ECAP-Conform was conducted, and at the second stage, rolling or drawing was performed. It is noted in all studies that a change in the type of loading had a beneficial effect on the properties of the produced materials. At the second stage of processing, an additional increase was observed in the microhardness and strength of UFG materials which had been produced by SPD at the first stage of processing. It is more difficult to unambiguously determine the effect of a change in the deformation type of producing UFG materials on the features of their microstructure. At the present time, the experimental data reported in the literature are not sufficient to summarize the results, especially for steels. Besides, of great importance is the microstructure formed immediately during SPD processing, as well as the nature of the material itself. After the rolling of even an equiaxed UFG structure, a structure was observed that was elongated in the direction of plastic straining. For copper, an increase in the structural homogeneity was revealed [16], and conversely, for an Al alloy, a separation of microstructure into two fractions was observed, one of which contained shear bands, and the other one contained equiaxed grains [17].

In this paper, we investigate the possibility of increasing the strength of an austenitic stainless steel through the use of combined strain processing. For this type of steel it is practically impossible to increase strength by thermal treatment, and thus microstructure refinement by deformation processing is an efficient means of strengthening.

2. Materials and Methods

Austenitic stainless steel was selected as an object of investigation. The chemical composition of the steel is given in Table 1. In order to produce a single-phase austenitic structure prior to SPD processing, the steel was water-quenched from a temperature of 1050 °C (exposure time 1 h). The SPD processing of rods with a diameter of 10 mm and a length of 100 mm was conducted by ECAP through 8 passes via route Bc at a temperature of 400 °C. The intersection angle of channels in the die-set was 120° (Figure 1).

Table 1. Chemical composition (wt. %) of the austenitic steel under investigation.

C	Cr	Ni	Ti	Si	S	P	Fe
0.08	16.19	9.13	0.3	0.58	0.03	0.08	bas.

The thermal conditions of ECAP processing were selected in accordance with earlier studies [6,19] that demonstrated the efficiency of SPD processing for microstructure refinement and enhancement of the mechanical properties of the austenitic stainless steel at the given temperature, as well as for the formation of grain-boundary segregations and nanotwins resulting in additional strengthening. The number of passes was selected in such a way as to be sufficiently large to impose such a strain under which the hardness and strength of a UFG billet reach saturation. The produced UFG state is further referred to as "ECAP".

Rolling was conducted in smooth rolls at the same temperature of billet heating, 400 °C, through 15 passes to a final strip thickness of 2.3 mm. The total reduction was 77% (Figure 1). The produced UFG state is further referred to as "ECAP + Rol". This regime was selected on the basis of the above-mentioned considerations, as well as to preserve the integrity of the billet.

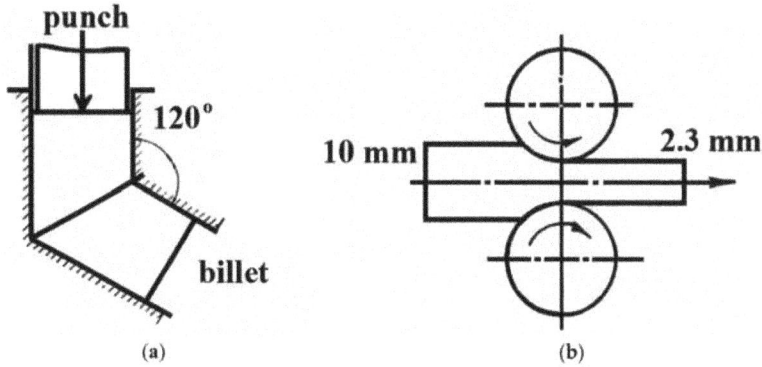

Figure 1. The principle of the combined processing of the steel (**a**) Stage I—ECAP, (**b**) Stage II—Rolling.

To study the effect of the combined processing on the microstructure and properties, we also investigated the billets subjected to rolling under the same conditions, but without a preliminary deformation processing by ECAP. This state is further referred to as "Rol".

The microstructure was studied in the longitudinal section of a rod and a strip. To investigate the microstructure, electrolytic etching was performed in a chemically pure nitric acid (the mass fraction of the acid was at least 65%). The etching time was from 5 to 10 s under a voltage of 13–20 V. Structural studies were performed using an Olympus GX51 optical microscope (Olympus Corp, Tokyo, Japan), a JEOL JSM-6490VL scanning electron microscope (Jeol Ltd, Tokyo, Japan) and a JEOL JEM-2100 transmission electron microscope (Jeol Ltd, Tokyo, Japan). The grain sizes were determined from the dark-field images of the microstructure. At least 300 grains were measured for each condition. The dislocation density ϱ_{xrd} was determined from the results of X-ray studies according to the expression [20]:

$$\varrho_{XRD} = \frac{2\sqrt{3}\left\langle \varepsilon^2 \right\rangle^{1/2}}{b \cdot d_{XRD}}$$

where $(\varepsilon^2)^{1/2}$ is the level of elastic microdistortions of the crystal lattice; b is the Burgers vector of dislocations; d_{xrd} is the size of coherent scattering domains.

Microhardness was measured on a Micromet-5101 device in the longitudinal direction. At least 30 measurements were made for each condition. Uniaxial tensile testing was performed on an INSTRON 8801 tensile testing machine (Instron Eng. Corp., High Wycomib, UK) at room temperature. For the tensile tests, flat samples with a gauge length of 4 mm were used, the strain rate was 10^{-3} s^{-1}.

3. Results

The microstructure of the steel in the as-received state was represented by equiaxed austenite grains with a mean size of (9 ± 2) μm (Figure 2a). In some grains, twins were observed. The volume fraction of grains containing twins was about 10%.

After quenching, the size of austenite grains increased up to an average value of (40 ± 11) μm. Practically all grains contained wide twins. At the boundaries of austenite grains and at twin boundary/grain boundary intersections, serrations were observed.

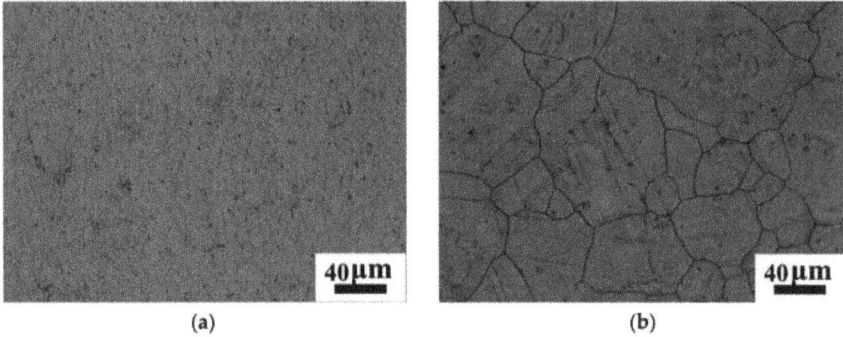

Figure 2. Microstructure of the austenitic steel: as-received condition (**a**); after quenching (**b**).

3.1. Microstructure of the Austenitic Steel after SPD Processing and Rolling

After SPD processing by ECAP, within the austenite grains we observed the formation of differently-directed shear bands (Figure 3a—the sample axis is vertical). As a result of the intersection of these bands, new boundaries form and grain refinement takes place. The microstructure is heterogeneous. At 10,000 times magnification (Figure 3b), relatively coarse grains with sizes of several µm and fine grains with sizes much smaller than 1 µm are visible. The coarse grains are elongated in the direction of the sample axis (Figure 3a). The volume fraction of the regions with relatively coarse grains amounts to about 10%.

When the structure was examined in detail by TEM, structural heterogeneity was also revealed (Figure 3c). A large volume of the structure (about 60%) is represented by shear bands with thin boundaries, within which a developed dislocation structure in the form of wide dislocation boundaries is observed. These boundaries divide the bands into non-equiaxed cells. The cell size amounts to, on average, about 180 nm in the transverse direction and 370 nm in the longitudinal direction (Figure 3c). Alongside shear bands, practically equiaxed grains with a reduced dislocation density and thin equilibrium boundaries are present in the structure. The grain size is about 350 nm. Separate deformation twins are observed in the grains (about 10 nm in thickness) (Figure 3d). The fraction of grains with twins does not exceed 5%. The average spacing between the twin boundaries is about 75 nm. The selected area electron diffraction pattern shown in the insert in Figure 3c reveals separate reflections located circumferentially, which indicates high-angle misorientations of grain boundaries.

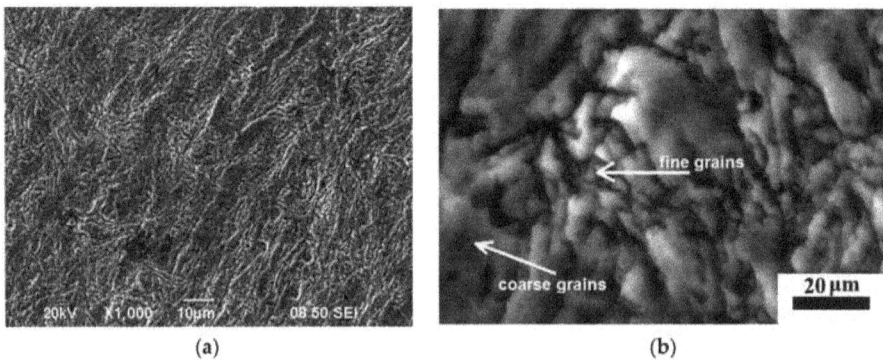

Figure 3. *Cont.*

Figure 3. Microstructure of the steel after SPD processing (via ECAP) in the longitudinal section ("ECAP" condition): (**a,b**) SEM; (**c,d**) TEM, the aperture size for diffraction patterns ~1 μm^2.

Thus, after SPD processing via ECAP, a heterogeneous austenitic UFG structure is formed. This structure consisted of grains/subgrains elongated in the direction of straining, with a small number of twins.

After rolling of the ECAP-processed steel, further grain refinement is observed (Figure 4). Individual grains are practically not identified by an optical microscope. The boundaries of the original austenite grains are not visible either (Figure 4a).

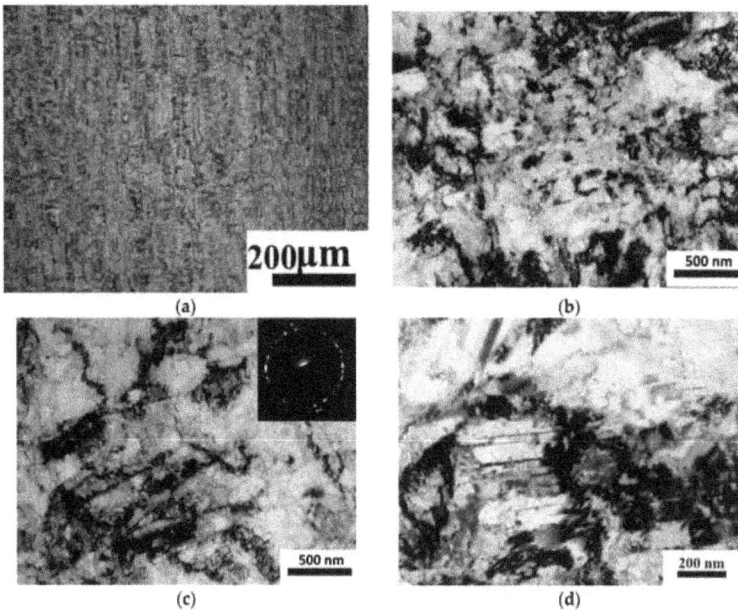

Figure 4. Structure of the austenitic steel after ECAP and subsequent rolling to a total reduction in area of 77% ("ECAP + Rol" condition): (**a**) optical microscopy; (**b–d**) TEM, the aperture size for diffraction patterns ~1 μm^2.

When the microstructure is examined by TEM, it can be seen that the microstructure has become more homogeneous (Figure 4b) as compared to the one observed in the "ECAP" state (Figure 3c).

The structure has a grain/cellular character. Shear bands are preserved in separate regions, but the fraction of banded structure is only about 10%. The dislocation density increases, while the size of structural elements decreases to 110 nm. Thin twins are observed in the grains (Figure 4d). The fraction of grains containing twins increases to 14%. The average twin spacing decreases to 30 nm. The electron diffraction pattern shown in the insert of Figure 4c has a ring-shaped form, which indicates high-angle misorientation between grains. Thus, combined loading leads to further microstructure refinement—the mean grain size decreases to 110 nm, and the fraction of nanotwins grows.

In the steel samples after rolling ("Rol" condition) the boundaries of original austenite grains (Figure 5a), elongated in the rolling direction are still observed. Formation of shear bands is distinctly observed within the grains there. At the boundaries of the original austenite grains and at the shear band/grain boundary intersections, ledges are seen. A banded structure (Figure 5b,c) is also observed in some areas. Inside the bands there are wide boundaries dividing grains into cells (Figure 4b–d). The average cell size amounts to 560 nm. The structure is characterized by an increased dislocation density. Twins are almost absent. Thus, in the "Rol" condition, the steel is characterized by a banded cellular structure with a cell size of 560 nm, which does not contain twins.

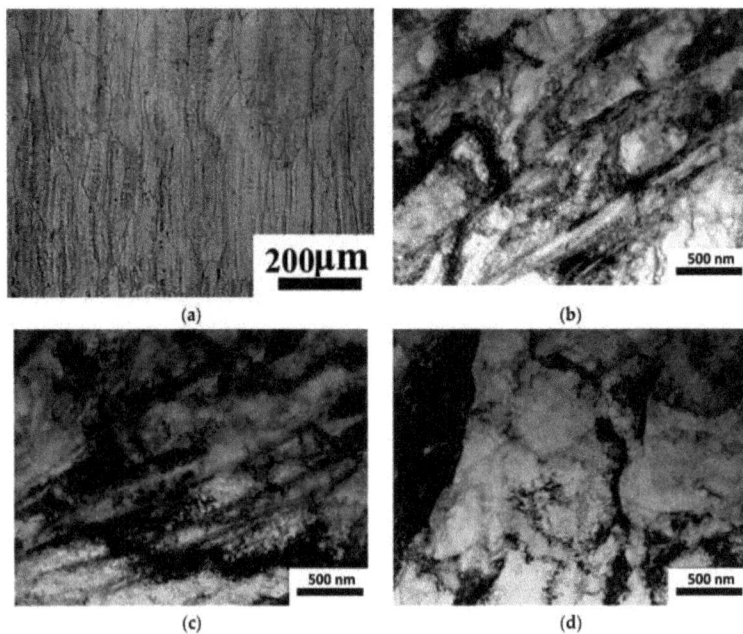

Figure 5. Microstructure of the austenitic steel after rolling to a total reduction in area of 77% ("Rol" condition): (**a**) optical microscopy; (**b–d**) TEM.

3.2. Mechanical Properties of the Austenitic Steel

The average microhardness value of the austenitic steel in the as-received state consists (1970 ± 60) MPa (Figure 6). After quenching microhardness declines slightly to a value of (1820 ± 30) MPa, which is related to the growth of austenite grains, as well as to a more complete dissolution of excess phases during heating prior to quenching.

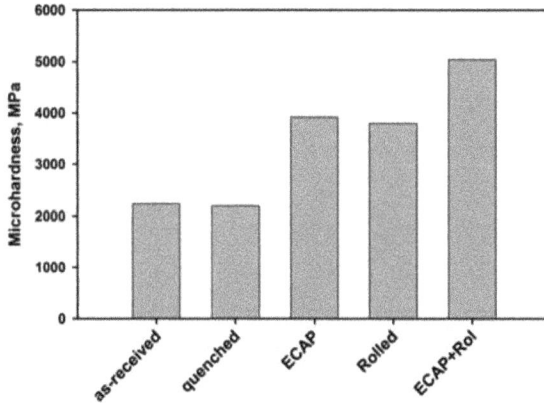

Figure 6. Microhardness of the austenitic steel after different types of processing.

As a result of microstructure refinement, in the "ECAP" state the average microhardness value of the steel grows two-fold, reaching (3920 ± 50) MPa. After rolling to 77% without a preliminary ECAP processing ("Rol"), there is also observed an increase in hardness, very similar to ECAP processing—to 3800 ± 50 MPa. A combination of ECAP and rolling results in an even greater increase in microhardness, namely 33%, reaching (5040 ± 40) MPa (Figure 6).

In a similar manner, straining has an effect on the steel's strength as well. Figure 7 shows the engineering stress-strain curves obtained during tensile tests of the steel samples in different state. It is obvious that the deformation behavior of the material changes depending of the type of processing of the steel. The quantitative data on the mechanical properties of the steel in different state are summarized in Table 2. Peculiar to the quenched condition there is a significant capability for strengthening: the yield stress is $\sigma_{0.2} = 200$ MPa, the ultimate tensile strength (UTS) is 2.5 times higher—$\sigma_{ult} = 720$ MPa. The elongation is $\delta = 65\%$.

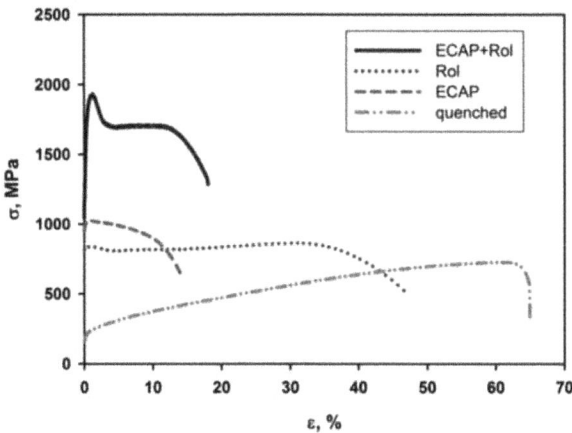

Figure 7. Engineering stress-strain curves of the steel after different types of processing.

Table 2. Mechanical properties of the steel after different types of processing.

Condition	Offset Yield Stress $\sigma_{0.2}$ (MPa)	Upper Yield Stress (Corresponds to Yield Drop) σ_{Bu} (MPa)	Lower Yield Stress (Corresponds to Yield Plateau) σ_{l} (MPa)	Ultimate Tensile Strength σ_{ult} (MPa)	Uniform Elongation $\delta_{uniform}$ (%)	Elongation to Failure δ (%)	Microhardness (MPa)
quenching	200	-	-	720	62	65	1820
ECAP	950	-	-	1020	1	14	3920
Rol	-	830	800	855	33	47	3800
ECAP + Rol	-	1925	1700	1720	11	18	5040

The steel in the "ECAP" state is characterized by a high value of yield stress $\sigma_{0.2} = 950$ MPa with a very short period of insignificant strengthening: the uniform elongation is only 1%, the UTS is $\sigma_{ult} = 1020$ MPa. After that, a rapid strain localization and necking take place, corresponding to the region with a stress decline in the diagram. The total elongation is $\delta = 14\%$. In spite of similar values of microhardness and yield stress in the "ECAP" and "Rol" states, the deformation behavior of those differs significantly. In the curve of the rolled sample, there appears a weakly expressed yield drop, corresponding to the upper yield stress $\sigma_u = 830$ MPa. After that there is observed a yield plateau, corresponding to the lower (physical) yield stress $\sigma_l = 800$ MPa and the region of weak strengthening. The UTS is $\sigma_{ult} = 855$ MPa, the elongation to failure is $\delta = 47\%$. The highest strength is exhibited by the samples after the "ECAP + Rol" combined loading. This state is displayed by the curve with a distinct yield drop. The upper yield stress is $\sigma_u = 1925$ MPa, and the lower yield stress, corresponding to the yield plateau, is $\sigma_l = 1700$ MPa. The curve does not demonstrate notable strengthening, strain localization starts as the elongation reaches the value $\delta = 11\%$, and the total elongation is $\delta = 18\%$.

Thus, the type of deformation processing determines not only the level of properties, but also the tensile mechanical behavior of the steel.

4. Discussion

Numerous studies on the SPD processing of bulk metallic billets via ECAP have demonstrated that the increase in hardness and strength is observed after the initial one or two passes, after which further strengthening becomes much slower [21–23]. Meanwhile, the possibilities for strength enhancement in a material have not yet been exhausted. This is confirmed by the fact that under processing by high-pressure torsion, as a rule, the observed hardness values are significantly larger than the ones that can be attained by ECAP processing [10].

Microstructural features of metallic materials provide activation of the related deformation mechanisms, which, in their turn, contribute to the strengthening of the given material. Varying SPD parameters one can purposefully form the targeted features in the produced UFG materials and put into action the corresponding strengthening mechanisms. In UFG materials produced by SPD, strengthening can be achieved due to several mechanisms [1–7,23]:

1. Grain-boundary and dislocation strengthening. During the grain refinement the volume fraction of grain boundaries, which are an efficient impediment for dislocation movement, significantly increases. For the formation of new strain-induced boundaries, dislocation generation in various slip systems is necessary.

   When analyzing the types of loading realized in the course of SPD, it is necessary to mention two distinctive features typical of SPD processing:

   a. a high hydrostatic constituent, which is especially significant in high-pressure torsion, but present in all deformation techniques;
   b. an essential non-monotony of strain, typical for most SPD techniques, such as ECAP or multiple forging.

   Both of these features enable activating additional slip systems, thus leading to an increase in dislocation density, formation of new interfaces and microstructure refinement.

2. Solid-solution strengthening and precipitation hardening. These mechanisms are competing ones, since as a result the alloying of a solid solution, the corresponding strengthening grows, but the amount of dispersed particles (providing precipitation hardening) decreases. The contribution of these constituents to strengthening is determined primarily by the deformation temperature. It has been shown [24,25] that at room temperature the dissolution of second-phase particles prevails, but as the deformation temperature is increased, precipitation is observed.

3. Formation of segregations at grain boundaries. This process is also connected with the solid solution decomposition during SPD and a transfer of solute atoms to the boundaries. The action of this mechanism is also thermally dependent: at room temperature no segregations were observed, and at elevated temperatures the formation of the grain boundary segregations was shown [5,26].

4. Formation of twins. For a number of materials, including austenitic steels, it is typical that nanotwins form during SPD processing. The high-angle boundaries of nanotwins are also impediments for dislocation movement and, consequently, they provide additional strengthening. Twinning may be activated when possibilities for slip are limited. When the scheme of the stress-strain state is changed (in this particular case, by changing the type of loading), the direction of action of the maximum tangential stresses changes with respect to the sample's axis. As a result of such a change, new slip systems should be activated, and the activation of twinning is also possible.

Thus, a change in the loading scheme may activate at least several of the above-mentioned factors of strengthening: an increase in dislocation density, grain refinement and an increase in the fraction of twins. These conclusions are confirmed by studies conducted on various materials. For instance, in [17], in the Al alloy 5083 after ECAP processing and additional compression, imitating rolling conditions, an increase in dislocation density was observed. An enhancement of strength after ECAP-Conform processing and compression of Ti [15], after ECAP processing and rolling of Cu [16], was accounted for by the formation of additional low-angle boundaries within grains and a transformation of the low-angle boundaries into high-angle ones. This conclusion is also consistent with the studies on the microstructure of austenitic steel in different states reported in the present study (see Table 3).

Table 3. Features of the structure of steel after straining.

Condition	Dislocation Density (m^{-2})	Grain/Cell Size (nm)	Fraction of Shear Bands	Fraction of Grains with Twins (%)	Twin Spacing (nm)
ECAP	1.28×10^{14}	350	60	5	75
ECAP + Rol	7.19×10^{14}	110	10	14	30
Rol	4.27×10^{14}	560	80	-	-

Comparison of the microstructural parameters of steel in different states demonstrates that the size of structural elements considerably decreases as compared with the "ECAP" and the "Rol" states as a result of combined loading. This leads to a considerable increase in the density of grain boundaries.

Let us consider a generalized dependence of yield stress on grain size in terms of the Hall-Petch relation, presented on the basis of literature data in Figure 8.

The results obtained in the present study are also presented in the graph. It can be seen that the points corresponding to the "ECAP" or "Rol" states have a certain deviation from the line summarizing literature data towards larger values of yield stress. Moreover, the point corresponding to the "ECAP + Rol" condition is located much higher than expected in accordance with the Hall-Petch relation.

As considered above, strengthening of nanostructured steels is provided not only by grain size. For austenitic steels, additional strengthening is introduced by the dislocation mechanism, as well as by twin boundaries, as was demonstrated in [6]. In the general case, the contributions of different mechanisms follow linear additivity [2,3,5–7,23]:

$$\Delta \sigma_y = \Delta \sigma_{FS} + \Delta \sigma_{SS} + \Delta \sigma_\varrho + \Delta \sigma_{GB}$$

where σ_{FS} is the friction stress of γ-iron's lattice; $\Delta \sigma_{SS}$ is solid-solution strengthening; $\Delta \sigma_\varrho$ is dislocation strengthening; $\Delta \sigma_{GB}$ is grain-boundary strengthening.

Let us estimate the contribution of these mechanisms into the yield stress of the investigated steel in each state.

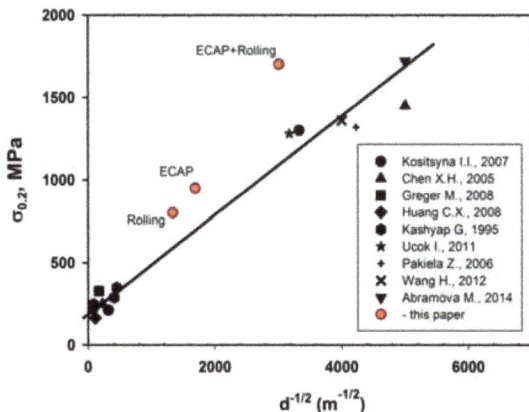

Figure 8. Hall-Petch relation for chromium-nickel austenitic steels, built on the basis of literature data [6,27–34] (back marks) and the results of the present study (red marks).

The lattice friction stress and solid solution hardening can be defined from the Hall-Petch relation, displayed in Figure 8, as the stress corresponding to the infinitely large grain size. In the given case, it is $\Delta\sigma_0 = 195$ MPa.

Dislocation strengthening can be estimated according to:

$$\Delta\sigma_\varrho = \alpha MbG\varrho^{1/2}$$

where $\alpha = 0.3$ is a constant; $M = 3.05$ is the Taylor factor; $G = 77$ GPa is the shear modulus and $b = \sqrt{2}a/2$ is the Burgers vector for the investigated steel.

Let us define additional grain-boundary strengthening, taking into account the presence of twins, as [6]:

$$\Delta\sigma_{GB} = (1-f)\,k_y d^{-1/2} + fk_y\lambda^{-1/2}$$

where f is the fraction of grains with twins; $k_y = 0.3$ MPa·m$^{1/2}$ is a constant derived from the dependence in Figure 8; d is the average grain/cell size; λ is the average twin spacing.

The results of the analysis are given in Table 4.

Table 4. Calculated results for the contribution of different mechanisms to the strengthening of the steel.

Condition	$\Delta\sigma_\varrho$	$\Delta\sigma_{GB}$			Calculated Value	Experimental Value
		$\Delta\sigma_d$	$\Delta\sigma_{tw}$	$\Delta\sigma_d + \Delta\sigma_{tw}$		
ECAP	202	481	55	536	933	950
ECAP + Rol	480	778	242	1020	1688	1700
Rol	370	401	-	401	771	800

The calculated yield stress values are very close to the experimental ones. Microstructural studies and the presented estimations show that strength enhancement of the steel under a combined loading is provided predominantly by the grain-boundary hardening contribution in accordance with the Hall-Petch equation. Besides, unlike in ECAP processing, no twins were observed in the structure of the steel after rolling at a temperature of 400 °C. After the combined loading, the fraction of twins increases even compared to the "ECAP" state, and this component also notably contributes to the steel's strengthening (see Table 4). The dislocation contribution into yield stress grows almost two-fold.

In addition to further strengthening, the combined "ECAP + Rol" loading also has an effect on the steel's deformation behavior, which is principally different not only from the quenched state, but also from the steel's behavior in the "ECAP" and "Rol" states. In the quenched state, the microstructure is characterized by a small density of grain boundaries and wide twins. After straining dislocation density increases, and as a result, an extensive region of strengthening and a high value of ductility are observed in the curve. In the "ECAP" state the stress-strain curve is typical for materials subjected to SPD—the maximum stress is achieved at early deformation stage, then rapid localization of strain and failure occur.

After "ECAP + Rol" treatment, a distinct yield drop is observed in the curve. Its appearance could be caused by segregations. The formation of segregations during elevated temperature SPD processing was found in recent years in SPD alloys, including austenitic steels [6]. In the samples after rolling, the appearance of a weakly expressed yield drop could indicate the formation of segregations or atmospheres pinning dislocations. Evidently, the formation of a UFG structure with a high density of grain boundaries during ECAP processing stimulates segregation formation during subsequent rolling, which is expressed in the yield phenomenon observed in the curve. The contribution of segregations can be estimated as the difference between the upper and the lower yield stresses, which amounts to 225 MPa for the steel in the "ECAP + Rol" condition. However, this issue requires an additional detailed study.

It should be noted that the steel after "ECAP + Rol" is characterized by rather high values of both uniform (11%) and total (18%) elongation. This may also be related to the pinning of dislocations by atmospheres or segregations of solutes: after the disruption of the blocking of a large quantity of dislocations, their free movement is possible, thus ensuring an additional deformation of the sample.

Thus, the application of the combined "ECAP + Rol" technique results in a considerable growth in the density of grain boundaries and increases the dislocation density and fraction of twins in the microstructure, which enables enhancement of the strength characteristics, while at the same time preserving the ductility of the UFG austenitic steel.

5. Conclusions

(1) A combination of SPD processing and a conventional metal forming technique for the rolling of austenitic steel leads to a further refinement of a homogeneous UFG cell-granular microstructure with a high density of grain boundaries and a large fraction of twins.

(2) As a result, the tensile mechanical behavior of the UFG steel samples produced by the combined "ECAP + Rol" loading changes—it exhibits a yield drop, to which corresponds the upper yield stress of 1925 MPa, as well as a yield plateau, and the yield stress amounts to 1700 MPa. The obtained values of strength are 1.5 times higher than the values of yield stress obtained when using only the ECAP technique (950 MPa) or only rolling (~815 MPa). Besides, in the UFG sheet produced by combined loading, a rather reasonable level of ductility is preserved: a uniform elongation of 11% and a total elongation of 18%.

(3) The enhancement of the strength characteristics is achieved as a result of a combined action of several strengthening mechanisms: grain-boundary strengthening, dislocation strengthening, twinning-induced strengthening and, presumably, strengthening due to the formation of solute segregations in grain boundaries.

Acknowledgments: The authors acknowledge the financial support from the Ministry of Education and Science of the Russian Federation under Grant Agreement No. 14.583.21.0012 (unique identification number RFMEFI58315X0012) and by the International Research & Development Program of the National Research Foundation of Korea (NRF) funded by the Ministry of Science, ICT and Future Planning (MSIP) of Korea (Grant No. K1A3A1A49.070466, FY2014). A part of investigations was conducted using the facilities of the

Common Use Center Nanotech (Ufa State Aviation Technical University) and Centre for X-ray Diffraction Studies (Research Park of Saint-Petersburg State University).

Author Contributions: M.V.K. fulfilled the general analysis of results, estimation of hardening contributions and she was responsible for writing the manuscript; M.M.A. was dealing with processing by rolling, specimen preparation, TEM measurements and mechanical testing of the samples; N.A.E. was responsible for XRD analysis and data processing as well as for discussion of hardening mechanisms; G.I.R. conducted ECAP-processing; R.Z.V. participated in task definition and performed general supervision of the conducted studies All authors contributed to discussion and summarizing of results as well as to correction and revisions of the manuscript.

Conflicts of Interest: The authors declare no conflict of interest.

References

1. Valiev, R.Z.; Horita, Z.; Langdon, T.G.; Zehetbauer, M.J.; Zhu, Y.T. Fundamentals of superior properties in bulk nano SPD materials. *Mater. Res. Lett.* **2016**, *4*, 1–21. [CrossRef]
2. Valiev, R.Z.; Enikeev, N.A.; Langdon, T.G. Towards superstrength of nanostructured metals and alloys, produced by SPD. *Met. Mater.* **2011**, *49*, 1–9.
3. Hasan, H.S.; Peet, M.J.; Avettand-Fénoël, M.-N.; Bhadeshia, H.K.D.H. Effect of tempering upon the tensile properties of a nanostructured bainitic steel. *Mater. Sci. Eng. A* **2014**, *615*, 340–347. [CrossRef]
4. Valiev, R.Z.; Estrin, Y.; Horita, Z.; Langdon, T.G.; Zehetbauer, M.J.; Zhu, Y.T. Producing Bulk Ultrafine-Grained Materials by Severe Plastic Deformation: Ten Years Later. *JOM* **2016**, *68*, 1216–1226. [CrossRef]
5. Kamikawa, N.; Abe, Y.; Miyamoto, G.; Funakawa, Y.; Furuhara, T. Tensile behavior of Ti, Mo-added low carbon steels with interphase precipitation. *ISIJ Int.* **2014**, *54*, 212–221. [CrossRef]
6. Abramova, M.M.; Enikeev, N.A.; Valiev, R.Z.; Etienne, A.; Radiguet, B.; Ivanisenko, Y.; Sauvage, X. Grain boundary segregation induced strengthening of an ultrafine-grained austenitic stainless steel. *Mater. Lett.* **2014**, *136*, 349–352. [CrossRef]
7. Ganeev, A.V.; Karavaeva, M.V.; Sauvage, X.; Courtois-Manara, E.; Ivanisenko, Y.; Valiev, R.Z. On the nature of high-strength of carbon steel produced by severe plastic deformation. *IOP Conf. Ser. Mater. Sci. Eng.* **2014**, *63*. [CrossRef]
8. Bylja, O.I.; Vasin, R.A.; Ermachenko, A.G.; Karavaeva, M.V.; Muravlev, A.V.; Chistjakov, P.V. The influence of simple and complex loading on structure changes in two-phase titanium alloy. *Scr. Mater.* **1997**, *36*, 949–954. [CrossRef]
9. Berdin, V.K.; Karavaeva, M.V.; Syutina, L.A. Effect of the type of loading on the evolution of microstructure and crystallographic texture in VT9 titanium alloy. *Met. Sci. Heat Treat.* **2003**, *45*, 423–427. [CrossRef]
10. Valiev, R.Z.; Zhilyaev, A.P.; Langdon, T.G. *Bulk Nanostructured Materials: Fundamentals and Applications*; John Wiley & Sons, Inc.: New York, NY, USA, 2014; p. 456.
11. Iwahashi, Y.; Horita, Z.; Nemoto, M.; Langdon, T.G. The process of grain refinement in equal-channel angular pressing. *Acta Mater.* **1998**, *46*, 3317–3331. [CrossRef]
12. Wetscher, F.; Pippan, R. Cyclic high-pressure torsion of nickel and ARMCO iron. *Philos. Mag.* **2006**, *86*, 5867–5883. [CrossRef]
13. Salischev, G.; Zaripova, R.; Galeev, R.; Valiahmetov, O. Nanocrystalline structure formation during severe plastic deformation in metals and their deformation behavior. *Nanostruct. Mater.* **1995**, *6*, 913–916. [CrossRef]
14. Belyakov, A.; Tsuzaki, K.; Kaibyshev, R. Nanostructure evolution in an austenitic stainless steel subjected to multiple forging at ambient temperature. *Mater. Sci. Forum* **2011**, *667–669*, 553–558. [CrossRef]
15. Polyakov, A.; Gunderov, D.; Sitdikov, V.; Valiev, R.; Semenova, I.; Sabirov, I. Physical simulation of hot rolling of ultra-fine grained pure titanium. *Metall. Trans. B* **2014**, *45B*, 2315–2326. [CrossRef]
16. Stepanov, N.D.; Kuznetsov, A.V.; Salischev, G.A.; Raab, G.I.; Valiev, R.Z. Effect of cold rolling on microstructure and mechanical properties of copper subjected to ECAP with various number of passes. *Mater. Sci. Eng. A* **2012**, *554*, 105–115. [CrossRef]
17. Murashkin, M.Y.; Enikeev, N.A.; Kazykhanov, V.U.; Sabirov, I.; Valiev, R.Z. Physical simulation of cold rolling of ultra-fine grained Al 5083 alloy to study microstructure evolution. *Rev. Adv. Mater. Sci.* **2013**, *35*, 75–85.
18. Sabbaghianrad, S.; Langdon, T.G. Microstructural saturation, hardness stability and superplasticity in ultrafine-grained metals processed by a combination of severe plastic deformation techniques. *Lett. Mater.* **2015**, *5*, 335–340. [CrossRef]

19. Vorhauer, A.; Kleber, S.; Pippan, R. Influence of processing temperature on microstructure and mechanical properties of high-alloyed single-phase steels subjected to severe plastic deformation. *Mater. Sci. Eng. A* **2005**, *410–411*, 281–284. [CrossRef]

20. Williamson, G.K.; Smallman, R.E. III. Dislocation densities in some annealed and cold-worked metals from measurements on the X-ray Debye-Scherrer spectrum. *Philos. Mag.* **1956**, *1*, 34–45. [CrossRef]

21. Dobatkin, S.V.; Rybal'chenko, O.V.; Raab, G.I. Structure formation, phase transformations and properties in Cr-Ni austenitic steel after equal-channel angular pressing and heating. *Mater. Sci. Eng. A* **2007**, *463*, 41–45. [CrossRef]

22. Pang, J.C.; Yang, M.X.; Yang, G.; Wu, S.D.; Li, S.X.; Zhang, Z.F. Tensile and fatigue properties of ultrafine-grained low-carbon steel processed by equal channel angular pressing. *Mater. Sci. Eng. A* **2012**, *553*, 157–163. [CrossRef]

23. Whang, S.H. *Nanoctructured Metals and Alloys. Processing, Microstructure, Mechanical Properties and Applications*; Woodhead Publishing Limited: Cambridge, UK, 2011.

24. Ivanisenko, Y.; Lojkwski, W.; Valiev, R.Z.; Fecht, H.-J. The mechanism of formation of nanostructure and dissolution of cementite in a pearlitic steel during high pressure torsion. *Acta Mater.* **2003**, *51*, 5555–5570. [CrossRef]

25. Karavaeva, M.V.; Nurieva, S.K.; Zaripov, N.G.; Ganeev, A.V.; Valiev, R.Z. Microstructure and mechanical properties of medium-carbon steel subjected to severe plastic deformation. *Met. Sci. Heat Treat.* **2012**, *4*, 1–5. [CrossRef]

26. Ganeev, A.V.; Karavaeva, M.V.; Sauvage, X.; Ivanisenko, Y.; Valiev, R.Z. The grain-boundary precipitates in ultrafine-grained carbon steels produced by HPT. In Proceedings of the XV International Conference on Intergranular and Interphase Boundaries in Materials, Moscow, Russia, 23–27 May 2016.

27. Kositsyna, I.I.; Sagaradze, V.V. Phase transformations and mechanical properties of stainless steel in the nanostructural state. *Bull. Russ. Acad. Sci. Phys.* **2007**, *71*, 293–296. [CrossRef]

28. Chen, X.H.; Lu, J.; Lu, L.; Lu, K. Tensile properties of a nanocrystalline 316L austenitic stainless steel. *Scr. Mater.* **2005**, *52*, 1039–1044. [CrossRef]

29. Greger, M.; Vodárek, V.; Dobrzański, L.A.; Kander, L.; Kocich, R.; Kuřetová, B. The structure of austenitic steel AISI 316 after ECAP and low-cycle fatigue. *J. Ach. Mater. Manuf. Eng.* **2008**, *28*, 151–158.

30. Huang, C.X.; Yang, G.; Gao, Y.L.; Wu, S.D.; Zhang, Z.F. Influence of processing temperature on the microstructures and tensile properties of 304L stainless steel by ECAP. *Mater. Sci. Eng. A* **2008**, *485*, 643–650. [CrossRef]

31. Kashyap, B.; Tangri, K. On the Hall-Petch relationship and substructural evolution in type 316L stainless steel. *Acta Mater.* **1995**, *43*, 3971–3981. [CrossRef]

32. Üçok, İ.; Ando, T.; Grant, N. Property enhancement in type 316L stainless steel by spray forming. *Mater. Sci. Eng. A* **1991**, *133*, 284–287. [CrossRef]

33. Pakieła, Z.; Garbacz, H.; Lewandowska, A.; Suś-Ryszkowska, M.; Zieliński, W.; Kurzydłowski, K. Structure and properties of nanomaterials produced by severe plastic deformation. *Nukleonika* **2006**, *51*, 19–25.

34. Wang, H.; Shuro, I.; Umemoto, M.; Kuo, H.-H.; Todaka, Y. Annealing behavior of nano-crystalline austenitic SUS316L produced by HPT. *Mater. Sci. Eng. A* **2012**, *556*, 906–910. [CrossRef]

metals

MDPI

Article

The Structural Evolution and Segregation in a Dual Alloy Ingot Processed by Electroslag Remelting

Yu Liu [1,2], Zhao Zhang [1,2], Guangqiang Li [1,2,3,*], Qiang Wang [1,2], Li Wang [1,2] and Baokuan Li [4]

[1] The State Key Laboratory of Refractories and Metallurgy, Wuhan University of Science and Technology, Wuhan 430081, China; liuyuwust@yeah.net (Y.L.); zhangzhaowust@163.com (Z.Z.); wangqiangwust@wust.edu.cn (Q.W.); wustwangli@163.com (L.W.)

[2] Key Laboratory for Ferrous Metallurgy and Resources Utilization of Ministry of Education, Wuhan University of Science and Technology, Wuhan 430081, China

[3] Collaborative Innovation Center of Steel Technology, University of Science and Technology Beijing, Beijing 100083, China

[4] School of Metallurgy, Northeastern University, Shenyang 110004, China; libk@smm.neu.edu.cn

* Correspondence: liguangqiang@wust.edu.cn; Tel./Fax: +86-27-6886-2665

Academic Editor: Robert Tuttle
Received: 28 November 2016; Accepted: 16 December 2016; Published: 21 December 2016

Abstract: The structural evolution and segregation in a dual alloy made by electroslag remelting (ESR) was investigated by various analytical techniques. The results show that the macrostructure of the ingot consists of two crystallization structures: one is a quite narrow, fine, equiaxed grain region at the edge and the other is a columnar grain region, which plays a leading role. The typical columnar structure shows no discontinuity between the CrMoV, NiCrMoV, and transition zones. The average secondary arm-spacing is coarsened from 35.3 to 49.2 μm and 61.5 μm from the bottom to the top of the ingot. The distinctive features of the structure are attributed to the different cooling conditions during the ESR process. The Ni, Cr, and C contents markedly increase in the transition zone (TZ) and show a slight increase from the bottom to the top and from the surface to the center of the ESR ingot due to the partition ratios, gravity segregation, the thermal buoyancy flow, the solutal buoyancy flow, and the inward Lorentz force. Less dendrite segregation exists in the CrMoV zone and the transition zone due to a stronger cooling rate (11.1 and 4.5 °C/s) and lower Cr and C contents. The precipitation of carbides was observed in the ingot due to a lower solid solubility of the carbon element in the α phase.

Keywords: structure; segregation; electroslag remelting; dual alloy; transition zone

1. Introduction

With the increase in power generation efficiency of single-cylinder steam turbines using a combined cycle, the rotor produced by the traditional bolted high/intermediate pressure-low pressure shaft has unmet needs. Compared with the traditional bolted shaft, the dual alloy shaft can provide a significant improvement in power generating efficiency. The manufacture of a dual alloy shaft via welding technology requires a long production cycle. Electroslag remelting (ESR) has the advantage of producing dual alloy ingots, which can improve the dual alloy single shaft yield and quality. During the production process, two pre-melted steel rods containing different alloy compositions are connected via welding to form a single electrode, and the single electrode is then remelted with ESR technology [1].

In the ESR process, a Joule heating created by the alternating current travelling through the highly resistive molten slag is sufficient to melt the electrode. Then, metal droplets are formed at the electrode tip. Due to the higher density, the metal droplets pass through the liquid slag layer and form a liquid metal pool in the water-cooled mold, which can purify liquid metal. With the heat transferring to the mold, the liquid metal is solidified to forming an ingot. Furthermore, a Lorentz force is created by the interaction between the self-induced magnetic field and the current [2]. The final quality and properties of the ingot strongly depend on the microstructure forming during the solidification process.

In order to obtain a high quality ESR dual alloy ingot, it is necessary to achieve a narrow chemical transition zone (TZ) of the ingot and maintain a continuous structure in TZ because a distorted chemical transition zone and discontinuous structure might increase the risk of running the rotor at high temperature due to thermal expansion mismatch [3]. The structure of an ESR ingot mainly consists of columnar grains and equiaxed grains. The grain growth direction and the secondary dendrite arm spacing (SDAS) are the most important macro- and microstructure factors for ingot quality. The grain growth direction is generally defined by the grain growth angle, which is the angle between the primary dendrites of the grain and the axis of the ingot [4]. Practice has shown that ESR ingots with a small grain growth angle can exhibit improved hot forging performance [5]. Enormous mathematical models and experiments [4–7] have confirmed that the local temperature gradients and solidification conditions have a vital effect on the structure of ESR ingots. However, the subtle differences of the structural evolution in different zones of ESR dual alloy ingot have rarely been reported. Macrosegregation is one of most common and serious defects in ESR ingots that occurs in the solidification process because of the uneven distribution of the solute in the liquid and solid phases [2]. Dendrite segregation is also common in ESR ingots due to interdendritic elemental enrichment [7]. Some researchers have used mathematical models to study the element redistribution in ingots [8,9]. The solute transport in the ESR process is dominated by the combined effect of the Lorentz force, the solutal buoyancy, and the thermal buoyancy. However, the experimental research of the segregation in different zones of the ESR dual alloy ingot is also relatively lacking.

Because of these factors, the authors were motivated to experimentally explore the underlying mechanism of the evolution of the structure and segregation of ESR dual alloy ingots in detail. The subtle differences of the structures and segregation in different zones (CrMoV zone, the TZ, and the NiCrMoV zone) of an ESR ingot were determined. This work is designed to provide fundamental research for the structure and segregation of ESR dual alloy ingots, providing basic knowledge for the manufacture of the dual alloy rotor for use in steam turbines.

2. Experimental Section

2.1. Experimental Apparatus and Method

The experiment was carried out using a laboratory-scale electroslag furnace (Herz, Shanghai, China) with a copper mold under an open air atmosphere. The inner diameter, the lateral wall thickness, and the height of the mold were 120, 65, and 600 mm, respectively. The weight of the ESR ingot was about 35.5 kg in the present experiment. The electrode was comprised of two pieces of pre-melted bars, and connected via welding. The upper part was an NiCrMoV alloy bar (Elec. NiCrMoV), and the lower one was an CrMoV alloy bar (Elec. CrMoV). The chemical composition of the consumable electrode is displayed in Table 1, which was examined by the ICP-AES (Tailun, Shanghai, China), the carbon and sulfur analyzer (Jinbo, Wuxi, China), and the oxygen and nitrogen analyzer (LECO, St. Joseph, MO, USA). The slag was composed of 70 mass pct calcium fluoride and 30 mass pct aluminum oxide. The weight and the thickness of the slag layer were 2.3 kg and about 60 mm, respectively. The root-mean-square value and frequency of the alternating current were constant at 1500 A and 50 Hz.

Table 1. Chemical composition of consumable electrode used in present experiment (wt %). Elec.: Electrode, T.: Total.

Electrode	C	Mn	Si	P	S	Cr	Ni	Mo	Al	Ti	T.[O]
Elec. NiCrMoV	0.106	1.67	0.37	0.018	0.039	16.28	7.45	0.117	0.009	0.012	0.0156
Elec. CrMoV	0.074	3.94	0.40	0.020	0.011	12.25	5.85	0.14	0.014	0.008	0.0121

2.2. Specimen Preparation and Analyzing Methods

The dual alloy ingot was equally split into two parts along the length using wire-electrode cutting. Steel filings were obtained by drilling along the longitudinal centerline of the section every 20 mm, and along the transverse radius every 15 mm for C, Cr, and Ni analysis, as shown in Figure 1. The Ni, Cr, and C contents in the ingot were analyzed with the ICP-AES and the carbon and sulfur analyzer. Three slices were taken from the upper (NiCrMoV), middle (TZ), and lower parts (CrMoV) of the ingot, and three $6 \times 6 \times 6$ mm^3 specimens A (NiCrMoV), B (TZ), and C (CrMoV) were then sampled from three slices, respectively (Figure 1). The NiCrMoV zone, the TZ (transition zone), and the CrMoV zone were decided according to the composition profile along the ESR ingot axial. Another part of the ingot was ground, polished, and finally etched via aqua regia for a certain time. Figure 2 illustrates the macrostructure of the dual alloy ingot. The three specimens were etched at 75 °C in a picric acid solution for metallographic observation by optical microscopy (OM, Carl Zeiss, Jena, Germany), then ground, polished, and etched via aqua regia for determining dendrite segregation via scanning electron microscopy (SEM, FEI, Hillsboro, OR, USA) coupled with energy dispersive spectrometer (EDS, Oxford Instruments, Oxford, UK). A carbon replica specimen was examined in a transmission electron microscopy (TEM, JEOL, Tokyo, Japan) operating at 200 kV for selected-area electron diffraction (SAED) analysis to observe the precipitates.

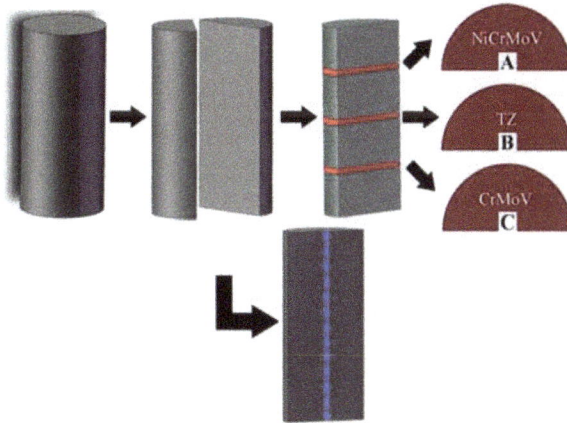

Figure 1. Schematic drawing of the dissection of the electroslag remelting (ESR) ingot.

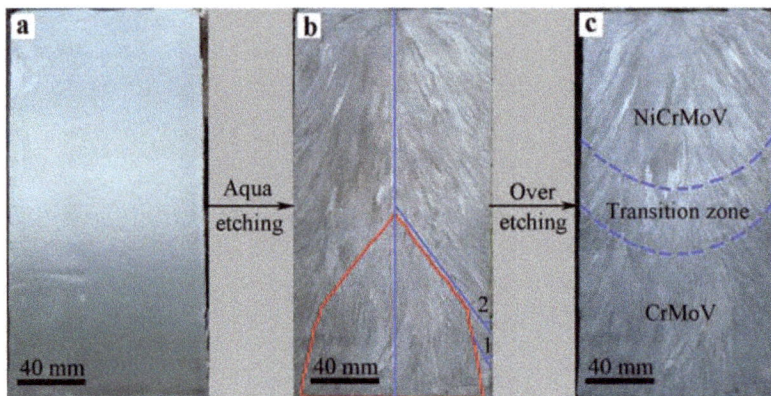

Figure 2. (a) The polished ESR ingot; (b and c) the macrostructure of the ESR ingot etched via aqua regia; (c) over-etched to show the macrostructure of the ESR ingot. The red frame and blue line in (b) mean vertical columnar grain and growth direction of inclined columnar grain, respectively.

3. Results and Discussions

3.1. Macro- and Microstructure Evolution of Steel Ingot Made via the Electroslag Remelting (ESR) Process

Two structures of the crystal are shown in Figure 2b—one is a quite narrow, fine, equiaxed grain region at the edge, and the other is a columnar grain region, which plays a leading role. The typical columnar structure shows no discontinuity between the CrMoV zone, the NiCrMoV zone, and the transition zone (Figure 2b,c). A thin layer of the equiaxed grain is located under the ingot surface, and the columnar grain grows inside the ingot. The columnar grain nucleates at the bottom and grows upward (inside the frame in Figure 2b). Meanwhile, the columnar grain nucleated at the lateral wall grows in a certain angle. After a period of growth, the two columnar grains would meet each other, and the inclined columnar grain hinders the growth of the vertical columnar grain. The inclined columnar grain continuously grows to the center of the ingot, which forms an inverted chevron structure with an angle ranging from 30° to 37° (from Angle 1 to Angle 2 in Figure 2b) with respect to the vertical axis. The columnar grain at the top is almost perpendicular to the vertical axis. After longer time etching, the CrMoV zone is darker than the TZ and the NiCrMoV zone due to the lower Cr and Ni contents (Figure 2c). The variation in optical microstructure at different positions (A, B, and C) throughout the cross section of the ESR ingot is shown in Figure 3a–c. The dendritic structure is formed throughout the cross section, which gradually becomes coarsened from the bottom (Figure 3c) to the top (Figure 3a) of the ingot. The secondary dendrite arm-spacing (SDAS) measured at the bottom, middle and top of the ingot is shown in Figure 3d. The average SDAS is the average of four SDAS's at three sampling points. The average SDAS throughout the section at the bottom, middle, and top parts of the ingot are 35.3, 49.2, and 61.5 μm, respectively. The cooling rate (C_R/ °C/s) can be calculated by Equation (1) as follows [10]:

$$\lambda_S(\mu m) = (169.1 - 720.9 \cdot [\%C]) \cdot C_R^{-0.4935} (0 < [\%C] \leq 0.15) \tag{1}$$

According to Equation (1), the cooling rates of the bottom, middle and top parts of the ingot are 11.1, 4.5, and 2.3 °C/s, respectively. This indicates that the cooling intensity decreases from the bottom to the top of the ingot.

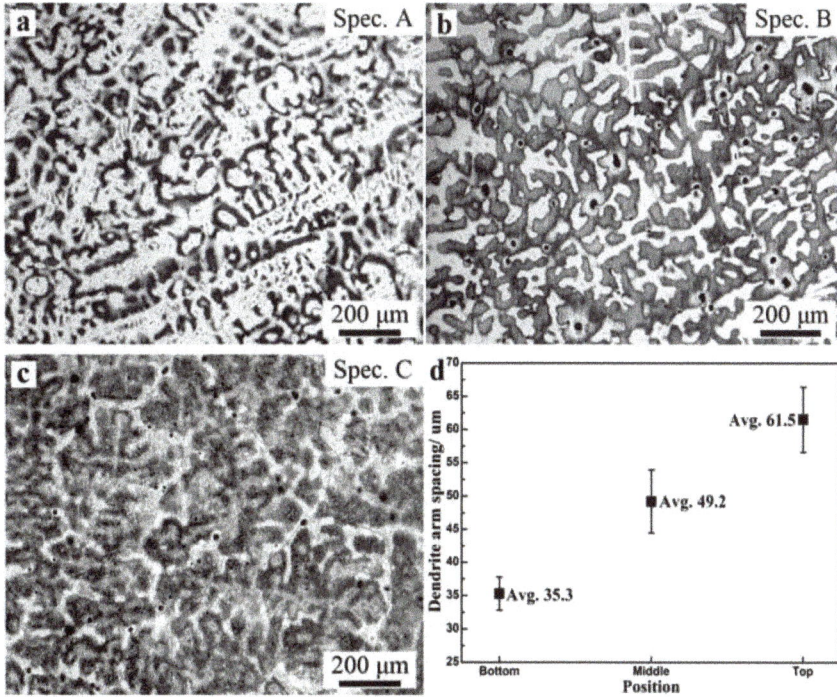

Figure 3. Optical micrographs of the (**a**) Spec. A, NiCrMoV, (**b**) Spec. B, transition zone (TZ), and (**c**) Spec. C, CrMoV in the ESR ingot; (**d**) secondary dendritic arm spacing at different positions (A, B, and C) of the ESR ingot.

3.2. Macrosegregation of the ESR Ingot

The distributions of Ni, Cr, and C concentrations along the vertical centerline and radius in the ESR ingot are shown in Figure 4. Figure 4a indicates that the Ni, Cr, and C contents markedly increase in the transition zone until they reaches the nominal concentration of Elec. NiCrMoV. It can be seen from Figure 4b that the Ni, Cr, and C contents at the middle of the ingot are higher than those at the surface.

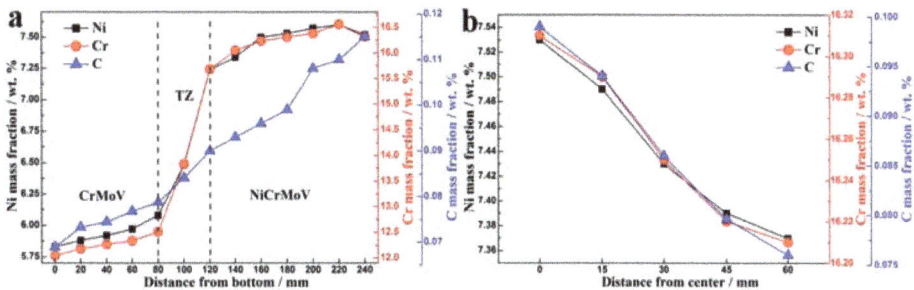

Figure 4. Ni, Cr, and C concentration distributions along (**a**) the longitudinal centerline and (**b**) the transverse radius in the ESR ingot.

In our previous work, we studied the fluid flow during the ESR process [1,9,11]. The metal droplets formed at the electrode tip sink through the slag layer to form a crescent-shaped molten metal pool in the mold, which is deep in the center of the ingot and becomes shallow gradually outward along the radius. Figure 5 illustrates the flows of the molten slag and metal pools during the ESR process. The flow in the metal pool is induced by the thermal buoyancy and the Lorentz forces [12,13]. The metal near the mold is cooled by the water through the mold, resulting in a lower temperature and higher density. Then, the hot metal will float up and the cold metal will sink down, and a circular flow is formed in the vicinity of the water-cooled mold due to the thermal buoyancy flow. The large temperature difference at the solidification front also gives rise to a clockwise circular flow. The cool metal moves down along the inclined solidification front and washes out the solidifying mushy zone. Around the base of the metal pool, the cooling intensity decreases and the hot metal turns up toward the slag–metal interface and then back to the mold wall. Furthermore, according to Faraday's law of electromagnetic induction, the downward current would induce a clockwise circular magnetic field (looking down from the top). The interaction between the downward current and the clockwise magnetic field creates an inward Lorentz force, which also pushes the metal from the periphery to the bottom.

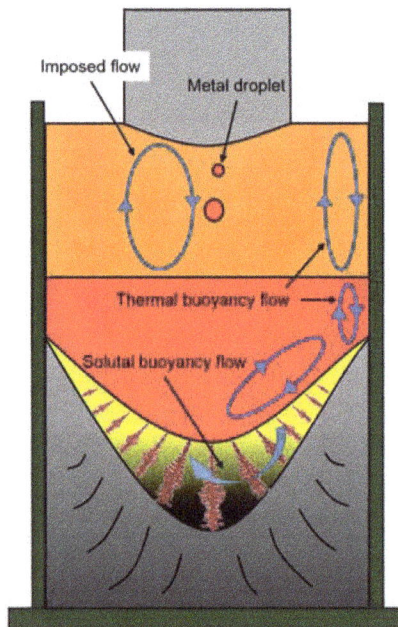

Figure 5. Schematic of the flows of the molten slag and metal pools in the ESR process.

The solute elements Ni, Cr, and C become enriched in the mushy zone in Ni, Cr, and C because of their partition ratios (k_{Ni} = 0.94, k_{Cr} = 0.76, and k_C = 0.34) [14] between solid and remained liquid alloy, which are less than 1. In addition, the density of C (ρ_C = 1800 kg/m^3) and Cr (ρ_{Cr} = 6900 kg/m^3) is lower than that of iron (ρ_{Fe} = 7500 kg/m^3) [15], resulting in the so-called gravity segregation, with higher Cr and C contents in the liquid. Due to the clockwise circular flow at the solidification front, the solute-poor (Ni, Cr, and C) metal in the pool displaces the solute-rich metal through washing the mushy region. Furthermore, the solute enrichment increases the metal density and thus promotes the sinking of the liquid metal [16,17]. As a result, the Ni, Cr, and C accumulate at the base of the pool and the concentration increases with increasing distance from the bottom of the ingot. There is a lower concentration region at the outer layers of the ingot, where solidification occurs first and Ni, Cr, and C are depleted. The solute transport during the ESR process is dominated by the Lorentz force, the thermal and solutal buoyancy forces, and the gravity segregation.

3.3. Microsegregation and Precipitates of the ESR Ingot

Figure 6 shows the SEM micrograph of the dendrite segregation in the ESR ingot. It indicates that severe dendrite segregation occurs in interdendritic areas. In order to analyze the solute-rich elemental distribution in interdendritic areas of the ESR ingot, the SEM micrograph and corresponding elemental line scanning image of the interdendritic areas are presented in Figure 7. It reveals that Cr and C elements are enriched in interdendritic areas, whereas the Ni element exhibits a slight decrease (Figure 7b). During the solidification of the ingot, the interdendritic areas enrich carbon and chromium [7], which always form larger carbides (Figure 7a) due to severe dendrite segregation. Furthermore, SEM analysis was conducted to reveal the dendrite segregation distribution in Specimens A (NiCrMoV), B (TZ), and C (CrMoV). The magnified micrographs of the specimens are presented in Figure 8a–c, indicating that dendrite segregation becomes severer from the bottom to the top of the ingot. This is the result of different cooling intensities and Cr and C contents. Less dendrite segregation exists in the CrMoV zone and the TZ due to stronger cooling rates (11.1 and 4.5 °C/s) and lower Cr and C contents.

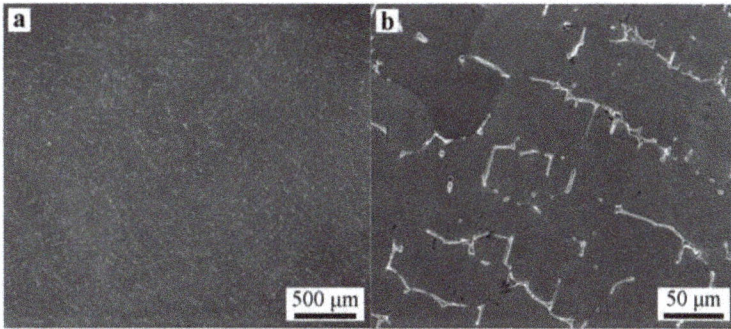

Figure 6. Scanning electron microscopy (SEM) (**a**) low magnification and (**b**) high magnification micrographs of interdendritic segregation in the ESR ingot.

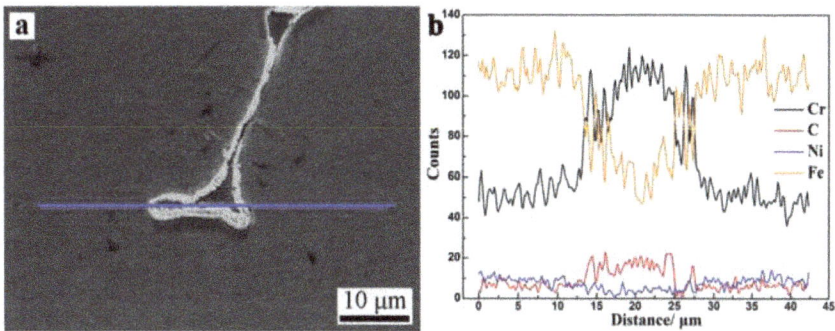

Figure 7. (**a**) SEM micrograph and (**b**) elemental line scanning images (Cr, C, Ni, and Fe) of interdendritic segregation in the ESR ingot. Blue line in (**a**) is the range of line scan.

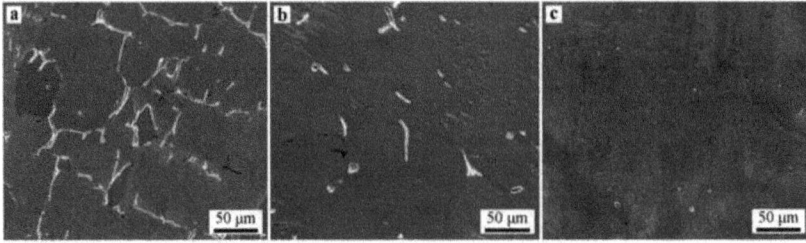

Figure 8. SEM images of dendrite segregation distribution in the (**a**) Spec. A, NiCrMoV, (**b**) Spec. B, TZ, and (**c**) Spec. C, CrMoV.

The SEM micrograph and EDS spectral image of precipitates observed in the ESR ingot are shown in Figure 9. Figure 9a shows that precipitates distribute randomly, and the size of the big precipitates is up to about 40 μm. The corresponding EDS analysis reveals the precipitates mainly consist of C, Cr, Ni, and Fe (Figure 9b). In order to analyze the precipitates in detail, TEM analysis were conducted to reveal the crystal structure of the precipitates. Figure 10 shows the TEM micrograph and the corresponding SAED patterns of the precipitates. According to the analysis of the SAED, the spots $(31\bar{1})$ and $(\bar{2}4\bar{2})$ correspond to $M_{23}C_6$ with an FCC crystal structure in Figure 10b. The $M_{23}C_6$ carbide is common in high chromium steel [18]. In the present work, the ingot contains 0.074%–0.106% carbon, and it belongs to hypoeutectoid steel. During the solidification of the ingot, the γ phase transformed to the α phase and the solid solubility of carbon in the α phase was lower, which resulted in carbon enrichment. Enriched carbon always combines with Cr or another alloying element to form large carbides. Fine $M_{23}C_6$ carbides can improve the strength and toughness of steel by the mechanism of precipitation strengthening [18]: (1) pinning the grain boundary and hindering the movement of grain boundaries; (2) confining the creep cavity between the precipitated phase, making it difficult to grow, prolonging the rupture time; and (3) changing the solid solution of the two sides of the grain boundary, improving the sliding ability of dislocation near the grain boundary, improving the plasticity of the grain boundary, and eliminating notch sensitivity. However, the massive precipitates easily form the crack source and reduce the strength of the matrix. In the present work, the some precipitates are massive, which is detrimental to the strength of the matrix. The dimensions of $M_{23}C_6$ in the ESR ingot were larger than those of the carbides generated by the thermal treatments as reported in the literature [19,20]. This indicates that an appropriate heat treatment process is expected to eliminate the severer dendrite segregation and dissolve the massive carbides. Otherwise, the massive carbides and severer dendrite segregation would stay and distribute in the structure.

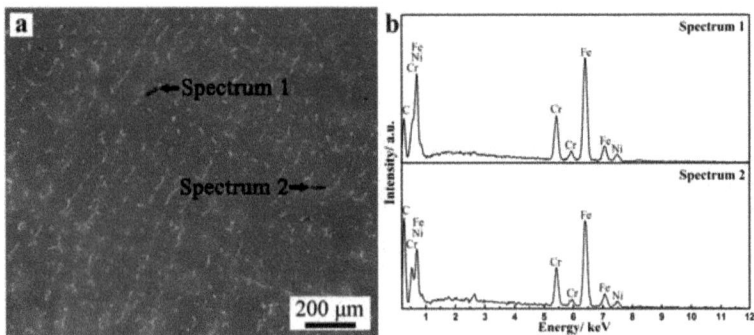

Figure 9. (**a**) SEM micrograph and (**b**) energy dispersive spectrometer (EDS) images of precipitates observed in the Spec. A.

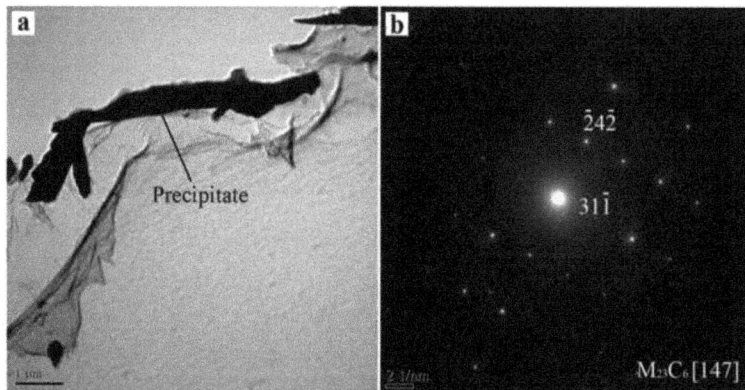

Figure 10. Transmission electron microscopy (TEM) micrograph and the corresponding selected-area electron diffraction (SAED) patterns of the precipitates: (**a**) TEM micrograph of the precipitate; (**b**) the diffraction pattern of $M_{23}C_6$ carbide.

4. Conclusions

The following conclusion can be drawn from the present study.

(1) Two crystallization structures were observed in the ESR ingot: one is a quite narrow, fine, equiaxed grain region at the edge of the ingot, and the other is a columnar grain region, which plays a leading role. The typical columnar structure shows no discontinuity between the CrMoV zone, the NiCrMoV zone, and the transition zone. The average second arm-spacing is coarsened from 35.3 to 49.2 μm and 61.5 μm from the bottom to the top of the ingot. The distinctive features of the structure are attributed to different cooling conditions during the ESR process.

(2) The Ni, Cr, and C show a slight increase from the end to the top and from the surface to the center of the ESR ingot due to the partition ratios, gravity segregation, the thermal buoyancy flow, the solutal buoyancy flow, and the inward Lorentz force. The Ni, Cr, and C contents markedly increase in the transition zone until it reaches the nominal concentration of Elec. NiCrMoV.

(3) Severe dendrite segregation was observed in the ESR ingot, which becomes severer from the bottom to the top of the ingot. Less dendrite segregation exists in the CrMoV zone and the transition zone due to a stronger cooling rate (11.1 and 4.5 °C/s) and lower Cr and C contents. The precipitation of carbides was observed in the ESR ingot due to a lower solid solubility of the carbon element in the α phase. The massive carbides are detrimental to the strength of the matrix. An appropriate heat treatment process is expected to eliminate the severer dendrite segregation and dissolve the massive carbides. The result of the present work provides a reference for the manufacture of the dual alloy rotor to be used in steam turbines using a combined cycle, which would provide significant improvements in power generating efficiency.

Acknowledgments: The authors gratefully acknowledge the support from the National Natural Science Foundation of China (Grant No. 51210007) and the Key Program of Joint Funds of the National Natural Science Foundation of China and the Government of Liaoning Province (Grant No. U1508214).

Author Contributions: Guangqiang Li and Yu Liu conceived and designed the experiments; Yu Liu, Zhao Zhang and Li Wang performed the experiments; Yu Liu and Qiang Wang analyzed the data; Baokuan Li contributed materials and tools; Yu Liu wrote the paper.

Conflicts of Interest: The authors declare no conflict of interest. The founding sponsors had no role in the design of the study; in the collection, analyses, or interpretation of data; in the writing of the manuscript, and in the decision to publish the results.

Abbreviations

The following abbreviations are used in this manuscript:

ESR	electroslag remelting
TZ	transition zone
ICP-AES	inductively coupled plasma-atomic emission spectroscopy
OM	optical microscopy
SEM	scanning electron microscopy
EDS	energy dispersive spectrometer
TEM	transmission electron microscopy

References

1. Wang, Q.; Yan, H.; Ren, N.; Li, B. Effect of current on solute transport in electroslag remelting dual alloy ingot. *Appl. Therm. Eng.* **2016**, *101*, 546–567. [CrossRef]
2. Wang, Q.; Yan, H.; Ren, N.; Li, B. Effect of slag thickness on macrosegregation and transition zone width of electroslag remelting dual alloy ingot. *JOM* **2016**, *68*, 397–400. [CrossRef]
3. Kajikawa, K.; Ganesh, S.; Kimura, K.; Kudo, H.; Nakamura, T.; Tanaka, Y.; Schwant, R.; Gatazka, F.; Yang, L. Forging for advanced trubine applications: Development of multiple alloy rotor forging for turbine application. *Ironmak. Steelmak.* **2007**, *34*, 216–220. [CrossRef]
4. Rao, L.; Zhao, J.H.; Zhao, Z.X.; Ding, G.; Geng, M.P. Macro-and microstructure evolution of 5CrNiMo steel ingots during electroslag remelting process. *J. Iron Steel Res. Int.* **2014**, *21*, 644–652.
5. Chang, L.Z.; Shi, X.F.; Yang, H.S.; Li, Z.B. Effect of low-frequency AC power supply during electroslag remelting on qualities of alloy steel. *J. Iron Steel Res. Int.* **2009**, *16*, 7–11. [CrossRef]
6. Li, B.; Wang, Q.; Wang, F.; Chen, M. A coupled cellular automaton-finite-element mathematical model for the multiscale phenomena of electroslag remelting H13 die steel ingot. *JOM* **2014**, *66*, 1153–1165. [CrossRef]
7. Ma, D.; Zhou, J.; Chen, Z.; Zhang, Z.; Chen, Q.; Li, D. Influence of thermal homogenization treatment on structure and impact toughness of H13 ESR steel. *J. Iron Steel Res. Int.* **2009**, *16*, 56–60. [CrossRef]
8. Fezi, K.; Yanke, J.; Krane, M.J.M. Macrosegregation during electroslag remelting of alloy 625. *Metall. Mater. Trans. B* **2015**, *46*, 766–779. [CrossRef]
9. Wang, Q.; Wang, F.; Li, B.; Tsukihashi, F. A three-dimensional comprehensive model for prediction of macrosegregation in electroslag remelting ingot. *ISIJ Int.* **2015**, *55*, 1010–1016. [CrossRef]
10. Won, Y.M.; Thomas, B.G. Simple model of microsegregation during solidification of steels. *Metall. Mater. Trans. A* **2001**, *32*, 1755–1767. [CrossRef]
11. Wang, Q.; Li, B. Numerical investigation on the effect of fill ratio on macrosegregation in electroslag remelting ingot. *Appl. Therm. Eng.* **2015**, *91*, 116–125. [CrossRef]
12. Dong, J.; Cui, J.; Zeng, X.; Ding, W. Effect of low-frequency electromagnetic field on microstructures and macrosegregation of Φ270 mm DC ingots of an Al-Zn-Mg-Cu-Zr alloy. *Mater. Lett.* **2005**, *59*, 1502–1506. [CrossRef]
13. Zhang, B.; Cui, J.; Lu, G. Effect of low-frequency magnetic field on macrosegregation of continuous casting aluminum alloys. *Mater. Lett.* **2003**, *57*, 1707–1711. [CrossRef]
14. Schneider, M.C.; Beckermann, C. Formation of macrosegregation by multicomponent thermosolutal convection during the solidification of steel. *Metall. Mater. Trans. A* **1995**, *26*, 2373–2388. [CrossRef]
15. Weber, V.; Jardy, A.; Dussoubs, B.; Ablitzer, D.; Ryberon, S.; Schmitt, V.; Hans, S.; Poisson, H. A comprehensive model of the electroslag remelting process: Description and validation. *Metall. Mater. Trans. B* **2009**, *40*, 271–280. [CrossRef]
16. Wang, X.; Ward, R.M.; Jacobs, M.H.; Barratt, M.D. Effect of variation in process parameters on the formation of freckle in inconel 718 by vacuum arc remelting. *Metall. Mater. Trans. A* **2008**, *39*, 2981–2989. [CrossRef]
17. Mitchell, A. Solidification in remelting processes. *Mat. Sci. Eng. A* **2005**, *413*, 10–18. [CrossRef]

18. Tian, Z.; Bao, H.; He, X.; Li, Q.; Liu, Z. Strengthening mechanisms of heat resistant alloys used for steam turbine rotor working at 700 °C. *J. Iron Steel Res.* **2015**, *27*, 1–6. (In Chinese)
19. Jiao, S.Y.; Zhang, M.C.; Zheng, L.; Dong, J.X. Investigation of carbide precipitation process and chromium depletion during thermal treatment of Alloy 690. *Metall. Mater. Trans. A* **2010**, *41*, 26–42. [CrossRef]
20. Angeliu, T.M.; Was, G.S. Behavior of grain boundary chemistry and precipitates upon thermal treatment of controlled purity alloy 690. *Metall. Trans. A* **1990**, *21*, 2097–2107. [CrossRef]

![metals logo] *metals*

MDPI

Article

The Influence of Austenite Grain Size on the Mechanical Properties of Low-Alloy Steel with Boron

Beata Białobrzeska *, Łukasz Konat and Robert Jasiński

Department of Materials Science, Welding and Strength of Materials, Wrocław University of Technology, 50-370 Wrocław, Poland; lukasz.konat@pwr.edu.pl (Ł.K.); robert.jasinski@pwr.edu.pl (R.J.)
* Correspondence: beata.bialobrzeska@pwr.edu.pl; Tel.: +48-713203845

Academic Editor: Robert Tuttle
Received: 21 November 2016; Accepted: 12 January 2017; Published: 17 January 2017

Abstract: This study forms part of the current research on modern steel groups with higher resistance to abrasive wear. In order to reduce the intensity of wear processes, and also to minimize their impact, the immediate priority seems to be a search for a correlation between the chemical composition and structure of these materials and their properties. In this paper, the correlation between prior austenite grain size, martensite packets and the mechanical properties were researched. The growth of austenite grains is an important factor in the analysis of the microstructure, as the grain size has an effect on the kinetics of phase transformation. The microstructure, however, is closely related to the mechanical properties of the material such as yield strength, tensile strength, elongation and impact strength, as well as morphology of occurred fracture. During the study, the mechanical properties were tested and a tendency to brittle fracture was analysed. The studies show big differences of the analysed parameters depending on the applied heat treatment, which should provide guidance to users to specific applications of this type of steel.

Keywords: boron steels; austenite grain size; impact strength; fracture; tensile strength

1. Introduction

The grain size has a measurable effect on most of the mechanical properties. For example, at room temperature, the hardness, yield strength, tensile strength, fatigue strength, and impact strength all increase with decreasing grain size. The influence of grain size on the mechanical properties of steel is most commonly expressed in a Hall-Petch Equation (Equations (1) and (2)) [1–4]. This classic equation can also be used to predict hardness. A similar equation applies to the brittle cracking (cleavage fracture stress, σ_f) of high-strength steels (Equation (3)). According to this criterion, increase in the grain size eases the cracking process, because the greater the number of dislocations that pile up, the less stress is required for crack development [5].

$$\sigma_y = \sigma_0 + \frac{k_y}{\sqrt{d}} \tag{1}$$

where:

σ_y yield stress;
σ_0 the friction resistance for dislocation movement within the polycrystalline grains;
k_y a measure of the local stress needed at a grain boundary for the transmission of plastic flow—unpinning constant;
d the average grain size.

The value of the k_y coefficient in Equation (1) has been described with the following relationship:

$$k_y \approx 3 \left[\frac{2Gb\tau_b}{q\pi} \right]^{1/2}$$

(2)

where:

G modulus of rigidity (shear);
b Burgers vector (dependent on the type of crystal lattice);
q geometrical factor (dependent on the type of crystal lattice);
τ_b critical stress required for passing the slide through the grain boundary [2].

$$\sigma_f = \sigma_{0f} \frac{k_f}{\sqrt{d}}$$

(3)

where:

σ_f fracture stress;
σ_{0f} and k_f the experimentally determined constants, and $k_f > k_y$ from Equation (2) has the following value:

$$k_f \geq \left(\frac{6\pi\gamma G}{1 - \nu} \right)^2$$

(4)

where:

γ surface energy of a crack;
ν Poisson's ratio.

However, Morris [4] noticed that the Hall-Petch equation is not unequivocal for martensitic steels. Some studies concerning the influence of the prior austenite grain size on the austenite to ferrite transformation temperature and different ferrite morphologies in Nb-micro-alloyed (HSLA) steel have been performed [6]. Similar studies have been performed to investigate the impact of the prior austenite grain size on the morphology and mechanical properties of martensite in medium carbon steel [3] and to predict the austenite grain growth in low-alloy steels [7], but there is still a lack of knowledge about the influence of the prior austenite grain size on the basic mechanical properties in low-alloy boron steels with high resistance to abrasive wear.

In recent times, in various industries, increasingly frequent attempts have been made to increase the durability of machine elements by applying boron to the construction of a low-alloy, high-strength steel. This group of materials includes, among others, constant Hardox. By using the advanced technology of their production, these steels achieve high abrasive wear resistance in combination with high strength and a sufficient toughness. Due to the fact that the technologies used in industrial conditions require the structural materials to be tolerant of complex thermal or thermal-plastic processes, an important factor determining the properties of these materials is the austenite grain size. So far, the knowledge of these advanced steels is not sufficient, and publicly available information about them is mere advertising. Therefore, an analysis of the growth of austenite grains in these steels according to the austenitizing temperature in order to prevent the degradation of their very favorable mechanical properties appears to be justified. These steels are exposed to the degradation of their favorable properties due to the thermal processes used for assembling components made of these steels. Therefore, it is important to control the grain size of these steels during their thermal treatment. This study analyzed the growth of austenite grains in, for example, Hardox 450, which is widely used in industry.

Another very important issue, from the point of view of the impact resistance level achieved in the low-alloy steels, is related to the type and morphology of the martensite structure. As a result of transformation in the low- and medium-carbon steels, martensite initially nucleates in random areas of austenite grain, and its growth takes place afterward [8]. A martensite area created in this way, after reaching the grain boundary, may initiate further development of martensitic areas in neighbouring grains, according to the analogous variant. The whole process is autocatalytic in character. It results from the way of passing stress (through the grain boundaries) caused by the greater specific volume of martensite in relation to austenite. Accordingly, it can be stated that the greater degree of heterogeneity of crystallographic orientation variants of the formed martensite corresponds to a larger number of initial nuclei [8–10]. A characteristic feature of martensite created in this way is the three-level hierarchy of morphology, consisting of laths, blocks, and packets. The laths of martensite, creating a block, have the same crystallographic orientation and thus represent the same variant of the martensite structure formed. By contrast, the packets are created by clusters of blocks of the same habitus plane, corresponding to the plane {111}γ of the primary austenite. In the simplified version of considerations (neglecting the chance of separate blocks of martensite of the same variant appearing in one austenite grain), it can be stated that, in the single packet blocks of martensite, a maximum of six variants may be present. This results from six possible crystallographic directions within one plane {111} of austenite, of which austenite has four. In relation to that, in a single grain of prior austenite, 24 different variants of martensite may appear simultaneously at the wide-angle inter-block boundaries [1,11].

From the point of view of the static strength and the impact resistance of the low- and medium-carbon steel, their direct dependence on the size of blocks and packets, which constitute the effective dimensions of the martensitic structure in the meaning of the Hall-Petch relationship, has to be underlined. From the data contained in the works [1–3], decrease in the grain size of the primary austenite (and thus the reduction of the martensite packets sizes) from 200 μm to 5 μm results in an increase in strength of 235 MPa and an eight-fold increase in steel impact resistance. In relation to the above, in the opinion of the authors of this work, it is worth analyzing the morphology and size of the packets of martensitic structure while considering the impact of the grain size of the prior austenite on the selected mechanical properties of low-alloy steels with boron.

2. Materials and Methods

For the tests, Hardox 450, a material from a group of low-alloy boron steels with a high resistance to abrasive wear, was selected [12–15]. All samples were taken from the longitudinal direction relative to the sheet rolling direction. The chemical composition and mechanical properties of the analyzed material (according to the manufacturer's data and the research data) are presented in Tables 1 and 2. The plate thickness was 30 mm. Chemical composition analysis was performed with a spectral method using a glow-discharge spectrometer. The content of oxygen and nitride was measured using a NO determinator.

Table 1. Selected mechanical properties of the investigated steel in the as-received condition. KCV: notched impact strength.

Mechanical Properties	$R_{p0.2}$ (MPa)	Hardness (HB)	KCV_{-20} (J/cm^2)
Manufacturer's data [12]	1100–1300	425–75	27

Table 2. Chemical composition of the investigated steel.

Element (wt %)	C	Mn	Si	P	S	Ni	Cr	V	Al	Ti
Content of elements	0.223	1.32	0.489	0.009	0.004	0.044	0.784	0.004	0.035	0.02

Element (wt %)	Nb	B	Cu	Co	Mo	As	Pb	O	N	
Content of elements	0.005	0.0011	0.015	0.016	0.012	0.009	0.002	0.049885	0.003805	

The metallographic studies were performed using a light microscope (Nikon Corporation, Tokyo, Japan). The microstructure in the as-received condition of the investigated steel is presented in Figure 1. The tested material in the as-received state showed a structure of low-carbon martensite. The martensitic microstructure exhibits a high homogeneity and some features that can be described as being similar to tempered martensite, with precipitates of some non-metallic inclusions like titanium nitrides.

Figure 1. The microstructure of the investigated low-alloy boron steel with a high resistance to abrasive wear in the as-received state. Etched state, light microscope.

Samples were austenitized for a holding time of 20 min at temperatures of 900, 1000, 1100, and 1200 °C and then quenched in water. After each heat treatment, the samples of the prior austenite grain were tempered at 250 °C for 30 min, in order to retain the detail of the austenite microstructure and to allow identification of the prior austenite grain boundaries. The samples were etched with 5% picric acid at a temperature of 55 °C in accordance with the standard PN-H-04503:1961P. The measurements of the austenite grain size were performed using the program NIS Elements. Each average austenite grain size was evaluated from 100 measurements.

Packet size was measured using the linear intercept method on SEM images. About 100 martensite packets in each sample were measured to obtain the average packet size.

In order to determine the impact of heat treatment on the basic mechanical and plastic properties of the tested steel, tensile testing was conducted at ambient temperature, based on the valid standard PN-EN ISO 6892-1:2010 (metallic materials—tensile testing). The research was carried out on an Instron 5982 machine (Instron, High Wycombe, UK) using an extensometer to measure elongation. Proportional rectangular samples were tested with an original gauge length of $L_0 = 35$ mm. Testing rates were based on stress rate (Method B according to the ISO Standard 6892). Within the elastic and plastic range up to the yield strength, the strain rate was 0.002 1/s; after the yield strength, the stress rate exceeded 25 MPa/s until fracture occurred. The following mechanical properties were determined: non-proportional extension (yield strength, $R_{p0.2}$), tensile strength (R_m), percentage elongation after fracture (A), and percentage reduction of area (Z).

In order to determine the value of the absorbed energy (KV), the notched impact strength (KCV), and the type of fracture related to the austenitic temperature, a Charpy impact test was performed. The study was performed in accordance with Standard PN-EN ISO 148-1:2010 (Metallic materials—Charpy pendulum impact test) on the Zwick Roell pendulum hammer RPK300 (Zwick Roell Gruppe, Ulm, Germany) using an initial energy of 300 J. Standard samples, V-notched to a depth of 2 mm, were tested. The tests were carried out after the samples were cooled to -40 ± 2 °C and conditioned for 15 min in a mixture of liquid nitrogen and isopropanol. The temperature was monitored using a digital thermometer Center (Center Technology Corp., New Taipei City, Taiwan), and the transfer time for all samples was less than 5 s. Fractographic analysis was performed using a stereo microscope (Nikon Corporation, Tokyo, Japan) and SEM (JEOL Ltd., Tokyo, Japan).

Samples for mechanical research were cut in the longitudinal direction to the rolling direction.

3. Results

3.1. Austenite Grain Growth Analysis

A comparison of the austenite grain size attained after changing the temperature from 900, 1000, 1100, and 1200 °C for 20 min is presented in Figures 2 and 3.

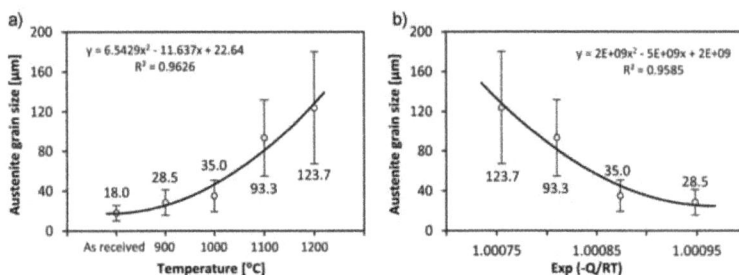

Figure 2. Austenite grain size (**a**) in the as-received state and at different austenitizing temperatures; (**b**) as an Arrhenius plot.

The austenite grain growth with increasing austenitizing temperature was approximated with a quadratic function at the level of correlation $R^2 = 0.96$. An approximation of the results with this model is consistent with the model shown in [8].

The measurement results of austenite grain size in the as-received state are shown in Figure 3a. The sample in the as-received state is characterized by the lowest grain size. The values vary within the range 6–38 μm. The average value is approximately 18.0 ± 7.8 μm. The manufacturer did not provide details of heat treatment in addition to information about hardening and tempering [12]. The distribution was approximated with a logarithmic normal model. This model is similar to the distribution of austenite grains size presented in [16]. Almost 50% of the measurements are grains with a diameter of 10–20 μm. Abnormal grains were not observed (Figure 4a).

The austenitizing process at a temperature of 900 °C caused an increase in the average grain size by nearly 60%, compared to the as-received state. The size varies between 12 and 58 μm. The average value is 28.5 ± 12.8 μm. Again, abnormal grains were not observed (Figure 4b). The measurement results are approximated with a logarithmic function. The most common diameters of grains were across the ranges 11–18 μm and 32–39 μm (Figure 3b).

During the austenitizing at 1000 °C, the minimum measured grain size was approximately 13.5 μm. This value is more than 2 times higher than that measured in the as-received state (i.e., 6 μm). The average grain size is 35.0 ± 15.7 μm. Grains from 10 to 19 μm constitute 24% (Figure 3c). However, a higher frequency of grains in the range of approximately 60–70 μm is found (Figure 3c). The frequency of occurrence of each austenite grain size cannot be approximated with any mathematical model.

Austenitizing at 1100 °C caused the appearance of clearly abnormal austenite grains. Their maximum size ranged from 211 to 296 μm (Figure 4d). They were not included in the calculation of the average grain size. Grain sizes in the ranges of 50–69 μm and 107–126 μm occurred more frequently than others (Figure 3d). The average size of the austenite grains is 93.3 ± 38.4 μm.

The structure obtained after austenitizing at 1200 °C is significantly different from the others (Figure 4e). There is a strong grain growth, and the fine-grained regions disappear. The average grain size is 123.7 ± 56.5 μm. This is about three times higher than that obtained at 900 °C. Comparing the results obtained for the sample in as-received state, there was a five-fold growth of the average austenite grain size. Seventy-one percent of the grain diameters are of a value exceeding 100 µm, of which 13% are grains with a diameter greater than 200 µm. The results are presented in graphical form and approximated with a logarithmic model (Figure 3e).

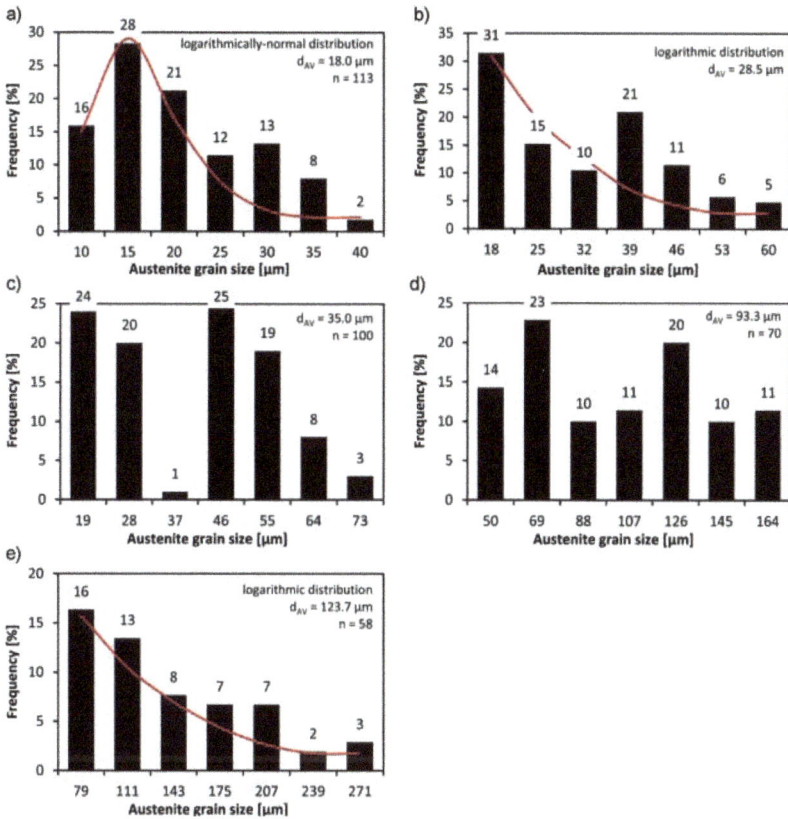

Figure 3. The frequency of occurrence of austenite grain size of the tested steel: (**a**) in the as-received state; (**b**) after austenitizing at 900 °C, 20 min; (**c**) after austenitizing at 1000 °C, 20 min; (**d**) after austenitizing at 1100 °C, 20 min; (**e**) 1200 °C, 20 min.

The migration of grain boundaries can be compared to a diffusion process. Therefore, the grain growth rate increases with an increase in the holding temperature. It can be seen in Figure 4b that the growth of austenite grains is gradual at a low austenitizing temperature. The abnormal grains are presented in the microstructure after heating at 1000 °C (Figure 4c), 1100 °C (Figure 4d), and 1200 °C (Figure 4e). The difference between these microstructures is in the number of fine grains. The decrease in the number of fine grains indicates that larger austenite grains can merge from smaller ones and grow gradually with the increase in holding temperature. The abnormal grains develop in areas where fine grains are embedded. These fine grains surround the abnormal grains. When the fine grain areas disappear, the normal growth of the coarse grains can take place.

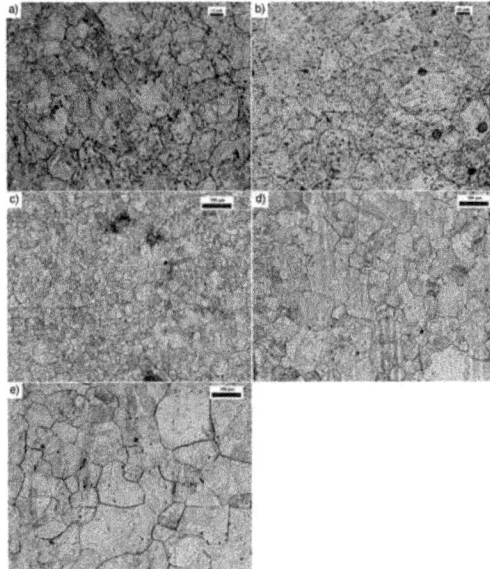

Figure 4. Micrographs of austenite grain boundaries in the as-received state and under different annealing conditions: (**a**) in the as-received state; (**b**) 900 °C, 20 min; (**c**) 1000 °C, 20 min; (**d**) 1100 °C, 20 min; (**e**) 1200 °C, 20 min. Etched state, light microscope.

3.2. The Tensile Test and the Impact Strength Test

The values of the mechanical properties determined during the tensile tests and impact strength test are presented in Table 3.

Table 3. Selected mechanical properties of tested steel after austenitizing at different temperature.

Austenitizing Temperature T (°C)	Tensile Strength R_m (MPa)	Yield Strength $R_{p0.2}$ (MPa)	Elongation A (%)	Reduction of Area Z (%)	Impact Strength KCV_{-40} (J/cm^2)
As received state	1433	1106	14.6	46	70
900	1445	1076	14.1	41	49
1000	1413	1016	13.2	37	38
1100	1425	1006	12.9	37	30
1200	1382	987	12.6	39	19

The results of the tensile tests show a decrease in percentage elongation after fracture (A) associated with increasing the austenitizing temperature from 900 to 1200 °C (Figure 5a). The relative change of its value between the austenitizing temperatures of 900 and 1000 °C is more than 6%, between the austenitization temperatures of 1000 and 1100 °C, 2%, and between the austenitization temperatures of 1100 and 1200 °C, also about 2%. The total relative change of percentage elongation between the temperatures of 900 and 1200 °C is about 11%. As can be seen, the greatest decrease in percentage elongation took place after austenitizing at 1000 °C, when the austenite grains growth was about 22%. However, a further rise in austenitizing temperature does not cause rapid degradation of the ductile values represented by percentage elongation, despite the appearance of abnormal grains and a sharp increase in the austenite grain size between the austenitization temperatures of 1100 and

1200 °C. After austenitization at 1200 °C, the austenite grain is the greatest, followed by a decrease in the percentage elongation values, but its value is still maintained at a satisfactory level.

In the case of the percentage reduction of area (Z), the situation is similar (Figure 5b). There has been a decline in the value of the reduction of area of about 5%. Worth particular mention is the fact that there is no difference between the percentage reduction of area after austenitization at 1000 and 1100 °C, although the relative increase in austenite grain size connected with the appearance of abnormal grains was then considerable and amounted to 166.6%.

The highest values of plastic properties (A = 15%; Z = 47%) were noted for the as-received state characterized with the finest grains of austenite.

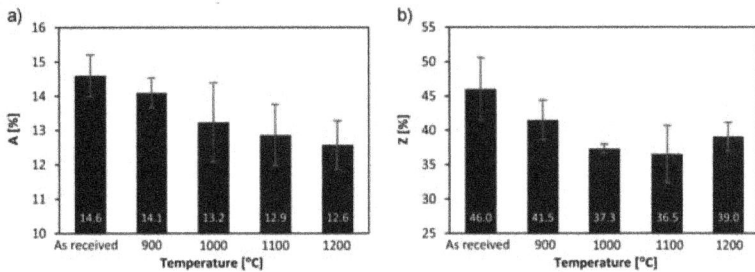

Figure 5. Percentage elongation (**a**) and percentage reduction of area (**b**) of the tested steel in the as-received state and after different austenitization temperatures.

The characteristic downward trend can also be noticed in the case of the values of the yield strength ($R_{p0.2}$) (Figure 6a) and the tensile strength (R_m) (Figure 6b), and is associated with an increase in austenitization temperature. It is worth noticing that, after austenitization at 900 °C, the tensile strength value is higher than in the as-received state, but the yield strength value is lower.

However, the continued downward trend in the case of tensile strength is not strong. Note that, after austenitizing at 1100 °C, the tensile strength was higher than the austenitizing at a lower temperature. The highest decrease in tensile strength of about 3% was noted between the austenitizing temperatures of 1100 and 1200 °C. The total decline in the value of tensile strength within the extreme range of austenitization temperatures was more than 4%.

A greater decrease was noticed in the case of the yield strength, even from the austenitization temperature of 1000 °C. The yield strength after austenitizing at this temperature decreased by about 6% in comparison to the yield strength after austenitizing at 900 °C. At higher temperature, the decrease in the yield strength value was not as significant and was about 1% and 2%, respectively, between samples austenitized at 1000 and 1100 °C and between samples austenitized at 1100 and 1200 °C. The overall decrease in yield strength within the extreme austenitization temperature range was more than 8% and two times higher than the decrease in the value of tensile strength.

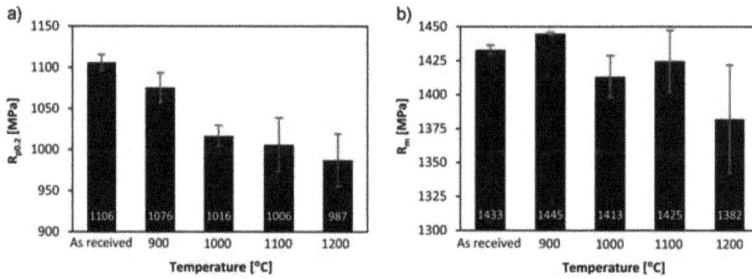

Figure 6. Tensile strength (**a**) and yield strength (**b**) of tested steel in the as-received state and after different austenitizing temperatures.

The tested material after austenitizing at different temperatures is characterized by the different resistance to brittle fracture, which in many applications has fundamental importance. Comparing the obtained values of impact strength, it is found that the impact strength decreases by nearly 61% between the average values for extreme austenitizing temperatures, confirming the trend determined in the tensile testing and unambiguously attesting to a drop in the plastic properties of the material (Figure 7). However, it can be seen that the impact of the austenitic temperature was stronger in the case of impact strength. Of interest is the very high impact strength value in the as-received state, which is characterized by the smallest austenite grain size. The difference in austenite grain size between the as-received state and the state after austenitizing at 900 °C is about 58%; however, the impact strength value drops by about 30%. This proves that the impact strength is very sensitive to the austenite grain size.

Assuming the criterion of a minimum impact strength equal to 35 J/cm^2, this steel meets this criterion until the austenitization temperature reaches 1000 °C. After austenitizing at a temperature of 1100 °C, the impact strength of this steel was 30 J/cm^2, which was not much lower than the criterion cited. The criterion of a minimum impact strength equal to 35 J/cm^2 is adopted for construction materials in many applications [17].

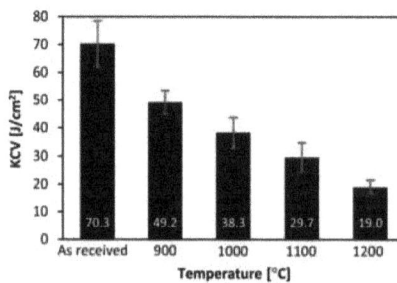

Figure 7. Impact strength KCV of tested steel in the as-received state and after different austenitizing temperatures.

3.3. Analysis of Martensite Morphology

Figure 8 presents the morphology of the martensitic structure of Hardox 450 steel in different states of heat treatment. In all the analyzed cases, the created martensite was in the form of laths laid in packets inside the prior austenite grains. The dependence between the size of the martensite packets and the size of the former austenite is shown in Figure 9. It has been shown that the size of the packets increased as a function of the heat treatment state of the steel (growth of the austenite grain). The average size of the packets increased from 3 µm to 16 µm, which corresponds to the former austenite grain growth within the range of 18–124 μm. The increase in the packet size growth represents approximately the linear function. Figure 10 illustrates the distribution of the corresponding fractions of the martensite packets for the samples of different sizes of grain of the former austenite. At the base of the performed normal distributions, there is a high correlation of the appearance of the martensite fraction with the distribution of the former austenite grain growth for the different heat treatment variants shown in Figure 4.

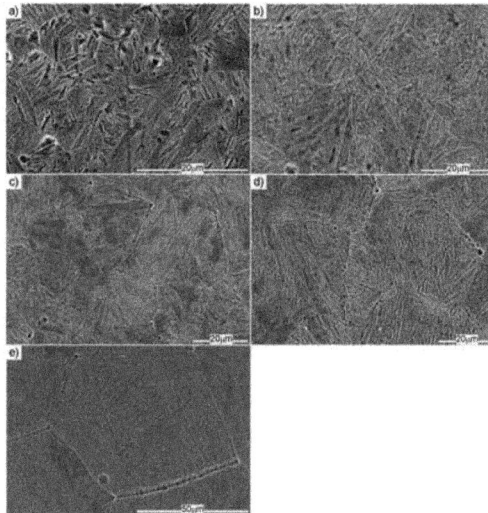

Figure 8. Micrographs showing martensitic packets in the Hardox 450 steel in the as-received state and under different annealing conditions: (**a**) in the as-received state; (**b**) 900 °C, 20 min; (**c**) 1000 °C, 20 min; (**d**) 1100 °C, 20 min; (**e**) 1200 °C, 20 min. Etched state, SEM.

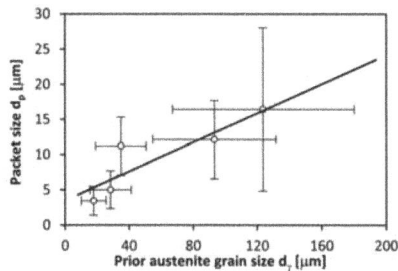

Figure 9. Relationship between the packet size and the prior austenite grain size in the Hardox 450 steel.

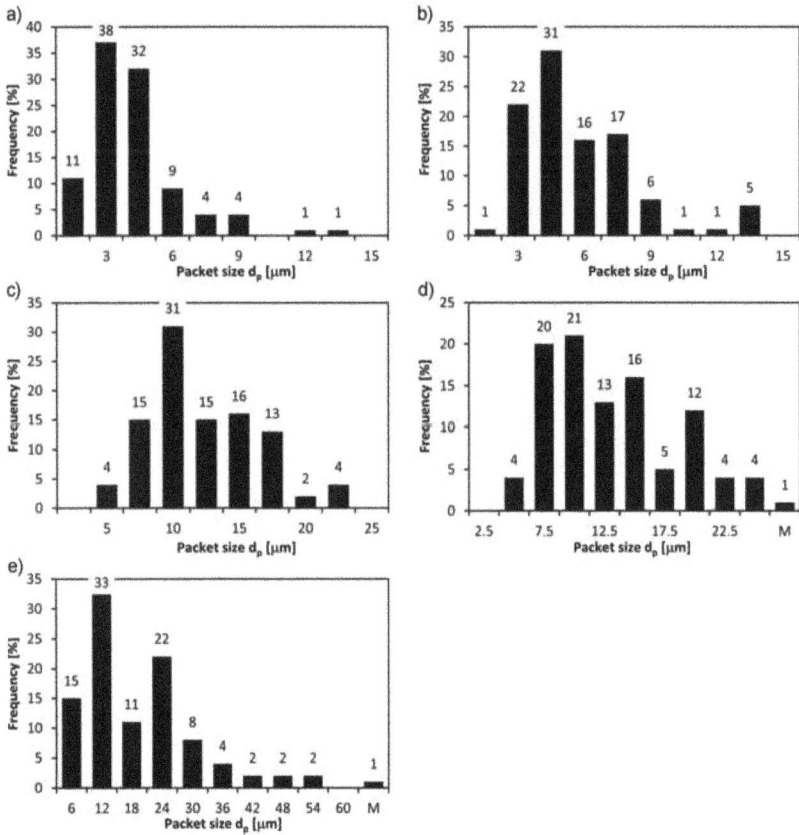

Figure 10. Distribution of martensite packet size in Hardox 450 steel with the prior austenite grain sizes: (**a**) 18 μm; (**b**) 28.5 μm; (**c**) 35 μm; (**d**) 93.3 μm, M—packet size more than 25 μm; (**e**) 123.7 μm, M—packet size more than 60 μm.

3.4. Fracture Analysis

Another way to determine the temperature of ductile-to-brittle transition is by direct analysis of the nature of the fracture. It is assumed that the criterion of impact strength (35 J/cm^2) corresponds to the occurrence of the mid-brittle and mid-ductile fracture [17].

Figure 11 summarizes the macroscopic views of fractures of the tested steel at temperature—40 °C in the as-received state and after different austenitizing temperatures. Figures 12–16 show representative SEM images of the fracture surfaces of the investigated steels.

Macroscopic analysis of the tested samples showed a significant divergence in the assessment of the embrittlement threshold based on the value of impact strength and direct analysis of the fracture. In all samples, there is a large central zone with typical facets for brittle fracture. In the case of the tested steel in the as-received state and after austenitizing at 900 °C, the plastic zones under the mechanical notch and in lateral edges shared respectively 28% and 30% of the whole fracture. In the case of samples austenitized at 1000 °C, the central brittle zone occupied more than 77%. Samples austenitized at higher temperatures were constantly characterized by nearly 100% fraction of brittle fracture—84% (for the sample austenitized at 1100 °C) and 89% (for the sample austenitized at 1200 °C). The fraction area of brittle fracture significantly increases by about 27% with increase in the austenitizing temperature.

Figure 11. Macroscopic view of the fractures that occurred after the Charpy impact strength test of the as-received state sample and the samples austenitized at different temperatures.

Analysis of the fractures in the individual samples austenitized at different temperatures showed some differences in structure. The change in fracture morphology as a function of austenitization temperature (growth of the austenite grain) may be characterized based on the morphology and the fraction of cleavage ridges appearing in the analyzed fractures. With an increase in the austenitizing temperature, the fraction of plastic areas inside the grains decreases, and typically cleavable areas even appear, after which the fraction of the main cleavage ridges increases.

The fracture in the as-received sample is shown in Figure 12a–d. The lateral zones, as well as that under the mechanical notch, constitute the ductile fracture with voids of various diameters (Figure 12a,d). The plastic zone under the mechanical notch of the pitting structure is shown in the figure. The fracture partly shows the "flaky" structure. It is considered that such a fracture is created as a result of the slides and the subsequent decohesion and appearance of microcracks in the {100} planes. The connection of microcracks through shear of the walls dividing them gives the characteristic appearance of the fracture in the form of overlapping flakes. As a result, the fracture is initiated with plastic deformation—a slide—but the cracking itself runs in principle along the specific crystallographic planes [5]. The pits are parabolic in shape, which proves the action of tangent forces in the process of fracture formation. Additionally, smooth areas can be distinguished, free of typical pitting relief. The central zone is occupied by a so-called quasi-cleavage fracture (Figure 12b). This is typical of steels with a martensitic structure, as well as a bainitic one, at temperatures below the brittle fracture transition temperature. This type of fracture forms through cleavage cracking in small local areas, after which their combination into one cracking surface takes place as a result of the plastic deformation. Despite the fact that the facets here are similar to the cleavable ones due to the presence of the carved "river patterns," identification of the crystallographic planes is almost impossible. This is not a typical quasi-cleavage fracture because the system of "river patterns", while meandering, creates pits at the large surface whose structure can resemble a ductile fracture. The ridges of the quasi-cleavable facets are characterized by extensive topography, which also indicates large plastic deformation during their formation. Moreover, numerous transverse cracks were observed. In the plastic zones of the pitted structure, inclusion precipitations are stuck.

Figure 12c shows the fault, which extends above the crack surface, having marked at the shoulder the ductile fracture of the pitted structure, which could also be evidence that the small column from the matrix was pulled (Figure 12c). One gets the impression that the columns joined the surfaces separated by cracking. If that was so, then we can speak of local breaking of the crack development. Thus, these photographs are proof of the very uneven and developed microrelief of the crack surfaces.

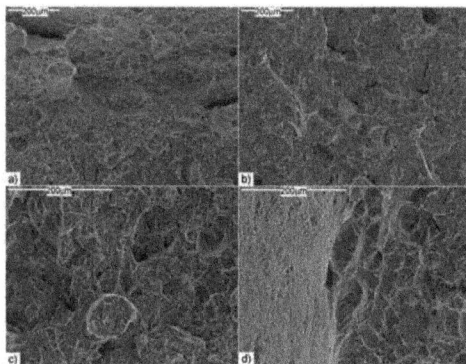

Figure 12. Fracture of the tested sample in the as-received state: (**a**) the zone under the mechanical notch; (**b**) the quasi-brittle central zone with visible ductile area; (**c**) the central zone with visible "small column"; (**d**) transition between the plastic lateral zone and the quasi-cleavage zone, SEM.

The fracture of the sample austenitized at a temperature of 900 °C is characterized by extensive topography; however, it does not differ in terms of quality from the fracture in the as-received state, although the value of the impact resistance varies greatly (Figure 13a–d). The characteristic feature is the appearance of a larger number of fine transverse cracks (Figure 13d). Under the mechanical notch, where the ductile fracture appears, the topography is very extensive (Figure 13a), and there is a large fracture of ductile fracture in the lateral zones, as well as in the central part of the sample, in the form of bands (Figure 13b). In the central part of the fracture, a large fracture of plastic zones appears with the characteristic pitted structure, which extends over the facets. The "river patterns" themselves combine into basins, with the characteristic plastic structure. Additionally, small columns were noticed, the lateral zones of which are built of the ductile fracture (Figure 13c). The central zone constitutes the quasi-cleavable fracture.

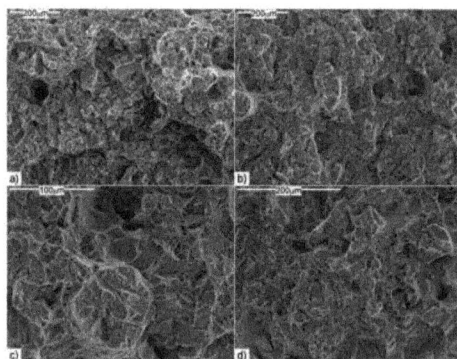

Figure 13. Fracture of the tested sample austenitized at 900 °C: (**a**) the zone under the mechanical notch; (**b**) the quasi-brittle central zone with visible ductile area; (**c**) the central zone with visible "small column"; (**d**) the quasi-cleavable zone with visible transverse cracks, SEM.

After austenitizing at a temperature of 1000 °C, despite a drop in the impact resistance value, the character of the fractures also changes insignificantly (Figure 14a–c). The topography is still extensive; in the central zone, the strained facets can be distinguished with characteristics of the "river patterns" relief. However, the cleavage ridges, where the typical ductile fracture appears, are narrower than in the case of the sample austenitized at a temperature of 900 °C (Figure 14a,b). Additionally, the characteristic small columns are visible (Figure 14c). Because of the characteristic "river patterns" relief, the surface of the fracture more closely resembles the quasi-cleavable fracture. The basins cross the whole grain. The appearance of these discontinuities means that the cracking ran not in a single crystallographic plane, but leapt from one plane to the other through shearing or the secondary cracking of the walls dividing them. It has been shown that this effect appears when the cracking front encounters screw dislocation, where the fault height is conditioned by the size of the Burgers vector [5]. The increase in energy absorbed during cracking is associated with the formation of the leaps, which is equivalent to a reduction of brittleness. The presence of the faults influences a change in direction of the cracking propagation. As a result, the growth of the crack is delayed in some sections, which entails a bending of its front, and this affects the merging of the adjacent faults, and their combination into the system of "river patterns".

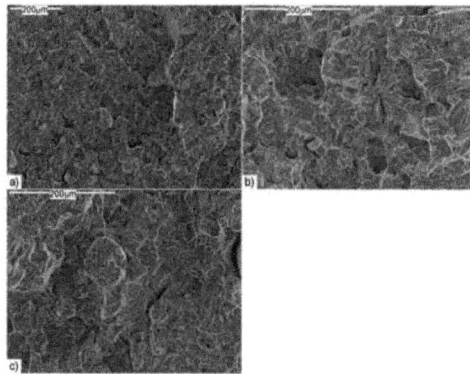

Figure 14. Fracture of the tested sample austenitized at 1000 °C: (**a**) the quasi-brittle central zone with visible ductile area; (**b**) the magnified central zone of the characteristic relief of "river patterns" with visible plastic ridges; (**c**) the central zone with visible "small column", SEM.

The appearance of abnormal grains and the increase in grain size as a result of austenitizing at a temperature of 1100 °C caused a drop in impact resistance below the criterion of 35 J/cm^2. This was reflected in minor qualitative changes of the character of the central zone of the fracture in comparison to the samples austenitized at lower temperatures (Figure 15a–c). It seems that the drop in impact resistance was influenced more by the smaller fracture of plastic zones under the mechanical notch and near the edges than by the fraction of these zones in the central part of the fracture. The plastic zone near the edge of the sample shows the "flaky" structure (Figure 15d). The surface of the fracture is characterized by extensive topography with larger hills and holes, and numerous transverse cracks were noticed. A large fraction of typically cleavable areas with the cleavage faults inside the grains was noticed (Figure 15b). The appearance of these areas caused a significant drop in the impact resistance value as compared to the samples austenitized at higher temperatures. The cracks continue to form the branching, but not as intensively as in the case of samples austenitized at lower temperatures, which could have been influenced by the increase in grain size of the tested steel. It is understood that the secondary slots propagate in the direct vicinity of the main slot, until they show the characteristic discontinuity (the slot is created by the whole system of small micro slots). One of the reasons for the formation of such micro slots is the appearance of micro faults in the fracture surface, which can be attributed to defects in the internal structure. In the case of crystalline bodies, such defects will be dislocations and the grain boundaries [5]. As shown in [3], the density of dislocations is higher in the case of steel with a finer grain.

After austenitizing at the highest temperature, the fraction of plastic zones is small; instead, in the central zone, we still have to deal with the quasi-cleavage fracture (Figure 16a,b). The facets created in this way are small and do not have clear boundaries. Moreover, the characteristic feature of quasi-cleavage is that the secondary cleavage faults are transformed into cleavage ridges of the surrounding facet. Pits can also be noticed, as characteristic features of ductile fracture (Figure 16b). The facets of the characteristic relief of "river patterns" combine through the so-called cleavage ridges with features characteristic of ductile cracking as a result of plastic deformation. Mainly, they are located at lateral surfaces of the faults. Besides the quasi-cleavable areas, a significant fraction of typically cleavable areas was noticed (Figure 16a). The fracture topography is still extensive, and the transverse cracks and "small columns" appear; however, the holes and hills in the central zone are larger than in the case of fractures in samples austenitized at lower temperatures, which could be influenced by the coarse grain structure.

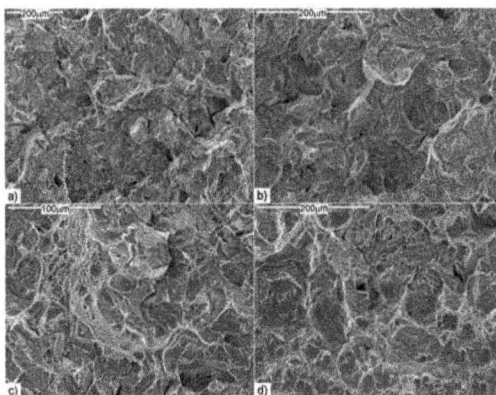

Figure 15. Fracture of the tested sample austenitized at 1100 °C: (**a**) the quasi-brittle central zone with visible ductile area; (**b**) the central zone with visible "small column" and cleavable zones; (**c**) the ductile band in the central zone; (**d**) plastic zone in the lower part of the sample (opposite to the mechanical notch), SEM.

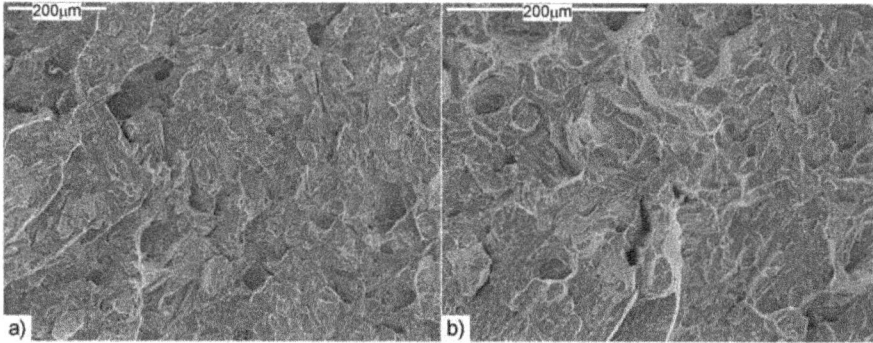

Figure 16. Fracture of the tested sample austenitized at 1200 °C: (**a**) the central zone of the fracture, visible small plastic areas and the typically cleavable areas; (**b**) the ductile bands and transverse crack in the central fracture zone, SEM.

4. Discussion

Selected mechanical properties for the analyzed steel in the as-received state and dependent on the austenitizing temperature have thus been shown, which has practical value. In this section, the values are to be dependent on the size of the austenite grains and packets of martensite. Figures 17 and 18 show the selected mechanical properties dependent on the size of the austenite grain in the analyzed steel and of the martensite packets. It can be noticed that, in the case of the impact resistance, the values were approximated with the exponential function with a good coefficient of correlation, while, in the case of the yield strength, the best correlation coefficient was obtained while approximating with the linear function.

Based on the achieved results, it can be stated that the plastic and not the strength properties are significantly more sensitive to a change in austenite grain size (especially when grain growth begins at lower austenitizing temperatures, even before the abnormal grains appear). In particular, the resistance of the tested steel to loads of dynamic character largely degrades, although the critical value of 35 J/cm^2 was not reached until the grains grew to about 90 μm. It can be concluded that the first phase of austenite grain growth is very critical in the process of reducing the mechanical properties.

In turn, the analysis of the impact of the martensite packet sizes on the yield strength has shown that, in the case of the lath martensite structure, the dependence described by the Hall-Petch relationship closely relates to the packet sizes as well as the effective martensite grain size in the low- and medium-carbon steel. It can be concluded that a change in the yield strength in the Hardox 450 steel corresponds with the inversion of the square root of the former austenite grain and the martensite packet size (Figure 17). However, it has to be underlined that the noted drop in the yield strength, as a result of the increase in grain size of the prior austenite from 18 μm to 124 μm (almost 700%), amounted only to 119 MPa (about 11%). In relation to the martensite, lowering the yield strength by the given value corresponded to an increase in the packet size of about 480% (3–16 μm). However, the impact of the increase in grain size of the former austenite and the packet size on the static tensile strength was not observed to be large.

Figure 18 presents the dependence between the grain size of the former austenite and martensite packets and the impact resistance value of the Hardox 450 steel. As a result of the increase in grain size of the former austenite from 18 μm to 124 μm, as well as the size of the packets, a more than 3.5-fold reduction in the impact resistance of the tested steel was noticed. This indicates a major influence of the packet size characteristic of martensitic steels on fracture work, especially at lower temperatures.

Undoubtedly, a change in the austenite grain size has an influence on impact resistance; thus, the martensite packets contributed to a decrease in the fraction of plastic zones near the fracture edge, as well as a change in character of the central quasi-cleavable zone, where a decrease in the fraction of the plastic ridges of cleavage occurred, the typical brittle facets appeared, and the means of propagation of the secondary cracks changed.

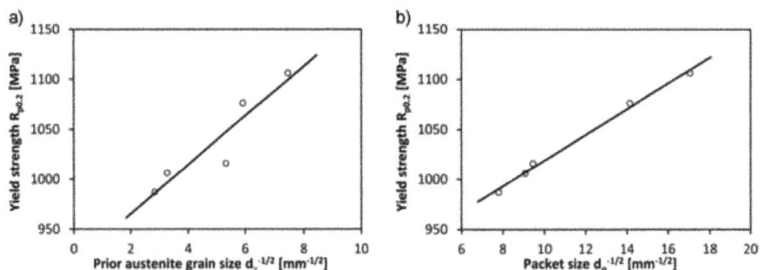

Figure 17. Dependence of the yield strength on the prior austenite grain size (**a**) and the martensite packet size (**b**) for the Hardox 450 steel.

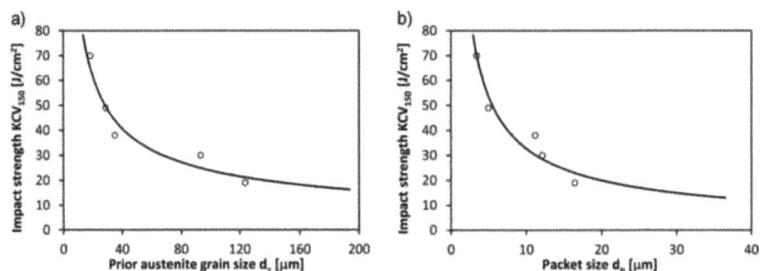

Figure 18. Influence of the prior austenite grain size (**a**) and martensite packet size (**b**) on the impact strength KCV_{150} at 233K ($-40\ ^\circ$C) for the Hardox 450 steel.

In the Hardox 450 steel (independent of the applied variant of heat treatment), the value of the yield strength value is related to the classic Hall-Petch rule. In the case of martensitic structures, the effective size of the grains considered in the above rule constitutes the martensite laths arranged in the form of packets of identical crystallographic orientation. However, it has to be mentioned that the size of the martensite packets, being closely related to the size of the prior austenite grains, is not taken into account in the Hall-Petch rule as the factor determining the plastic properties of steels. Again, it has to be noted that, in the case of Hardox 450 steel, the main decisive factor on the plastic properties is the value of Charpy impact energy, the value of which (in contrast to percentage elongation A and the percentage reduction of area Z) decreased as a function of the state of heat treatment, despite maintaining high-strength parameters. It has to be stressed, however, that, for the analyzed group of steels, the plastic properties, in most cases, are analyzed on the basis of the fracture morphology obtained.

The greater impact of the grain size on the impact resistance than on strength properties can be explained by the greater sensitivity of the impact resistance not so much on the austenite grain size as on the size of the martensite packets. Similar conclusions were reached by the authors of [18,19], who developed the heat treatment of steel, enabling a reduction in austenite grain size that would result in high resistance combined with high ductility in minus temperatures. According to the author of [4], "grain refinement is accomplished by controlling the martensitic transformation to break up the crystallographic alignment between adjacent martensite laths, interrupting the cleavage fracture path. In this case, grain refinement does not ordinarily cause a substantial increase in strength, probably because {110} planes lie along the long axis of the laths, which are not significantly refined."

In summary, this study forms part of the current research on modern steel groups with higher resistance to abrasive wear. The growth of austenite grains is an important factor in the analysis of the microstructure, as the grain size has an effect on the kinetics of phase transformation. The microstructure, however, is closely related to the mechanical properties of the material such as yield strength, tensile strength, percentage elongation, and impact strength, as well as the morphology of the fractures that occur. Therefore, it is important to control the grain size of metals and metal alloys during processes connected with thermal treatment such as welding or cutting.

5. Conclusions

1. The steel analyzed in the as-received state is characterized by high-strength properties combined with high impact resistance at lower temperatures and by ductility.
2. The lowest average size of prior austenite grains noted for the as-received state was 18.0 ± 7.8 μm. Increasing the austenitizing temperature causes an increase in the average grain size, which, at 1200 °C, is 123.7 ± 56.5 μm. These changes can be approximated by a quadratic function.
3. The austenite grain size has a strong influence on the mechanical properties of the investigated steel. However, in comparison to the significant changes in impact resistance, the austenite grain growth is accompanied by a small change in strength properties, i.e., the tensile strength and the yield strength.
4. As a result of increasing the grain size of the prior austenite from 18 μm to 124 μm, as well as the packet sizes, a more than 3.5-fold drop in the impact resistance of the tested steel was found. This indicates a major influence of the packet size characteristic of martensitic steels on the value of the fracture work, especially at lower temperatures.
5. Assuming the criterion of minimum impact strength is equal to 35 J/cm^2, this steel meets this criterion until the austenitization temperature reaches 1000 °C. However, after austenitizing at a temperature of 1100 °C, the impact strength of this steel was 30 J/cm^2, which is not much lower than the criterion cited.
6. The analysis of the influence of the martensite packets size on the yield strength showed that, in the case of the lath martensite structure, the dependence described by the Hall-Petch relation is closely related to packet size as well as the effective size of martensite grain in the low- and medium-carbon steel. It can be concluded that a change in the yield strength value of the Hardox 450 steel corresponds with the inverse of the square root from the grain size of the prior austenite and the packet size of the martensite.
7. Undoubtedly, the change in the size of the austenite grains and the martensite packets has an influence on the impact resistance, which contributed to a decrease in the fraction of plastic zones near the fracture edge and a change in the character of the central zone of the quasi-cleavable zone, where a decrease in the share of plastic cleavage ridges appeared, typical brittle facets appeared, and the means of propagation of the secondary cracks changed.

Author Contributions: Beata Białobrzeska contributed reagents/materials/analysis tools, conceived and designed the experiments, performed experiments, analysed data, wrote the paper; Łukasz Konat contributed reagents/materials/analysis tools, analysed data; Robert Jasiński conceived and designed the experiments performed experiments, analysed data.

Conflicts of Interest: The authors declare no conflict of interest.

References

1. Morito, S.; Tanaka, H.; Konishi, R.; Furuhara, T.; Maki, T. The morphology and crystallography of lath martensite in Fe-C alloys. *Acta Mater.* **2003**, *51*, 5323–5331. [CrossRef]
2. Wang, C.; Wang, M.; Shi, J.; Hui, W.; Dong, H. Effect of Microstructure Refinement on the Strength and Toughness of low alloy martenitic steel. *J. Mater. Sci. Technol.* **2007**, *23*, 659–664.

3. Prawoto, Y.; Jasmawati, N.; Sumeru, K. Effect of Prior Austenite Grain Size on the Morphology and Mechanical Properties of Martensite in Medium Carbon Steel. *J. Mater. Sci. Technol.* **2012**, *28*, 461–466. [CrossRef]
4. Morris, J.W., Jr. The Influence of Grain Size on the Mechanical Properties of Steel. In Proceedings of the International Symposium on Ultrafine Grained Steels, Fukuoka, Japan, 20–22 September 2001; Takaki, S., Maki, T., Eds.; Iron and Steel Institute of Japan: Tokyo, Japan, 2001; pp. 34–41.
5. Maciejny, A. *Kruchość Metali*; Wydawnictwo "Śląsk" Katowice: Katowice, Poland, 1973.
6. Esmailian, M. The effect of cooling rate and austenite grain size on the austenite to ferrite transformation temperature and different ferrite morphologies in microalloyed steels. *Iranian J. Mater. Sci. Eng.* **2010**, *7*, 7–14.
7. Lee, S.-J.; Lee, Y.-K. Prediction of austenite grain growth during austenitization of low alloy steels. *Mater. Des.* **2008**, *29*, 1840–1844. [CrossRef]
8. Dobrzański, L.A. *Podstawy Nauki o Materiałach i Metaloznawstwo*; WNT: Warsaw, Poland, 2003.
9. Krauss, G. *Steels, Processing, Structure, and Performance*; ASM International: Novelty, OH, USA, 2005.
10. Guimaraes, J.R.C.; Rios, P.R. Quantitative Interpretation of Martensite Microstructure. *Mater. Res.* **2011**, *14*, 97–101. [CrossRef]
11. Kitahara, H.; Ueji, R.; Tsuji, N.; Minamino, Y. Crystallographic features of lath martensite in low-carbon steel. *Acta Mater.* **2006**, *54*, 1279–1288. [CrossRef]
12. Manufacturer's Data. Available online: http://www.ssab.com/Global/HARDOX/Datasheets/en/168_HARDOX_450_UK_Data%20Sheet.png (accessed on 2 March 2015).
13. Dudziński, W.; Konat, Ł.; Pękalski, G. Structural and strength characteristics of wear-resistant martensitic steels. *Arch. Foundry Eng.* **2008**, *8*, 21–26.
14. Dudziński, W.; Konat, Ł.; Pękalska, L.; Pękalski, G. Structures and properties of Hardox 400 and Hardox 500 steels. *Inżynieria Materiałowa* **2006**, *3*, 139–142.
15. Konat, Ł. The Structure and Properties of Steel Hardox and Their Potential Application in the Conditions of Abrasive Wear and Dynamic Loads. Ph.D. Thesis, Wrocław University of Technology, Wrocław, Poland, 2007.
16. Gutérrez, N.Z.; Luppo, M.I.; Danon, C.A.; Caraballo, I.T.; Capdevila, C.; de Andrés, C.G. Heterogeneous austenite grain growth in martensitic 9Cr steel. Coupled influence of the initial metallurgical state and the heating rate. *Mater. Sci. Technol.* **2013**, *29*, 1254–1266. [CrossRef]
17. Wyrzykowski, J.W.; Pleszakow, E.; Sieniawski, J. *Odkształcenie i Pękanie Metali*; WNT: Warszawa, Poland, 1999.
18. Kim, H.J.; Kim, Y.H.; Morris, J.W., Jr. Thermal Mechanisms of Grain and Packet Refinement in a Lath Martensitic Steel. *ISIJ Int.* **1998**, *38*, 1277–1285. [CrossRef]
19. Morris, J.W., Jr.; Guo, Z.; Krenn, C.R. *Heat Treating: Steel Heat Treating in the New Millennium*; Midea, S.J., Pfaffmann, G.D., Eds.; ASM: Metals Park, OH, USA, 2000.

metals

MDPI

Article

Effects of Austenitizing Conditions on the Microstructure of AISI M42 High-Speed Steel

Yiwa Luo [1,2], Hanjie Guo [1,2,*], Xiaolin Sun [1,2,3], Mingtao Mao [1,2,3] and Jing Guo [1,2]

[1] School of Metallurgical and Ecological Engineering, University of Science and Technology Beijing, Beijing 100083, China; lyw918@126.com (Y.L.); b20140121@xs.ustb.edu.cn (X.S.); composure_guy@163.com (M.M.); guojing@ustb.edu.cn (J.G.)
[2] Beijing Key Laboratory of Special Melting and Preparation of High-End Metal Materials, Beijing 100083, China
[3] Central Iron & Steel Research Institute, Beijing 100083, China
[*] Correspondence: guohanjie@ustb.edu.cn; Tel.: +86-138-0136-9943

Academic Editor: Robert Tuttle
Received: 4 December 2016; Accepted: 12 January 2017; Published: 18 January 2017

Abstract: The influences of austenitizing conditions on the microstructure of AISI M42 high-speed steel were investigated through thermodynamic calculation, microstructural analysis, and in-situ observation by a confocal scanning laser microscope (CSLM). Results show that the network morphology of carbides could not dissolve completely and distribute equably in the case of the austenitizing temperature is 1373 K. When the austenitizing temperature reaches 1473 K, the excessive increase in temperature leads to increase in carbide dissolution, higher dissolved alloying element contents, and unwanted grain growth. Thus, 1453 K is confirmed as the best austenitizing condition on temperature for the steel. In addition, variations on the microstructure and hardness of the steel are not obvious when holding time ranges from 15 to 30 min with the austenitizing temperature of 1453 K. However, when the holding time reaches 45 min, the average size of carbides tends to increase because of Ostwald ripening. Furthermore, the value of M_s and M_f decrease with the increase of cooling rate. Hence, high cooling rate can depress the martensitic transformation and increase the content of retained austenite. As a result, the hardness of the steel is the best (65.6 HRc) when the austenitizing temperature reaches 1453 K and is held for 30 min.

Keywords: high-speed steel; austenitizing temperature; cooling rate; carbides; martensite

---

1. Introduction

High-speed steels (HSS) have been widely used to make engineering cutting tools in quenched and high-temperature tempered conditions due to their high hardness, wear resistance, and favorable high-temperature properties [1–3]. Among them, AISI M42 HSS is one of the most popular one owing to its excellent combination of hardness and toughness.

The mechanical properties of AISI M42 HSS are determined by the martensitic matrix and distribution of carbides. The cast structure of high-speed steel can be improved by subsequent heat treatment processes such as annealing, quenching, and tempering. Under annealing conditions, high-speed steels have a ferrite matrix with plenty of undissolved carbides. Following quenching, they contain martensite, retained austenite, and undissolved carbides. The final microstructure after tempering, which may occur several times, mainly consists of tempered martensite and well-distributed hard carbides [1,4–6]. It is known that microstructural factors like distribution of carbides, as well as characteristics of the martensitic matrix, play important roles in optimizing the properties of high-speed steel such as hardness, wear resistance, fracture toughness, and thermal-fatigue behavior. It is also well known that the martensite morphology of HSS depends on austenitizing temperature and its holding time, prior deformation of the austenite matrix, chemical composition of alloys, and the cooling rate [7,8]. A lot of studies have been performed on the effects of austenitizing conditions on the microstructure of high-speed steel. Sarafianos [9] reported that the mechanical properties of high-speed steel can be improved by austenitizing heat treatment at 1463 K for 30 s. The microstructure is altered to a mixture of martensite and retained austenite, and the types of carbides after austenitizing are mainly Mo and W enriched carbides. Hashimoto et al. [10] noticed that the austenitizing time did not significantly change the morphology, size, and distribution of eutectic carbides. Fu et al. [11] investigated the effects of quenching and tempering treatment on the microstructure, mechanical properties, and abrasion resistance of high-speed steel. Their results showed that the hardness of high-speed steel increased with the increase of austenitizing temperature, but decreased while the austenitizing temperature exceeds 1323 K. When the austenitizing temperature reached 1273 K, the metallic matrix all transformed into martensite. Afterwards, the eutectic carbides dissolve into the metallic matrix and their continuous network distribution changed into the broken network. However, Kang et al. [12] reported that the HSS showed the best wear resistance when the austenitizing temperature was 1473 K, because M_6C phase was dissolved at such a high temperature, only MC carbide could be stable for its property of hindering crack propagation. Accordingly, opinions are divergent on the effective austenitizing temperature and its holding time, which are beneficial to the mechanical property of high-speed steel.

The present work aims to identify the correlation between microstructure and hardness of AISI M42 HSS. It also intends to provide essential conditions for heat treatment by investigating the effects of austenitizing temperature, holding time, and cooling rate on the microstructure of AISI M42 HSS.

2. Materials and Methods

Material used in this investigation was produced and developed at the pilot plant of the Central Irom & Steel Research Institute (Beijing, China). The AISI M42 high-speed steel scrap was melted in vacuum induction furnace and refined through electroslag remelting (ESR) furnace. The ESR ingot was 16 cm in diameter and 100 cm in length. The chemical compositions of the produced ingot were measured by spectrograph and summarized in Table 1. According to this composition, the equilibrium phase diagram was calculated by Thermal-Calc software (Thermal-Calc Software, Stockholm, Sweden) with the TCFE7 database. A thermal dilatometer (DIL, Fuji Electric, Tokyo, Japan) was employed to measure martensite start temperature (M_s), martensite finish temperature (M_f), and simulate the continued cooling transformation (CCT) curve of steel. To examine the correlation between martensitic transformation and cooling rate, the as-cast samples were machined into 4 mm in diameter and 10 mm in length and heated up to 1473 K (1200 °C) with the heating rate of 10 K/s, maintained for 5 min and then cooled down with a series of cooling rates, such as 20 K/s, 10 K/s, 5 K/s, 1 K/s, 0.5 K/s, 0.1 K/s.

Table 1. Chemical compositions of AISI M42 high-speed steel. (wt. %).

C	Si	Mn	P	S	Cr	W	Mo	V	Co
1.11	0.7	0.39	0.016	0.002	4.58	1.39	9.46	1.37	8.04

The specimens machined into 20 mm × 20 mm × 20 mm were heated to 1373, 1423, 1453, and 1473 K (1100, 1150, 1180, and 1200 °C) at a rate of 10 K/s, and then held for 15 min followed by oil quenching in a box-type furnace. The specimens which heated to 1453 K (1180 °C) were held for 15, 30, and 45 min. Figures 1 and 2 show the detailed quenching process. In order to avoid the deformation and cracking due to excessive internal stress of the steel, the specimens were preheated at 1123 K for 10 min. The as-cast steel, as well as the quenched specimens, were machined into 10 mm × 10 mm × 10 mm and polished for microstructural observation using scanning electron microscope (SEM, Carl-Zeiss, Oberkochen, Germany). The average grain size of the specimens and volume fraction of carbides were counted by Image-Pro Plus software. The final value was the average by measuring 10 photos of each condition. Specimens for grain size measurement were polished and etched using a Nital etchant (Ethanol + 4 vol. % Nitric acid) to reveal the grain boundary, and then measured according to GB/T 6394-2002. Rockwell hardness tests were conducted on these specimens (10 mm × 10 mm × 10 mm) using a Hardness Tester (TIME, Beijing, China). Impact toughness of the specimens with different austenitizing conditions were measured by Charpy Tester (SANS, Shenzhen, China), the specimens were machined into 10 mm × 10 mm × 55 mm without notch.

In order to observe the evolution of microstructure in situ, a confocal scanning laser microscope (CSLM, Lasertec, Yokohama, Japan) equipped with an infrared image furnace was employed in this study. The specimens were heated up to 1473 K (1200 °C) in an alumina crucible, thermal insulated for 15 min, and cooled to room temperature at a cooling rate of 5 K/s and 20 K/s, respectively. The whole process of the experiment was under an ultra-high purity argon gas and the cooling rate was controlled by adjusting nitrogen gas flowrate.

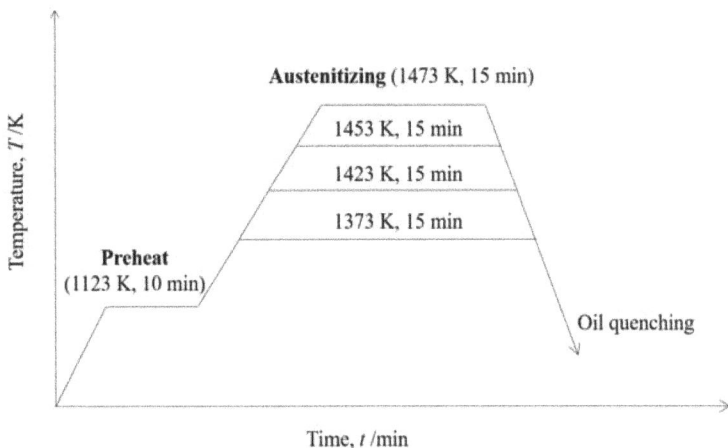

Figure 1. Experimental process for the investigation of effect of austenitizing temperature on the microstructure of AISI M42 HSS.

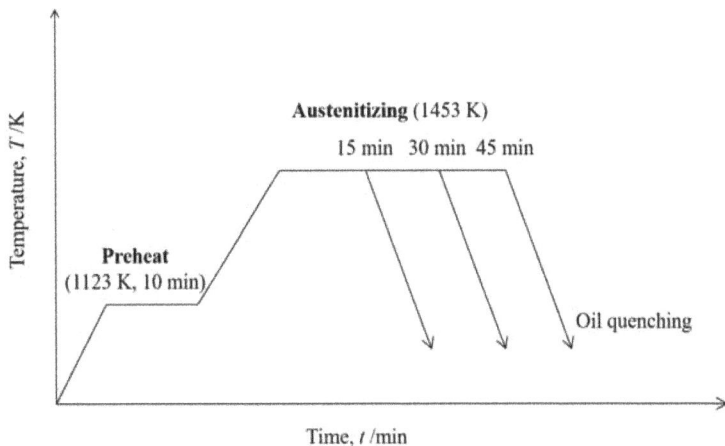

Figure 2. Experimental process for the investigation of effect of holding time on the microstructure of AISI M42 HSS.

3. Results and Discussion

3.1. Phase Transformation and Precipitation Behavior of Carbides

Because the microstructure of as-cast HSS usually consists of ferrite and pearlite, austenitizing and subsequent rapid cooling were employed to improve hardness of the steel. The phase stability diagram of AISI M42 HSS obtained from Thermo-Calc was shown in Figure 3.

As shown in Figure 3, the liquidus temperature and the solidus temperature of present AISI M42 HSS are 1680 K (1407 °C) and 1500 K (1227 °C), respectively. When the steel is slow cooling to room temperature, the carbides of the present steel consisted of M_6C, M_7C_3, and MC. MC starts to precipitate at 1550 K (1277 °C) in the solid-liquid zone, while M_6C and M_7C_3 begin to precipitate in the solid phase at 1500 K (1227 °C) and 1140 K (867 °C), respectively. That means when the specimens are heated to the experimental temperatures and rapid cooling, M_7C_3 have been dissolved into the matrix and the carbides of the present specimens consisted of MC and M_6C.

The determined CCT diagram of AISI M42 HSS austenitized at 1473 K (1200 °C) was shown in Figure 4. The dotted lines means different cooling rates range from 100 K/h to 30 K/s. The martensite point (M_s), critical temperatures of A_{c1} and A_{c3} (the start and end of austenitic transformation on heating) were determined corresponding to a heating (10 K/s) up to 1473 K (1200 °C) and subsequent cooling down (5 K/s) to room temperature. As shown in Figure 4, the pearlitic transformation is only possible when the cooling rate is as slow as 100 K/h. Bainitic transformation occurs at the range from 100 to 1000 K/h. For the cooling rate of 1 to 30 K/s, martensite would precipitate directly and cross the area of pearlite and bainite.

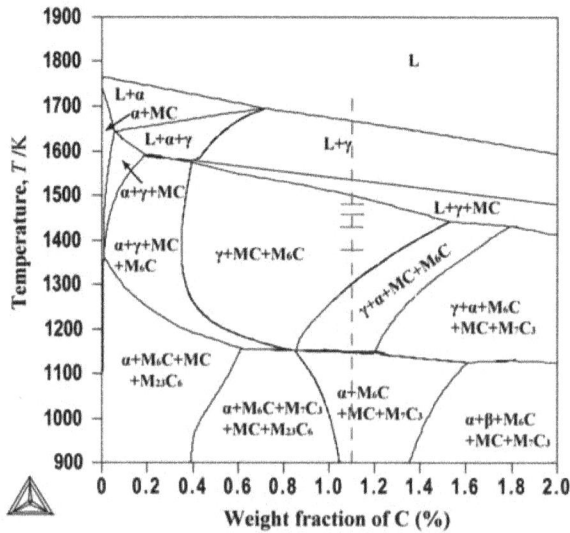

Figure 3. Phase equilibrium diagram of AISI M42 HSS calculated using Thermo-Calc software. A dotted vertical line means the C concentration (1.11 wt. %) of the present steel. The blue lines show the region of four different temperatures.

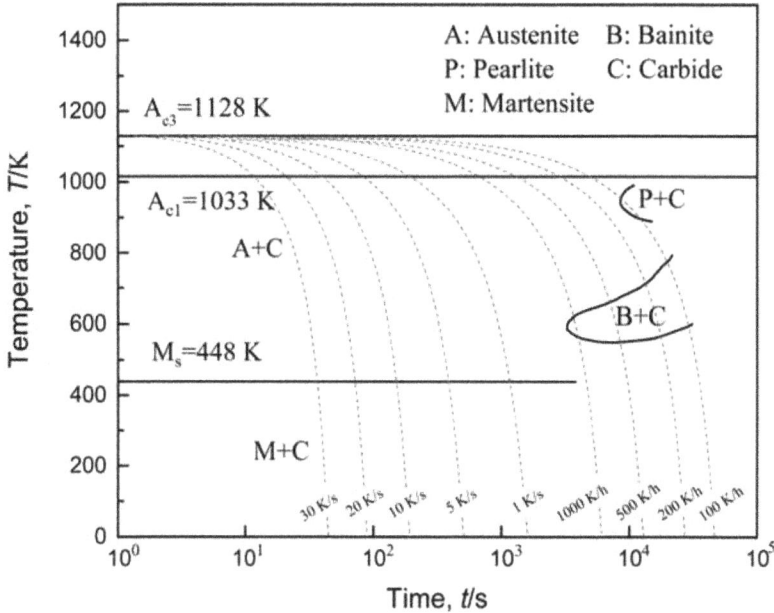

Figure 4. CCT diagram for AISI M42 HSS austenitized at 1473 K (1200 °C). The dotted lines mean different cooling rates range from 100 K/h to 30 K/s.

3.2. Effects of Austenitizing Temperature on Microstructure and Properties

SEM micrograph of as-cast HSS specimen was shown in Figure 5a. As indicated by arrows, network carbides M_2C, and graininess-shaped carbides MC (red circles) were found in the micrograph. However, M_2C did not appear in the phase stability diagram because M_2C particles are in the metastable phase. MC carbides were dispersed in the matrix and the average size of them was much smaller than M_2C carbides. The grains in the as-cast HSS were completely surrounded by the layers of M_2C, namely, grain boundaries (GBs) were completely wetted by the second solid phase M_2C carbides. Microstructure of AISI M42 HSS specimens oil cooled at different temperatures were shown in Figure 5b–e. The population of the carbides was being steadily reduced with the increase of temperature during austenitization. When the austenitizing temperature was 1373 K (1100 °C), fibrous M_2C particles decomposed into M_6C and MC particles [13,14]. However, the network morphology of carbides still exists as shown in Figure 5b. With the increasing of austenitizing temperature, the continuous network distribution of carbides turn to be broken and scattered as shown in Figure 5c,d. The portion of completely wetted GBs decreased as well. When GBs wetted by a liquid phase (melt), the amount of completely wetted GBs always increases with the increase of temperature due to higher entropy of a melt [15,16]. However, when GBs are wetted by a second solid phase, the completely wetted GBs will decrease with the increase of temperature [17,18]. Moreover, M_s point will decrease and the stability of austenite will increase because the high-temperature austenite dissolves more carbon and alloying elements with the increase of austenitizing temperature [19,20]. When the austenitizing temperature reaches 1473 K (1200 °C), more carbides dissolve into the metallic matrix so that the quantity of carbides reduced rapidly as shown in Figure 5e. No sufficient number of carbides were available to pin down the grain boundaries to prevent grain coarsening, so it is more probable for cracks to be formed [4,9]. Besides, austenitizing temperature approaching the melting point of the experimental AISI M42 HSS (1500 K) leads to partial melting and agminate carbide precipitation of the specimen. Figure 6 shows the relationship between the average grain size and austenitizing temperature of the specimens. The average grain size presents a smooth increase with the increase of temperature. However, it shows an impressive promotion (35 μm) during a period from 1453 to 1473 K. Consequently, an excessive increase in temperature results in increased carbide dissolution and unwanted grain growth.

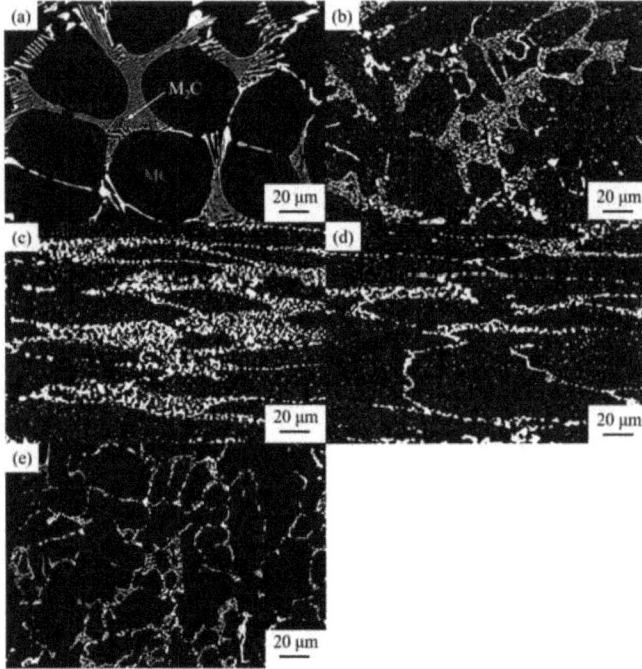

Figure 5. SEM microstructures of the as-cast AISI M42 HSS specimen (**a**); the oil cooling specimens at 1373 K (**b**); 1423 K (**c**); 1453 K (**d**); and 1473 K (**e**).

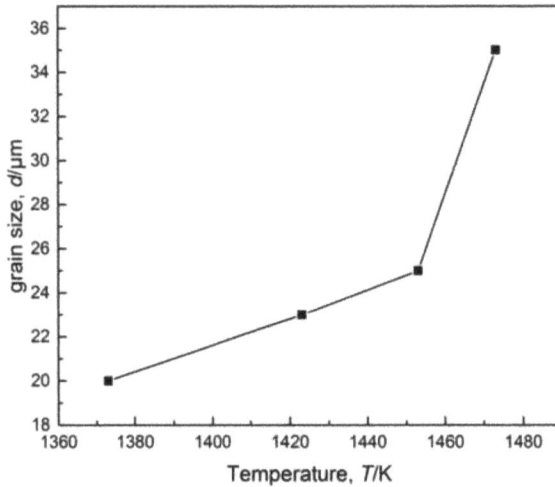

Figure 6. Relationships between grain size and austenitizing temperature.

Table 2 presents the corresponding hardness and impact toughness of AISI M42 HSS specimens with different austenitizing temperatures and holding time. The hardness increased with the austenitizing temperature between 1373 and 1453 K, but decreased when the austenitizing temperature reached 1473 K. This phenomenon could be explained by the reducing of well-distribution hard carbides. The hardness of quenched HSS is affected by the microstructure, the quantity of carbides, and alloy elements in the martensite and the retained austenite [21,22]. Carbon and alloying elements dissolved into the austenite, and then transferred to martensite by oil cooling, and the quantity increased with the increase of austenitizing temperature. However, when the austenitizing temperature reached 1473 K, there was an excess dissolution of carbon and alloying elements in the austenite. Besides, M_s tends to decrease when alloying elements dissolved in the steel at high temperature [21]. It lead to excess retained austenite remaining in the microstructure after oil cooling which reduced the hardness of the steel. On the contrary, the impact toughness increased with the increase of austenitizing temperature for the increasing of retained austenite. Consequently, in the experimental temperature range, the tendency of the hardness of the specimens presented a peak, and the impact toughness increased with increasing temperature.

Table 2. Effects of austenitizing conditions on hardness and impact toughness of AISI M42 HSS.

Austenitizing Temp. (K)	Holding Time (min)	Hardness (HRc)	Impact Toughness (J)
1373	15	58.5	8.7
1423	15	59.8	10.4
1453	15	65.5	10.9
1453	30	65.7	11.2
1453	45	64.8	11.9
1473	15	63.7	12.5

3.3. Effects of Austenitizing Time on Microstructure and Properties

The series in Figure 7 represent SEM micrographs of the specimens austenitized at 1453 K for 15 to 45 min. From Figure 7a,b, change on the microstructure of the specimens was not obvious under the austenitizing time ranges between 15 and 30 min, as well as the hardness (Table 2). However, the hardness of the specimen decreased when the holding time up to 45 min. Besides, the average size of the carbides tends to increase but the quantity of carbides seems decrease as shown in Figure 7c. This phenomenon could be explained by the fact of Ostwald ripening. Since carbides with large sizes have a lower interfacial concentration. The resulting concentration gradients give rise to diffusional transport from small carbides to the matrix then to large carbides. Thus, large carbides tend to grow at the expense of the small carbides, the average particle size increases, and the total number of carbides decreases [23,24]. The impact toughness of the specimens increased with the increase of holding time. Because more carbides and alloy elements dissolved in the matrix promoted solution strengthening, and made retained austenite stable. Figure 8 is the volume fraction of carbides in the specimens with different holding time and austenitizing temperatures. The blue line means cooling at different austenitizing temperature but with the same holding time—i.e., 15 min. The red line represents cooling with different holding time at the same austenitizing temperature, i.e., 1453 K. Compared with that of the as-cast specimen, the quantity of carbides in AISI M42 HSS after austenitizing reduced greatly. At 1453 K, the quantity of carbides showed a smooth decrease with the increase of holding time. It could be thought that this decrease in the amount of the carbides is due to the dissolution of carbides into austenite matrix phase. Moreover, the decrease of carbides with different holding time is much weaker than that with different austenitizing temperatures. Consequently, the effect of austenitizing temperature on the microstructure and hardness of AISI M42 HSS is more influential than the austenitizing time.

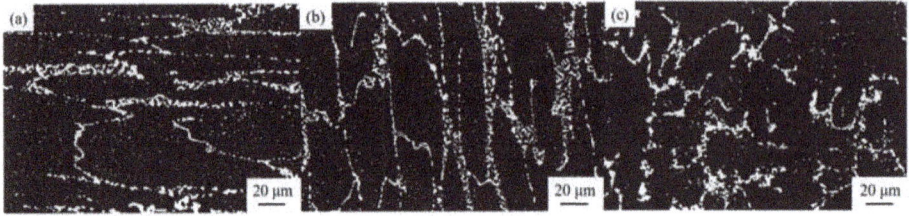

Figure 7. Microstructures of the oil cooling AISI M42 HSS specimens at 1453 K for 15 min (**a**); 30 min (**b**); and 45 min (**c**).

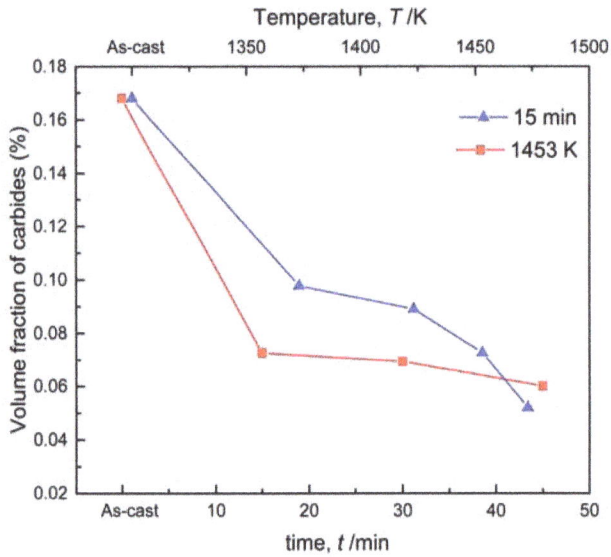

Figure 8. The volume fraction of carbides in the specimens with different holding time and austenitizing temperatures.

3.4. Effects of Cooling Rate on Microstructure and Properties

In order to investigate the effects of cooling rate on the microstructure and observe the phase transformation and precipitation of AISI M42 HSS in situ, CSLM was employed in combination with thermal dilatometer experimental determination. M_s and M_f were measured by a thermal dilatometer and shown in Figure 9.

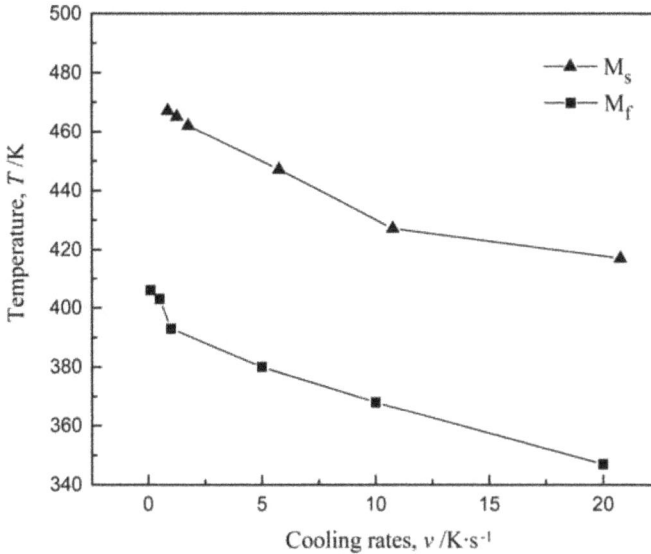

Figure 9. M_s and M_f of as-cast AISI M42 HSS under different cooling rates.

As the cooling rate increases, the value of the M_s and M_f points decrease consequently (Figure 9). When the cooling rate is 5 K/s, the value of M_s and M_f are 448 K (175 °C) and 380 K (107 °C), respectively. When the cooling rate reaches 20 K/s, the value of M_s and M_f decrease to 419 K (146 °C) and 347 K (74 °C), respectively.

Figure 10 shows the sequential image of the in-situ observation on AISI M42 HSS cooling from 1473 K (1200 °C) to room temperature by the rate of 5 K/s. Each image is arranged in the order of the corresponding temperature. Figure 10a,b is magnified 500 times and Figure 10c–f is magnified 1000 times. In the heat up stage, some small carbides precipitated because of the volume contraction caused by austenite transformation (Figure 10a). When the temperature reached 1473 K (1200 °C), an outburst of precipitation and growth of carbides appeared (Figure 10b). At the beginning of cooling, the specimen consisted of austenite and carbides which precipitate on the grain boundaries, and the grain size was about 15 to 25 μm (Figure 10c). When the temperature decreased to 1281 K (1008 °C), acicular ferrite started to transform from austenite (Figure 10d). The martensitic transformation began at about 446 K (173 °C) with a saltatory mode along the grain boundary and the undissolved carbides, as shown in Figure 10e. During the increase of marensite, a position of grain boundaries were replaced by martensite. No pearlite appears in the range from 1281 K (1008 °C) to 446 K (173 °C), which indicates that the cooling rate of 5 K/s is greater than the critical cooling rate of pearlitic transformation (Figure 4). When the temperature dropped to room temperature, martensitic transformation was still underway and there was a small group of retained austenite in the microstructure (Figure 10f). It is worth noting that martensitic transformation is usually finished rapidly in most of low-carbon steels [25,26]. However, for the high-carbon steel AISI M42, martensite transformed with a saltatory mode. Martensite still emerged sporadically during in-situ observation when the temperature dropped to room temperature. Because martensitic transformation is affected by shear resistance in kinetics, it could be worth considering that a higher content of carbon would increase the shear resistance of martensitic transformation and slow down the transformation rate.

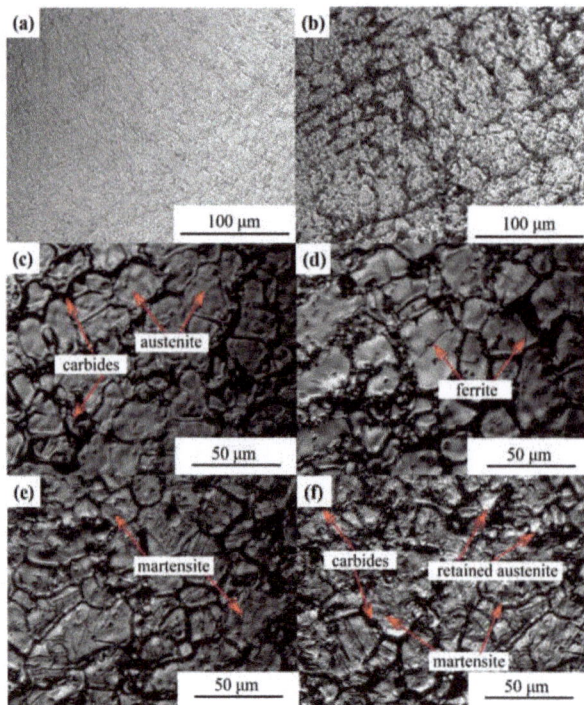

Figure 10. In-situ observation of phase transformation during austenitizing process of AISI M42 HSS with the cooling rate of 5 K/s: (**a**) 1423 K (×500); (**b**) 1473 K (×500); (**c**) 1473 K (×1000); (**d**) 1281 K (×1000); (**e**) 446 K (×1000); (**f**) room temperature (×1000).

The phase transformation and precipitation behavior of AISI M42 HSS austenitizing at 1473 K (1200 °C) with the cooling rate of 20 K/s was shown in Figure 11. It seems to be very different from that with the cooling rate of 5 K/s. At the beginning of cooling, the specimen also consisted of austenite and carbides (Figure 11a). Acicular and triangular ferrite transformation began at 1323 K (1050 °C) (Figure 11b). The grain size was uneven between 20 and 50 μm. Because the cooling rate (20 K/s) is much greater than the critical cooling rate of pearlitic transformation, no pearlite appeared in the process of cooling (Figure 4). When the temperature decreased to 421 K (148 °C), martensite started to transform from the austenite (Figure 11c). Because there was so little time for martensitic transformation and carbide dissolution, there was still much retained austenite that had not finished the transformation in the microstructure (Figure 11d). Moreover, increasing the cooling rate could decrease M_s and M_f point which narrow the martensitic transformation range (Figure 9). The carbides likewise were hardly dissolved and presented network, which could provide routes for crack propagation and make cracks deeper [27,28]. Therefore, it might be worth considering that a high cooling rate can depress the martensitic transformation and increase the content of retained austenite. Consequently, an exorbitant cooling rate prevents less retained austenite, better microstructure, and higher hardness in AISI M42 HSS.

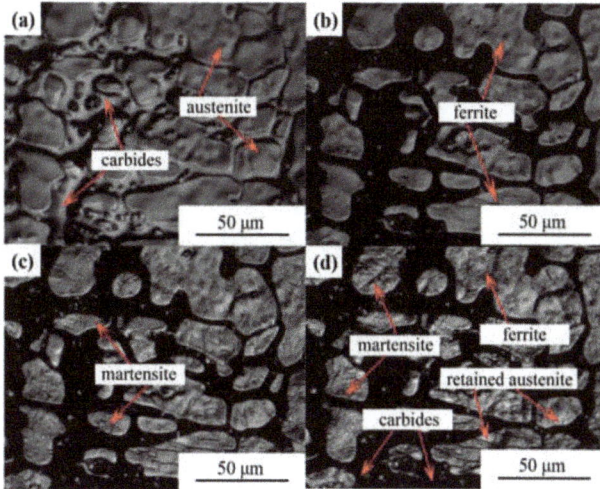

Figure 11. In-situ observation of phase transformation during austenitizing process of AISI M42 HSS with the cooling rate of 20 K/s: (**a**) 1473 K (×1000); (**b**) 1323 K (×1000); (**c**) 421 K (×1000); (**d**) room temperature (×1000).

4. Conclusions

In this study, effects of austenitizing temperature, holding time, and cooling rate on the microstructure and hardness of AISI M42 HSS were investigated. The conclusions have been drawn as follows:

1. When the austenitizing temperature is 1373 K, the network morphology of carbides could not dissolve completely and distribute equably. When the austenitizing temperature reaches 1473 K, the excessive increase in temperature leads to increased carbide dissolution and unwanted grain growth. 1453 K gives the best austenitizing condition on temperature for AISI M42 HSS.

2. The amount of hardness increases with the increase of austenitizing temperature between 1373 and 1453 K, but decreases when the austenitizing temperature reaches 1473 K. The hardness of the AISI M42 HSS does not change much when the holding time is less than 30 min, but decreases when the holding time reaches 45 min because of Ostwald ripening. The impact toughness of the specimens increases with the increase of austenitizing temperature and holding time.

3. The value of M_s and M_f decrease with the increase of cooling rate. High cooling rate can depress the martensitic transformation and increase the content of retained austenite, which prevents less retained austenite, better microstructure, and higher hardness in AISI M42 HSS.

Acknowledgments: This study was financially supported by the National Natural Science Foundation of China (NSFC) by a Grant of No. U1560203 and No. 51274031.

Author Contributions: Yiwa Luo and Hanjie Guo conceived and designed the experiments; Mingtao Mao and Jing Guo performed the experiments; Yiwa Luo and Xiaolin Sun analyzed the data; Xiaolin Sun contributed analysis tools; Yiwa Luo wrote the paper.

Conflicts of Interest: The authors declare no conflict of interest.

References

1. Kim, C.K.; Park, J.I.; Lee, S.; Kim, Y.C.; Kim, N.J.; Yang, J.S. Effects of alloying elements on microstructure, hardness, and fracture toughness of centrifugally cast high-speed steel rolls. *Metall. Mater. Trans. A* **2005**, *36*, 87–97. [CrossRef]

2. Sano, Y.; Hattori, T.; Haga, M. Characteristics of high-carbon high speed steel rolls for hot strip mill. *ISIJ Int.* **1992**, *32*, 1194–1201. [CrossRef]

3. Zhou, X.F.; Fang, F.; Li, F.; Jiang, J.Q. Morphology and microstructure of M_2C carbide formed at different cooling rates in AISI M2 high speed steel. *J. Mater. Sci.* **2011**, *46*, 1196–1202. [CrossRef]

4. Yu, T.H.; Yang, J.R. Effect of retained austenite on GPM A30 high-speed steel. *J. Mater. Eng. Perform.* **2007**, *16*, 500–507. [CrossRef]

5. Cambronero, L.E.G.; Gordo, E.; Torralba, J.M.; Ruiz-Prieto, J.M. Comparative study of high speed steels obtained through explosive compaction and hot isostatic pressing. *Mater. Sci. Eng. A* **1996**, *207*, 36–45. [CrossRef]

6. Xu, L.; Xing, J.; Wei, S.; Chen, H.; Long, R. Comparative investigation to rolling wear properties between high-vanadium high-speed steel and high-chromium cast iron. *J. Xi'an Jiaotong Univ.* **2006**, *40*, 275–278.

7. Himuro, Y.; Kainuma, R.; Ishida, K. Martensitic transformation and shape memory effect in ausaged Fe-Ni-Si alloys. *ISIJ Int.* **2002**, *42*, 184–190. [CrossRef]

8. Durlu, T.N. Effects of high austenitizing temperature and austenite deformation on formation of martensite in Fe-Ni-C alloys. *J. Mater. Sci.* **2001**, *36*, 5665–5671. [CrossRef]

9. Sarafianos, N. The effect of the austenitizing heat-treatment variables on the fracture toughness of high-speed steel. *Metall. Mater. Trans. A* **1997**, *28*, 2089–2099. [CrossRef]

10. Hashimoto, M.; Kubo, O.; Matsubara, Y. Analysis of carbides in multi-component white cast iron for hot rolling mill rolls. *ISIJ Int.* **2004**, *44*, 372–380. [CrossRef]

11. Fu, H.; Qu, Y.; Xing, J.; Zhi, X.; Jiang, Z.; Li, M.; Zhang, Y. Investigations on heat treatment of a high-speed steel roll. *J. Mater. Eng. Perform.* **2008**, *17*, 535–542. [CrossRef]

12. Kang, M.; Lee, Y.K. The effects of austenitizing conditions on the microstructure and wear resistance of a centrifugally cast high-speed steel roll. *Metall. Mater. Trans. A* **2016**, *47*, 3365–3374. [CrossRef]

13. Zhou, X.; Fang, F.; Li, G.; Jiang, J. Morphology and properties of M_2C eutectic carbides in AISI M2 steel. *ISIJ Int.* **2010**, *50*, 1151–1157. [CrossRef]

14. Hwang, K.C.; Lee, S.; Lee, H.C. Effects of alloying elements on microstructure and fracture properties of cast high speed steel rolls part I: Microstructural analysis. *Mater. Sci. Eng. A* **1998**, *254*, 282–295. [CrossRef]

15. Straumal, B.B.; Kucheev, Y.O.; Efron, L.I.; Petelin, A.L.; Majumdar, J.D.; Manna, I. Complete and incomplete wetting of ferrite grain boundaries by austenite in the low-alloyed ferritic steel. *J. Mater. Eng. Perform.* **2012**, *21*, 667–670. [CrossRef]

16. Modica, L. Critical point wetting. *J. Chem. Phys.* **1977**, *66*, 359–361.

17. Straumal, B.B.; Baretzky, B.; Kogtenkova, O.A.; Straumal, A.B.; Sidorenko, A.S. Wetting of grain boundaries in Al by the solid Al_3Mg_2 phase. *J. Mater. Sci.* **2010**, *45*, 2057–2061. [CrossRef]

18. Straumal, B.B.; Gornakova, A.S.; Kucheev, Y.O.; Baretzky, B.; Nekrasov, A.N. Grain boundary wetting by a second solid phase in the Zr-Nb alloys. *J. Mater. Eng. Perform.* **2012**, *21*, 721–724. [CrossRef]

19. Zhou, B.; Shen, Y.; Chen, J.; Cui, Z.S. Evolving mechanism of eutectic carbide in as-cast AISI M2 high-speed steel at elevated temperature. *J. Shanghai Jiaotong Univ. Sci.* **2010**, *15*, 463–471. [CrossRef]

20. Deng, Y.K.; Chen, J.R.; Wang, S.Z. *High Speed Tool Steel*; Guo, G.C., Ed.; Metallurgical Industry Press: Beijing, China, 2002; pp. 199–200.

21. Yamamoto, K.; Kogin, T.; Harakawa, T.; Murai, N.; Kuwano, M.; Ogi, K. Effects of alloying elements in hardenability for high C high speed steel type alloy. *J. Jpn. Foundry Eng. Soc.* **2000**, *72*, 90–95.

22. Barlow, L.D.; Toit, M.D. Effect of austenitizing heat treatment on the microstructure and hardness of martensitic stainless steel AISI 420. *J. Mater. Eng. Perform.* **2012**, *21*, 1327–1336. [CrossRef]

23. Snyder, V.A.; Akaiwa, N.; Alkemper, J.; Voorhees, P.W. The influence of temperature gradients on Ostwald ripening. *Metall. Mater. Trans. A* **1999**, *30*, 2341–2348. [CrossRef]

24. Garay-Reyes, C.G.; Hernandez-Santiago, F.; Cayetano-Castro, N.; Martinez-Sanchez, R.; Hernandez-Rivera, J.L.; Dorantes-Rosales, H.J.; Cruz-Rivera, J. Analysis of Ostwald ripening in Ni-rich Ni-Ti alloys by diffusion couples. *J. Bull. Mater. Sci.* **2014**, *37*, 823–829. [CrossRef]

25. Alvarado-Meza, M.A.; Garcia-Sanchez, E.; Covarrubias-Alvarado, O.; Salinas-Rodriguez, A.; Guerrero-Mata, M.P.; Colas, R. Effect of the high-temperature deformation on the M_s temperature in a low C martensitic stainless steel. *J. Mater. Eng. Perform.* **2013**, *22*, 345–350. [CrossRef]

26. Morito, S.; Yoshida, H.; Maki, T.; Huang, X. Effect of block size on the strength of lath martensite in low carbon steels. *Mater. Sci. Eng. A* **2006**, *438–440*, 237–240. [CrossRef]

27. Nogueira, R.A.; Riberiro, O.C.S.; Das-Neves, M.D.M.; de Lima, L.F.C.P.; Filho, F.A.; Friedrich, D.N.; Boehs, L. Influence of the heat treatment on the microstructure of AISI T15 high speed steel. *Mater. Sci. Forum* **2003**, *416–418*, 89–94. [CrossRef]

28. Hwang, K.C.; Lee, S.; Lee, H.C. Effects of alloying elements on microstructure and fracture properties of cast high speed steel rolls part II: Fracture behavior. *Mater. Sci. Eng. A* **1998**, *254*, 296–304. [CrossRef]

metals

MDPI

Article

The Effects of Cr and Al Addition on Transformation and Properties in Low-Carbon Bainitic Steels

Junyu Tian, Guang Xu *, Mingxing Zhou, Haijiang Hu and Xiangliang Wan

The State Key Laboratory of Refractories and Metallurgy, Hubei Collaborative Innovation Center for Advanced Steels, Wuhan University of Science and Technology, 947 Heping Avenue, Qingshan District, Wuhan 430081, China; 13164178028@163.com (J.T.); kdmingxing@163.com (M.Z.); hhjsunny@sina.com (H.H.); wanxiangliang@wust.edu.cn (X.W.)
* Correspondence: xuguang@wust.edu.cn; Tel.: +86-15697180996

Academic Editor: Robert Tuttle
Received: 23 December 2016; Accepted: 27 January 2017; Published: 31 January 2017

Abstract: Three low-carbon bainitic steels were designed to investigate the effects of Cr and Al addition on bainitic transformation, microstructures, and properties by metallographic method and dilatometry. The results show that compared with the base steel without Cr and Al addition, only Cr addition is effective for improving the strength of low-carbon bainitic steel by increasing the amount of bainite. However, compared with the base steel, combined addition of Cr and Al has no significant effect on bainitic transformation and properties. In Cr-bearing steel, Al addition accelerates initial bainitic transformation, but meanwhile reduces the final amount of bainitic transformation due to the formation of a high-temperature transformation product such as ferrite. Consequently, the composite strengthening effect of Cr and Al addition is not effective compared with individual addition of Cr in low-carbon bainitic steels. Therefore, in contrast to high-carbon steels, bainitic transformation in Cr-bearing low-carbon bainitic steels can be finished in a short time, and Al should not be added because Al addition would result in lower mechanical properties.

Keywords: bainitic transformation; microstructure; property; Cr; Al

1. Introduction

Low-carbon bainitic steels are commonly designed with alloying elements added to achieve favorable properties [1–11]. It is well known that the mechanical properties of bainitic steel are significantly influenced by the volume fraction of bainite, the amount of retained austenite (RA), the precipitation of cementite, among other factors [2]. The main purpose of the addition of alloying elements is to promote bainitic transformation and control the microstructures. For example, the addition of manganese, copper, and nickel, among others, are strong austenite stabilizers and can result in high stability of austenite and higher amounts of RA [3–5]. Silicon is used to restrain the formation of cementite during bainitic transformation [6]. In addition, a higher strength can be achieved due to precipitation hardening and grain refinement effects by adding vanadium, titanium, molybdenum, or niobium [7–11].

Chromium (Cr) and aluminium (Al) are also very important alloying elements, which are commonly added to low-carbon bainitic steels. Some researchers have investigated the effect of Al addition on bainitic transformation [5,12–17]. For example, Jimenez-Melero et al. [5] and Zhao et al. [12] reported that the addition of Al significantly increases the chemical driving force for the formation of bainitic ferrite plates and shortens the austenite-to-bainite transformation time. Similar results were also obtained by Garcia-Mateo et al. [13] and Hu et al. [14,15]. They claimed that the addition of Co and Al accelerates the bainitic transformation by increasing the free energy change accompanying the austenite-to-bainite ferrite transformation. Moreover, Monsalve et al. [16] and Meyer et al. [17] proved that Al promotes the formation of ferrite.

As to Cr, You et al. [18] and Chance et al. [19] studied the effect of Cr on continuous cooling transformation (CCT) diagrams, indicating that a single C-curve is separated by a bay of austenite stability due to the presence of Cr. They claimed that the addition of Cr may delay the bainitic transformation. Kong et al. [20] and Zhang et al. [21] investigated the influence of Cr on the transformation kinetics, demonstrating that the addition of Cr enhances the hardenability of metastable austenite. Moreover, Zhou et al. [22] investigated the effect of Cr on transformation and microstructure, and showed that Cr appreciably restrains ferrite transformation.

In summary, some investigations have been conducted on the effects of individual Cr or Al addition on microstructure and properties of bainitic steels. Cr and Al, as important alloying elements, are often compositely added to many bainitic steels [12–15]. However, few studies have reported on the composite effects of the combined addition of Cr and Al on the transformation, microstructure, and properties in low-carbon bainitic steels. Therefore, three kinds of low-carbon bainitic steels were designed in the present study to investigate the effects of Cr and Al addition on bainitic transformation, microstructure, and properties. Heat treatment experiments were performed on ThermecMaster-Z hot thermal–mechanical simulator followed by microstructure and property analyses, as well as quantitative characterization of bainitic transformation with dilation data. The purpose of the present study is to clarify the effects of the combined addition of Cr and Al on the transformation, microstructure, and properties in low-carbon bainitic steels. The results are useful for optimizing the composition design of Cr–Al alloying low-carbon steels.

2. Experimental Procedure

Three low-carbon bainitic steels with different chemical compositions were designed and their compositions are given in Table 1. Silicon (Si), manganese (Mn), and molybdenum (Mo), as important alloying elements, are often added in bainitic steels. The addition of Si prevents the formation of carbide, as carbide is detrimental to mechanical properties. The addition of Mn and Mo can enhance the hardenability of metastable austenite, which increases the bainite amount. Therefore, some Si, Mn, and Mo were added to the three steels. In addition, only Cr addition to steel B was used to study the effects of Cr on transformation and properties in the low-carbon bainitic steel, while the combined addition of Cr and Al in steel C was used to investigate the effects of Al and the composite addition of Cr and Al. The three steels were refined using a laboratory-scale 50 kg vacuum furnace. Cast ingots were heated at 1280 °C for 2 h and then hot-rolled to 12 mm thick plates in seven passes. The start and finish rolling temperatures were 1180 °C and 915 °C, respectively. After rolling the plates, they were cooled to 550 °C at a cooling rate of 20 °C/s followed by final air-cooling to room temperature.

Table 1. The chemical compositions of steels (wt %).

Steels	C	Si	Mn	Cr	Mo	Al	N	P	S
A (base)	0.218	1.831	2.021	/	0.227	/	<0.003	<0.006	<0.003
B (Cr)	0.221	1.792	1.983	1.002	0.229	/	<0.003	<0.006	<0.003
C (Cr + Al)	0.219	1.824	2.041	1.021	0.230	0.502	<0.003	<0.006	<0.003

Samples for dilatometric tests were machined to a cylinder of 8 mm diameter and 12 mm height. The top and bottom surfaces of samples were polished conventionally to keep the measurement surface level. The experiments were conducted according to the procedure shown in Figure 1 on a ThermecMaster-Z hot thermal–mechanical simulator equipped with a light-emitting diode (LED) dilatometer to quantitatively analyze the bainitic transformation of the three steels. The specimens were heated to 1000 °C at a rate of 10 °C/s and held for 900 s to achieve a homogeneous austenitic microstructure, followed by cooling to 350 °C at a rate of 10 °C/s. The austenitizing temperature of 1000 °C is larger than Ac_3 (the temperature at which transformation of ferrite to austenite is completed during heating). The small grain size can be obtained at a lower heating temperature. In addition, there are few inclusions and precipitates in steels, so the holding time of 900 s was chosen. The cooling rate of 10 °C/s refers to the cooling ability in the central area of thick plates in industrial production. After isothermal holding for 3600 s at 350 °C for bainite precipitation, the samples were air-cooled to room temperature.

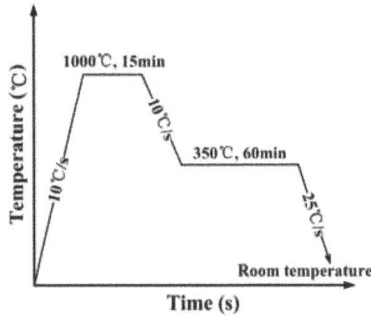

Figure 1. Experimental procedure.

According to the empirical Equations (1) in Ref. [23] and (2) in Ref. [24], the bainite starting temperature (B_S) and martensite starting temperature (M_S) of base steel are 471 °C and 336 °C, respectively. Therefore, the isothermal transformation temperature is designed to be 350 °C. It can be seen in Equation (2) that the Ms decreases by the addition of Cr, so the M_S temperature for steels B and C are smaller than 350 °C. It is known that finer bainite laths—and more of them—can be obtained at the lower phase transition temperature, which is beneficial for the mechanical properties of bainitic steel [11]. Moreover, martensite transformation will occur at the lower isothermal temperature (e.g., 300 °C). Bainite laths become coarse at the higher isothermal temperature (e.g., 400 °C), which is harmful to the mechanical properties of bainitic steel.

$$B_S \, (^\circ C) = 630 - 45Mn - 40V - 35Si - 30Cr - 25Mo - 20Ni - 15W \tag{1}$$

$$M_S \, (^\circ C) = 498.9 - 333.3C - 33.3Mn - 27.8Cr - 16.7Ni - 11.1(Si + Mo + W) \tag{2}$$

The time–temperature-transformation (TTT) curves of the three steels are plotted by MUCG83 software developed by Bhadeshia at Cambridge University (Figure 2). Chromium can enhance the stability and hardenability of austenite, so that the TTT curve moves to bottom-right with Cr addition, indicating that it is easy to obtain bainite at a certain cooling rate. In addition, higher temperature transformation may occur by Al addition, which is harmful to bainitic transformation.

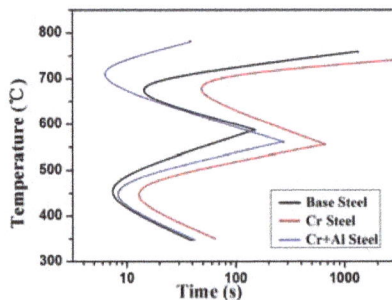

Figure 2. The time–temperature-transformation (TTT) curves of three steels.

Additionally, in order to investigate the properties of the tested steels, 140 × 20 × 10 mm blocks were cut from hot-rolled sheets and heat-treated using the same procedure shown in Figure 1. The specimens were mechanically polished and etched with a 4% nital solution for microstructure examination. Both bainite morphology and fracture surfaces were examined using a Nova 400 Nano field-emission scanning electron microscope (FE-SEM) operated at an accelerating voltage of 20 kV. The volume fraction of bainite in the three specimens was calculated by Image-Pro Plus software, and tensile tests were carried out on UTM-4503 electronic universal tensile tester at room temperature. Tensile specimens were prepared according to the ASTM standard and the beam displacement rate was 1 mm/min. Four tensile tests were conducted for each kind of tested steel and the corresponding average values were calculated in this work. In order to determine the volume fraction of RA, X-ray diffraction (XRD) experiments were conducted on BRUKER D8 ADVANCE diffractometer, using unfiltered Cu Kα radiation and operating at 40 kV and 40 mA.

3. Results

3.1. Microstructure

Figure 3 presents the microstructures of the three steels before heating treatment. It can be observed that the microstructures of the three steels mainly consist of lath-like bainite. The grain sizes of prior-austenite before heating treatment are measured by Image-Pro Plus software to be 34.6 ± 8.5 μm, 31.5 ± 8.5 μm, and 39.2 ± 8.5 μm for steels A, B, and C, respectively. In addition, few inclusions and precipitates are observed in the three steels due to very small amounts of N, P, and S (Table 1 and Figure 3). At the same time, the same processing routes are utilized for all three steels. Therefore, the influences of inclusions and precipitates in the three steels are small and similar.

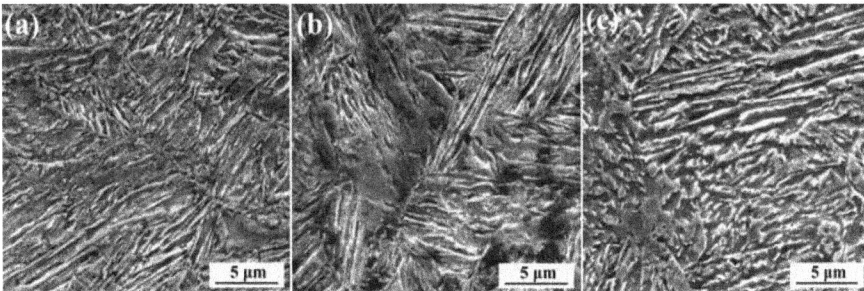

Figure 3. SEM micrographs of three low-carbon bainitic steels before heating treatment: (**a**) base, (**b**) Cr addition, and (**c**) Cr and Al addition.

Figure 4 shows the typical SEM microstructures of the three steels after isothermal holding for 3600 s at 350 °C following austenization at 1000 °C for 900 s. The classification method proposed in Ref. [11,25] is used in the present work to identify the microstructure in low-carbon bainitic steels: the microstructure is classified as ferrite (F), bainite ferrite (BF), and martensite (M). It can be observed that the microstructures of the three specimens mainly consist of lath-like BF and martensite/austenite (M/A) islands as shown in Figure 4. Prior-austenite grain boundaries (AGB) are well defined, as shown by arrows in Figure 4b. The original austenite grain size of the three tested steels, which influences the bainite morphology [11], was calculated by Image-Pro Plus software (Table 2). The prior-austenite grain sizes of the three steels have no significant difference. In addition, some ferrite is observed in steel A (base steel) and steel C (Cr + Al steel), as marked by the arrow in Figure 4c, but there is no ferrite in steel B (Figure 4b). According to the micrographs, the volume fraction of bainite was calculated by Image-Pro Plus software using the method in Ref. [11,26]. Further, the volume fractions of RA were calculated based on XRD results using the method in Ref. [4,27]. The results are shown in Table 3.

It reveals that the sample with only Cr addition has the largest amount of bainite, while the base steel without Cr and Al addition has the smallest percentage of bainite. It is clear that, compared with base steel, Cr addition obviously increases the amount of bainite. However, the combined addition of Cr and Al decreases the bainite amount to the level of base steel, indicating that Al addition in Cr-bearing low-carbon bainitic steel obviously retards the bainitic transformation. It should be pointed out that the M_S of base steel is calculated to be 374 °C and 386 °C by empirical Equation (3) in Ref. [28] and (4) in Ref. [29], respectively. However, the microstructures of the three steels after isothermal transformation at 350 °C mainly consist of bainite rather than martensite (Figure 4). It indicates that the empirical Equations (3) and (4) for M_S are not suitable for the three tested steels in the present study. The calculated results of B_S and M_S with different equations are not the same. The M_S temperatures of the base steel calculated by some equations are higher than 350 °C and others are lower than 350 °C, indicating that the different equations were obtained with different steel grades and experimental conditions. Moreover, the B_S and M_S in these equations may be corresponding to equilibrium conditions. The B_S and M_S are also affected by cooling rate. Therefore, B_S and M_S calculated by theoretical equations can only be used as theoretical reference. The real B_S and M_S at a certain cooling rate for a steel grade should be measured by experiments.

$$M_S \ (°C) = 537.8 - 361.1C - 38.9(Mn + Cr) - 19.4Ni - 27.8Mo \tag{3}$$

$$M_S \ (°C) = 561.1 - 473.9C - 33Mn - 16.7(Ni + Cr) - 21.1Mo \tag{4}$$

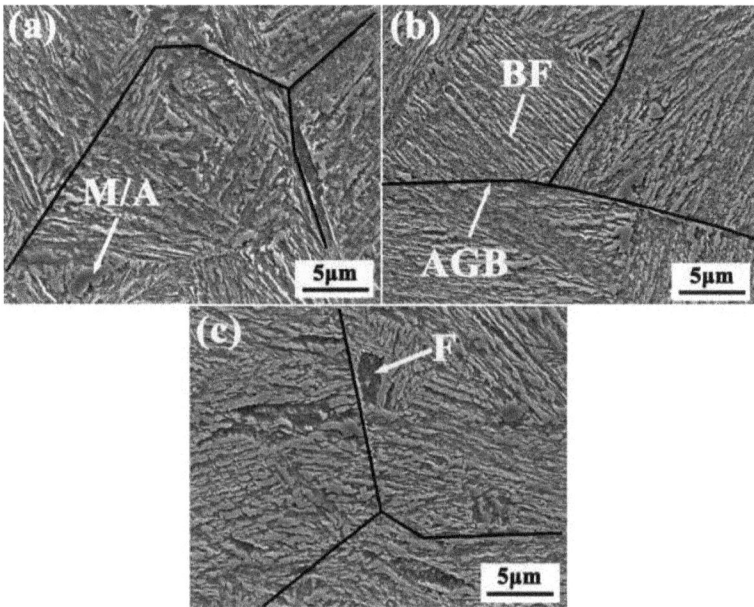

Figure 4. SEM micrographs of three low-carbon bainitic steels after isothermal holding at 350 °C for 3600 s: (**a**) base, (**b**) Cr addition, and (**c**) Cr and Al addition.

Table 2. Prior-austenite grain sizes of three steels (μm).

	Base	Cr	Cr + Al
Prior-austenite grain size	30.8 ± 9.4	29.4 ± 9.1	32.6 ± 8.5

Table 3. The volume fractions of bainite ferrite (BF) and retained austenite (RA) in three steels.

Steels	$V_{(BF)}$ (%)	$V_{(RA)}$ (%)
A (base)	45.6	3.5
B (Cr)	68.4	11.5
C (Cr + Al)	48.6	10.7

3.2. Mechanical Properties

The engineering strain–stress curves of the three tested steels are presented in Figure 5, and the corresponding mechanical properties are given in Table 4. Both the strength and the elongation are improved by only Cr addition, while the combined addition of Cr and Al has no significant effect on strength and elongation. Compared with steel A (base steel), the ultimate tensile strength (UTS) of Cr addition steel (steel B) increases by 135 MPa, while the UTS and yield strength (YS) increments of Cr and Al additional steel (steel C) are only 21 MPa and 12 MPa, respectively. Additionally, the strength and total elongation (TE) of steel C (Cr + Al) is unexpectedly smaller than that of steel B (only Cr), suggesting that no further improvement of mechanical properties occurs by Al addition in Cr-bearing bainitic steel.

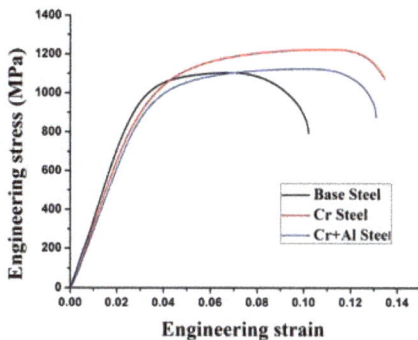

Figure 5. Engineering strain–stress curves of three tested steels with different compositions.

Table 4. Mechanical Properties of steels with different compositions.

Steels	UTS (MPa)	YS (MPa)	TE (%)	UTS × TE (GPa%)
A (base)	1103 ± 18	867 ± 22	10.2 ± 0.4	11.25 ± 0.007
B (Cr)	1238 ± 21	889 ± 18	13.1 ± 0.8	16.22 ± 0.017
C (Cr + Al)	1124 ± 15	873 ± 16	12.8 ± 0.5	14.39 ± 0.008

UTS: ultimate tensile strength; YS: yield strength; TE: total elongation.

Figure 6 displays the tensile fracture morphologies of the three steels. The mix of quasi-cleavage and dimples is exhibited in steel A (base steel), as shown by arrows in Figure 6a. A small number of quasi-cleavage fractures with a river pattern are observed in steel A, without Cr and Al addition (Figure 6a), indicating a portion of brittle fracture [19]. Nevertheless, this kind of brittle fracture microstructure rarely appears in steels B (only Cr) and C (Cr + Al). Additionally, it is observed that the diameters of dimples in the fracture of the steels with only Cr addition and combined addition of Cr and Al are larger than that of steel A without Cr and Al addition. This means that the toughness of steel is improved by Cr addition. The result is consistent with the mechanical properties of steels listed in Table 4.

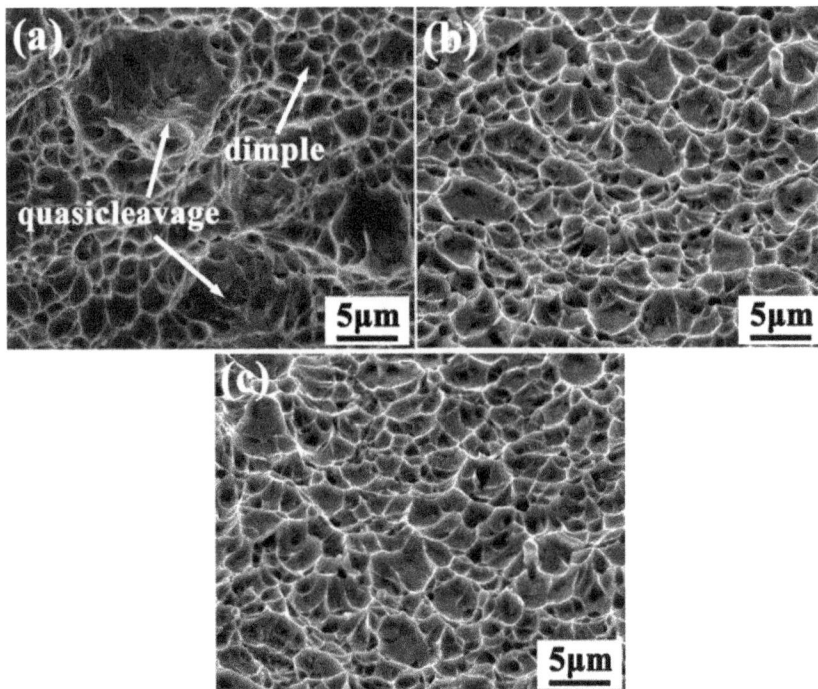

Figure 6. Fractographs of three tested steels with different compositions: (**a**) base, (**b**) Cr addition, and (**c**) Cr and Al addition.

3.3. Thermal Dilatometry

In order to quantitatively and accurately investigate the influence of combined Cr and Al addition on bainite transformation, dilatometric experiments were conducted on the thermo–mechanical simulator. According to the recorded dilatometric data, dilatation curves of the three steels are plotted. Figure 7 shows the recorded dilatation curves and transformation rates of the three steels during isothermal holding at 350 °C. Figure 7a shows dilatations as a function of holding time during isothermal holding at 350 °C, where the beginning of isothermal holding was selected as the zero point of abscissa and ordinate axes. The transformation temperature was constant and no extra force was applied on the sample during isothermal holding, thus the dilatation in Figure 7a represents the real bainite transformation amount. It can be observed that compared with the base steel, the final amount of bainite obviously increases with the addition of Cr, but it merely slightly rises with the combined addition of Cr and Al. Moreover, the amount of bainite transformation of steel C (Cr + Al steel) is obviously smaller than that of steel B (Cr steel), indicating that the addition of Al in Cr-bearing bainitic steel has a negative effect on the amount of bainite.

In addition, dilatation rates versus holding time during isothermal holding at 350 °C are given in Figure 7b. It shows that bainite transformation in steel C with combined Cr + Al addition and steel A without Cr and Al is completed prior to steel B with individual Cr addition. Although the maximum amount of bainite appears in steel B (Cr steel), the fastest transformation rate shows up in steel C (Cr + Al steel). It indicates that the addition of Al accelerates initial bainitic transformation, while the addition of Cr delays bainitic transformation. Besides, it is noticed that there is no distinguishable difference in the transformation processes between steel A (base steel) and steel C (Cr + Al steel). The time consumed to complete bainitic transformation is basically equal for the two steels, indicating that the combined Cr and Al addition has an ignorable effect on bainitic transformation rate compared with the base steel. Moreover, bainitic transformation completes quickly in the three low-carbon steels in the present study. This is contrast to high-carbon steels in which several hours or days are needed to finish bainitic transformation [13,15].

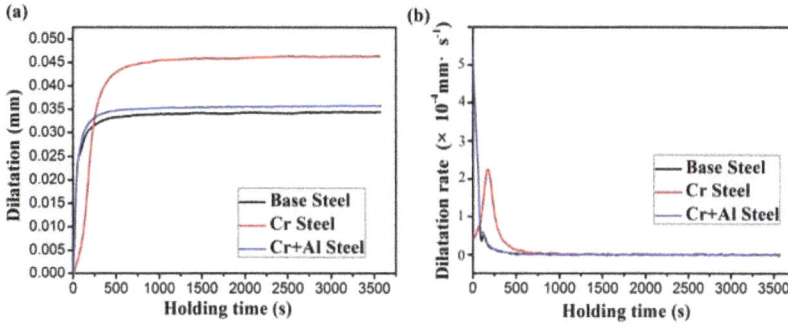

Figure 7. (a) Dilation curves and (b) transformation rates of three steels recorded by dilatometer on thermal simulator during bainitic transformation at 350 °C.

4. Discussions

4.1. Influence of Cr Addition

SEM micrographs (Figure 4) show that steel A mainly consists of bainite sheaves, a small amount of ferrite, and M/A, while no ferrite exists in steel B. As reported by some researches [30–32], bainitic transformation is characterized by incomplete reaction, thus some untransformed austenite after isothermal holding could transform into martensite. The results of dilatometric test (Figure 7) and microstructural determination (Table 3) both indicate that the bainite amount obviously increases with about 1% Cr addition. For bainitic steels, more bainite amount and RA fractions can improve mechanical properties [3,33,34]. In the present work, compared with steel A, the volume fractions of bainite and RA in steel B increase by 22.8% and 8%, respectively, resulting in an increment of 135 MPa in UTS and about 3% in TE. It can also be explained from the viewpoint of wetting of grain boundaries. It can be observed that the M/A particles with lath morphology clearly wet the austenite grain boundary (AGB) as shown in the bottom-right in Figure 4b, while they wet few AGBs in Figure 4a,c. It was reported by Straumal et al. [35] that the transition from incomplete to complete surface wetting is a phase transformation. It indicates that more M/A particles distribute in steel B (Cr steel) than steel A (base steel) and steel C (Cr + Al steel). The increased surface area contact of martensite particles with ferrite facilitates stress transfer from ductile to hard phase, which contributes to its high strength [36]. It demonstrates that a small amount of Cr addition in low-carbon bainitic steel improves strength of steel with a better total elongation. The existence of ferrite in steel A indicates that high-temperature phase transition happens during the cooling process before bainite transformation. However, with the Cr addition, the ferrite transformation is avoided, resulting in an increased bainite fraction. It is reported that Cr causes a separation of the bainite C-curve and extends the bainite formation field [18,19]. Similar results can be obtained from Equations (1) and (2). The addition of Cr decreases the B_S and M_S, which contributes to the greater amount of bainite. Moreover, Cr addition enhances the hardenability of metastable austenite [22] and increases the stability of austenite to ferrite [37]. Therefore, more undercooled austenite can transform into lath-like BF in the Cr addition steel, which leads to the improvement of the mechanical properties of steel B.

4.2. Influence of Al Addition

Figure 7 shows that bainite transformation is accelerated by Al addition, which is consistent with the results by Hu et al. [14,15], Caballero et al. [37] and others. SEM micrographs (Figure 4) indicate that steel B (Cr steel) mainly consists of bainite sheaves without ferrite, while a small amount of ferrite presents in steel C (Cr +Al steel), showing that Al addition promotes the formation of ferrite. According to the dilation diagram (Figure 7), the total dilatation of steel C (Cr + Al steel) is only 0.0368 mm, reduced by 19.5% compared to steel B (Cr steel, 0.0457 mm), which is consistent with the result listed in Table 3. It indicates that 0.5% Al addition in Cr-bearing steel obviously reduces the final bainite amount. In addition, the product of tensile strength and total elongation for steel C (Cr + Al steel) slightly decreases from 16.22 GPa% for steel B (Cr steel) to 14.39 GPa%. Fonstein et al. [38] and Meyer et al. [17] reported that Al addition can promote the formation of a high-temperature transformation product such as ferrite. Therefore, the amount of bainite transformation decreases because of less supercooled austenite. On the other hand, the results by Caballero and Bhadeshia [37] indicate that Al addition has no significant effect on final bainite amount. This depends on whether the ferrite transformation occurs. There is no ferrite transformation in their study, while ferrite appears in the present study, which leads to the decrease of bainite amount. Therefore, the mechanical properties of steel C (Cr + Al steel) are slightly smaller compared to steel B (Cr steel).

4.3. Influence of Composited Addition of Cr and Al

It can be observed from Table 3 that the volume fractions of bainite in steels A and C are 45.6% and 48.6%, respectively. Also, from the dilation curves, the dilatation of steel C (Cr and Al) is 0.0368 mm, increased by 0.0022 mm compared to the 0.0346 mm expansion of base steel, demonstrating that the effect of combined addition of Cr and Al on bainite transformation is very small. In addition, the UTS of steel C (Cr and Al) increases by only 21 MPa compared with base steel, indicating that the strengthening effect of combined Cr and Al addition has no significant improvement over that of steel A (base steel). As mentioned previously, although the only Cr addition increases the final bainite transformation amount, Al addition in Cr-bearing low-carbon steel reduces the amount of bainite transformation because of the formation of high-temperature transformation product (F). This means that the addition of Al weakens the promotion function of Cr on bainite transformation. Therefore, the composite strengthening effect of Cr and Al addition has no significant improvement over that of base steel. Al addition is very effective at shortening the bainitic transformation time in high-carbon bainitic steel [13,37]. However, for the low-carbon bainitic steels, bainitic transformation can finish in a short time even without Al addition (Figure 7), and Al should not be added in Cr-bearing low-carbon bainitic steels because Al addition would result in lower mechanical properties.

5. Conclusions

Three low-carbon bainitic steels were designed in the present work. Metallographic method and dilatometry were used to investigate the effects of Cr and Al addition on bainitic transformation, microstructures, and properties. The results show that the individual addition of Cr is effective for improving the strength of low-carbon bainite steel by increasing the amount of bainite transformation. In addition, in Cr-bearing low-carbon steel, Al addition leads to lower mechanical properties due to decreased amount of bainite transformation, although the addition of Al accelerates the initial bainitic transformation and shortens the austenite-to-bainite transformation time. Moreover, the addition of Al can effectively shorten the bainitic transformation time in high-carbon bainitic steels. However, bainitic transformation in Cr-bearing low-carbon bainitic steels can finish in a short time. Therefore, it is not necessary to add Al for the acceleration of bainitic transformation because Al addition would result in lower mechanical properties.

Acknowledgments: The authors gratefully acknowledge the financial supports from National Natural Science Foundation of China (NSFC) (No. 51274154), the National High Technology Research and Development Program of China (No. 2012AA03A504). State Key Laboratory of Development and Application Technology of Automotive Steels (Baosteel Group).

Author Contributions: Guang Xu, supervisor, conceived and designed the experiments; Junyu Tian conducted experiments, analyzed the data and wrote the paper; Mingxing Zhou, conducted experiments; Haijiang Hu conducted experiments; Xiangliang Wan, conducted experiments. All authors participated in the discussion of experimental results.

Conflicts of Interest: The authors declare no conflict of interest. The founding sponsors had no role in the design of the study; in the collection, analyses, or interpretation of data; in the writing of the manuscript, and in the decision to publish the results.

References

1. Zhou, M.X.; Xu, G.; Wang, L.; Yuan, Q. The varying effects of uniaxial compressive stress on the bainitic transformation under different austenitization temperatures. *Metals* **2016**, *6*, 119. [CrossRef]
2. He, J.G.; Zhao, A.M.; Yao, H.; Zhi, C.; Zhao, F.Q. Effect of ausforming temperature on bainite transformation of high carbon low alloy steel. *Mater. Sci. Forum.* **2015**, *817*, 454–459. [CrossRef]
3. Hu, H.J.; Xu, G.; Wang, L.; Zhou, M.X.; Xue, Z.L. Effect of ausforming on the stability of retained austenite in a C-Mn-Si bainitic steel. *Met. Mater. Int.* **2015**, *21*, 929–935. [CrossRef]
4. Zhou, M.X.; Xu, G.; Wang, L.; He, B. Effects of austenitization temperature and compressive stress during bainitic transformation on the stability of retained austenite. *T. Indian. I. Metals.* **2016**, 1–7. [CrossRef]
5. Jimenez-Melero, E.; Dijk, N.H.V.; Zhao, L.; Sietsma, J.; Offerman, S.E.; Wright, J.P. The effect of aluminium and phosphorus on the stability of individual austenite grains in trip steels. *Acta Mater.* **2009**, *57*, 533–543. [CrossRef]
6. Girault, E.; Mertens, A.; Jacques, P.; Houbaert, Y.; Verlinden, B.; Humbeeck, J.V. Comparison of the effects of silicon and aluminium on the tensile behaviour of multiphase trip-assisted steels. *Scripta Mater.* **2001**, *44*, 885–892. [CrossRef]
7. Heller, T.; Nuss, A. Effect of alloying elements on microstructure and mechanical properties of hot rolled multiphase steels. *Ironmak. Steelmak.* **2005**, *32*, 303–308. [CrossRef]
8. Pereloma, E.V.; Timokhina, I.B.; Russell, K.F.; Miller, M.K. Characterization of clusters and ultrafine precipitates in Nb-containing C–Mn–Si steels. *Scripta Mater.* **2006**, *54*, 471–476. [CrossRef]
9. Shi, W.; Li, L.; Yang, C.X.; Fu, R.Y.; Wang, L.; Wollants, P. Strain-induced transformation of retained austenite in low-carbon low-silicon trip steel containing aluminum and vanadium. *Mater. Sci. Eng. A* **2006**, *429*, 247–251. [CrossRef]
10. Wang, X.D.; Huang, B.X.; Wang, L.; Rong, Y.H. Microstructure and mechanical properties of microalloyed high-strength transformation-induced plasticity steels. *Metall. Mater. Trans. A* **2008**, *39*, 1–7. [CrossRef]
11. Hu, H.J.; Xu, G.; Wang, L.; Xue, Z.L.; Zhang, Y.L.; Liu, G.H. The effects of Nb and Mo addition on transformation and properties in low carbon bainitic steels. *Mater. Des.* **2015**, *84*, 95–99. [CrossRef]
12. Zhao, J.; Wang, T.S.; Lv, B.; Zhang, F.C. Microstructures and mechanical properties of a modified high-C–Cr bearing steel with nano-scaled bainite. *Mater. Sci. Eng. A* **2015**, *628*, 327–331. [CrossRef]
13. Garcia-Mateo, C.; Caballero, F.G.; Bhadeshia, H.K.D.H. Acceleration of low-temperature bainite. *ISIJ Int.* **2003**, *43*, 285–288. [CrossRef]
14. Hu, F.; Wu, K.M.; Zheng, H. Influence of Co and Al on pearlitic transformation in super bainitic steels. *Ironmak. Steelmak.* **2012**, *39*, 535–539. [CrossRef]
15. Hu, F.; Wu, K.M.; Zheng, H. Influence of Co and Al on bainitic transformation in super bainitic steels. *Steel Res. Int.* **2013**, *84*, 1060–1065. [CrossRef]
16. Monsalve, A.; Guzmán, A.; Barbieri, F.D.; Artigas, A.; Carvajal, L.; Bustos, O. Mechanical and microstructural characterization of an aluminum bearing trip steel. *Metall. Mater. Trans. A* **2016**, *47*, 3088–3094. [CrossRef]
17. Meyer, M.D.; Mahieu, J.; Cooman, B.C.D. Empirical microstructure prediction method for combined intercritical annealing and bainitic transformation of trip steel. *Mater. Sci. Technol.* **2002**, *18*, 1121–1132. [CrossRef]
18. You, W.; Xu, W.H.; Liu, Y.X.; Bai, B.Z.; Fang, H.S. Effect of chromium on CCT diagrams of novel air-cooled bainite steels analyzed by neural network. *J. Iron. Steel Res. Int.* **2007**, *14*, 39–42.

19. Chance, J.; Ridley, N. Chromium partitioning during isothermal transformation of a eutectoid steel. *Metall. Mater. Trans. A* **1981**, *12*, 1205–1213. [CrossRef]
20. Kong, L.; Liu, Y.; Liu, J.; Song, Y.; Li, S.; Zhang, R. The influence of chromium on the pearlite-austenite transformation kinetics of the Fe–Cr–C ternary steels. *J. Alloy. Compd.* **2015**, *648*, 494–499. [CrossRef]
21. Zhang, G.H.; Chae, J.Y.; Kim, K.H.; Dong, W.S. Effects of Mn, Si and Cr addition on the dissolution and coarsening of pearlitic cementite during intercritical austenitization in Fe-1mass%C alloy. *Mater. Charact.* **2013**, *81*, 56–67. [CrossRef]
22. Zhou, L.Y.; Liu, Y.Z.; Yuan, F.; Huang, Q.W.; Song, R.B. Effect of Cr on transformation of ferrite and bainite dual phase steels. *J. Iron Steel Res. Int.* **2009**, *21*, 37–41.
23. Zhao, Z.; Cheng, L.; Liu, Y.; Northwood, D.O. A new empirical formula for the bainite upper temperature limit of steel. *J. Mater. Sci.* **2001**, *36*, 5045–5056. [CrossRef]
24. Rowland, E.S.; Lyle, S.R. The application of M_S points to case depth measurement. *ASM Trans.* **1946**, *37*, 26–47.
25. Xiao, F.; Liao, B.; Ren, D.; Shan, Y.; Yang, K. Acicular ferritic microstructure of a low-carbon Mn–Mo–Nb microalloyed pipeline steel. *Mater. Charact.* **2005**, *54*, 305–314. [CrossRef]
26. Zhou, M.X.; Xu, G.; Wang, L.; Hu, H.J. Combined effect of the prior deformation and applied stress on the bainite transformation. *Met. Mater. Int.* **2016**, *22*, 956–961. [CrossRef]
27. Lindström, A. Austempered High Silicon Steel: Investigation of Wear Resistance in A Carbide Free Microstructure. Master's Thesis, Luleå Tekniska Universitet, Sweden, 2006.
28. Grange, R.A.; Stewart, H.M. The temperature range of martensite formation. *Trans. AIME* **1946**, *167*, 467–472.
29. Steven, W.; Haynes, A.G. The temperature formation of martensite and bainite in low-alloy steels. *J. Iron Steel Inst.* **1956**, *183*, 349–359.
30. Wang, X.; Zurob, H.S.; Xu, G.; Ye, Q.; Bouaziz, O.; Embury, D. Influence of microstructural length scale on the strength and annealing behavior of pearlite, bainite, and martensite. *Metall. Mater. Trans. A* **2013**, *44*, 1454–1461. [CrossRef]
31. Wang, X.L.; Wu, K.M.; Hu, F.; Yu, L.; Wan, X.L. Multi-step isothermal bainitic transformation in medium-carbon steel. *Scripta Mater.* **2014**, *74*, 56–59. [CrossRef]
32. Cornide, J.; Garcia-Mateo, C.; Capdevila, C.; Caballero, F.G. An assessment of the contributing factors to the nanoscale structural refinement of advanced bainitic steels. *J. Alloy. Compd.* **2012**, *577*, S43–S47. [CrossRef]
33. Hu, H.J.; Xu, G.; Zhou, M.X.; Yuan, Q. Effect of Mo content on microstructure and property of low-carbon bainitic steels. *Metals* **2016**, *6*, 173. [CrossRef]
34. Shi, J.; Sun, X.; Wang, M.; Hui, W.; Dong, H.; Cao, W. Enhanced work-hardening behavior and mechanical properties in ultrafine-grained steels with large-fractioned metastable austenite. *Scripta Mater.* **2010**, *63*, 815–818. [CrossRef]
35. Straumal, B.B.; Baretzky, B.; Kogtenkova, O.A.; Straumal, A.B.; Sidorenko, A.S. Wetting of grain boundaries in Al by the solid Al_3Mg_2 phase. *J. Mater. Sci.* **2010**, *45*, 2057–2061. [CrossRef]
36. Ahmad, E.; Manzoor, T.; Ziai, M.M.A.; Hussain, N. Effect of martensite morphology on tensile deformation of dual-phase steel. *J. Mater. Eng. Perform.* **2012**, *21*, 1–6. [CrossRef]
37. Caballero, F.G.; Bhadeshia, H.K.D.H. Very strong bainite. *Curr. Opin. Solid State Mater. Sci.* **2004**, *8*, 251–257. [CrossRef]
38. Fonstein, N.; Yakubovsky, O.; Bhattacharya, D.; Siciliano, F. Effect of niobium on the phase transformation behavior of aluminum containing steels for trip products. *Mater. Sci. Forum.* **2005**, *500–501*, 453–460. [CrossRef]

Article

Effect of Low-Temperature Sensitization on Hydrogen Embrittlement of 301 Stainless Steel

Chieh Yu [1,†], Ren-Kae Shiue [1], Chun Chen [1] and Leu-Wen Tsay [2,*]

1   Department of Materials Science and Engineering, National Taiwan University, Taipei 106, Taiwan;
    d02527001@ntu.edu.tw (C.Y.); rkshiue@ntu.edu.tw (R.-K.S.); gchen@ntu.edu.tw (C.C.)
2   Institute of Materials Engineering, National Taiwan Ocean University, Keelung 20224, Taiwan
*   Correspondence: b0186@mail.ntou.edu.tw; Tel.: +886-2-2462-5324
†   Current Address: National Chung-Shan Institute of Science and Technology, Materials & Electro-Optics
    Research Division, Lung-Tan, Tao-Yuan 325, Taiwan.

Academic Editor: Robert Tuttle
Received: 7 January 2017; Accepted: 10 February 2017; Published: 15 February 2017

Abstract: The effect of metastable austenite on the hydrogen embrittlement (HE) of cold-rolled (30% reduction in thickness) 301 stainless steel (SS) was investigated. Cold-rolled (CR) specimens were hydrogen-charged in an autoclave at 300 or 450 °C under a pressure of 10 MPa for 160 h before tensile tests. Both ordinary and notched tensile tests were performed in air to measure the tensile properties of the non-charged and charged specimens. The results indicated that cold rolling caused the transformation of austenite into α' and ε-martensite in the 301 SS. Aging at 450 °C enhanced the precipitation of $M_{23}C_6$ carbides, G, and σ phases in the cold-rolled specimen. In addition, the formation of α' martensite and $M_{23}C_6$ carbides along the grain boundaries increased the HE susceptibility and low-temperature sensitization of the 450 °C-aged 301 SS. In contrast, the grain boundary α'-martensite and $M_{23}C_6$ carbides were not observed in the as-rolled and 300 °C-aged specimens.

Keywords: stainless steel; hydrogen embrittlement; hydrogen charging; G and σ phases; α' martensite

---

1. Introduction

Austenitic stainless steels (SSs) have been extensively used in industrial applications due to their good combination of corrosion resistance and mechanical properties. Metastable austenitic SSs may undergo phase transformation from austenite (γ) into ferromagnetic α'-martensite during plastic deformation [1–5]. The induced martensite during tensile tests at room temperature enhances the elongation to fracture, hardness, and tensile strength [5]. However, the induced martensite increases the susceptibility of austenitic stainless steels to hydrogen embrittlement (HE) [2–8]. In hydrogen-containing environments, the more austenite transforms into martensite, the more ductility loss of the material occurs [6]. Hydrogen suppresses the transformation of austenite to martensite, leading to hydrogen softening and localized brittle fracture [5]. Moreover, the suppressed formation of α'-martensite in the highly strained region of a 304L specimen at 80 °C accounts for its lowered HE susceptibility [7]. It has been reported that strain-induced α'-martensite acts as a "hydrogen diffusion highway" in hydrogen-charged 304 SS [8], resulting in an increased hydrogen concentration at the crack tip and accelerated crack growth.

Austenitic stainless steels are susceptible to stress corrosion cracking (SCC) in chloride-containing solutions [9–15]. In $MgCl_2$ solution, the mechanism of environment-induced cracking of austenitic stainless steels can be either SCC or HE, depending on the test temperature [10,11]. A suitable amount of cold work can lower the SCC susceptibility of austenitic stainless steels, whereas excessive cold working reverses that trend [16,17]. The machined 304L SS showed a marked increase in SCC susceptibility as compared with the solution-annealed, unmachined, and cold worked samples [18,19]. Unlike martensite-free 304 SS, 304 SS with machining-induced martensite is greatly embrittled and undergoes premature failure in 40 MPa hydrogen [20].

As austenitic stainless steels are heated in the temperature range between 500 °C and 850 °C, chromium carbides form along the grain boundary, leaving an adjacent chromium depletion zone. This phenomenon is called sensitization and it is responsible for intergranular corrosion or intergranular SCC (IGSCC) [21–24]. Cold-working austenitic stainless steels accelerates carbide precipitation, even in a Ti-stabilized (AISI 321) SS [24]. Cathodic hydrogen-charging greatly reduces the ductility of tensile specimens and decreases the time-to-failure of 304L SS and 308L weld metals under constant load tests, especially for the sensitized 304L (650 °C/24 h aging) [22]. Furthermore, the nucleation of tiny carbides in many austenitic SSs is enhanced by welding, thermo-mechanical processing, or slow cooling from the solution-annealed temperature. Such thermo or thermo-mechanical processes might not immediately induce IGSCC of the alloys. The number of carbide precipitates remains unchanged, but they grow in size during subsequent long term service below 500 °C. Sensitization of austenitic SSs at temperatures below the classic sensitization range (500 °C–800 °C) is referred to as low-temperature sensitization (LTS).

Cold deformation increases the degree of 304 SS sensitization up to 65 times that of undeformed 304 SS tested at 500 °C [25]. In a previous study, 304 SS welds exposed to a temperature of 450 °C for 6600 h suffered from IGSCC due to the presence of $(Fe,Cr)_{23}C_6$ carbides along grain boundaries [26]. Cold working of the steel increases the sensitization kinetics of austenitic SSs by up to 15%, while further cold working shows less effect on sensitization [23]. The induced martensite in cold-worked 304 [27] or 304L [28] SSs causes low-temperature sensitization at 380 °C [27] and 500 °C [28], thereby increasing the SCC susceptibility in a BWR (boiling water reactor) simulated environment [28]. Furthermore, sensitization causes α'-martensite transformation preferentially along the grain boundaries of 304 and 316 SSs, which provides a high diffusivity path of hydrogen to the crack tip [29] and enhances the intergranular fracture and HE susceptibility of the steel [29].

In this study, cold-rolled 301 SS was hydrogen-charged in an autoclave at 300 °C or 450 °C at the pressure of 10 MPa for 160 h before straining. The effects of low-temperature sensitization during aging/hydrogen-charging at 300 °C or 450 °C on the microstructure were investigated. The HE susceptibility of various specimens was correlated with the corresponding microstructures, particularly the induced martensite and fine precipitates in the specimens.

2. Experimental Procedures

The chemical composition of the AISI 301 SS used in this study was 16.71 Cr, 6.89 Ni, 0.08 C, 1.16 Mn, 0.54 Si, 0.02 P, 0.003 S, and the balance Fe in wt %. The 301 SS in the plate form with a thickness of 4.5 mm was solution-annealed at 1050 °C for 30 min and had a hardness of Hv 176. Cold rolling of the 301 SS plate with 30 % in thickness reduction was performed at room temperature, and designated as CR (cold-rolled) specimen. For comparison, CR specimens were aged at 300 °C or 450 °C for 160 h, and these specimens were respectively designated as CR-300 and CR-450.

Figure 1 shows the dimensions of the double-edge notched tensile and standard tensile specimens, which had a thickness of 3 mm. Cold-rolled specimens were hydrogen-charged in an autoclave at 300 °C or 450 °C at a pressure of 10 MPa for 160 h before straining. They were designated as CR-300H and CR-450H, respectively. Standard tensile specimens, according to ASTM E8 specification with a gauge length of 25 mm, were wire-cut directly from the cold-rolled plates along the rolling direction. Ordinary tensile tests were carried out at a strain rate of 6×10^{-4} s^{-1} (crosshead displacement rate of 0.9 mm/min) in laboratory air to determine the tensile properties of the non-hydrogen-charged specimens. Notched tensile tests were performed to evaluate the HE susceptibility of the hydrogen-charged specimens at a crosshead displacement rate of 0.72 mm/min at room temperature.

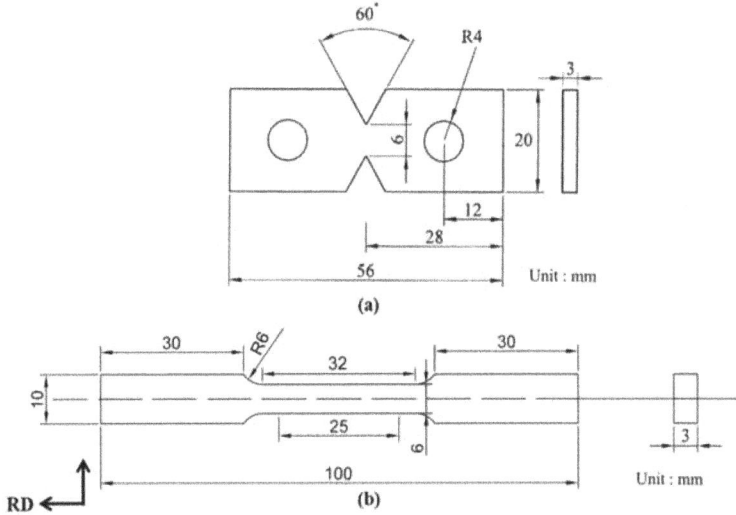

Figure 1. Schematic dimensions of (**a**) a double-edge notched specimen and (**b**) a standard tensile specimen. (RD: Rolling Direction).

A Fischer MP30 ferritescope (Windsor, CT, USA) can measure the ferrite contents precisely in austenitic and duplex SSs. The ferritescope was used in this study to determine the amounts of strain-induced α'-martensite in all specimens [30,31]. The hydrogen-charged samples (CR-300H and CR-450H) with the dimensions of $1 \times 1 \times 1$ cm^3 were ground by No. 2000 SiC paper before cleaning. The hydrogen contents of the charged samples (CR-300H and CR-450H) were measured by using the LECO-TCH 600 (Saint Joseph, MI, USA). After hydrogen-charging, they were melted in a crucible. The amounts of H and O were calculated from the H$_2$O, CO, and CO$_2$ mixture. Fractographs of each specimen after tensile straining were inspected with a NOVA-450 scanning electron microscope (SEM, Hillsboro, OR, USA). Selected specimens were inspected by using a JEOL 2000EX transmission electron microscope (TEM, Akishima, Japan). A high resolution TEM (HRTEM), FEI Tecnai G2 F20 (Hillsboro, OR, USA), was applied to inspect the nano-sized precipitates in the specimen.

3. Results

3.1. Microstructural Observation

Figure 2 shows optical metallographs of cold-rolled specimens with different aging conditions. Figure 2a is a composite photograph of the CR specimen revealing the microstructures along three perpendicular directions. No severe texture was observed in the CR specimen after 30% thickness reduction. The CR specimen contained basket-weaved slip bands within equiaxial austenite grains. In the case of the specimens aged at 300 (CR-300) and 450 °C (CR-450), slip bands could be still observed clearly, as shown in Figure 2b,c. The ferritescope was used to determine the amount of strain-induced α' martensite formed in the tested specimens. The ferrite contents of the CR, CR-300, and CR-450 specimens were 26%, 26%, and 29%, respectively. It is well known that heating strain-induced α' martensite to an appropriate temperature will cause it to revert into austenite, leading to a decrease in ferrite content. It was deduced that increased amount of α' martensite in the CR-450 specimen was associated with a specific phase transformation in the specimen.

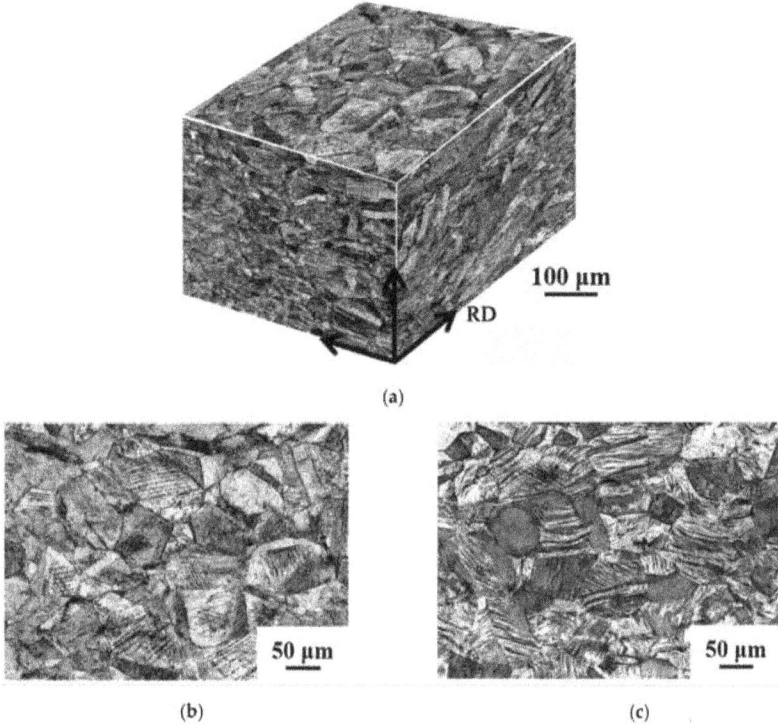

Figure 2. Optical metallographs of: (**a**) three mutually perpendicular planes of the CR, (**b**) CR-300, and (**c**) CR-450 specimens.

3.2. Mechanical Properties

Table 1 lists the mechanical properties of the test specimens. The 30% cold rolling obviously increased the surface hardness of CR specimen to Hv 455, as compared with the Hv 176 of the annealed 301 SS. Dislocation recovery during aging at 300 °C was expected to be responsible for the slight decrease in surface hardness of CR-300 specimen to Hv 445. The surface hardness of CR-450 specimen further reduced to Hv 414 due to the overage at 450 °C. The results of ordinary tensile tests revealed that all three tested specimens had the same elongation of 24%. The ordinary tensile strengths of the CR and CR-300 specimens were similar. However, the YS and UTS of the CR-450 specimen were lower than those of the CR and CR-300 ones.

Table 1. Mechanical properties of tested specimens with 30% reduction in thickness.

Specimen	YS [a] (MPa)	UTS [b] (MPa)	EL [c] (%)	Hardness(Hv)	NTS [d] (MPa)
CR	1350	1506	24	455	1737
CR-300 [e]	1400	1544	24	445	1743
CR-450 [f]	1200	1296	24	414	1584
CR-300H [g]	-	-	-	-	1339
CR-450H [h]	-	-	-	-	843

[a] YS: offset yield strength; [b] UTS: ultimate tensile strength; [c] EL: elongation; [d] NTS: notched tensile strength; [e] CR-300: CR specimen aged at 300 °C for 160 h in vacuum; [f] CR-450: CR specimen aged at 450 °C for 160 h in vacuum; [g] CR-300H: CR specimen hydrogen-charged at 300 °C for 160 h; [h] CR-450H: CR specimen hydrogen-charged at 450 °C for 160 h.

Notched tensile tests were performed in air to evaluate the notch brittleness of the specimens. The NTSs of the CR, CR-300, and CR-450 specimens were 1737, 1743, and 1584 MPa, respectively. The NTS of the CR-450 specimen was lower than those of the CR and CR-300 ones. This is consistent with the YS and UTS results for the CR, CR-300, and CR-450 specimens. It is noted that the NTS was significantly higher than the corresponding UTS for all specimens, indicating that the notched specimens possessed enough toughness to resist localized brittle fracture before rupture in air.

Notched tensile strengths tested in air versus displacement curves of the non-charged (solid lines) and hydrogen-charged (dotted lines) specimens are shown in Figure 3. The NTSs of the CR and CR-300 specimens were similar, but the CR-300 specimen had a slightly higher notch displacement (notch ductility) than the CR one. In contrast, the NTS and notch displacement of the CR-450 specimen were obviously lower than those of the CR and CR-300 ones tested in air. Subsequently, aging the cold-rolled 301 SS at 450 °C was found to deteriorate its notched tensile properties. As shown in Figure 3, hydrogen-charging caused a significant drop in NTSs and notch displacements of both the CR-300H and CR-450H specimens. HE was responsible for such a severe degradation of notch tensile properties of the hydrogen-charged specimens. The results also indicated that the CR-450H specimen was more damaged by hydrogen-charging than the CR-300H one. Aging the cold worked specimen at higher temperature was expected to improve its ductility and reduce brittleness, but these improvements did not occur in the CR-450 specimen. This result would be addressed in the later discussion of the mechanism of deterioration in CR-450.

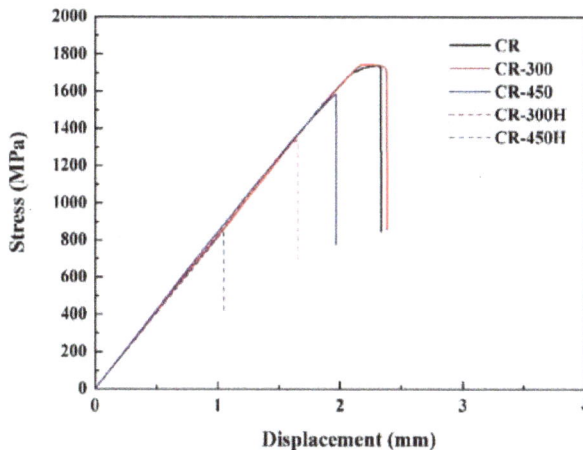

Figure 3. Notched tensile tests of non-hydrogen-charged (solid lines) and hydrogen-charged (dash lines) specimens.

3.3. Fractographic Examinations

The macroscopic fracture appearances of selected notched tensile specimens are shown in Figure 4. The fractographs of the CR and CR-300 specimens were alike, comprising extensive shear fractured (SF) regions (shear lips) on the lateral surfaces and triangular flat fracture (FF) zones ahead of the notch fronts, as shown in Figure 4a. A noticeable reduction of the SF regions and an increase in the size of the FF zone were observed in the CR-450 specimen after the notched tensile test (Figure 4b). Moreover, the fractograph shown in Figure 4b exhibited lamellar tears and many fine secondary cracks. Even without hydrogen-charging, the CR-450 specimen showed inherent notch brittleness in comparison with the CR and CR-300 specimens.

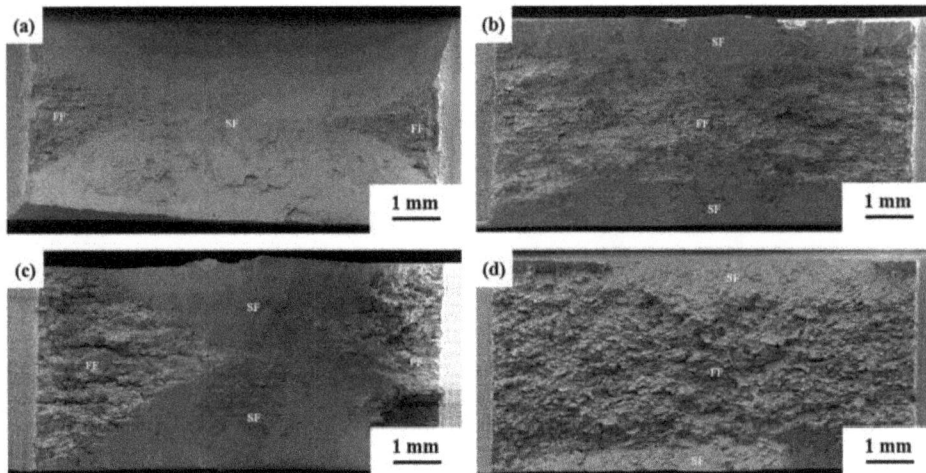

Figure 4. Macro-fracture appearances of (**a**) CR-300, (**b**) CR-450, (**c**) CR-300H, and (**d**) CR-450H specimens.

It is known that hydrogen tends to accumulate and embrittle the elastic-plastic boundaries of high-strength steels [32], and this embrittlement becomes more severe at the locations of stress concentration, such as a notch [33,34]. Moreover, a notch can produce a triaxial stress state, and limits the plastic deformation ahead of the notch tip. The marked decreases in NTSs and notch displacements of the hydrogen-charged specimens (CR-300H and CR-450H in Figure 3) were associated with significant changes in their fractographs as illustrated in Figure 4c,d. In the CR-300H specimen, most SF zones were replaced by FF regions with parallel secondary cracks (Figure 4c). For the CR-450H specimen, the fractured surface was dominated by FF regions (Figure 4d) due to the high HE susceptibility.

Failure analyses of all fractured surfaces after tensile tests were performed using an SEM. Ductile dimple fracture was observed in all ordinary tensile specimens (not shown here). SEM fractographs of selected specimens after notched tensile tests are presented in Figure 5. The fracture modes of the CR-300 and CR-450 specimens differed slightly. In the CR-300 specimen, predominant fine shallow dimples mixed with sparsely flat cleavage-like fracture were observed within the FF zones ahead of notch front (Figure 5a). In contrast, mainly cleavage fracture with secondary cracks along austenite grain boundaries was found ahead of the notch front of the CR-450 specimen (Figure 5b), indicating the brittle nature of the sample. The CR-450 specimen, even without hydrogen-charging, was likely to suffer intergranular fracture under strain. Regarding the hydrogen-charged specimens, extensive cleavage-like fracture together with numerous secondary cracks was observed in the CR-300H specimen (Figure 5c), while the CR-450 specimen mainly exhibited intergranular fracture (Figure 5d).

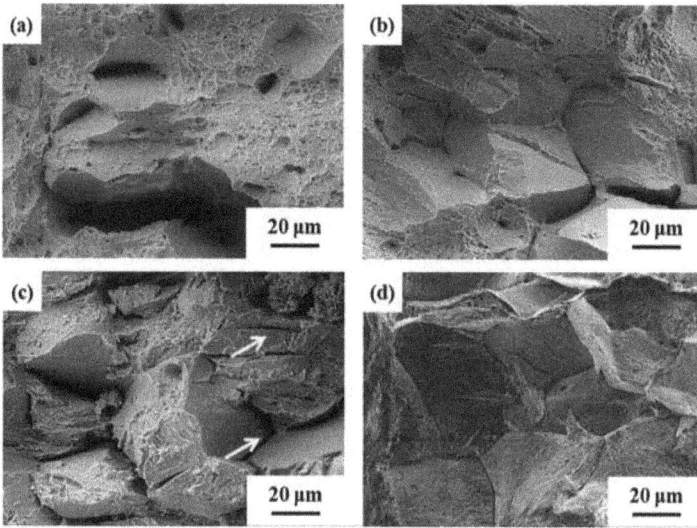

Figure 5. SEM fractographs of (**a**) CR-300, (**b**) CR-450, (**c**) CR-300H, and (**d**) CR-450H specimens after notched tensile test.

3.4. TEM Examinations

Figure 6 displays detailed TEM micrographs of the CR specimen. There were three distinct structures—ε-martensite, α′-martensite, and austenite—in the CR specimen. The microstructure of the CR specimen was primarily comprised of parallel strips of ε-martensite (hcp) in the austenite (γ, fcc) matrix, as illustrated in Figure 6a. The ε-martensite formed in slip bands due to the 30% cold rolling of the 301 SS. The α′-martensite (bcc) had also been found mainly at the intersections of ε-martensite strips. Similar results were obtained in prior study of the SCC of cold-rolled 304L SS [4,12], consistent with the location of α′-martensite, as shown in Figure 6b.

Figure 6. TEM micrographs of the CR specimen illustrating (**a**) ε-martensite strips in the austenite matrix, (**b**) α′-martensite at the intersections of ε-martensite strips.

In the CR-450 specimen, further aging at 450 °C for 160 h enhanced the transformation of metastable austenite into α′-martensite and the precipitation of $M_{23}C_6$ carbides along the grain boundaries, as displayed in Figure 7. In the CR and CR-300 specimens, the grain boundary α′-martensite and $M_{23}C_6$ carbides were not observed. Due to the presence of fresh grain boundary α′-martensite, the ferrite contents of the CR-450 specimen were higher than those of the CR and

CR-300 specimens. Both the precipitation of $M_{23}C_6$ carbides and increased α'-martensite along the grain boundaries resulted in increasing HE susceptibility of the CR-450H specimen, as compared with the CR-300H specimens. It was obvious that cold rolling the 301 SS induced partial austenite transformation into α'- and ε-martensite. The subsequent aging at 450 °C promoted the formation of grain boundary α'-martensite and $M_{23}C_6$ carbides, which were responsible for the low-temperature sensitization of the SS.

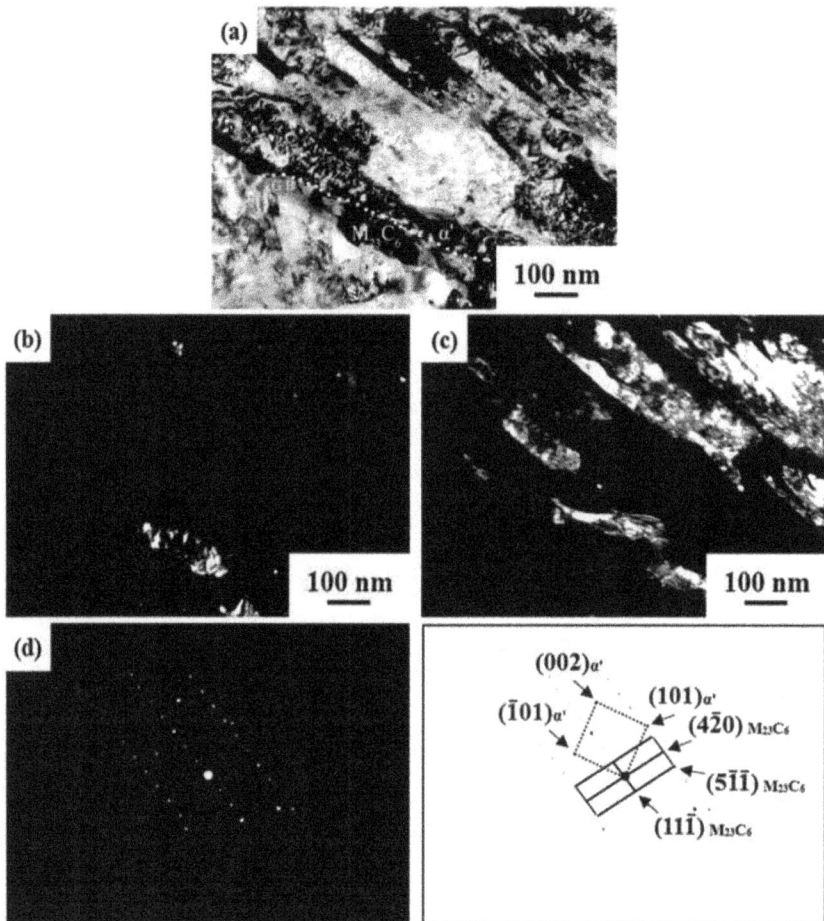

Figure 7. TEM images of the CR-450 specimen: (**a**) bright field image, (**b**) dark field image of carbides using $(11\bar{1})$ diffraction spot, (**c**) dark field image of α' martensite using $(\bar{1}01)$ spot, (**d**) selected area diffraction pattern with two identified structures.

Figure 8 shows TEM images of the G and σ phases in the CR-450 specimen. Nano-sized precipitates, G and σ phases, were found in the α' martensite of the CR-450 specimen, as shown in Figure 8a–c. The average size of the G phase precipitates was greater than that of the σ phase precipitates. The lattice image was transformed into a diffraction pattern by fast Fourier transform (FFT) for identification of the crystal structure of nano-sized precipitates by high resolution TEM. Figure 8b is a lattice image of the G phase formed in the α' matrix with a [100] zone axis. A σ precipitate in the α' matrix with a [013] zone axis is shown in Figure 8d.

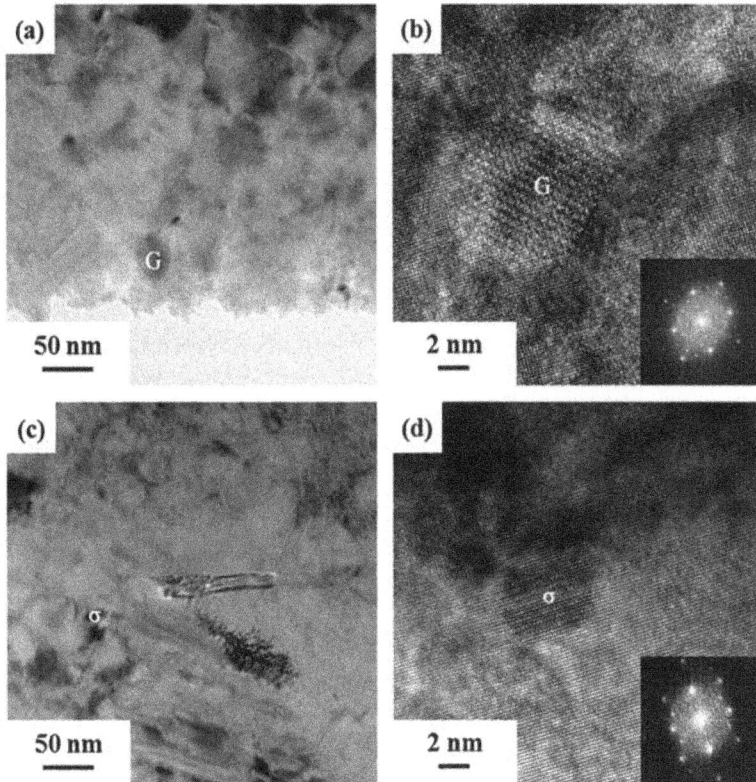

Figure 8. TEM images and high resolution lattice images with fast Fourier transform diffraction patterns of the CR-450 specimen: (**a**,**b**) G phase; and (**c**,**d**) σ phase.

4. Discussion

In general, an increase in annealing temperature reduces the strain-hardening effect of cold-rolled steels. Annealing treatment can improve the resistance of work-hardened steel to HE. As shown in Figure 6, cold work enhanced the formation of ε- and α′-martensite in the metastable austenitic SS. Moreover, metastable austenitic SSs undergo traditional sensitization when heated above 500 °C, and low-temperature sensitization occurs below 500 °C, if the cold work is applied. The formation of induced α′-martensite is responsible for its easy sensitization. In this work, aging the cold-rolled 301 SS at 450 °C (CR-450) assisted the formation of grain boundary α′-martensite and $M_{23}C_6$ carbides. In contrast, grain boundary α′-martensite and $M_{23}C_6$ carbides were not observed in the CR and CR-300 specimens.

As illustrated in Table 1, both UTS and YS of the CR-300 specimen were slightly higher than those of the CR one. In contrast, UTS and YS of the CR-450 specimen were obviously lower than those of CR and CR-300 ones. Moreover, all specimens showed equivalent tensile elongation tested in air. The results indicated increasing the annealing temperature to 450 °C did not improve the ductility but caused a decline in strength of the cold worked 301 SS. The results of notched tensile tests in air also revealed that the NTS of CR-450 specimen was lower than those of the others. As shown in Figure 4, the CR-450 specimen comprised of extensive flat fracture region (Figure 4b), in contrast, shear fracture dominated the fractographs of CR and CR-300 specimens (Figure 4a). It demonstrated the fact of high notch brittleness of the CR-450 specimen, and could be related to the brittle microstructures in the specimen. SEM fractographs of the CR-450 specimen after the notched tensile test in air showed cleavage-like fracture along with intergranular fracture (Figure 5b), which could be associated with the formation of grain boundary α'-martensite and $M_{23}C_6$ carbides, as shown in Figure 7. Without the presence of brittle microstructures along grain boundaries, the notched fracture appearance of the CR and CR-300 specimens showed predominantly shallow dimple fracture (Figure 5a).

It was reported that the HE susceptibility of the austenitic SSs is affected by the compositions [35,36] and prestrained level of the SS [37,38]. Dynamic interactions between hydrogen and the induced martensite play important roles in the HE of the hydrogen-charged 304 SS [37]. The hydrogen enhances the formation of α'-martensite, which facilitates HE and dominates the fracture process of the metastable austenitic SS [7]. Moreover, the hydrogen source influences the location of crack initiation and propagation in austenitic SSs. An external hydrogen source promotes crack initiation and propagation at the surface. Similar to surface cracks, the propagation of internal cracks is accelerated by the presence of internal hydrogen in the 304 and 316 SSs [35].

In Table 1, both CR-300H and CR-450H specimens were highly susceptible to internal HE, especially for the CR-450H one. It was deduced that high sensitivity to HE of the cold worked 301 SS could be attributed to its low austenite stability under straining. The CR-450H specimen suffered from a greater loss of ductility than the CR-300H specimen, as displayed in Figure 3. Both macro- and micro-fractographs, as shown in Figures 4 and 5, also confirmed that the CR-450H specimen was more susceptible to internal HE than the CR-300H specimen. High extent of intergranular fracture in the CR-450H specimen was associated with the brittle characteristics of grain boundaries.

Nano-sized G and σ particles were observed in α'-martensite of CR-450 specimen (Figure 8). The precipitation of G phase in the ferrite matrix has also been reported in the duplex SS after prolonged aging above 350 °C [39]. The precipitation of nano-sized G and σ phases further strengthened the interiors of the grains of the CR-450 sample. The strengthening of the grain interior by nano-sized precipitates highlighted the weakness of the embrittled grain boundaries and favored crack propagation therein. Actually, hydrogen-charging caused more hydrogen to diffuse into the specimen at 450 °C than that at 300 °C. The hydrogen concentrations of the CR-300H and CR-450H were 12 and 20 ppm, respectively. A high hydrogen concentration is usually accompanied by high HE susceptibility. Both the deteriorated microstructure and the high hydrogen content accounted for the high HE susceptibility and extensive intergranular fracture of the CR-450H specimen.

5. Conclusions

The effects of cold rolling and subsequent hydrogen-charging at 300 °C or 450 °C for 160 h on the microstructure, tensile properties, and HE susceptibility of 301 SS were investigated. Cold rolling caused ε-martensite to form in parallel strips of the slip bands, and α'-martensite was found mainly at the intersections of ε-martensite. Hydrogen-charging at 300 °C and 450 °C led to severe HE of the cold-rolled 301 SS, particularly in the 450 °C charged specimens. In specimens that were not hydrogen-charged, the notched tensile fracture of the 450 °C-aged one showed brittle fracture appearance, which comprised of cleavage-like fracture together with intergranular separations. In the CR-450 specimen, α'-martensite and $M_{23}C_6$ carbides were formed along the grain boundary. Because of this fresh grain boundary of α'-martensite, the ferrite contents of the CR-450 specimen were higher than those of the CR and CR-300 specimens. Moreover, very fine precipitates including the G and σ phases were found in the α'-martensite. The formation of grain boundary α'-martensite and $M_{23}C_6$ carbides together with nano-sized precipitates in the α'-martensite were responsible for the high HE susceptibility and low-temperature sensitization of the CR-450 specimen. In contrast, those of grain boundary precipitates and nano-sized phases were not observed in the CR and CR-300 specimens.

Acknowledgments: The authors gratefully acknowledge the partial financial support of this research by the National Science Council, Republic of China, Taiwan.

Author Contributions: L.W. Tsay and R.K. Shiue designed and planned the experiment. C. Chen contributed to the discussion of the results. C. Yu performed the tests and TEM examinations. All co-authors contributed to the manuscript proof and submissions.

Conflicts of Interest: The authors declare no conflict of interest.

References

1. Böhner, A.; Niendorf, T.; Amberger, D.; Höppel, H.W.; Göken, M.; Maier, H.J. Martensitic transformation in ultrafine-grained stainless steel AISI 304L under monotonic and cyclic loading. *Metals* **2012**, *2*, 56–64. [CrossRef]
2. Michler, T.; Naumann, J. Microstructural aspects upon hydrogen environment embrittlement of various bcc steels. *Int. J. Hydro. Energy* **2010**, *35*, 821–832. [CrossRef]
3. Michler, T.; Naumann, J.; Hock, M.; Berreth, K.; Balogh, M.P.; Sattler, E. Microstructural properties controlling hydrogen environment embrittlement of cold worked 316 type austenitic stainless steels. *Mater. Sci. Eng. A* **2015**, *628*, 252–261. [CrossRef]
4. Gey, N.; Petit, B.; Humbert, M. Electron backscattered diffraction study of ε/α′ martensitic variants induced by plastic deformation in 304 stainless steel. *Metall. Mater. Trans. A* **2005**, *36*, 3291–3299. [CrossRef]
5. Bak, S.; Abro, M.; Lee, D. Effect of hydrogen and strain-induced martensite on mechanical properties of AISI 304 stainless steel. *Metals* **2016**, *6*, 169. [CrossRef]
6. Pan, C.; Chu, W.Y.; Li, Z.B.; Liang, D.T.; Su, Y.J.; Gao, K.W.; Qiao, L.J. Hydrogen embrittlement induced by atomic hydrogen and hydrogen-induced martensites in type 304L stainless steel. *Mater. Sci. Eng. A* **2003**, *351*, 293–298. [CrossRef]
7. Lai, C.L.; Tsay, L.W.; Chen, C. Effect of microstructure on hydrogen embrittlement of various stainless steels. *Mater. Sci. Eng. A* **2013**, *584*, 14–20. [CrossRef]
8. Mine, Y.; Narazaki, C.; Murakami, K.; Matsuoka, S.; Murakami, Y. Hydrogen transport in solution-treated and pre-strained austenitic stainless steels and its role in hydrogen-enhanced fatigue crack growth. *Int. J. Hydro. Energy* **2009**, *34*, 1097–1107. [CrossRef]
9. Alyousif, O.M.; Nishimura, R. Stress corrosion cracking and hydrogen embrittlement of sensitized austenitic stainless steels in boiling saturated magnesium chloride solutions. *Corros. Sci.* **2008**, *50*, 2353–2359. [CrossRef]
10. Alyousif, O.M.; Nishimura, R. The stress corrosion cracking behavior of austenitic stainless steels in boiling magnesium chloride solutions. *Corros. Sci.* **2007**, *49*, 3040–3051. [CrossRef]
11. Alyousif, O.M.; Nishimura, R. The effect of test temperature on SCC behavior of austenitic stainless steels in boiling saturated magnesium chloride solution. *Corros. Sci.* **2006**, *48*, 4283–4293. [CrossRef]
12. Lai, C.L.; Tsay, L.W.; Kai, W.; Chen, C. Notched tensile tests of cold-rolled 304L stainless steel in 40 wt. % 80 °C MgCl$_2$ solution. *Corros. Sci.* **2009**, *51*, 380–386. [CrossRef]
13. Lai, C.L.; Tsay, L.W.; Kai, W.; Chen, C. The effects of cold rolling and sensitisation on hydrogen embrittlement of AISI 304L welds. *Corros. Sci.* **2010**, *52*, 1187–1193. [CrossRef]
14. Li, W.J.; Young, M.C.; Lai, C.L.; Kai, W.; Tsay, L.W. The effects of rolling and sensitization treatments on the stress corrosion cracking of 304L stainless steel in salt-spray environment. *Corros. Sci.* **2013**, *68*, 25–33. [CrossRef]
15. Raman, R.; Siew, W. Stress corrosion cracking of an austenitic stainless steel in nitrite-containing chloride solutions. *Materials* **2014**, *7*, 7799–7808. [CrossRef]
16. García, C.; Martín, F.; Tiedra, P.D.; Heredero, J.A.; Aparicio, M.L. Effects of prior cold work and sensitization heat treatment on chloride stress corrosion cracking in type 304 stainless steels. *Corros. Sci.* **2001**, *43*, 1519–1539. [CrossRef]
17. García, C.; Martín, F.; de Tiedra, P.; Alonso, S.; Aparicio, M.L. Stress corrosion cracking behavior of cold-worked and sensitized type 304 stainless steel using the slow strain rate test. *Corros. Sci.* **2002**, *58*, 849–857. [CrossRef]
18. Ghosh, S.; Kain, V. Effect of surface machining and cold working on the ambient temperature chloride stress corrosion cracking susceptibility of AISI 304L stainless steel. *Mater. Sci. Eng. A* **2010**, *527*, 679–683. [CrossRef]
19. Ghosh, S.; Kain, V. Microstructural changes in aisi 304L stainless steel due to surface machining: Effect on its susceptibility to chloride stress corrosion cracking. *J. Nucl. Mater.* **2010**, *403*, 62–67. [CrossRef]

20. Martin, M.; Weber, S.; Izawa, C.; Wagner, S.; Pundt, A.; Theisen, W. Influence of machining-induced martensite on hydrogen-assisted fracture of AISI type 304 austenitic stainless steel. *Int. J. Hydro. Energy* **2011**, *36*, 11195–11206. [CrossRef]

21. Domankova, M.; Marek, P.; Magula, V. Effect of deformation and heat treatment on grain boundary sensitization in austenitic stainless steels. *Metal* **2006**, *23*.

22. Nasreldin, A.M.; Gad, M.M.A.; Hassan, I.T.; Ghoniem, M.M.; El-sayed, A.A. Effect of hydrogen charging on the tensile and constant load properties of an austenitic stainless steel weldment. *J. Mater. Sci. Technol.* **2001**, *17*, 444–448.

23. Parvathavarthini, N.; Dayal, R.K. Influence of chemical composition, prior deformation and prolonged thermal aging on the sensitization characteristics of austenitic stainless steels. *J. Nucl. Mater.* **2002**, *305*, 209–219. [CrossRef]

24. Silva, M.J.G.; Souza, A.A.; Sobral, A.V.C.; de Lima-Neto, P.; Abreu, H.F.G. Microstructural and electrochemical characterization of the low-temperature sensitization of AISI 321 stainless steel tube used in petroleum refining plants. *J. Mater. Sci.* **2003**, *38*, 1007–1011. [CrossRef]

25. Singh, R. Influence of cold rolling on sensitization and intergranular stress corrosion cracking of AISI 304 aged at 500 °C. *J. Mater. Process. Technol.* **2008**, *206*, 286–293. [CrossRef]

26. Prasanthi, T.N.; Sudha, C.; Parameswaran, P.; Punniyamoorthy, R.; Chandramouli, S.; Saroja, S.; Rajan, K.K.; Vijayalakshmi, M. Failure analysis of a 304 steel component aged at 623 K. *Eng. Fail. Anal.* **2013**, *31*, 28–39. [CrossRef]

27. Lv, J.; Luo, H. Temperature dependence of sensitization on tensile pre-strained AISI 304 stainless steels. *J. Alloy. Compd.* **2014**, *588*, 509–513.

28. Kain, V.; Chandra, K.; Adhe, K.N.; De, P.K. Effect of cold work on low-temperature sensitization behaviour of austenitic stainless steels. *J. Nucl. Mater.* **2004**, *334*, 115–132. [CrossRef]

29. Han, G.; He, J.; Fukuyama, S.; Yokogawa, K. Effect of strain-induced martensite on hydrogen environment embrittlement of sensitized austenitic stainless steels at low temperatures. *Acta Mater.* **1998**, *46*, 4559–4570. [CrossRef]

30. Talonen, J.; Hänninen, H. Formation of shear bands and strain-induced martensite during plastic deformation of metastable austenitic stainless steels. *Acta Mater.* **2007**, *55*, 6108–6118. [CrossRef]

31. Talonen, J.; Aspegren, P.; Hänninen, H.J. Comparison of different methods for measuring strain induced α'-martensite content in austenitic steels. *Mater. Sci. Technol.* **2004**, *20*, 1506–1512.

32. Yokobori, A.T.; Nemoto, T.; Satoh, K.; Yamada, T. Numerical analysis on hydrogen diffusion and concentration in solid with emission around the crack tip. *Eng. Fract. Mech.* **1996**, *55*, 47–60. [CrossRef]

33. Hardie, D.; Liu, S.E. The effect of stress concentration on hydrogen embrittlement of a low alloy steel. *Corros. Sci.* **1996**, *38*, 721–733. [CrossRef]

34. Liu, S.E.; Ziyong, Z.; Wei, K. Notch severity effect on hydrogen embrittlement of type 4340 steel. *J. Mater. Sci. Technol.* **1996**, *12*, 51–56.

35. San Marchi, C.; Michler, T.; Nibur, K.A.; Somerday, B.P. On the physical differences between tensile testing of type 304 and 316 austenitic stainless steels with internal hydrogen and in external hydrogen. *Int. J. Hydro. Energy* **2010**, *35*, 9736–9745. [CrossRef]

36. Neuharth, J.J.; Cavalli, M.N. Investigation of high-temperature hydrogen embrittlement of sensitized austenitic stainless steels. *Eng. Fail. Anal.* **2015**, *49*, 49–56. [CrossRef]

37. Yoshioka, Y.; Yokoyama, K.I.; Sakai, J.I. Role of dynamic interactions between hydrogen and strain-induced martensite transformation in hydrogen embrittlement of type 304 stainless steel. *ISIJ Int.* **2015**, *55*, 1772–1780. [CrossRef]

38. Matsuo, T.; Yamabe, J.; Matsuoka, S. Effects of hydrogen on tensile properties and fracture surface morphologies of type 316L stainless steel. *Int. J. Hydro. Energy* **2014**, *39*, 3542–3551. [CrossRef]

39. Lo, K.H.; Lai, J.K.L. Microstructural characterization and change in a.c. magnetic susceptibility of duplex stainless steel during spinodal decomposition. *J. Nucl. Mater.* **2010**, *401*, 143–148. [CrossRef]

metals

MDPI

Article

Precipitation Behavior of Carbides in H13 Hot Work Die Steel and Its Strengthening during Tempering

Angang Ning [1], Wenwen Mao [2], Xichun Chen [3], Hanjie Guo [2,*] and Jing Guo [2]

[1] School of Material Science and Engineering, Taiyuan University of Technology, Taiyuan 030024, China; ningangang1986@163.com
[2] School of Metallurgical and Ecological Engineering, University of Science and Technology Beijing, Beijing 100083, China; hrc7087097@163.com (W.M.); guojingzq@163.com (J.G.)
[3] Department of High Temperature Materials, Central Iron and Steel Research Institute, Beijing 100094, China; chenxichun@189.cn
* Correspondence: guohanjie@ustb.edu.cn; Tel.: +86-10-6233-4964

Academic Editor: Robert Tuttle
Received: 6 December 2016; Accepted: 18 February 2017; Published: 23 February 2017

Abstract: The properties of carbides, such as morphology, size, and type, in H13 hot work die steel were studied with optical microscopy, transmission electron microscopy, electron diffraction, and energy dispersive X-ray analysis; their size distribution and quantity after tempering, at different positions within the ingot, were analyzed using Image-Pro Plus software. Thermodynamic calculations were also performed for these carbides. The microstructures near the ingot surface were homogeneous and had slender martensite laths. Two kinds of carbide precipitates have been detected in H13: (1) MC and M_6C, generally smaller than 200 nm; and (2) $M_{23}C_6$, usually larger than 200 nm. MC and M_6C play the key role in precipitation hardening. These are the most frequent carbides precipitating at the halfway point from the center of the ingot, and the least frequent at the surface. From the center of the ingot to its surface, the size and volume fraction of the carbides decrease, and the toughness improves, while the contribution of the carbides to the yield strength increases.

Keywords: H13 steel after tempering; thermodynamic calculation; carbides; precipitation strengthening

1. Introduction

As a typical hot work die steel, H13 (4Cr5MoSiV1) (where "4" means that the carbon content is about 0.4%), has excellent resistance to heat, wear, and thermal mechanical fatigue [1]; thus, it is widely employed in high temperature applications, such as die casting molds, hot rolling, hot extrusion, and hot forging, where the operating tool is repeatedly subjected to high temperatures and loads [2]. Toughness and yield strength, determined by the martensite and carbides [3,4], are the most important mechanical properties of H13 [5].

Most carbides dissolve in H13 during the quenching process and precipitate from martensite as uniformly dispersed nanoparticles during tempering, resulting in secondary hardening. Thus, the heat treatment process has significant effects on the properties of the carbides and hence on the quality of the steel. Extensive research has been performed on both the solidification process in H13 and the effects of heat treatment; changes in the microstructure and categories of carbides have been studied in the laboratory and also at the industrial plant scale [6–8]. By electrolytically extracting precipitates after tempering H13, Song et al. [9] discovered that the carbides in H13 were mainly V_8C_7, which not only includes pseudo-eutectic carbides not dissolved in H13 during forging, but also includes the secondary hardening carbides precipitated from martensite. These secondary carbides are regarded as the main strengthening phase because of their fine particle size and well-dispersed distribution. In addition, M_2C, M_6C, and M_7C_3-type carbides also precipitate during tempering. Zhang et al. [10] measured the particle distribution of MC by using small-angle X-ray scattering and reported that the precipitation strengthening increment of tempering at 873 K was 171.72 MPa. Fu et al. [11] demonstrated that the precipitation strengthening increments of Fe_3C and $Ti(C,N)$ during rapid cooling after rolling were 194.4 MPa and 130 MPa, respectively, in Ti microalloyed high-strength weathering steel. These reports, however, did not precisely describe the shape, size, and distribution of the carbides. The influence of precipitates on the mechanical properties—especially toughness and yield strength—of hot work die steel is also rarely reported.

A previous study by our group [12] investigated the effect of different heat treatments on the precipitates and strengthening mechanism in H13, and a root-mean-square summation law, which was confirmed as the most applicable for H13 steel, was summarized as $\sigma = \sigma_g + \sigma_s + (\sigma_d^2 + \sigma_p^2)^{\frac{1}{2}}$, where σ_g, σ_s, σ_d and σ_p represent the contributions of fine-grain strengthening, solid-solution strengthening, dislocation strengthening, and precipitates' strengthening.

In this article, the types of carbides formed in H13 during tempering are described, their characteristics are clarified, and then thermodynamic calculations are performed. The type, size, and volume of carbides are analyzed in three different regions of the H13 ingot, and their contribution to the yield strength is discussed. The influence of precipitates on yield strength and impact properties is also discussed.

2. Experimental Materials and Methods

The chemical composition of H13 hot work die steel is shown in Table 1. The smelting process included electric arc furnace (EAF) melting, ladle refining (LF), vacuum degassing (VD), and electro-slag remelting (ESR). The ingot diameter was 220 mm and its available weight was up to 5 t.

Table 1. Chemical composition of H13 steel, %.

C	Si	Mn	P	S	Cr	Ni	Cu	Mo	V	Al	N	T [O]
0.39	0.88	0.34	0.0064	0.0005	5.13	0.086	0.054	1.5	0.99	0.047	0.0015	0.0017

The heat treatment was simulated industrial processing and was performed as follows:

First, the ESR ingot was forged at 1373 K to bar stock with a diameter of 105 mm. Then the forged ingot was annealed at 1133 K for 10 h, cooled to 773 K in the furnace, and further cooled to room temperature in air, as shown in Figure 1a.

Next, a 105 mm diameter × 60 mm thick sample was cut from the central part of the ingot. This sample was preheated to 1113 K at a rate of 13 K/min, maintained for 20 min, heated to 1303 K at a rate of 6 K/min, maintained for an additional 30 min, and then cooled in oil at 33 K/min, as shown in Figure 1b.

Finally, the sample was tempered for 2 h at 863 K, and then cooled in air to room temperature, as shown in Figure 1c.

Smaller samples were then cut from the main sample for transverse and longitudinal impact value (Charpy V-notch) tests, tensile strength measurements, and metallographic studies. The dimensions of the Charpy test samples were $10 \times 10 \times 55$ mm^3, the tensile test samples were $\Phi 8 \times 120$ mm^2 (the gripping end was $\Phi 12 \times 50$ mm^2), and the metallographic samples were $10 \times 10 \times 10$ mm^3. The positions of sampling were: near the center (#1), halfway between the center and the outer surface (hereafter referred to as "half-radius") (#2), and near the outer surface (#3) of the H13 ingot, as shown in Figure 2.

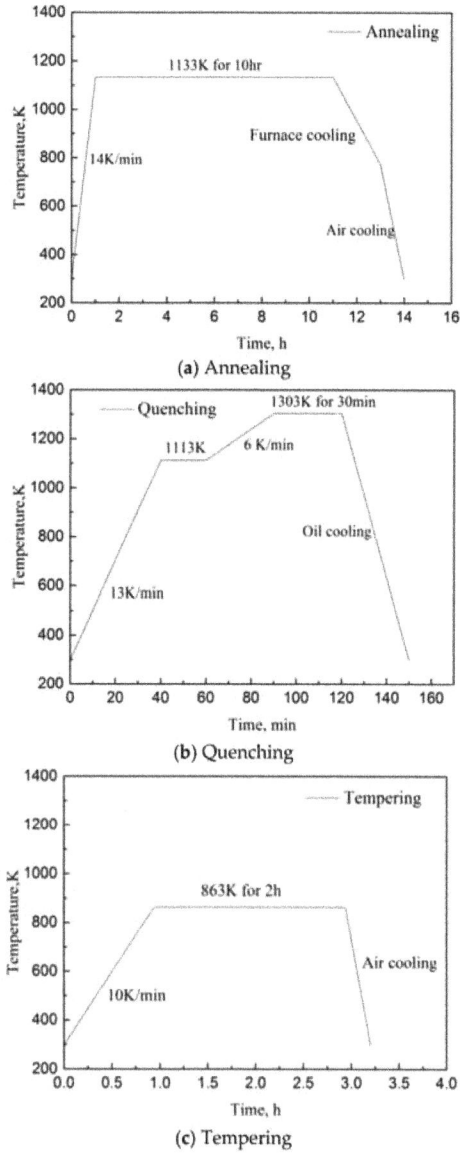

Figure 1. Heat treatment for the H13 ingot after forging. (**a**) Annealing; (**b**) Quenchin; (**c**) Temperin.

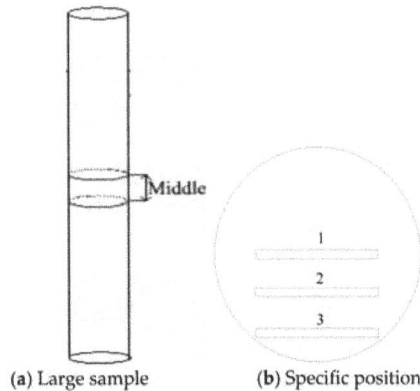

(a) Large sample   (b) Specific position

Figure 2. Positions of the H13 steel samples: (**a**) Large sample taken from the middle of the ingot after annealing; (**b**) Individual test samples taken from the center (#1), half-radius (#2), and outer surface (#3) of (**a**) after tempering.

The impact values of the three samples were tested by a ZBC2452-B Pendulum impact testing machine (MTS System Corporation, Shenzhen, China). Tensile strength, yield strength, elongation, and area reduction were examined with a CMT4105 electronic universal testing machine (MTS System Corporation, Shenzhen, China). A 500MRA Rockwell hardness tester was used to measure the hardness of the samples. The microstructure of the steel was observed with a 9XB-PC optical microscope (Shanghai Optical Instrument Factory, Shanghai, China). The morphology of the carbides in the three samples was examined with an F30 high resolution transmission electron microscope (HR TEM) (FEI Company, Hillsboro, OR, USA). The carbon extraction replica method was used to prepare the TEM samples; the specific steps were as follows: first etch the polished metallographic samples in 8% nitric acid alcohol solution, then coat them with a layer of evaporated carbon film, approximately 20–30 nm thick, and finally extract the precipitates by using 10% nitric acid alcohol solution and mount the carbon membrane on a copper grid. The morphology of the precipitates was analyzed by TEM after the samples were dried.

3. Experimental Results

3.1. Metallographic Structure after Tempering

As shown in Figure 3, the microstructure of H13 after tempering includes martensite, fine-grained pearlite, and a large number of carbides. The heterogeneity of the microstructures is obvious both at the center and at the half-radius of the ingot (Figure 3a,b). Light field represents the alloy-rich and carbide-rich field, while the situation is opposite in the dark field where the carbides dissolve and martensite forms [13]. Especially in the center, there exist segregation bands composed of eutectic carbides, as shown in Figure 3a. Compared with the center and the half-radius of the ingot, the microstructures near the outer surface (Figure 3c) are much more homogeneous and appear to have the narrowest martensite laths.

(a) Center

(b) Half-radius

(c) Outer surface

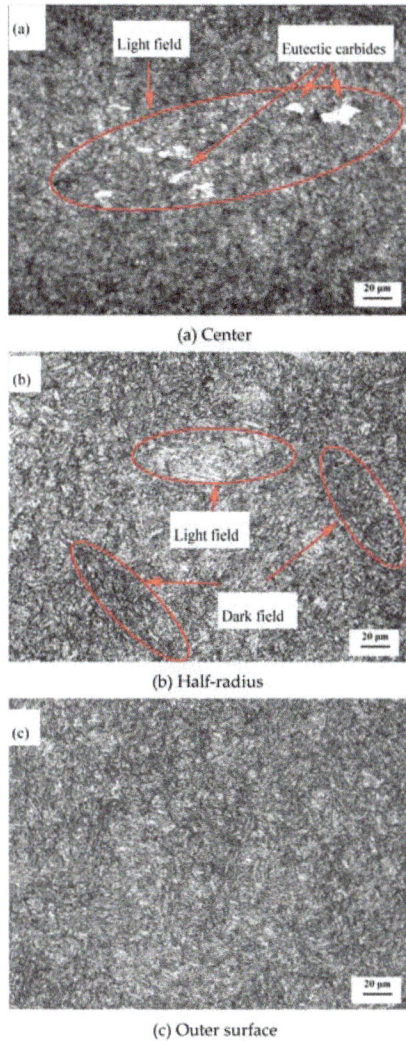

Figure 3. Microstructure of the H13 ingot at different positions after tempering. (**a**) Center; (**b**) Half-radius; (**c**) Outer surface.

It is well known that the cooling speed is lowest in the center and fastest at the surface during ESR [14]. It is difficult to eliminate eutectic carbides and segregation during ESR by conventional heat treatment. Thus, the formation of eutectic carbides consumes alloying elements and carbon in H13, influencing the formation of nanocarbides; meanwhile, the segregation that causes the heterogeneous distribution of elements in H13 can in some cases also have an effect on the distribution, sizes, and amounts of nanocarbides.

3.2. Types of Precipitates during Tempering

After observing 300 photomicrographs (14.5 × 14.5 μm^2 each) as well as the EDS results, it was determined that most of the fine particles are V-rich MC-type carbides along with some Mo-rich carbides, usually squares or elongated in shape and smaller than 200 nm, as shown in Figures 4a and 5a.

Some Cr-rich carbides were also observed; these are usually larger than the V- and Mo-rich carbides—200 nm or more—and are irregularly spherical in shape. As shown in Figures 4 and 6, V_8C_7 was found near the surface of the H13 ingot, and $M_{23}C_6$ was also detected [15]. However, most of the $M_{23}C_6$ was found at the center of the H13 ingot in this experiment. V_8C_7 belongs to the family of cubic MC-type carbides, and its lattice constant is $a = b = c = 0.833$ nm. The particles found near the surface of our H13 ingot were approximately spherical in shape, with a diameter of about 100 nm.

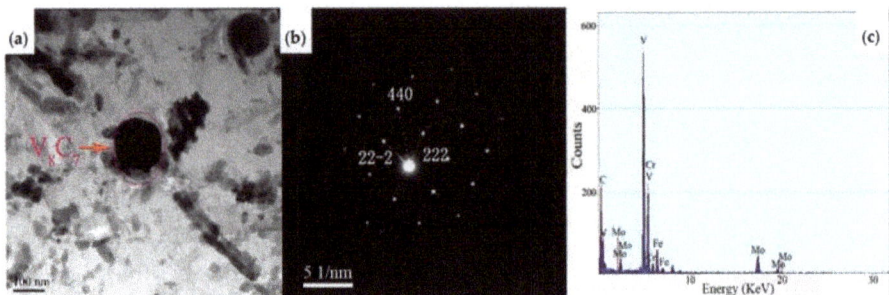

Figure 4. Precipitates of V_8C_7 near the surface of the H13 ingot after tempering: (a) Morphology by TEM; (b) SAED (Selected Area Electron Diffraction) analysis; (c) EDS (Energy Dispersive X-ray Spectrum) analysis.

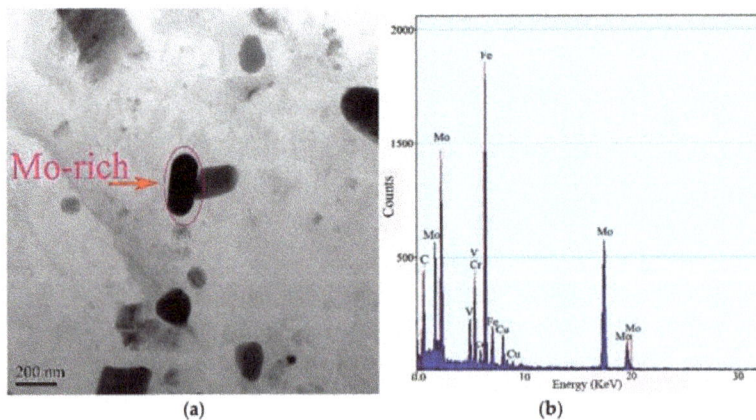

Figure 5. Precipitates of Mo-rich carbides at half-radius of the H13 ingot after tempering: (a) Morphology by TEM; (b) EDS analysis.

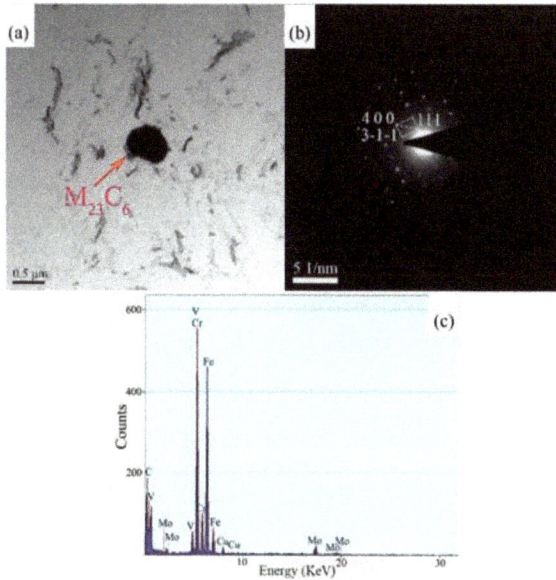

Figure 6. Precipitates of $M_{23}C_6$ at the center of the H13 ingot after tempering: (**a**) Morphology by TEM; (**b**) SAED analysis; (**c**) EDS analysis.

According to the morphology and EDS analysis in Figure 5, the indicated nanoparticle is an M_6C-type carbide, which is detected at the half-radius of the H13 ingot, rich in Mo and Fe, with an elongated shape, about 200×90 nm^2.

The types of precipitates that were found in H13 by using the carbon extraction replica method were consistent with the published literature [7,9]. It was also found in this experiment that although the size of the Mo-rich M_6C particles was larger than that of the V-rich MC, both of them are widely and uniformly distributed throughout the ingot, and both work as secondary hardening and contribute to tempering resistance. V-rich MC, mixed with M_2C and M_6C in steel, can improve the tempering resistance [16].

3.3. Analysis of the Size and Volume of Precipitate Carbides

In the present trials, 15 large fields of view (6000×) and 15 small fields of view (30,000×) were randomly selected from points #1, #2, and #3 in order to analyze the volumes and size distribution of the carbides. Figure 7a,b shows the micro-morphology and distribution of carbides in H13 after tempering for 2 h.

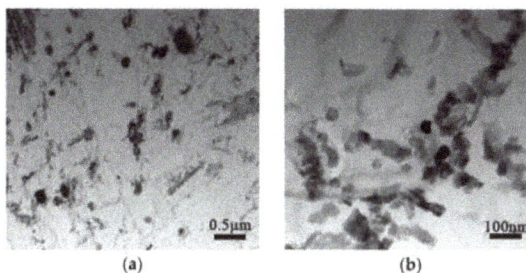

Figure 7. Distribution of carbides at the half-radius of the H13 ingot after tempering. (**a**) Large field of view; (**b**) Small field of view.

115

The amount of precipitates and their size distribution were obtained by using Image-Pro Plus software (Media Cybernetics Inc., Rockville, MD, USA). Image-Pro Plus is designed for processing, enhancing, and analyzing pictures; it has exceptionally rich measurement and customization features. The statistical results are listed in Table 2.

Table 2. Statistics of precipitates at different positions in H13 after tempering.

Sample Number	Visual Field Area/μm^2	Field Number	Number of Precipitates	Average Size/nm
1	3.8 × 3.8	15	711	82.6
	0.8 × 0.8	15	394	
2	3.8 × 3.8	15	815	81.6
	0.8 × 0.8	15	327	
3	3.8 × 3.8	15	683	60.7
	0.8 × 0.8	15	261	

Table 2 shows the quantity and size distribution of the carbides. It can be seen that the sizes of the precipitation carbides decrease gradually from the center to the surface of the H13 ingot, while the precipitation quantity is largest at the half-radius. The average size of the nanocarbides at the center is almost the same as that at the half-radius. Under the influence of eutectic carbides and the segregation of metallic elements, nanocarbides precipitate insufficiently; therefore, the precipitation quantity at the center is smaller than that at the half-radius after tempering. Many fine carbides on the surface, such as M_2C, do not have enough time to precipitate because of the rapid oil cooling speed after tempering; therefore there are very few carbides near the surface. However, near the surface, the average size of carbides is 60.7 nm; this size is finer than the corresponding sizes at either the center or at the half radius. In order to obtain H13 steel with sufficiently high strength and hardness, the appropriate holding time and temperature should be set to allow for sufficient nanocarbides to precipitate.

The precipitation quantities in different particle size ranges (<150 nm) of these three positions is shown in Figure 8, and it can be seen that:

(1) The carbides less than 150 nm in size precipitate most frequently at the half-radius of the H13 ingot. Therefore, the quantity of precipitates at this position is the largest.
(2) The greatest number of carbides are of sizes less than 50 nm, occurring near the surface of the ingot. Therefore, the average size is smallest at this position.
(3) The particle size distribution deviates from a Gaussian distribution, possibly because short-time tempering leads to insufficient growth of the precipitates and a large number of small newly-formed carbides precipitate during tempering.

It can be concluded that, after tempering for 2 h, the carbides had an average size of 76 nm, which is finer than the size after ESR (90 nm), forging annealing (206 nm), and quenching (137 nm) [17,18]. Although the residual carbides grow and coarsen at lower temperatures during quenching [7], martensite decomposes and secondary phases are precipitated during tempering; thus a large number of nanocarbides, which have a significant effect on precipitation strengthening, precipitate from the grain boundaries and inside the grains of martensite. Therefore, the quantity increases and the average size decreases significantly, as shown in Table 2.

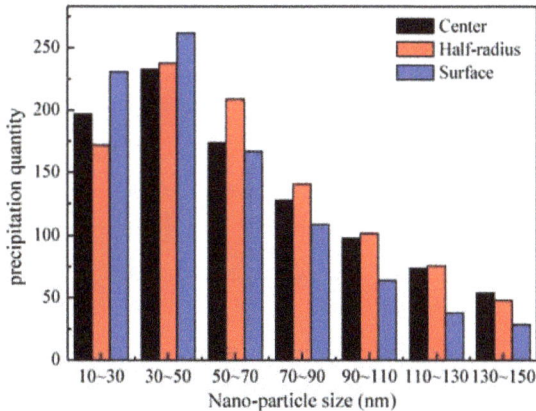

Figure 8. Size distribution of carbides in H13 at the three different locations after tempering.

4. Discussion

4.1. Thermodynamic Calculation of Carbides during Tempering

As mentioned above, MoC, VC, and V_8C_7 belong to the MC family of carbides; and $M_{23}C_6$ is a Cr-rich carbide. The following study focuses on the precipitation temperatures of VC, V_8C_7, $Cr_{23}C_6$, and MoC, and their evolution during tempering is also discussed.

According to the relevant chemical reaction and Gibbs free energy [19–21], a precipitation reaction of $Cr_{23}C_6$ and MoC is obtained.

$$23[Cr]_\gamma + 6[C]_\gamma = Cr_{23}C_6(s), \Delta G^{\varnothing} = -959797.2 + 1172.7T \tag{1}$$

$$[Mo]_\gamma + [C]_\gamma = MoC(s), \Delta G^{\varnothing} = -149038 + 94.97T \tag{2}$$

The equilibrium solubility product in austenite is obtained from Equations (1) and (2).

$$\ln\left(w[Cr]_\%^{23} \cdot w[C]_\%^6\right)_\gamma = 141.05 - \frac{115443.49}{T} \tag{3}$$

$$\ln\left(w[Mo]_\% \cdot w[C]_\%\right)_\gamma = 11.42 - \frac{17926.15}{T} \tag{4}$$

Owing to the very low solubility of MC-type carbides in austenite, it is difficult to test the solubility of MC-type carbides experimentally, and the only result available at the present time has been deduced from thermodynamic data. According to the published literature [21], the solubility products of V_8C_7 and VC in ferrite and austenite are given by the following:

$$\ln\left(w[V]_\% \cdot w[C]_\%^{0.875}\right)_\alpha = 13.01 - \frac{21510.02}{T} \tag{5}$$

$$\ln\left(w[V]_\% \cdot w[C]_\%\right)_\gamma = 15.48 - \frac{21878.5}{T} \tag{6}$$

The precipitation temperatures of these carbides can be obtained by inserting the following values into Equations (3)–(6): $w[Cr]_\% = 5.13$, $w[Mo]_\% = 1.5$, $w[V]_\% = 0.99$, and $w[C]_\% = 0.39$; the results are listed in Table 3.

Table 3. Precipitation temperatures of carbides in H13 during solidification.

Austenite Region			
Carbides	V_8C_7	MoC	VC	$Cr_{23}C_6$
Temperature	1553.5 K	1499.0 K	1331.8 K	1058.2 K

Only when the temperature is below the precipitating temperatures do these carbides precipitate. The austenite region of H13 is between the liquid line at 1755 K and the A_{r1} line at 1048 K [1]. From Table 3, the precipitation sequence is MC > $M_{23}C_6$.

According to experimental results combined with theoretical calculations, it can be concluded that V_8C_7 precipitates at a higher temperature (1553.5 K), but it has low coarsening speed (Figure 4). V-rich carbides are more widely dispersed and finer than Cr carbides. Another conclusion is that $M_{23}C_6$ precipitates at a lower temperature (1058.2 K), but has a faster coarsening speed (Figure 5), because the diffusion rate of Cr in austenite is greater than that of V [21]. $M_{23}C_6$ has high Cr content and thus Cr-rich carbides are easy to grow at the grain boundaries [22]. Cr carbides have a larger precipitation volume fraction and are not so widely distributed in the form of fine particles. MoC precipitates are not stable at high temperature (<1040 K), and they are easier to combine with Fe and Cr to form M_6C [1].

In actual circumstances, carbon and alloy segregation always exists, so the precipitation temperature of carbides will fluctuate in different regions.

4.2. Effect of Precipitates in H13 during Tempering on Mechanical Properties

The mechanical properties of H13 after 2 h tempering are shown in Table 4.

Table 4. Tensile property and hardness of different positions in H13 after tempering.

Sample Number	Tensile Strength, R_m (MPa)	Yield Strength, R_p (MPa)	Elongation, A (%)	Area Reduction Rate, Z (%)	HRC
Center, #1	1764.6	1456.0	7.0	30.6	49.2
Half-radius, #2	1743.1	1426.4	8.9	28.9	47.5
Surface, #3	1750.2	1436.0	8.0	35.1	46.2

According to the theory of precipitation strengthening [21], when a large number of fine nanocarbides precipitate during tempering, the strength of the steel increases because of the decomposition of martensite. Although a large number of precipitates may cause a decrease in plasticity, the reduction in alloying element content and decrease in dislocation density may improve the plasticity [23–25]. As shown in Table 4, the strength and plasticity after tempering are almost independent of the sample position. However, the hardness seems to be lower from the center to the surface, showing that the ingot was heated unevenly.

From the size distribution of Figure 8, it can be seen that the average size of precipitates is larger than 10 nm; therefore, the bypass mechanism has the main effect on precipitation strengthening [11]. In order to calculate the contribution of the precipitates to yield strength, we employed segment calculation and summation of the results (Table 5). According to the methods of McCall-Boyd [26] and the Ashby-Orowan correction model [27], the formula for the volume fraction and precipitation strengthening of the precipitates in H13 is obtained as Formulas (12) and (13). McCall and Boyd [28] analyzed the characteristics of precipitates in ThO_2 by employing carbon extraction replica methods in the 1960s. The McCall-Boyd method is an accurate way to calculate the volume fraction of precipitates with uneven distribution in alloys.

Table 5. Size, volume, and calculation of precipitation strengthening in H13 of different positions after tempering for 2 h.

Size (nm)	Amount			Average Diameter, D (nm)			Volume Fraction, f			Average Radius, r (nm)			Yield Strength Increment, τ_P (MPa)		
	1#	2#	3#	1#	2#	3#	1#	2#	3#	1#	2#	3#	1#	2#	3#
10–30	197	172	231	21.79	21.73	21.74	0.00015	0.00013	0.00035	10.90	10.87	10.87	26.78	25.01	40.88
30–50	233	238	262	40.27	40.44	39.31	0.00061	0.00063	0.0013	20.14	20.22	19.66	33.86	34.25	50.36
50–70	174	209	167	59.22	59.12	58.66	0.00099	0.0012	0.0019	29.61	29.56	29.33	31.82	34.87	43.89
70–90	128	141	109	79.67	79.1	79.67	0.0013	0.0014	0.0022	39.84	39.55	39.84	28.99	30.38	37.73
90–110	98	102	64	98.43	99.01	99.86	0.0015	0.0016	0.0021	49.22	49.51	49.93	26.42	26.99	30.20
110–130	74	76	38	118.74	118.15	120.02	0.0017	0.0017	0.0018	59.37	59.08	60.01	23.77	24.07	24.08
130–150	54	48	29	139.54	138.94	139.2	0.0017	0.0015	0.0018	69.77	69.47	69.60	20.91	19.70	21.60
150–170	38	53	17	160.42	159.11	157.63	0.0016	0.0022	0.0014	80.21	79.56	78.82	17.97	21.19	16.90
170–190	26	35	9	178.54	177.69	180.89	0.0013	0.0018	0.00095	89.27	88.85	90.45	15.14	17.55	12.59
190–210	24	26	6	200.26	198.01	199.23	0.0016	0.0017	0.00077	100.13	99.01	99.62	14.83	15.41	10.45
210–230	15	11	4	218.68	220.87	222.11	0.0012	0.00087	0.00064	109.34	110.44	111.06	11.90	10.20	8.69
230–250	11	11	4	242.57	239.75	238.34	0.0011	0.0010	0.00074	121.29	119.88	119.17	10.36	10.34	8.79
250–270	12	9	1	260.18	259.54	254.88	0.0013	0.00099	0.00021	130.09	129.77	127.44	10.94	9.47	4.44
270–290	5	6	2	275.09	278.86	278.88	0.00062	0.00076	0.00050	137.55	139.43	139.44	7.13	7.82	6.37
290–310	5	1	0	294.86	307.76	0	0.00071	0.00015	0	147.43	153.88	0	7.21	3.24	0
310–330	4	1	1	320.86	316.7	313.36	0.00067	0.00016	0.00032	160.43	158.35	156.68	6.53	3.26	4.59
330–350	5	2	0	340.15	334.18	0	0.00094	0.00036	0	170.08	167.09	0	7.37	4.65	0
350–370	2	1	0	357.39	353.43	0	0.00042	0.00020	0	178.70	176.72	0	4.69	3.31	0
Total	1105	1142	944				0.0193	0.0184	0.0169				279.84	301.72	321.56

$$f_i = \left(\frac{1.4\pi}{6}\right) \cdot \left(\frac{N_i D_i^2}{A}\right) \tag{7}$$

$$\sigma_s = \sum_{i=1}^{n} \sigma_i = \sum_{i}^{n} [\frac{10\mu b}{5.72\pi^{3/2} r_i} f_i^{\frac{1}{2}} \ln(\frac{r_i}{b})] \tag{8}$$

Here, A represents the area of the photos in μm^2; N_i represents the amounts of precipitates within a certain range; D_i represents the average diameter in nanometers for precipitates within a certain range; r_i represents the average radius in nanometers for precipitates within a certain range; f_i represents the volume fraction, in % of precipitates within a certain range; μ represents a shearing factor (80.26×10^3 MPa for steel); and b represents the Burgers Vector, with a value of 2.48×10^{-4} μm.

The volume fraction of precipitates and the increment of precipitation strengthening in the three different positions of H13 after tempering are calculated according to Formulas (7) and (8). The calculation process is listed in Table 5, and the results of the calculations are listed in Table 6.

Table 6. Contribution to yield strength of the precipitates in H13 after tempering.

Item	Center (1#)	Half-Radius (2#)	Surface (3#)
Average size (nm)	82.6	81.6	60.7
Volume fraction (%)	1.9	1.8	1.7
Contribution to the yield strength (MPa)	279.8	301.7	321.6
Actual yield strength (MPa)	1456.0	1426.4	1436.0
The proportion of yield strength attributable to precipitation strengthening (%)	19.2	21.2	22.4

From Table 6, combined with Figure 9, it can be seen that the contribution of precipitates to the yield strength increases as the average size of the precipitates becomes finer, and the volume fraction decreases from the center to the surface of the H13 ingot after 2 h tempering.

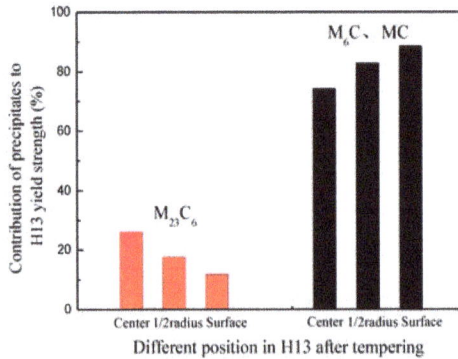

Figure 9. Contribution to yield strength of different types of carbides in different positions in H13 after tempering.

After tempering for 2 h, the contribution of nanoprecipitates to the yield strength of H13 remains at about 300 MPa. The average diameter and total volume fraction of precipitates are 60–83 nm and 1.7%–1.9%, respectively. The measured results of yield strength show little difference between the three positions. It was also found that the surface of H13 has the finest precipitates, and experiences the largest precipitation strengthening of H13. Combining the information from Section 3.2 and Table 5, the contribution of different types of carbides to yield strength is summarized in Figure 9.

From Figure 9, it can be seen that MC and M_6C have a stronger strengthening effect than $M_{23}C_6$ on H13 yield strength, contributing 74%–88% of precipitation strengthening. From the center to the surface of the H13 ingot, the contribution of $M_{23}C_6$ to the yield strength decreases, while the contribution of MC and M_6C to the yield strength increases. It can be concluded that MC and M_6C are the main carbides in precipitation strengthening. These carbides have good tempering stability as their average size is below 200 nm.

The transverse and longitudinal impact values at the center, half-radius, and surface of H13 after tempering for 2 h are shown in Figure 10.

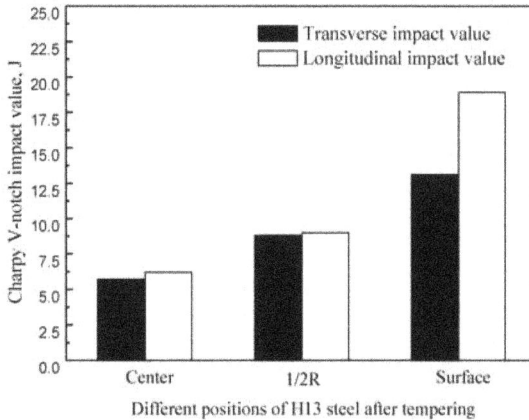

Figure 10. Impact values in different positions of H13 after tempering.

From Figure 10, the impact values after tempering become better from the center of the ingot to its outer surface. It can be seen that the toughness was the lowest at the center of the ingot. One reason is the formation of ribbon segregation and eutectic carbides in the center during ESR [12]. The other reason is that the volume fraction of the precipitates decreases, especially the dissolution of $M_{23}C_6$ [22] and the precipitation of MC, M_6C during tempering (Figure 9).

In conclusion, it is the homogeneous structure and fine precipitates with moderate distribution that improve the toughness of H13.

5. Conclusions

1. Microstructures near the surface of the H13 ingot are homogeneous and have the narrowest martensite laths.
2. V-rich MC, Mo-rich M_6C, and Cr-rich $M_{23}C_6$ are the main kinds of carbides that precipitate in H13 during tempering. The shapes of MC and M_6C are approximately square or elongated and their sizes are less than 200 nm. The shape of $M_{23}C_6$ is irregularly spherical and its size is greater than 200 nm. From thermodynamic calculations, the precipitation sequence is MC > $M_{23}C_6$.
3. Carbide size and volume fraction decrease from the center to the periphery of H13 during tempering. There are more precipitates at the half-radius of H13 than that at the center. The surface of the H13 ingot has the fewest precipitates.
4. The strength and plasticity after tempering are almost independent of the sample position. However, the hardness seems to be lower from the center to the surface, showing that the ingot was heated unevenly. The impact values after tempering become better from the center of the

ingot to its outer surface, because the compositions and structures are uniform near the surface of the ingot, and the precipitates are uniformly dispersed with the finest size.

5. The contribution of precipitation strengthening to yield strength increases from the center to the outer surface of the H13 ingot. The fraction of yield strength attributable to precipitation strengthening is 19.2%–22.4% after tempering. With a decreased contribution of $M_{23}C_6$ to precipitation strengthening, the contribution of MC and M_2C increases from the center to the surface. MC and M_6C have the main effect on precipitation strengthening in H13 after tempering, increasing the strength by up to 74.2%–88.4%.

Acknowledgments: The authors would like to express gratitude and appreciation to the National Natural Science Foundation of China (Grant No. 51274031) titled "Investigation on nanoscale precipitates in hot work die steel and comprehensive strengthening mechanism of steel". This study is also supported by the Beijing Key Laboratory of Special Melting and Preparation of High-end Metal Materials. Foundation Item: Supported by National Natural Science Foundation of China (No. 51274031)

Author Contributions: Angang Ning and Hanjie Guo conceived and designed the experiments; Angang Ning and Wenwen Mao performed the experiments; Angang Ning analyzed the data and wrote the paper; Jing Guo improved the English writing.

Conflicts of Interest: The authors declare no conflict of interest.

References

1. Feng, X.Z. *Mold Steel and Heat Treatment*; Machinery Industry Press: Beijing, China, 1984.
2. Sjöström, J.; Bergström, J. Thermal fatigue testing of chromium martensitic hot-work tool steel after different austenitizing treatments. *J. Mater. Process. Technol.* **2004**, *153–154*, 1089–1096. [CrossRef]
3. Pan, X.H.; Zhu, Z.C. The study of chemical composition and improvement and development for the H13 hot work die & mold steel. *Mold Manuf.* **2006**, *4*, 78–85.
4. Fuchs, K.D. Hot-work tool steels with improved properties for die casting applications, The use of tool steels: Experience research. In Proceedings of the 6th International Tooling Conference, Karlstad, Sweden, 10–13 September 2002; pp. 15–22.
5. Kheirandish, S.; Noorian, A. Effect of niobium on microstructure of cast AISI H13 hot work tool steel. *J. Iron Steel Res. Int.* **2008**, *15*, 61–66. [CrossRef]
6. Ma, D.S.; Zhou, J.; Chen, Z.Z.; Zhang, Z.K.; Chen, Q.A.; Li, D.H. Influence of thermal homogenization treatment on structure and impact toughness of H13 ESR steel. *J. Iron Steel Res. Int.* **2009**, *16*, 56–60. [CrossRef]
7. Hu, X.B.; Li, L. Changes of morphology and composition of carbides in H13 steel after thermal fatigue. *Trans. Mater. Heat Treat.* **2007**, *28*, 82–87.
8. Mebarki, N.; Delagnes, D.; Lamesle, P.; Delmas, F.; Levaillant, C. Relationship between microstructure and mechanical properties of a 5% Cr tempered martensitic tool steel. *Mater. Sci. Eng. A* **2004**, *387–389*, 171–175. [CrossRef]
9. Song, W.W.; Min, Y.A.; Wu, X.C. Study on carbides and their evolution in H13 hot work steel. *Trans. Mater. Heat Treat.* **2009**, *30*, 122–126.
10. Zhang, K.; Yong, Q.L.; Sun, X.J.; Li, Z.D.; Zhao, P.L.; Chen, S.D. Effect of tempering temperature on microstructure and mechanical properties of high Ti microalloyed directly quenched high strength steel. *Acta Metall. Sin.* **2014**, *50*, 913–920.
11. Fu, J.; Li, G.Q.; Mao, X.P.; Fang, K.M. Nanoscale cementite precipitates and comprehensive strengthening mechanism of steel. *Metall. Mater. Trans. A* **2011**, *42A*, 3797–3812. [CrossRef]
12. Mao, W.; Ning, A.; Guo, H. Nanoscale precipitates and comprehensive strengthening mechanism in AISI H13 steel. *Int. J. Miner. Metall. Mater.* **2016**, *23*, 1056–1064. [CrossRef]
13. Zhou, J.; Ma, D.S.; Liu, B.S.; Kang, A.J.; Li, X.Y. Research of band segregation evolution of H13 steel. *J. Iron Steel Res.* **2012**, *24*, 47–57.
14. Li, Z.B. *Electroslag Metallurgy Equipment and Technology*; Metallurgical Industry Press: Beijing, China, 2012.
15. Guo, K.X.; Ye, H.Q.; Wu, Y.K. *Application of Electron Diffraction Pattern in Crystallography*; Science Press: Beijing, China, 1983.
16. Gong, X.M. *Fundamentals and Applications of Phase Transition Theory*; Wuhan University of Technology Press: Wuhan, China, 2004.

17. Ning, A.G.; Guo, H.J.; Chen, X.C.; Sun, X.L. Precipitation behaviors of carbides in H13 steel during ESR, forging and tempering. *J. Univ. Sci. Technol. Beijing* **2014**, *36*, 895–902.
18. Ning, A.G.; Mao, W.W.; Guo, H.J.; Chen, X.C. Precipitation behaviors and strengthening of carbides in H13 steel during quenching. *Chin. J. Process Eng.* **2014**, *14*, 86–92.
19. Ye, D.L.; Hu, J.H. *Practical Inorganic Thermodynamic Data Manual*, 2nd ed.; Metallurgical Industry Press: Beijing, China, 2002.
20. Chen, J.X. *Steelmaking Common Chart Data Manual*, 2nd ed.; Metallurgical Industry Press: Beijing, China, 2010.
21. Yong, Q.L. *The Second Phase of the Steel Materials*; Metallurgical Industry Press: Beijing, China, 2006.
22. Klimiankou, M.; Lindau, R.; Moslang, A. Direct correlation between morphology of $(Fe,Cr)_{23}C_6$ precipitates and impact behavior of ODS steels. *J. Nucl. Mater.* **2007**, *367–370*, 173–178. [CrossRef]
23. Oikawa, T.; Zhang, J.J.; Enomoto, M.; Adachi, Y. Influence of carbide particles on the grain growth of ferrite in an Fe-0.1C-0.09V alloy. *ISIJ Int.* **2013**, *53*, 1245–1252. [CrossRef]
24. Garcia-Mateo, C.; Capdevila, C.; Caballero, F.G.; Andres, C.G.D. Influence of V precipitates on acicula ferrite transformation Part 1: The role of nitrogen. *ISIJ Int.* **2008**, *48*, 1270–1275. [CrossRef]
25. Cui, Z.Q.; Chun, Q.Y. *Metal Science and Heat Treatment*; Machinery Industry Press: Beijing, China, 2007.
26. Seher, R.J.; James, H.M.; Maniar, G.N. *Stereology and Quantitative Metallography, West Conshohocken*; Pellissier, G.E., Purdy, S.M., Eds.; ASTM International: West Conshohocken, PA, USA, 1972; pp. 119–137.
27. Kneissl, A.C.; Garcia, C.I.; Deardo, A.J. *HSLA Steels: Processing, Properties, and Applications*; Geoffrey, T., Zhang, S., Eds.; The Minerals, Metals and Materials Society: San Diego, CA, USA, 1992; pp. 99–102.
28. McCall, J.L.; Boyd, J.E. Proceedings International Metallographic Society, PIMTB. In Proceedings of the 1968 Annual Meeting, Boston, MA, USA, 21 July 1968.

metals

MDPI

Article

Assessment of the Residual Life of Steam Pipeline Material beyond the Computational Working Time

Marek Sroka [1,*], **Adam Zieliński** [2], **Maria Dziuba-Kałuża** [2], **Marek Kremzer** [1], **Magdalena Macek** [1] and **Artur Jasiński** [3]

1 Division of Material Processing Technology, Management and Computer Techniques in Materials Science, Institute of Engineering Materials and Biomaterials, Silesian University of Technology, Konarskiego 18a, 44-100 Gliwice, Poland; marek.kremzer@polsl.pl (M.K.); magdalena.macek@gmail.com (M.M.)
2 Institute for Ferrous Metallurgy, K. Miarki 12-14, 44-100 Gliwice, Poland; azielinski@imz.pl (A.Z.); mkaluza@imz.pl (M.D.-K.)
3 Energopomiar, J. Sowińskiego 3, 44-100 Gliwice, Poland; ajasinski@energopomiar.com.pl
* Correspondence: marek.sroka@polsl.pl; Tel.: +48-32-237-18-47

Academic Editor: Robert Tuttle
Received: 19 December 2016; Accepted: 28 February 2017; Published: 6 March 2017

Abstract: This paper presents the evaluation of durability for the material of repair welded joints made from (13HMF) 14MoV6-3 steel after long-term service, and from material in the as-received condition and after long-term service. Microstructure examinations using a scanning electron microscope, hardness measurements and creep tests of the basic material and welded joints of these steels were carried out. These tests enabled the time of further safe service of the examined repair welded joints to be determined in relation to the residual life of the materials. The evaluation of residual life and disposable life, and thus the estimation and determination of the time of safe service, is of great importance for the operation of components beyond the design service life. The obtained test results are part of the materials' characteristics developed by the Institute for Ferrous Metallurgy for steels and welded joints made from these steels to work under creep conditions.

Keywords: creep; degradation; welded joint; Cr-Mo-V steel; residual life

1. Introduction

Pressure components working at an elevated temperature are designed for a definitive working time. This time is based on temporary creep strength used for calculations. It is 100,000 h for old units, while, for those with supercritical working parameters designed and operated at present, it is 200,000 h. Most of the units operated in Poland have significantly exceeded the design service life of 100,000 h, reaching the actual operation time of more than 200,000 h. The extension of the operation time beyond the design one of 100,000 h is made by using the calculation methods based on data concerning the average temporary creep strength for 200,000 h and positive results of comprehensive investigations and diagnostic measurements. Usually, the critical components in the pressure part of boilers and turbines are subject to these investigations and evaluation. Out of these components, those working above the limit temperature, i.e., under creep conditions, are crucial (Figure 1).

The above-mentioned operation of steam boilers for much more than 200,000 h requires a new approach in the materials diagnostics. For safety reasons, a particularly important issue to be solved is creep strength of the welded joints of the steam pipelines working under creep conditions [1–6].

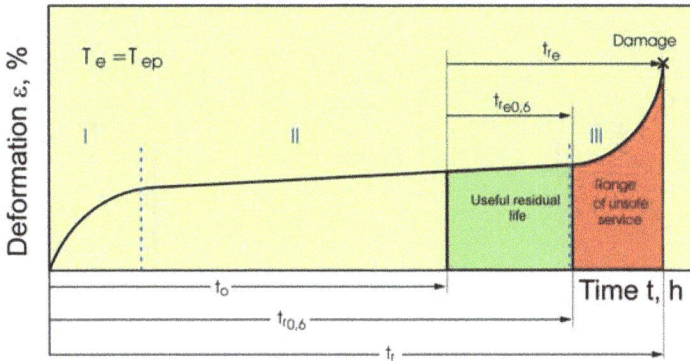

Figure 1. Schematic approach of the definition of residual and disposable durability.

In the evaluation of these components, it is important and necessary to evaluate the condition of their material [7–13]. This evaluation is carried out based on non-destructive or destructive materials tests. In the case of components working for more than 150,000 h, the estimate of residual life by non-destructive testing is not sufficient. It needs to be determined based on destructive tests performed on a sampled representative test specimen [14].

As part of the diagnostics, not only the basic material of the operated component but also the material of welded joints is subject to evaluation [15]. It is necessary to evaluate the condition of the material of welded joints to determine the component's ability to carry the required operating loads during its further service. If there is a need for repair to or replacement of a part or the entire component with a new one, the ability of the basic material under operation to carry out such a repair or replacement must be determined. When the condition of such material after service allows a repair to be made, it is necessary to develop a technology for its performance. The repair welded joint is defined as a new weld made to join a material after service with another material after service, and also to join a new material with a material after service (for replacement of a part of a structural component with a new one). Such repair welded joints are made during the renovation and modernisation works on pressure elements including, but not limited to, steam pipelines.

The subject-matter of the investigations, including the materials and their repair welded joints after long-term service made with materials after service or new materials, is an important issue overseen by the Institute for Ferrous Metallurgy. The selected results of investigations with regard to condition evaluation of the material of repair welded joints are the subject of this study. They mainly concern the elements of primary steam pipelines made from 13HMF (14MoV6-3) steel, which in the majority of Polish power plants exceeded the design service life of 100,000 h. Therefore, an important issue in the evaluation of the safe operation of these devices is to provide a numerical value of the time of their further operation and determine creep strength not for the material pipeline itself, but rather for the welded joints of these materials made during repairs.

2. Material for Investigations

The material for investigations was tested specimens of repair welded joints made from (13HMF) 14MoV6-3 steel after long-term service, and of material in the as-received condition and after long-term service. The summary of the material for investigations, including their steel grades, geometrical dimensions, working parameters, the current time of service and macrophotography of the test specimen is presented in Table 1.

Table 1. Material for investigations.

Repair Welded Joint Made from Pipeline Sections after Long-Term Service, Marked ZS1	
Steel grade: 14MoV6-3 Service time: material after 169,000 h service, marked ZS1	◪
Dimensions: 273 × 32 ($D_n \times g_n$)	
Working parameters of sections after service $t_0 = 538\ °C$; $p_0 = 13.0\ Mpa$	
Material for investigations: repair circumferential welded joint made under industrial conditions	
Repair Welded Joint Made from Pipeline Sections in the As-Received Condition (before Service) and after Long-Term Service, Marked ZS2	
Steel grade: 14MoV6-3 Service time: material in the as-received condition/material after 169,000 h service, marked ZS2	▬
Dimensions: 273 × 32 ($D_n \times g_n$)	
Working parameters of sections after service $t_0 = 538\ °C$; $p_0 = 13.0\ Mpa$	
Material for investigations: repair circumferential welded joint made under industrial conditions	

The check analysis of chemical composition of the examined materials of repair welded joints from low-alloy Cr-Mo-V steels after long-term service and a material in the as-received condition and after long-term service was performed in accordance with the following procedures: 3/CHEM,4 "Determination of C, Mn, Si, P, S, Cr, Ni, Cu, Mo, V, Ti, Al, Nb, B and Sn contents in low- and medium-alloy carbon steel by the spark optical emission spectrometry method using natural standards" with the optical emission spectrometer Magellan Q8 by Bruker, Germany. For the chemical composition of the examined steels with regard to the requirements of standard specification [16], see Table 2.

Table 2. Check analysis of chemical composition.

Grade of Material	Content of Elements (%)									
	C	Mn	Si	P	S	Cu	Cr	Ni	Mo	Others
14MoV6-3 according to [16]	0.10 0.18	0.40 0.70	0.15 0.35	max 0.04	max 0.04	max 0.25	0.30 0.60	max 0.30	0.50 0.65	V 0.22–0.35 Al max 0.02
14MoV6-3 169,000 h service Designation ZS1-PM1	0.16	0.58	0.35	0.017	0.018	0.20	0.46	0.23	0.62	V 0.29 Al 0.026
14MoV6-3 169,000 h service Designation ZS1-PM2	0.16	0.58	0.35	0.017	0.020	0.20	0.46	0.23	0.63	V 0.29 Al 0.024
14MoV6-3 in the as-received condition Designation ZS2-PM1	0.17	0.51	0.22	0.008	0.006	0.11	0.53	0.11	0.52	V 0.26 Al 0.013
14MoV6-3 169,000 h service Designation ZS2-PM2	0.16	0.59	0.34	0.018	0.018	0.21	0.48	0.22	0.59	V 0.28 Al 0.023

The analysis results of the check of chemical composition show that the materials of the examined test specimens of repair welded joints meet the requirements of the standard with regard to the chemical composition of the examined steel grade, i.e., 13HMF (14MoV6-3) [16].

3. Research Scope and Methodology

As part of the investigations, the properties of the material of the repair welded joints were evaluated. In the evaluation of the material condition and the level of required utility properties for repair welded joints, the following was subject to investigation:

- The microstructure of repair circumferential welded joints of components in the pressure part of a boiler was examined, with tests were carried out using a scanning electron microscope (SEM, FEI, Hillsboro, OR, USA) Inspect F on nital-etched metallographic microsections;
- Analysis of precipitation processes was carried out using X-ray analysis of isolated carbides, with the use of a Empyrean diffractometer (XRD, Panalytical, Almelo, Netherlands) and selective diffraction of electrons;
- The level of hardness for individual joint components and its nature in the course from the parent material through the heat-affected zone and weld was obtained, taken by Vickers method using a Future—Tech FM—7 machine (Kawasaki, Japan) at the indenter load of 10 kG;
- The material's residual life was determined based on abridged creep tests at a constant test stress corresponding to the operating one $\sigma_b = \sigma_r = $ const and at a constant test temperature T_b for each test. The tests were performed using Instron single-sample machines (Norwood, MA, USA) with an accuracy of temperature during the test of $\pm 1\,^{\circ}C$.

The obtained results of the investigations are part of the study, which is under preparation for verification of the proposed method for evaluating and predicting the time of further safe service of homogeneous circumferential welded joints from low-alloy Cr-Mo-V steels. In the case of its positive result, this test method will be used in materials diagnostics to be performed for the power industry.

4. Results

4.1. Microstructure Investigations: Structure of Steel in the As-Received Condition

The microstructure of 14MoV6-3 steel in the as-received condition is a mixture of bainite and ferrite, sometimes with a slight amount of pearlite. Moreover, very fine MC carbide precipitates that occur inside the ferrite grains are observed within the structure. In the bainite areas, there are small spheroidal cementite precipitates, while in the pearlite colonies, cementite lamellas exist. An example of the characteristic microstructure of 14MoV6-3 steel in the as-received condition is shown in Figure 2.

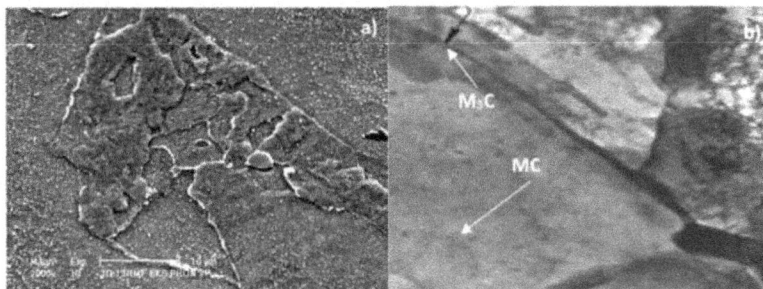

Figure 2. Structure image of 14MoV6-3 ferritic-bainitic steel in the as-received condition (**a**) SEM; (**b**) TEM.

4.2. Evaluation of the Microstructure of Repair Welded Joints

The investigations of microstructure were carried out on metallographic microsections. The microsections were made on the cross-section of test specimens of the examined components in the area of the weld and prepared by mechanical grinding and polishing as well as etching. The observations were performed with magnifications of 500 to 5000×. For the repair welded joint made from materials after long-term service, marked ZS1, the results of the investigations are presented as photographs of the microstructure of the materials of the circumferential welded joint components, in particular: parent material, heat affected zone of the joint and weld, respectively, in Figure 3.

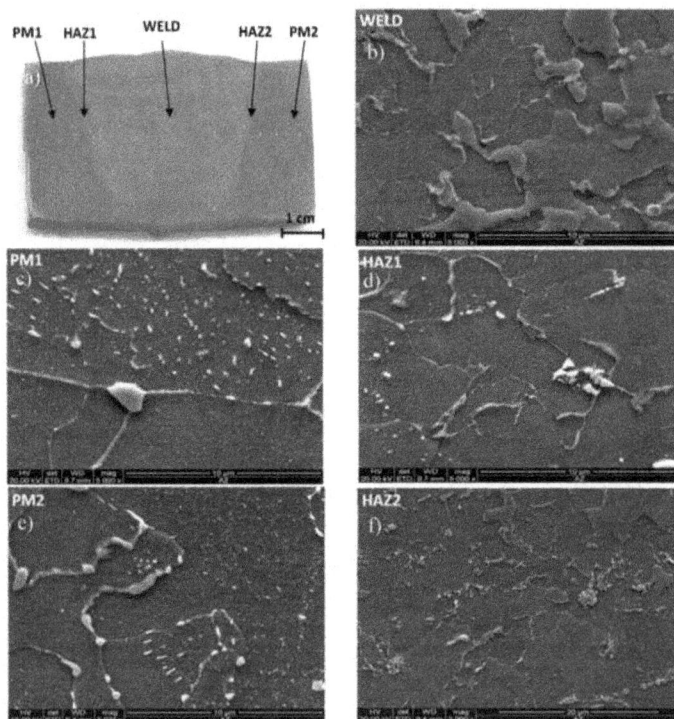

Figure 3. Structure of the material of components of the repair welded joint marked ZS1 made from 14MoV6-3 steel after 169,000 h service; microstructure investigation locations (**a**): parent material marked (**c**) PM1, (**e**) PM2; heat affected zone marked (**d**) HAZ1, (**f**) HAZ2; weld marked (**b**) WELD.

The results of the microstructure investigations for the components of the repair welded joint made from a material in the as-received condition and after long-term service, marked ZS2, in particular: parent material, heat affected zone and weld, are provided in Figure 4.

The classification of the microstructure including the evaluation and exhaustion extent t_e/t_r estimated based on the Institute for Ferrous Metallurgy's own classification [1] is provided in Table 3.

The parent material of the welded joints marked ZS1 (PM1, PM2) and ZS2 (PM2) after service was characterised by a ferritic microstructure with partially coagulated bainite areas. At the ferrite grain boundaries, there are precipitates of different size, mostly fine ones, whereas inside the ferrite grains, mostly very fine precipitates distributed evenly within the structure were observed.

The microstructure of the parent material of the welded joint marked ZS2 (PM1) in the as-received condition was characterised by ferritic-bainitic microstructure, which is typical for this type of steel.

In the microstructure of the materials of the examined repair welded joints, no discontinuities or microcracks, nor initiation of internal damage processes due to creep, were observed.

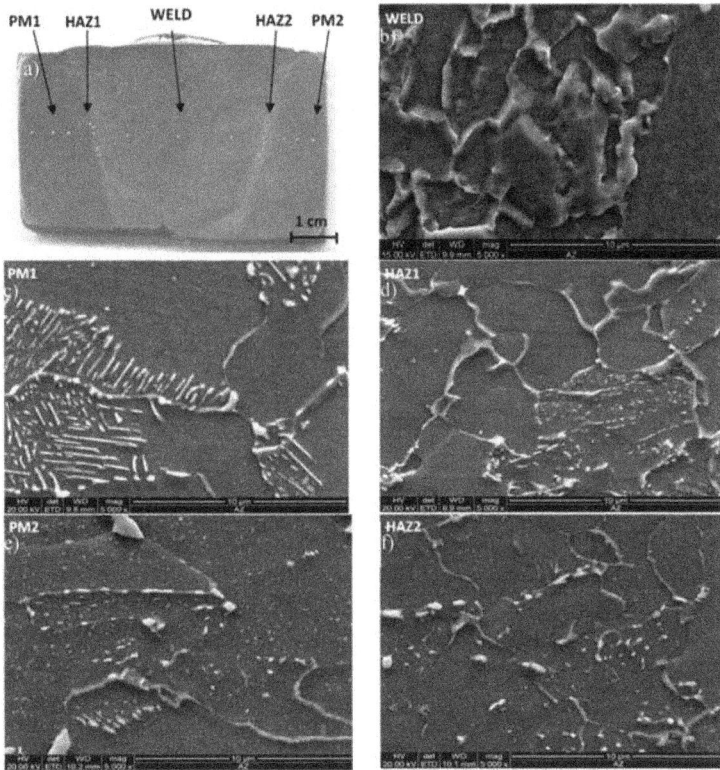

Figure 4. Structure of the material of components of the repair welded joint marked ZS2 made from 14MoV6-3 steel in the as-received condition and after 169,000 h service; investigation performance locations (**a**): parent material marked (**c**) PM1, (**e**) PM2; heat affected zone marked (**d**) HAZ1, (**f**) HAZ2; weld marked (**b**) WELD.

Table 3. Review of the results of microstructure investigations and hardness tests on the material of components of repair welded joints.

Material for Investigations	Figure No.	Description of Microstructure Material Condition—Exhaustion Degree	Hardness HV10
Repair welded joint Designation ZS1	PM1	Ferritic-bainitic structure. No discontinuities or micro-cracks are observed in the structure.	173
	PM2	Bainitic areas: class I/II, precipitates: class A. Damaging processes: class 0. CLASS 2, EXHAUSTION DEGREE: approx. 0.3 ÷ 0.4.	169
	HAZ1	Figure 3	247
	WELD	No discontinuities or micro-cracks are found in the structure.	240
	HAZ2		247
Repair welded joint Designation ZS2	PM1	Ferritic-bainitic structure. No discontinuities or micro-cracks are observed in the structure. Bainitic areas: class 0; precipitates: class 0; Damaging processes: class 0. MATERIAL CONDITION: CLASS 0; EXHAUSTION DEGREE: ~0.	160
	PM2	Ferritic-bainitic structure. No discontinuities or micro-cracks are observed in the structure. Bainitic areas: class I/II, precipitates: class A. Damaging processes: class 0. CLASS 2, EXHAUSTION DEGREE: approx. 0.3 ÷ 0.4.	168
	HAZ1	Figure 4	247
	WELD	No discontinuities or micro-cracks are found in the structure.	242
	HAZ2		168

4.3. X-ray Analysis of Phase Composition of Precipitated Carbide Isolates

As a result of dissolving the matrix of the material of the examined test specimens of repair welded joints by the electrolytic method, the existing carbides were isolated. The X-ray phase analysis was carried out on the obtained carbide isolate, and the existing carbides were identified. The obtained results of the investigations of the material of test repair welded joints are summarised in Table 4.

Table 4. Phase composition of carbides in repair welded joints.

Material Condition	Phase Composition of Carbides	Precipitation Sequence
As-received condition 14MoV6-3 steel	M_3C MC	$M_3C + MC$
14MoV6-3 steel 169,000 h service Designation ZS1-PM1	Isovit $Cr_{23}C_6$—main phase Cementite Fe_3C VC	$M_{23}C_{6\ main\_ph.} + M_3C_{av} + MC_{nw}$
14MoV6-3 steel 169,000 h service Designation ZS2-PM2	Isovit $Cr_{23}C_6$—main phase Cementite Fe_3C VC	$M_{23}C_{6\ main\_ph.} + M_3C_{av} + MC_{nw}$

The type and contribution of the revealed precipitates correspond to the exhaustion degree estimated based on the microstructure image of the examined materials of repair welded joints from low-alloy steels after operation beyond the design service time (Table 3).

The sequences of carbides (Table 4) within the examined materials formulated based on the X-ray diffraction analysis of electrolytically isolated carbide deposits confirm the class of microstructure as determined based on the analysis of its observed images.

4.4. Hardness Evaluation of Repair Welded Joints

Hardness measurement was taken by Vickers HV10 method on the transverse metallographic microsection of the examined repair circumferential welded joints made from materials after long-term service, marked ZS1, and material in the as-received condition and after long-term service marked ZS2. The HV10-hardness measurement results against the background of the macro photograph showing a cross-section of the repair welded joints in the examined test specimens are presented in Figure 5.

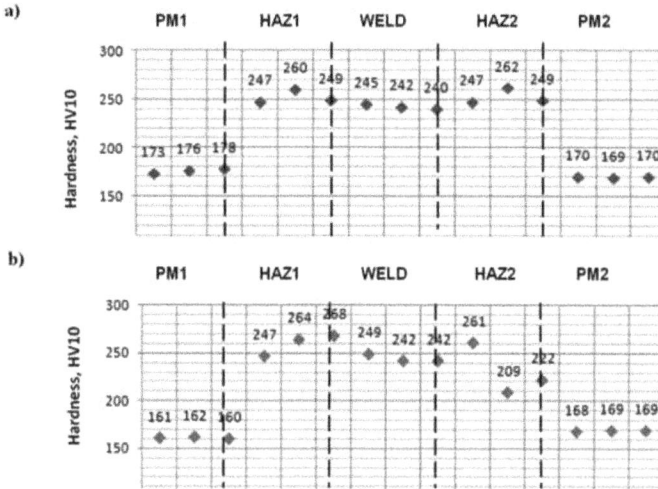

Figure 5. Distribution of the results of HV10 hardness measured on transverse microsections of the repair welded joint made from 14MoV6-3 steel. Test location—transverse microsection: (**a**) ZS1; (**b**) ZS2.

Hardness for all the examined components of repair welded joints is lower than the maximum permitted one, which is 350 HV10 for joints in the as-received condition and ranges from 160 to 179 HV10 for the parent material, from 209 to 268 for the heat-affected zone and from 240 to 249 HV10 for the weld material. This suggests that the examined welded joints were properly heat-treated after welding and will be able to transfer the required considerable loads, including those that occur during water pressure tests, shut-downs and start-ups. Hardness measurements have also shown no sudden changes when passing through the individual zones of the joint. Hardness for the 14MoV6 steel repair circumferential welded joint made from materials after 169,000 h service is, on average, 173 HV10 for the parent material, while in the weld it increases up to 262 HV10. Hardness for the 14MoV6 steel repair circumferential welded joint made from a material in the as-received condition and after 169,000 h service is, on average, 165 HV10 for the parent material, while in the weld it increases up to 268 HV10.

4.5. Abridged Creep Tests

The abridged creep tests were carried out for five test temperature levels ranging between 600 °C and 680 °C at 20 °C intervals with constant test stress σ_b = const corresponding to the operating one, which allows for obtaining test results within several months. This provides a good estimate of residual life t_{re} as it was verified in [17,18].

The method used to reduce the duration of creep tests involves accelerating the creep process by increasing the test temperature T_b well over the temperature level T_e suitable for operation in the samples maintained at a constant test stress corresponding to the operating one $\sigma_b = \sigma_r$ = const. They allow for the plotting of a straight line inclined at the time to rupture the t_r axis. The residual life is determined by extrapolation of the obtained straight line towards a lower temperature corresponding to the operating one T_e.

The results of creep tests for the examined repair welded joints marked ZS1 and ZS2 are summarised in Table 5 and presented in comparative graphs (Figure 6) as $\log t_z = f(T_b)$ at $\sigma_b = $ const, where t_r is the time to rupture in the creep test.

Table 5. Results of abridged creep tests.

Test Specimen Designation	Working Parameters		Test Stress σ_b (Mpa)	Test Temperature, T_b (°C)				
	Pressure P_r (MPa)	Temperature T_r (°C)		600	620	640	660	680
				Time to Rupture, t_r (h)				
Repair welded joint made from materials after 169,000 h service Designation ZS1	-	-	50	(3127)	1197	559	234	120
Repair welded joint made from material in the as-received condition and material after 169,000 h service Designation ZS2	-	-		(3161)	1178	822	179	103
Parent material after 169,000 h service Designation PM	13.0	538		(286)	(1365)	559	429	196
Repair welded joint made from materials after 169,000 h service Designation ZS1	-	-	55	2834	672	373	189	97
Repair welded joint made from material in the as-received condition and material after 169,000 h service Designation ZS2	-	-		(2592)	1297	481	191	84

(-)—tests in progress.

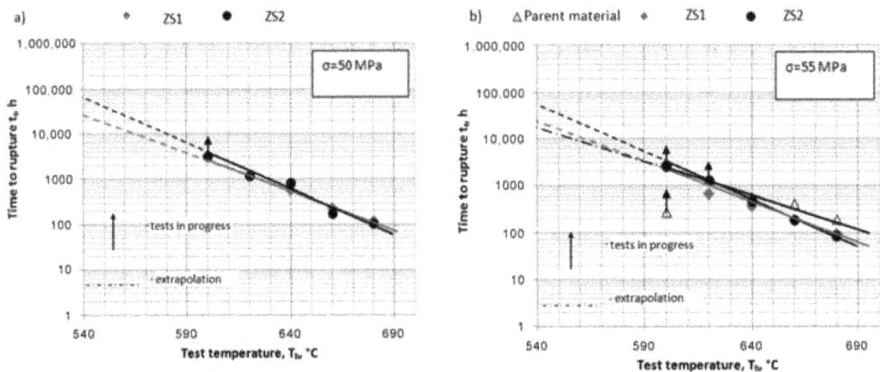

Figure 6. Result of abridged creep tests on the examined ZS1 and ZS2 repair joints and parent material in the form of $\log t_{re} = f(T_b)$ for the adopted stress level of further service (**a**) $b = 50$ MPa; (**b**) $b = 55$ MPa.

The comparison of the results of abridged creep tests in the form of $\log t_r = f(T_b)$ at $\sigma_b \approx \sigma_r = 50$ MPa for the repair welded joint made from 14MoV6-3 steel after 169,000 h service and the repair welded joint made from 14MoV6-3 material in the as-received condition and 14MoV6-3 material after 169,000 h service is presented in Figure 6a.

The comparison of the results of abridged creep tests in the form of $\log t_r = f(T_b)$ at $\sigma_b \approx \sigma_r = 55$ MPa for the parent material of 14MoV6-3 after 169,000 h service and the repair welded joint made from materials of 13HMF (14MoV6-3) after 169,000 h service and the repair welded joint made from 14MoV6-3 steel in the as-received condition and 14MoV6-3 steel after 169,000 h service is presented in Figure 6b.

On the basis of the previously completed creep tests, based on the extrapolation method used, the residual life (interpreted as the time to failure) was determined and the disposable residual life (being the safe time of service, which is about 0.6 of the residual life, Figure 1) was estimated as the safe time of service for the examined parent material, repair welded joint made from materials after long-term service and repair welded joint made from material in the as-received condition and after long-term service. The obtained results of extrapolation based on creep tests are summarised in Table 6 for two values of stress—50 and 55 MPa.

Table 6. Residual life determined and disposable residual life estimated by abridged creep tests of the parent material, repair welded joint made from materials after long-term service and repair welded joint made from material in the as-received condition and after long-term service.

Test Specimen Designation	Adopted Operating Stress σ_r (MPa)	Adopted Further Operation Temperature T_r (°C)	Estimated Life Time (h)	
			Residual t_{re}	Disposable Residual Life t_b (about 0.6 t_{re})
Joint from materials after long-term service Designation ZS1	50	538	25,000	15,000
Joint from material in the as-received condition and material after long-term service Designation ZS2			60,000	36,000
Native material Designation PM1	55		20,000	12,000
Joint from materials after long-term service Designation ZS1			23,000	13,800
Joint from material in the as-received condition and material after long-term service Designation ZS2			58,000	34,800

The residual life determined by extrapolation of creep results obtained in abridged tests, for the temperature of further service and the adopted stress level of further operation of the parent materials and repair welded joints, has allowed the disposable residual life, which is the time of further safe service, to be determined.

The residual life t_{re} determined for the adopted stress level of 50 MPa for the repair welded joint of the materials after service, marked ZS1, is 25,000 h and its estimated disposable life t_b is 15,000 h (Figure 6, Table 4), while the residual life t_{re} determined for the repair welded joint of the material after service and the material in the as-received condition, marked ZS2, is 60,000 h and the estimated disposable life is 36,000 h.

The residual life t_{re} determined for the adopted stress level of 55 MPa for the repair welded joint of the materials after service marked ZS1, is 23,000 h and the estimated disposable life t_b is approx. 14,000 h, while the residual life t_{re} determined for the repair welded joint of the material after service and the material in the as-received condition, marked ZS2, is 58,000 h and the estimated disposable life is approx. 35,000 h. For the parent material after service marked PM1, the residual life is 20,000 h, and the estimated disposable life is t_b 12,000 h.

The time of further safe service of the examined new repair welded joints may be assumed to be 15,000 h for the ZS1 joint and 36,000 h for the ZS2 joint at the further service stress σ_e = 50 MPa, while for the adopted further service stress σ_e = 55 MPa the time of further safe service of the examined new repair welded joints may be assumed to be approximately 14,000 h for the ZS1 joint and approximately 35,000 h for the ZS2 joint.

5. Conclusions

1. The set of destructive materials tests presented in this paper allows for the evaluation of material condition and determination of suitability for service of repair. It is of particular importance for the operation of steam pipelines beyond the design service time.

2. The evaluation of the material condition of repair welded joints is made based on a comprehensive summary of the results of investigations on mechanical properties, microstructure and abridged creep tests. These results are in turn a part of the database of the materials' characteristics for steels and their welded joints with materials showing varying degrees of degradation. This database is used in diagnostic tests for pressure parts of boiler elements.

3. The quantitative dimension of suitability for service of the material of repair welded joints is achieved by extrapolating the straight line obtained in abridged creep tests from $\log t_r = f(T_b)$ at σ_b = const towards the temperature of assumed operation, which allows residual life t_{re} and disposable residual life t_b to be determined for the working temperature.

4. The knowledge of the share of disposable residual life t_{be} in residual life t_{re} (t_{be}/t_{re}) allows the safe time of service of the examined joints to be determined for the required performance parameters.

5. The examined repair welded joints are suitable for operation for a limited time resulting from the disposable residual life determined for defined temperature and stress parameters of further service.

The completed tests of the material of steam pipeline and welded joints have shown that long-term operation beyond the design service time does not disqualify the material from service. It has been demonstrated that the modernisation and repair works carried out on the steam pipeline materials by making welded joints show lower creep strength than the basic material. The lower strength of repair welded joints in relation to the parent material should be taken into account in design calculations while extending the service time beyond the design service life.

It has also been demonstrated that, in contrast to the microstructural investigations and the basic investigations of mechanical properties, the abridged creep tests allow the real determination of the time of further safe operation of the elements of power equipment working beyond the design service life to be obtained.

The analysis of the research results of abridged creep tests shows that, independently, of the values of the stress, creep resistance of repair welded joint ZS2 is twice as high as welded joints marked ZS1. This difference is probably related to the higher creep resistance of the parent material resulting in a higher creep resistance of joints marked ZS2.

Acknowledgments: The publication was partially financed by the statutory grant from Faculty of Mechanical Engineering, the Silesian University of Technology for the year 2016. The results in this publication were obtained as a part of research co-financed by the National Centre for Research and Development under contract PBS3/B5/42/2015—Project: "Methodology, evaluation and forecast of operation beyond the analytical operation of welded joints in pressure elements of power boilers beyond the design work time".

Author Contributions: M.S., A.Z. and M.D.-K. conceived and designed the experiments; M.D.-K. and A.Z. performed the experiments; M.M., A.J. and M.K. analysed the data; M.S. wrote the paper.

Conflicts of Interest: The authors declare no conflict of interest.

References

1. Dobrzański, J. Materials science interpretation of the life of steels for power plants. *Open Access Libr.* **2011**, *3*, 1–228.
2. Golański, G.; Zieliński, A.; Słania, J.; Jasak, J. Mechanical properties of Vm12 steel after 30,000 hrs of ageing at 600 °C temperature. *Arch. Metall. Mater.* **2014**, *59*, 1351–1354.
3. Zieliński, A.; Golański, G. The influence of repair welded joint on the life of steam pipeline made of Cr-Mo steel serviced beyond the calculated working time. *Arch. Metall. Mater.* **2015**, *60*, 1045–1049. [CrossRef]
4. Cao, J.; Gong, Y.; Yang, Z.G.; Luo, X.M.; Gu, F.M.; Hu, Z.F. Creep fracture behavior of dissimilar weld joints between T92 martensitic and HR3C austenitic steels. *Int. J. Press. Vessels Pip.* **2011**, *88*, 94–98. [CrossRef]
5. Laha, K. Integrity Assessment of Similar and Dissimilar Fusion Welded Joints of Cr-Mo-W Ferritic Steels under Creep Condition. *Procedia Eng.* **2014**, *86*, 195–202. [CrossRef]
6. Sawada, K.; Tabuchi, M.; Hongo, H.; Watanabe, T.; Kimura, K. Z-Phase formation in welded joints of high chromium ferritic steels after long-term creep. *Mater. Charact.* **2008**, *59*, 1161–1167. [CrossRef]
7. Zieliński, A.; Sroka, M.; Miczka, M.; Śliwa, A. Forecasting the particle diameter size distribution in P92 (X10CrWMoVNb9-2) steel after long-term ageing at 600 and 650 °C. *Arch. Metall. Mater.* **2016**, *61*, 753–760.
8. Golański, G.; Zieliński, A.; Zielińska-Lipiec, A. Degradation of microstructure and mechanical properties in martensitic cast steel after ageing. *Materialwiss. Werkst.* **2015**, *46*, 248–255. [CrossRef]
9. Zieliński, A.; Golański, G.; Sroka, M.; Tański, T. Influence of long-term service on microstructure, mechanical properties, and service life of HCM12A steel. *Mater. High Temp.* **2016**, *33*, 24–32. [CrossRef]
10. Zieliński, A.; Golański, G.; Sroka, M.; Dobrzański, J. Estimation of long-term creep strength in austenitic power plant steels. *Mater. Sci. Technol.-Lond.* **2016**, *32*, 780–785. [CrossRef]
11. Dobrzański, J.; Hernas, A.; Moskal, G. Microstructural degradation in power plant steels. In *Power Plant Life Management and Performance Improvement*; Oakey, J.E., Ed.; Woodhead Publishing Limited: Sawston, UK, 2011.
12. Falat, L.; Svoboda, M.; Výrostková, A.; Petryshynets, I.; Sopko, M. Microstructure and creep characteristics of dissimilar T91/TP316H martensitic/austenitic welded joint with Ni-based weld metal. *Mater. Charact.* **2012**, *72*, 15–23. [CrossRef]
13. Kim, M.-Y.; Kwak, S.-C.; Choi, I.-S.; Lee, Y.-K.; Suh, J.-Y.; Fleury, E.; Jung, W.-S.; Son, T.-H. High-temperature tensile and creep deformation of cross-weld specimens of weld joint between T92 martensitic and Super304H austenitic steels. *Mater. Charact.* **2014**, *97*, 161–168. [CrossRef]
14. Sroka, M.; Zieliński, A.; Mikuła, J. The service life of the repair welded joint of Cr-Mo/Cr-Mo-V. *Arch. Metall. Mater.* **2016**, *61*, 969–974. [CrossRef]
15. Zieliński, A.; Golański, G.; Sroka, M. Influence of long-term ageing on the microstructure and mechanical properties of T24 steel. *Mater. Sci. Eng. A* **2017**, *682*, 664–672.
16. PN-EN 10216-2. *Pipes for Pressure Purposes with Specified Elevated Temperature Properties*; Boiler Tubes: Brussels, Belgium, 2014.
17. Zieliński, A.; Dobrzański, J.; Purzyńska, H.; Golański, G. Properties, structure and creep resistance of austenitic steel Super 304H. *Mater. Test.* **2015**, *57*, 859–865. [CrossRef]
18. Zieliński, A.; Sroka, M.; Hernas, A.; Kremzer, M. The effect of long-term impact of elevated temperature on changes in microstructure and mechanical properties of HR3C steel. *Arch. Metall. Mater.* **2016**, *61*, 761–766. [CrossRef]

metals

MDPI

Article

Effect of Heat Treatment on the Microstructure and Mechanical Properties of Nitrogen-Alloyed High-Mn Austenitic Hot Work Die Steel

Yi Zhang, Jing Li *, Cheng-Bin Shi, Yong-Feng Qi and Qin-Tian Zhu

State Key Laboratory of Advanced Metallurgy, University of Science and Technology Beijing (USTB),
Beijing 100083, China; 15801455355@163.com (Y.Z.); shicb09@163.com (C.-B.S.); yongfeng_qi@126.com (Y.-F.Q.);
zqtustb@163.com (Q.-T.Z.)
* Correspondence: lijing@ustb.edu.cn; Tel.: +86-189-1108-2675

Academic Editor: Robert Tuttle
Received: 27 January 2017; Accepted: 10 March 2017; Published: 14 March 2017

Abstract: In view of the requirements for mechanical properties and service life above 650 °C, a high-Mn austenitic hot work die steel, instead of traditional martensitic hot work die steel such as H13, was developed in the present study. The effect of heat treatment on the microstructure and mechanical properties of the newly developed work die steel was studied. The results show that the microstructure of the high-Mn as-cast electroslag remelting (ESR) ingot is composed of γ-Fe, V(C,N), and Mo_2C. V(C,N) is an irregular multilateral strip or slice shape with severe angles. Most eutectic Mo_2C carbides are lamellar fish-skeleton-like, except for a few that are rod-shaped. With increasing solid solution time and temperature, the increased hardness caused by solid solution strengthening exceeds the effect of decreased hardness caused by grain size growth, but this trend is reversed later. As a result, the hardness of specimens after various solid solution heat treatments increases first and then decreases. The optimal combination of hardness and austenitic grain size can be obtained by soaking for 2 h at 1170 °C. The maximum Rockwell hardness (HRC) is 47.24 HRC, and the corresponding austenite average grain size is 58.4 μm. When the solid solution time is 3 h at 1230 °C, bimodality presented in the histogram of the austenite grain size as a result of further progress in secondary recrystallization. Compared with the single-stage aging, the maximum impact energy of the specimen after two-stage aging heat treatment was reached at 16.2 J and increased by 29.6%, while the hardness decreased by 1–2 HRC. After two-stage aging heat treatment, the hardness of steel reached the requirements of superior grade H13, and the maximum impact energy was 19.6% higher than that of superior grade H13, as specified in NADCA#207-2003.

Keywords: austenite hot work die steel; heat treatment; microstructure; mechanical properties; nitrogen

1. Introduction

In recent years, with the development of manufacturing industry and advanced metal materials, the working conditions of tools and dies become increasingly severer. The problem of raising service temperatures of high-load tools for hot deformation of metals and extending their service life will remain current for as long as this kind of process exists in metal treatment. This problem is especially difficult to solve when the working surface of the tool is heated to 700–900 °C or higher temperatures. Current commercial martensitic die steels such as AISI H13 and AISI H21—which are widely used for hot extrusion dies, forging dies, and casting dies—have a limited working capacity above a working temperature of 650 °C because of the decreasing strength [1]. The impossibility to overcome such a barrier is explainable by the occurrence of a polymorphic $\alpha \rightarrow \gamma$ transformation in these steels, which limits their service temperature [2]. It would be natural to assume that die materials—which do not possess the mentioned barrier within the service temperature range and have a strength exceeding that of highly heat-resistant martensitic steels at a temperature exceeding 700 °C—should have an austenitic structure or belong to high-temperature alloys. Due to the high production cost of high-temperature alloys, highly heat-resistant austenitic steels may have good research and application prospects.

In contrast with martensitic hot work die steel, austenitic hot work die steel has no matrix phase transformation and is strengthened by precipitation hardening with various intermetallics, carbides, and nitrides [3]. The heat treatment gives austenitic steel excellent strength at elevated temperatures. In particular, when the temperature exceeds 650 °C, the strength of austenitic hot work die steel is much higher than that of martensitic type. Another effective method for increasing steel quality is alloying with nitrogen. It has been pointed out that the addition of an appropriate amount of nitrogen can control the amount of carbonitrides and the austenite grain size to improve the mechanical properties [4]. The addition of nitrogen in steel, even in small quantities, makes it possible to obtain highly complex properties, such as strength, impact elasticity, and corrosion resistance. In addition, nitrogen is an effective and economic substitute for some expensive elements: nickel, manganese, molybdenum, and tungsten. According to these characteristics, high-nitrogen steels, especially high-nitrogen austenitic stainless steels, have been studied intensively in recent years [5–7]. In addition, it also reported that the effects of nitrogen element increase the impact toughness of steel and high-temperature strength of hot work die steels [8–10].

The current commercial austenitic hot work die steels, such as $Cr_{14}Ni_{25}Co_2V$, $14Ni_{14}W_2Mo$, $5Mn_{15}Cr_8Ni_{25}Mo_3V$, $Mn_{10}Cr_8Ni_{10}Mo_3V$, and $Mn_{15}Cr_2Al_3V_2WMo$, contain a large amount of alloying elements, leading to high steel production costs. In the present work, a new nitrogen-alloyed Fe–Mn–Cr austenitic hot work die steel was developed by reducing Cr, Ni, and Mo contents and adding a certain amount of nitrogen. The studies on the development of austenitic hot work die steel serving above the temperature of 650 °C have been rarely reported. Zhang et al. [11] reported that high manganese–vanadium austenitic steel, which is used for copper alloyed hot press die, has a great future in the development of hot work die steel used at a high temperature over 700 °C. Grabovskii et al. [3] also reported that new austenitic alloys KhN35V6TYu (ÉK39) and KhN30VMYu (ÉK40) based on the Fe–Cr–Ni system can be recommended for manufacturing tools for hot deformation of metals serving at a temperature exceeding 700 °C. In this study, the effect of heat treatment on the microstructures and mechanical properties in nitrogen-microalloyed high-Mn austenitic hot work die steel (HMAS-N) were examined first. An alternative hot work die steel—which is used for dies, needles, and washers—for the hot pressing of copper alloys is expected to serve above 700 °C.

2. Experimental Section

2.1. Experimental Materials

HMAS-N steel was melted in a vacuum-induction furnace. Nitrogen-containing ferrochromium alloy was added in the smelting process. After vacuum-induction, the liquid steel was cast into the ingots of 120 mm diameter, which were thereafter used as consumable electrodes in electroslag remelting (ESR) experiments. The chemical composition of the steel is shown in Table 1. The experimental process includes forging, solid solution heat treatment (SSHT), and aging heat treatment. The as-cast ESR ingots were held at 1140 °C for 360 min and forged. The forging start and finish temperatures were $T_s = 1200$ °C and $T_f = 980$ °C. The ESR ingot was forged into a rod of Ø160 mm × 170 mm. For solid solution heat treatment and aging heat treatment, the specimens taken from forged sample were heat treated in an electric resistance furnace.

Table 1. Chemical composition of tested steel (mass %).

C	Mn	Cr	Mo	V	Si	P	S	N	Fe
0.56	14.5	3.192	1.641	1.723	0.52	0.015	0.029	0.15	Bal.

2.2. Microstructure and Precipitates of As-Cast ESR Ingot

Thermo-Calc software (Thermo-Calc Software Inc., Solna, Sweden) was used to investigate phase equilibrium diagrams and identify the chemistry of various precipitates at different temperatures. The specimens were subjected to grinding, polishing, and etching. The microstructure and precipitates in these specimens were analyzed by scanning electron microscope (SEM, FEI Quanta-250, FEI Corporation, Hillsboro, OR, USA) equipped with energy-dispersive X-ray spectrometer (EDS, Xflash 5030, Bruker, Germany). The specimens taken from ESR ingots were machined to a bar of Ø15 mm × 80 mm. The precipitates were electrolytically extracted from steel in organic solution (methanol, tetramethylammonium chloride, glycerin, diethanol amine). The collected precipitates were analyzed using XRD (Rigaku Dmax-RB, Rigaku, Tokyo, Japan) technique to determine their types, and observed by SEM.

2.3. Solid Solution Heat Treatment

The specimens of the HMAS-N were solid solution treated at the temperatures of 1170 °C, 1200 °C, and 1230 °C for 0.5, 1, 2, and 3 h. Thereafter, the specimens were quenched in water, and prepared for optical microscope (Leica DM4M, Leica Microsystems, Wetzlar, Germany) analysis. The average grain size was determined by the standard linear intercept method [12] through SISC IAS V8.0 image software [13]. About 400 austenite grains were selected randomly for the grain size measurement in each specimen. After various SSHT processes, the specimens were subjected to the same aging heat treatment at 720 °C for 2 h. Then, the specimens' Rockwell hardness values were measured.

2.4. Aging Heat Treatment

The optimum SSHT process can be selected by the above experimental scheme. After the specimens were subjected to the optimum SSHT, the following two kinds of aging heat treatments (single-stage aging and two-stage aging) were carried out. Single-stage aging and two-stage aging schemes are shown in Tables 2 and 3, respectively. For single-stage aging, the specimens, after optimal SSHT, were subjected to an aging heat treatment at 680–760 °C for 2 h. For two-stage aging, the specimens, after optimal SSHT, were subjected to pre-aging treatment at 650 °C for 1 h, and then re-aging was subsequently carried out in a temperature range from 700 °C to 800 °C for 1 h.

Table 2. Single-stage aging heat treatment process after solid solution heat treatment (SSHT).

The Specimen Number	Aging Temperature	Aging Time
S-680	680 °C	
S-700	700 °C	
S-710	710 °C	
S-720	720 °C	2 h
S-740	740 °C	
S-760	760 °C	

Table 3. Two-stage aging heat treatment process after SSHT.

The Specimen Number	Pre-Aging Temperature	Pre-Aging Time	Re-Aging Temperature	Re-Aging Time
T-700			700 °C	
T-720			720 °C	
T-740	650 °C	1 h	740 °C	1 h
T-760			760 °C	
T-780			780 °C	
T-800			800 °C	

2.5. Mechanical Properties

The impact energy of specimens was tested at room temperature, using a JB-30B testing machine, with the Charpy V-notch test method according to ASTM 370 [14]. The value of impact energy presented in this paper is an average of three measurements. The Rockwell hardness (HRC) measurements with an applied load of 150 kg were performed on the heat-treated specimens. The average value of five tests for each specimen were recorded as the standard hardness.

3. Results and Discussion

3.1. Precipitation and Microstructure of As-Cast ESR Ingot

The equilibrium formation of precipitates in the studied steel was investigated using Thermo-Calc software (TCFE7 database). Figure 1 shows the relationship between the amount of precipitates and their precipitation temperatures in the temperature range from 600 °C to 1600 °C. The transformation (precipitation) temperatures of precipitates in the investigated steel are listed in Table 4.

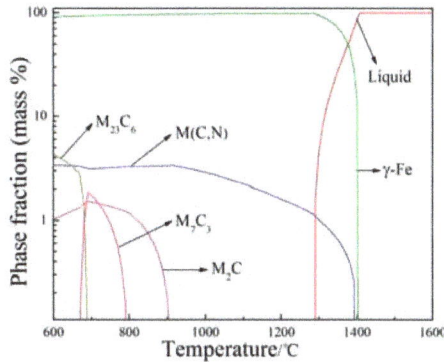

Figure 1. Equilibrium phase precipitations in nitrogen-microalloyed high-Mn austenitic hot work die steel (HMAS-N) calculated using Thermo-Calc (M represents metallic element atom; C represents carbon atom).

Table 4. Calculation results of transformation temperatures of precipitates in steel, calculated using Thermo-Calc.

	T_s/°C				T_f/°C	
γ-Fe	M(C,N)	M_2C	$M_{23}C_6$	M_7C_3	M_2C	M_7C_3
1400	1392	910	690	790	435	670

T_s represents phase transformation starting temperature, T_f represents phase transformation finish temperature.

It can be seen from Figure 1 and Table 4 that M(C,N), M_2C, $M_{23}C_6$, and M_7C_3 precipitate in sequence with the decrease of temperature. The precipitation temperature of M(C,N) is above 1300 °C. This indicates that M(C,N) is a highly stable phase and remains undissolved during heat treatment process.

To predict phase precipitation during liquid steel solidification in a practical ESR refining process, the Scheil–Gulliver model included in Thermo-Calc software was employed to calculate the nonequilibrium phase precipitation in HMAS-N, as shown in Figure 2. M(C,N) precipitates from liquid steel directly. As the austenite continues to precipitate from liquid steel, carbon and alloying element contents keep increasing. Primary carbides M_2C precipitate directly from liquid steel when the solid fraction of liquid steel exceeds 0.93. As shown in Figure 3a,b, partial primary carbides precipitate along preexisting austenite due to the composition of residual liquid steel reaching eutectic point. There are two types of precipitates (i.e., white- and gray-colored ones), which are represented by I and II, in as-cast ESR ingot as shown in Figure 4. The gray-colored and white-colored precipitates both distributed along grain boundaries. It can be seen in Figure 4 that some M(C,N) carbonitrides are mixed with M_2C carbides on the grain boundary. This is because the M(C,N) phase precipitates at a higher temperature prior to the precipitation of the M_2C phase.

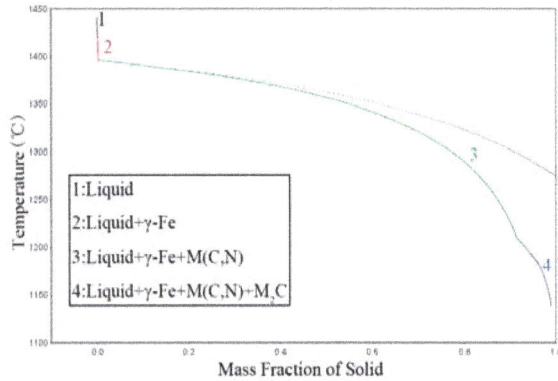

Figure 2. Nonequilibrium phase precipitation in HMAS-N, calculated using Thermo-Calc.

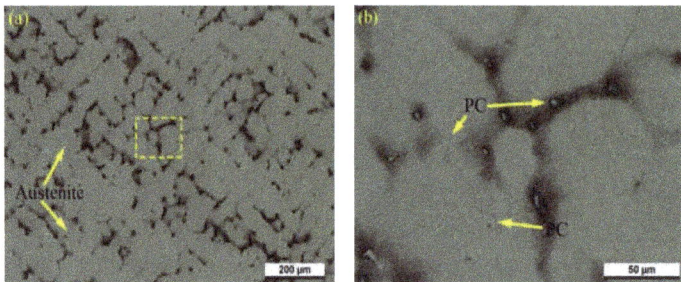

Figure 3. (a) Optical microscope (OM) images of as-cast electroslag remelting (ESR) ingot microstructure; (b) highly magnified images taken at the region marked by the yellow box in (a). PC represents primary carbides or carbonitrides.

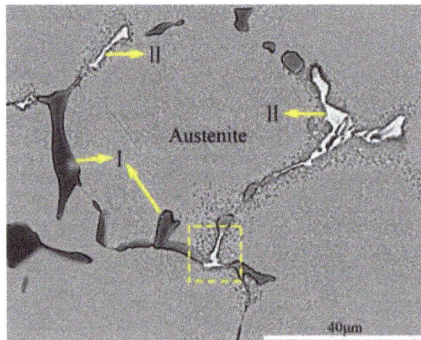

Figure 4. Scanning electron microscopy (SEM) images of as-cast ESR ingot microstructure.

It can be confirmed from the microstructure in Figure 4 and the XRD results in Figure 6a that the matrix is a single austenite phase. Table 5 presents the EDS results of the carbides or carbonitrides shown in Figure 5. As shown in Table 5 and Figure 6b, M(C,N) (namely V(C,N)) contains a certain amount of the elements Cr and Mo. In addition, M_2C is Mo_2C, which contains more Cr and Mo atoms. As shown in Figure 5, the morphology of V(C,N) is an irregular multilateral strip or slice shape with severe angles. The morphology of Mo_2C is rod-shaped or lamellar fish-skeleton-like, which is typical eutectic carbide according to eutectic reaction (L → γ + M_2C).

Table 5. Energy-dispersive X-ray spectrometer EDS analyzed results of carbides or carbonitrides (mass %).

Point	Element						
	C	V	Mo	Cr	N	Fe	Mn
1	14.68	41.60	0.64	1.68	33.11	6.31	1.98
2	56.69	6.88	17.35	7.94	0.05	5.55	5.54

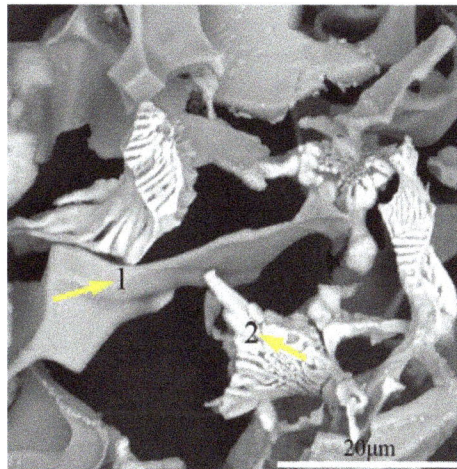

Figure 5. SEM micrographs showing the three-dimensional morphology of carbides or carbonitrides extracted from as-cast ESR ingot.

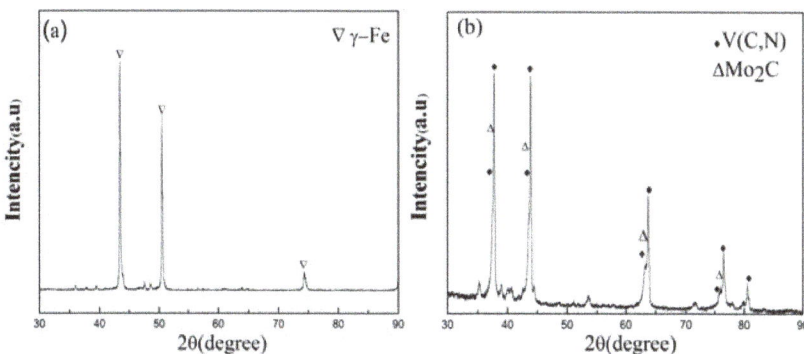

Figure 6. XRD patterns of steel block and precipitates powder: (**a**) steel block after etching of as-cast ESR ingot; (**b**) precipitate powders that were electrolytically extracted from as-cast ESR ingot.

3.2. Effect of Solid Solution Heat Treatment on Hardness and Microstructure

Solid solution heat treatment (SSHT) was carried out to dissolve coarse primary precipitates in the matrix, which were then reprecipitated from supersaturated solid solution during the subsequent aging process with the purpose of precipitation strengthening. If the solution temperature is too low, large primary precipitates are difficult to dissolve into matrix. However, if the solid solution temperature is too high, austenitic grain size (AGS) will grow up excessively. The AGS tremendously influences diffusive and diffusionless phase transformations, precipitation, and mechanical properties such as strength, hardness, toughness, and ductility [15]. Hence, the SSHT process is quite important for microstructures and mechanical properties. The effects of SSHT on the austenite grain size, microstructure, and Rockwell hardness were studied in detailed.

The hardness value of the specimens, which were subjected to various SSHT and the same aging treatment at 720 °C for 2 h, is shown in Figure 7a. The hardness value increased from 45.34 HRC to 47.24 HRC at a solution temperature of 1170 °C with increasing solution time from 0.5 h to 2 h. However, the hardness value decreased rapidly to 39 HRC when the solid solution time was 3 h at 1170 °C. When the solid solution temperature reached 1200 °C and 1230 °C, the hardness decreased linearly with the increase of solution holding time from 0.5 h to 3 h.

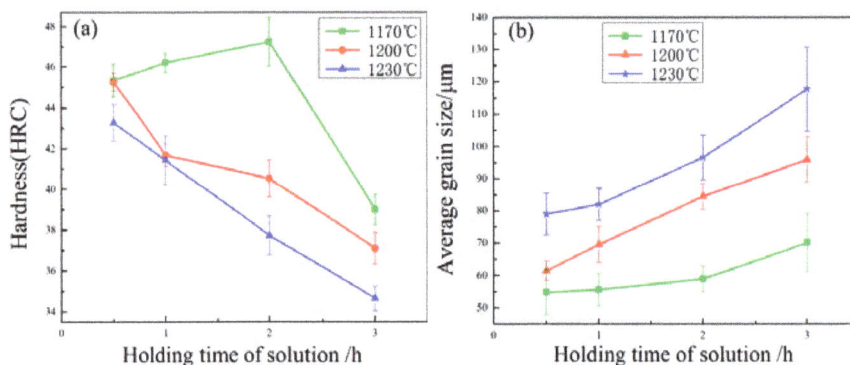

Figure 7. Rockwell hardness (**a**) and austenite average grain size (**b**) after various holding time at different solution temperatures.

The AGS after different solution heat treatments is shown in Figure 7b. The austenite grain grew gradually with increasing solution temperature and solution time. The austenite grains grew slowly at a solution temperature below 1200 °C. A tangible grain growth was detected at 1170 °C for soaking times over 3 h due to the coupled effects of the soaking time and temperature on the grain growth. The results presented in Figure 7b also implied that the effect of solution temperature on grain growth was more noticeable than solution time.

Figures 8 and 9 show the microstructure evolution at different solution temperatures and solution times with corresponding austenitic grain-size distribution histograms, respectively. As shown in Figure 9, the grain size distribution of austenite is close to lognormal distribution, which is consistent with the results presented by Han et al. [16] and Kurtz et al. [17]. When the solid solution time is 0.5 h at 1170 °C, the fine austenite grains distribute inhomogeneously as coarse austenite grains and contact each other, as shown in Figure 8a. Figure 9a indicates the grain size of austenite is mainly in the range of 20–50 μm. The percentage of AGS smaller than 20 μm is 14.9%. When the specimens were solid solution treated at 1170 °C for an aging time between 0.5 h and 2 h, AGS ranged from 54.8 μm to 58.4 μm. No obvious grain growth was observed in Figure 8a,b. Those primary carbonitride particles that precipitated along grain boundaries affected the diffusion of iron and carbon atoms, and finally prevented the growth of austenite grain in the heating process [15,18]. Compared with the results shown in Figure 9a, AGS distributes more uniform after SSHT for 2 h, as shown in Figure 9b. The grain size of austenite shown in Figure 9c was mostly in the range of 60–80 μm, and the number of austenite grains larger than 50 μm increased with rising solution temperature. When the solid solution time is 3 h at 1230 °C, the bimodality of the histogram presented as a result of further progress in the second recrystallization, presented in Figure 9d.

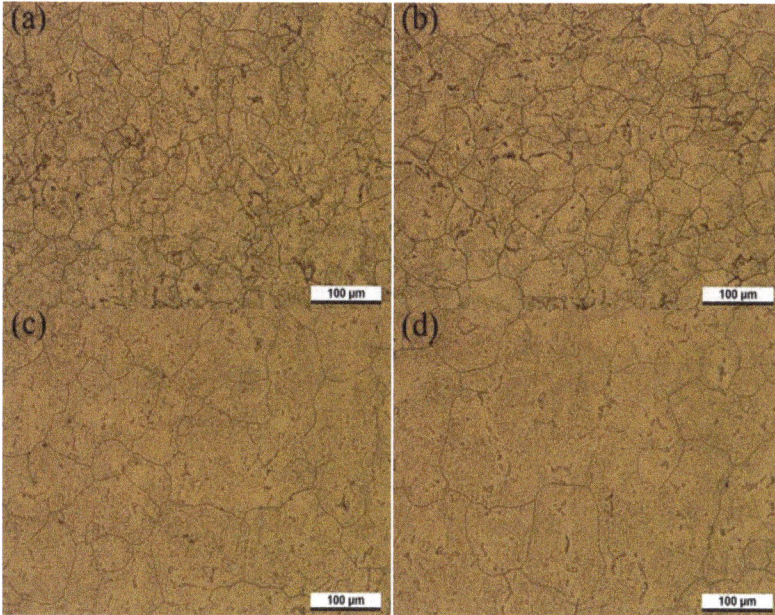

Figure 8. Effect of solution temperature and solution time on microstructure evolution at (**a**) 1170 °C for 0.5 h; (**b**) 1170 °C for 2 h; (**c**) 1200 °C for 2 h; and (**d**) 1230 °C for 3 h.

As shown in Figure 8a, the precipitates were not completely dissolved at 1170 °C, which is consistent with thermodynamic calculation using Thermo-Calc software. Thermodynamic calculation illustrates that precipitation temperature of the primary V(C,N) reaches up to 1390 °C, as shown in Table 4. N addition in steel increased the melting point and stability of carbonitrides. In addition, most primary V(C,N) were distributed along the austenite grain boundary and were not significantly influenced by solution heat treatment [19]. Hence, primary V(C,N) cannot completely dissolve into the matrix when the solution temperature is below 1390 °C. Therefore, it can be concluded that the pinning effect of undissolved V(C,N) [20] suppressed austenite grain growth with increasing solution time from 0.5 h to 2 h at 1170 °C. Meanwhile, more carbides, carbonitrides, and alloying elements dissolved into the austenite matrix with increasing solid solution time at 1170 °C. Owing to solution strengthening and the second phase strengthening during the process of aging heat treatment, the hardness of steel reached the maximum value at 2 h aging time when solution temperature was at 1170 °C. When aging time exceeded 2 h at 1170 °C, the austenite grain grew rapidly, which led to the decrease of hardness value. When the solution temperature is more than 1170 °C, the AGS grows rapidly, despite a significant amount of undissolved secondary phases. This is attributed to coagulation process during the SSHT process, leading to the growth of large particles at the expense of small ones and the reduction in the amount of precipitates. A small number of large precipitates do not limit grain boundary mobility, and pinning effects disappear [21]. Grain growth highly depends on the grain boundary migration and the atomic diffusion. Moreover, the dissolution/coarsening of precipitates occurs at higher temperatures [22,23]. Furthermore, the solute drag effect of alloying elements is not effective enough at very high temperatures due to their higher frequency of vibration and mobility [24]. Therefore, the grain growth kinetics is intensified by increasing solid solution temperature. At a very high solution temperature of 1230 °C, the austenite grains coarsened significantly. The coarse austenite grains accelerated the development and growth of fatigue-induced cracks, which will lead to hot work steel failure. After solid solution at 1230 °C for 3 h, grain coarsening became more evident as a result of further progress in the second recrystallization, shown in Figures 8d and 9d. Hence, austenitic grain-coarsening speed is accelerated, and undissolved V(C,N) pinning effect was weakened, resulting in the decrease of hardness in a linear trend when the solution temperature exceeds 1200 °C. Hence, the optimal combination of hardness and AGS can be obtained by soaking for 2 h at 1170 °C.

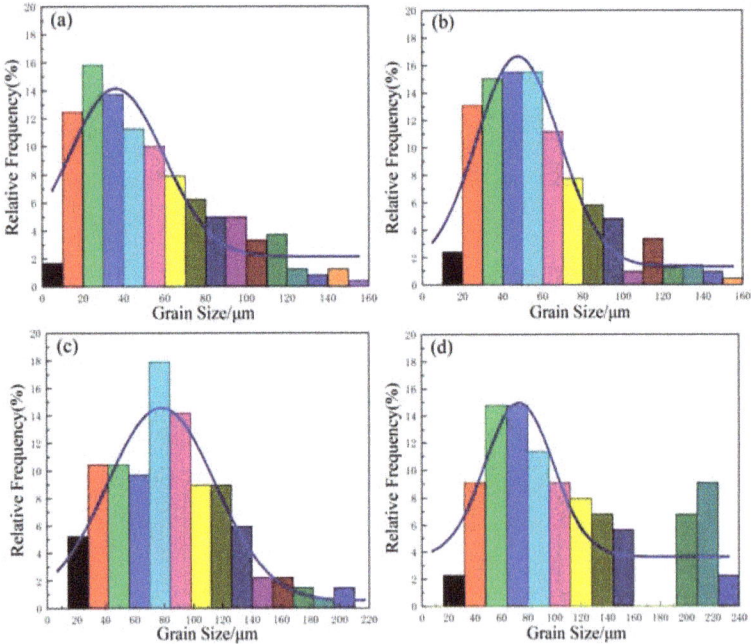

Figure 9. Grain-size distribution histograms obtained upon solution treatment at (**a**) 1170 °C for 0.5 h; (**b**) 1170 °C for 2 h; (**c**) 1200 °C for 2 h; and (**d**) 1230 °C for 3 h.

3.3. Effect of Aging Heat Treatment on Mechanical Properties and Microstructure

To remove residual stress caused by SSHT and dispersed small secondary phase particles, high-temperature aging heat treatment was carried out. The aging heat treatment process has a significant effect on comprehensive performance of HMAS-N. According to Grabovskii et al. and Lanskaya's [3,25] reports, two-stage aging instead of single-stage aging can substantially increase impact toughness at the same level of high-temperature strength, which positively affects the crack resistance of the tools and increases their service life. However, there are few literatures on the effect of multiple aging on the microstructure and properties of austenitic hot work die steels. Al–Zn–Mg alloys and some high-temperature alloys such as iron–nickel superalloys are subjected to multistage aging to enhance their comprehensive performance [26,27]. To improve the comprehensive performance of HMAS-N, the effects of single-stage aging and two-stage aging on the microstructure and properties of steel were studied.

As for single-stage aging, experiments were carried out in a temperature range from 680 °C to 760 °C for 2 h, as shown in Table 2. The hardness improves gradually with aging temperature increasing from 680 °C to 720 °C, as shown in Figure 10. When the aging temperature was at 720 °C, significant secondary hardening occurred, and the hardness reached the maximum value of 47.24 HRC. It can be attributed to the enrichment and segregation of precipitate-forming elements at 720 °C, such as V, Mo, Cr, C, and N [8]. As the aging temperature continued to increase, the hardness decreased to 45.04 HRC at 760 °C. As for the two-stage aging, pre-aging and re-aging heat treatment experiments were carried out, shown in Table 3. The hardness increases with the re-aging temperature increasing from 700 °C to 760 °C. When the two-stage aging temperature was at 760 °C, the hardness reached the peak value of 46.06 HRC. As the re-aging temperature continued to increase, the hardness decreased and dropped to 43.84 HRC at 800 °C. Compared to single-stage aging, the hardness value of two-stage aging is slightly less than single-stage aging at the same aging temperature. In addition, the temperature for maximum hardness delayed. To ensure that the steel has higher hardness and impact toughness to restrain crack initiation and growth in the process of service, the Charpy impact tests were carried out for the specimens with Rockwell hardness values greater than 45 HRC after aging heat treatment. The impact energies after different aging heat treatments are shown in Figure 11. For single-aging at 720 °C, the hardness value reached the maximum value of 47.24 HRC, whereas the impact energy reached the minimum value at 9.2 J. The impact energy reached the maximum value of 12.5 J and the corresponding hardness value was 46.05 HRC at 740 °C. For two-stage aging, the impact energy gradually increased and reached the maximum value of 16.2 J at 780 °C with increasing temperature from 740 °C to 780 °C. As for two-stage aging heat treatment, impact energy greatly improved despite the hardness decreasing by 1–2 HRC compared with single-stage aging.

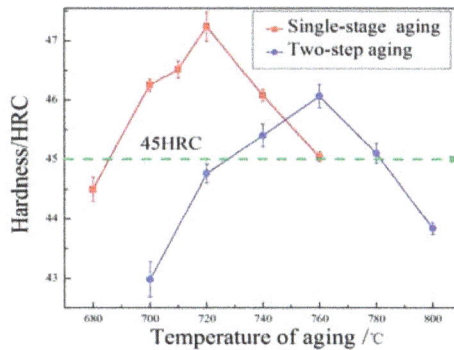

Figure 10. Hardness of specimens with different aging heat treatment processes.

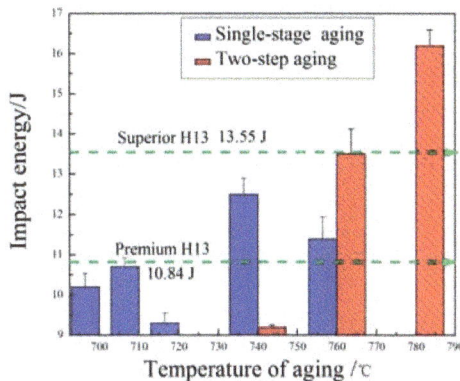

Figure 11. Impact energy of specimens aged with different aging heat treatment processes.

145

The maximum impact energy value of 16.2 J is not only higher than premium H13 (10.84 J), but also almost exceeds superior H13 (13.55 J) [28]. The newly prepared HMAS-N, in terms of the toughness, is considerably outstanding when taking H13 mechanical properties into consideration. Compared with the maximum impact energy of single-stage aging, two-stage aging impact energy increased 29.6%. Despite the lowest hardness value of 45.1 HRC at 780 °C re-aging temperature, it still meets the service requirements of hot work steel. Hence, in order to improve the comprehensive performance and service life, the optimal aging process is re-aging at 780 °C.

It is well known that structure and fractograph of a specimen are related to the mechanisms involved in the fracture process. The microstructure determines the performance of the steel. Figure 12a is a SEM micrograph of specimen S-720 at low magnification. It can be seen that there are massive, undissolved primary precipitates. Figure 12b–d are the scanning electron micrographs of specimens S-720, S-740, and S-760 at high magnification, respectively. When the aging temperature increased from 720 °C to 740 °C, the number of fine dispersed secondary carbonitrides increased, as shown in Figure 12b,c. As the aging temperature increased to 760 °C, the number of secondary carbonitrides decreased, but the secondary carbonitrides appeared to be aggregated, and the grain boundary was widened, as shown in Figure 12d. This is attributed to growth of large particles at the expense of small ones, resulting in the reduction in the total number of particles. Precipitation of carbides along the grain boundary will result in deterioration of toughness, which can be attributed to the decrease of grain boundary cohesion.

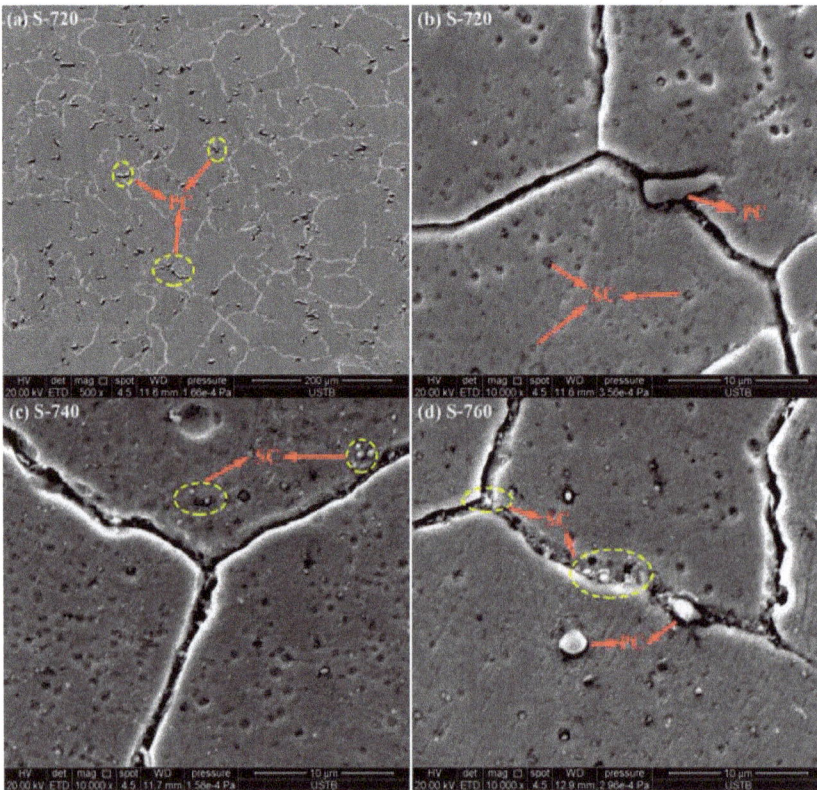

Figure 12. SEM microstructure of specimens after different aging heat treatment processes: (**a,b**) S-720, (**c**) S-740, (**d**) S-760. Note: PC represents primary precipitates. SC represents secondary precipitates.

The SEM fractograph of specimens after the Charpy test for a fixed solid solution temperature (1170 °C) and various aging processes are shown in Figure 13a–f. Both quasi-cleavage facets and ductile dimples were observed, which indicated the mixed mechanism of fracture. The specimen S-720 showed mainly brittle-fractured surface zones characterized by transgranular quasi-cleavage and intergranular fracture (Figure 13a). It can be noticed that quasi-cleavage fracture facets encircled by ridges are much smaller than the austenite grain (58.4 μm), and this kind of morphology forms when a crack propagates along the primary precipitates. The size of intergranular fracture grains (~60 μm) was similar to the size of austenite grains (58.4 μm). This implies that intergranular fracture along the boundaries of austenite grains operated partially for specimen S-720. As shown in Figure 13b, quasi-cleavage facets show massive primary precipitates, which is consistent with the observation in Figure 11a. It indicated that quasi-cleavage nucleated along the precipitates. Some portions of the fracture surface exhibited ductile-fractured surface with fine dimples (Figure 13c). The dimple percentage increases with increasing aging temperature (from 720 °C to 760 °C) for single-stage aging treatment. The specimens S-740 and S-760 showed mainly ductile-fractured areas with fines dimples and some quasi-cleavage features (Figure 13d,e). It is clear that the fracture surface of specimen S-760 contained microcracks and some spherical secondary precipitates precipitated from the aging process, as shown in Figure 13e. The EDS results of the precipitates shown in Figure 13b,e are presented in Table 6. Figure 14 shows fractograph EDS element distribution mappings of specimen S-760. Figure 14 and the EDS results of point 1, shown in Figure 13b, indicate that elements N, V, C were enriched in the same area, confirming that undissolved coarse primary carbonitrides are V(C,N), which also contains a small amount of Cr. Figure 14 and the EDS results of point 2, shown in Figure 13e, indicate that spherical secondary carbides are Mo-rich carbides. Specimen T-780 presented more ductile dimples and fewer quasi-cleavage facets (Figure 13f).

Table 6. The EDS results for precipitates (mass %).

Point	Element						
	C	V	Mo	Cr	N	Fe	Mn
1	28.48	40.50	0.24	0.92	27.57	1.72	0.57
2	37.07	0.26	16.57	1.65	0.00	43.62	0.83

The undissolved primary precipitates, austenite grain size, strength of the matrix, secondary precipitates, and the strength of grain boundary have great effects on the impact energy, while aging heat treatment has little effect on primary precipitates and austenite grain size. For specimen S-720, the massive undissolved primary V(C,N) had a higher hardness value (47.2 HRC) and is not easy to deform, which is responsible for embrittlement for the lower impact energy (9.3 J). A lower binding force among precipitates and steel matrix resulted from the existence of primary and secondary precipitates. It is much easier to generate cracks at the regions where primary V(C,N) are enriched.

With increasing aging temperature from 720 °C to 740 °C for single-stage aging, the amount of secondary precipitates increased, which meant the interstitial atoms dissolved in the matrix precipitated and the degree of supersaturation decreased, resulting in the softening of the steel matrix [29]. The increasing number of secondary precipitates has a negative effect on impact energy. The softer the base metal, the larger the plastic district around the crack tip when the stress concentration is high enough and plastic deformation occurs. A larger plastic district will result in greater crack propagation and larger impact energy [30]. For specimen S-720 and S-740, the base metal plays the main role on impact energy compared to the negative effect of secondary precipitates. Hence, impact energy increased from 9.3 J to 12.5 J with increasing temperature, from 720 °C to 740 °C, for single-stage aging temperature. When increasing temperature to 760 °C for the single-stage aging, the amount of secondary precipitates increases and the coarsening of the austenite grain boundary has a major effect on the impact energy, which led to impact energy decreasing from 12.5 J to 11.4 J.

For two-stage aging, the low-temperature pre-aging of the two-stage aging process is equivalent to the nucleation stage, and the high-temperature aging is the stabilization stage [26]. The first stage in aging of the alloy is to promote the formation of the GP (Guinier-Preston) zone, and then to obtain a large number of dispersed GP areas, and to make the GP areas extend to a certain scale without dissolution in the second stage of aging. In the second high-temperature aging process, the aging precipitation sequence of the alloy changes from GP zone to secondary precipitates, and the alloy enters the aging stage. This improves the working capacity of tools for hot deformation. The two-stage aging may be more conducive to distributed homogeneous precipitation of secondary precipitates and reducing growth of secondary precipitates, which will weaken the reduction of the impact energy compared to single-stage aging. As a result, the impact energy increases from 9.2 J to 16.2 J with a slight decrease of the strength when increasing the re-aging temperature from 740 °C to 780 °C.

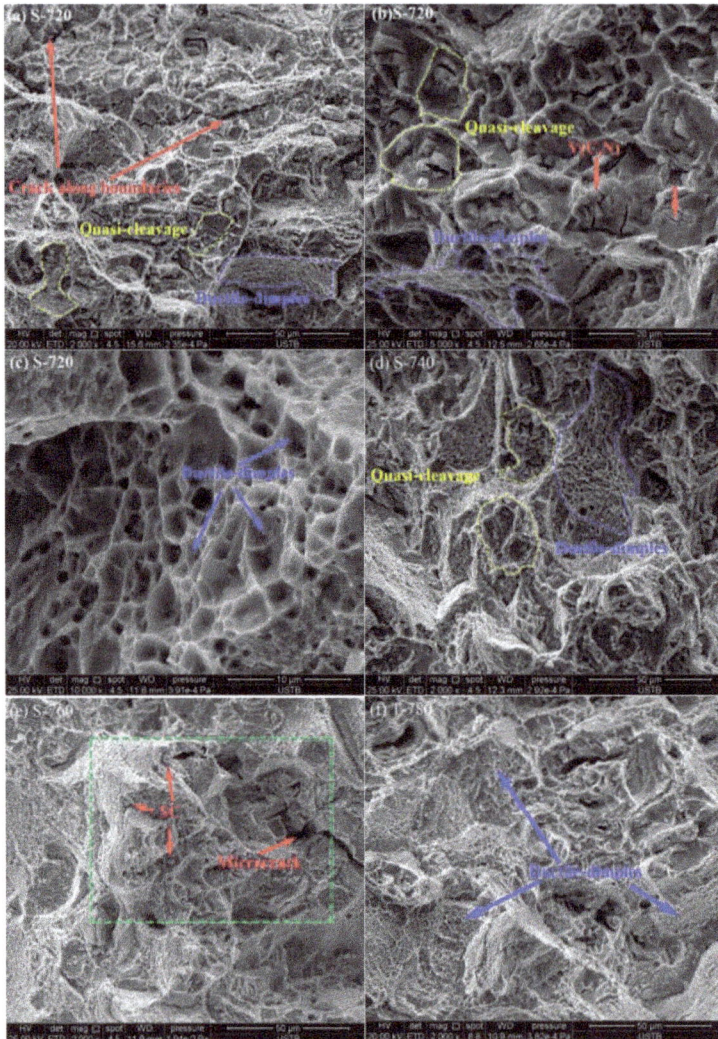

Figure 13. SEM fractographs of specimens after different aging heat treatment processes: (**a–c**) S-720; (**d**) S-740; (**e**) S-760; (**f**) T-780. (**b**) and (**c**) are highly magnified images taken at the region enriched with quasi-cleavage facets and dimples of specimen S-720, respectively. SC represents secondary precipitates.

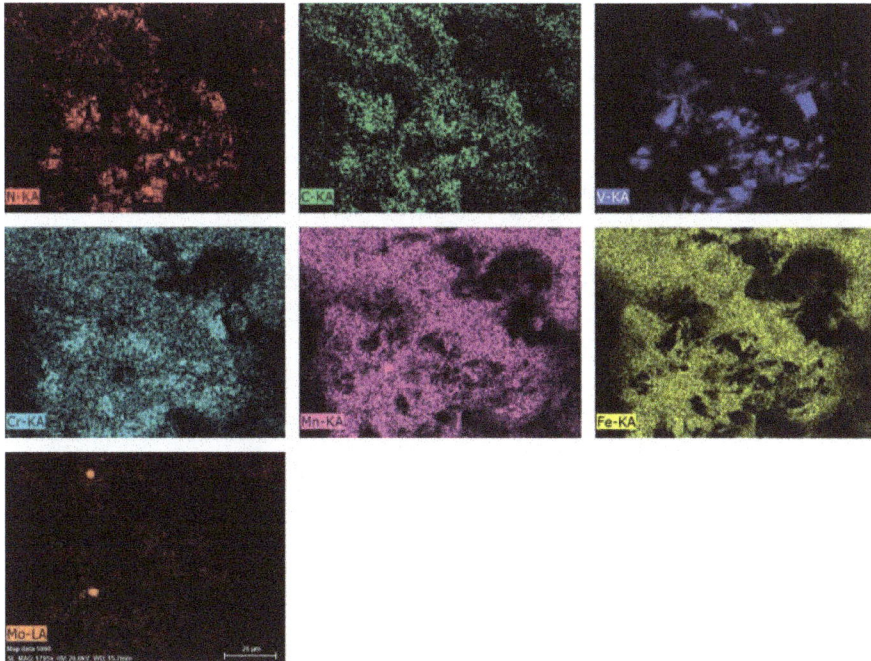

Figure 14. Energy-dispersive X-ray spectrometer (EDS) element mappings of S-760 fractograph, scale bar: 20 μm.

4. Conclusions

The microstructure of as-cast high-Mn ingot is composed of γ-Fe + M(C,N) + M_2C. M(C,N) precipitated from liquid steel directly, which is a highly stable phase. M(C,N) is V(C,N) containing Cr, Mo elements. M_2C is Mo_2C, containing more Cr and Mo elements, which formed by eutectic reaction (L → γ + M_2C). The morphology of V(C,N) is an irregular multilateral strip or slice shape with severe angles. Most of Mo_2C morphology are lamellar fish-skeleton-like and a small number of Mo_2C morphology are rod-shaped.

When increasing solid solution time from 0.5 h to 2 h at 1170 °C, AGS ranges from 54.8 μm to 58.4 μm and no significant grain growth was observed, which can be attributed to the pinning effect of undissolved V(C,N). Additionally, the hardness increased from 45.34 HRC to 47.24 HRC due to solid solution strengthening. When the solid solution temperature reached 1200 °C and 1230 °C, the effect of grain growth, which decreases hardness, exceeds the effect of solid solution strengthening, which increases hardness. The hardness value decreased linearly with the increase of solution temperature from 1200 °C to 1230 °C.

The AGS distribution of austenite is close to lognormal distribution. The austenite grains grew slowly at a solution temperature below 1200 °C. A tangible grain growth was detected at 1170 °C for soaking times over 3 h due to the synergic effect of the soaking time and temperature on the grain growth. When the solid solution time is 3 h at 1230 °C, the bimodality of histogram is presented as a result of further progress in secondary recrystallization.

For single-stage aging and two-stage aging, Rockwell hardness shows a trend of first increasing and then decreasing with increasing aging temperature. In addition, hardness reach the maximum at 740 °C (47.24 HRC) and 760 °C (46.06 HRC), respectively, which is attributed to significant secondary hardening. Compared to single-stage aging, impact energy greatly improved despite the hardness decreasing by 1–2 HRC. In addition, the temperature for maximum hardness delayed.

Both quasi-cleavage facets and ductile dimples were observed in the fractograph, which indicates the mixed mechanisms of fracture. The two-stage aging may be more conducive to homogeneously distributed precipitation of secondary precipitates and reduced growing tendency of secondary precipitates, which will weaken the reduction of the impact energy compared to single-stage aging. Compared with the maximum impact energy of single-stage aging, maximum impact energy reached 16.2 J and increased by 29.6% when using two-stage aging. Thus, the optimum heat treatment process is subjected to solution treatment for 2 h at 1170 °C, followed by pre-aging treatment for 1 h at 650 °C, and, finally, by re-aging for 1 h at 780 °C.

Acknowledgments: This work was financially supported by the National Natural Science Foundation of China (Grant No. 51504019 and No. 51374022).

Author Contributions: Yi Zhang and Jing Li conceived and designed the experiments; Yi Zhang and Yong-Feng Qi performed the experiments; Yi Zhang and Cheng-Bin Shi analyzed the data; Cheng-Bin Shi and Qin-Tian Zhu were responsible for language modification; Jing Li contributed reagents/materials/analysis tools; Yi Zhang wrote the paper.

Conflicts of Interest: The authors declare no conflict of interest.

References

1. Norström, L.Å.; Öhrberg, N. Development of hot-work tool steel for high-temperature applications. *Metals Technol.* **2013**, *8*, 22–26. [CrossRef]
2. Kremnev, L.S.; Brostrem, V.A. Heat resistance of tool steels and alloys. *Metal Sci. Heat Treat.* **1973**, *15*, 225–230. [CrossRef]
3. Grabovskii, V.Y.; Kanyuka, V.I. Austenitic Die Steels and Alloys for Hot Deformation of Metals. *Metal Sci. Heat Treat.* **2001**, *43*, 402–405. [CrossRef]
4. Adrian, H. Thermodynamic model for precipitation of carbonitrides in high strength low alloy steels containing up to three microalloying elements with or without additions of aluminium. *Mater. Sci. Technol.* **1992**, *8*, 406–420. [CrossRef]
5. Li, J.; Huang, H.Y. *High Nitrogen Steels and Stainless Steels: Manufacturing, Properties and Applications*; Chemical Industry Press: Beijing, China, 2006.
6. Stein, G.; Hucklenbroich, I. Manufacturing and Applications of High Nitrogen Steels. *Mater. Manuf. Process.* **2006**, *19*, 7–17. [CrossRef]
7. Rashev, T. Development of Laboratory and Industrial Installations for One Stage Production of HNS. *Mater. Manuf. Process.* **2004**, *19*, 31–40. [CrossRef]
8. Li, J.Y.; Zhao, P.; Yanagimoto, J. Effects of heat treatment on the microstructures and mechanical properties of a new type of nitrogen-containing die steel. *Int. J. Miner. Metall. Mater.* **2012**, *19*, 511–517. [CrossRef]
9. Li, J.Y.; Chen, Y.L.; Huo, J.H. Mechanism of improvement on strength and toughness of H13 die steel by nitrogen. *Mater. Sci. Eng. A* **2015**, *640*, 16–23. [CrossRef]
10. Jiang, H.; Wu, X.C.; Shi, N.N. Effects of nitrogen on microstructure and properties of austenitic hot work die steel. *Mater. Mech. Eng.* **2012**, *36*, 58–61.
11. Zhang, Z.Y.; Rong, J.K. A study of austenitic hot work die steel. *CISRIC Tech. Bull.* **1985**, *5*, 47–53.
12. ASTM E 112–10. *Standard Methods for Estimating the Average Grain Size*; ASTM International: West Conshohocken, PA, USA, 2010.
13. SISC IAS V8.0. Scientific Instrument Software Corporation Limited: Beijing, China, 2003.
14. ASTM A370. *Standard Test Methods and Definitions for Mechanical Testing of Steel Products*; ASTM International: West Conshohocken, PA, USA, 2010.
15. Lee, S.J.; Lee, Y.K. Prediction of austenite grain growth during austenitization of low alloy steels. *Mater. Des.* **2008**, *29*, 1840–1844. [CrossRef]
16. Han, L.Z.; Chen, R.K.; Gu, J.F.; Pan, J.S. Behavior of austenite grain growth in X12CrMoWVNbN10-1-1 ferrite heat-resistant steel. *Acta Metall. Sin.* **2009**, *45*, 1446–1450.

17. Kurtz, S.K.; Carpay, F.M.A. Microstructure and normal grain growth in metals and ceramics. Part I. Theory. *J. Appl. Phys.* **1980**, *51*, 5725–5744. [CrossRef]

18. Staśko, R.; Adrian, H.; Adrian, A. Effect of nitrogen and vanadium on austenite grain growth kinetics of a low alloy steel. *Mater. Charact.* **2006**, *56*, 340–347. [CrossRef]

19. Speer, J.G.; Michael, J.R.; Hansen, S.S. Carbonitride precipitation in niobium/vanadium microalloyed steels. *Metall. Mater. Trans. A* **1987**, *18*, 211–222. [CrossRef]

20. Matsuo, S.; Ando, T.; Grant, N.J. Grain refinement and stabilization in spray-formed AISI 1020 steel. *Mater. Sci. Eng. A* **2000**, *288*, 34–41. [CrossRef]

21. Adamczyk, J.; Kalinowska-Ozgowicz, E.; Ozgowicz, W.; Wusatowski, R. Interaction of carbonitrides V(C,N) undissolved in austenite on the structure and mechanical properties of microalloyed V-N steels. *J. Mater. Process. Technol.* **1995**, *53*, 23–32. [CrossRef]

22. Humphreys, F.J.; Hatherly, M. *Chapter 6–Recovery after Deformation. Recrystallization & Related Annealing Phenomena*; Elsevier Limited: Amsterdam, The Netherlands, 2004; pp. 169–213.

23. Dutra, J.C.; Siciliano, F., Jr.; Padilha, A.F. Interaction between Second-Phase Particle Dissolution and Abnormal Grain Growth in an Austenitic Stainless Steel. *Mater. Res.* **2002**, *5*, 379–384. [CrossRef]

24. Abbaschian, R.; Reedhill, R.E. *Physical Metallurgy Principles*, 4th ed.; Cengage Learning: Stamford, CT, USA, 2009.

25. Lanskaya, K.A. *High-Temperature Steels [in Russian]*; Metallurgiya: Moscow, Russia, 1969.

26. Werenskiold, J.C.; Deschamps, A.; Bréchet, Y. Characterization and modeling of precipitation kinetics in an Al–Zn–Mg alloy. *Mater. Sci. Eng. A* **2000**, *293*, 267–274. [CrossRef]

27. Yang, R.X.; Liu, Z.Y.; Ying, P.Y.; Li, J.L.; Lin, L.H.; Zeng, S.M. Multistage-aging process effect on formation of GP zones and mechanical properties in Al–Zn–Mg–Cu alloy. *TNMSC* **2016**, *26*, 1183–1190. [CrossRef]

28. NADCA Die Material Committee. *NADCA Recommended Procedures for H13 Tool Steel*; NADCA #207-2003; North America Die Casting Association: Arlington Heights, IL, USA, 2003.

29. Yan, P.; Liu, Z.; Bao, H.; Weng, Y.; Liu, W. Effect of tempering temperature on the toughness of 9Cr–3W–3Co martensitic heat resistant steel. *Mater. Des.* **2014**, *54*, 874–879. [CrossRef]

30. Gao, J.M. *Mechanical Properties of Materials*, 1st ed.; Wuhan University of Technology Press: Wuhan, China, 2004. (In Chinese)

metals | MDPI

Article

Corrosion Behavior of API X100 Steel Material in a Hydrogen Sulfide Environment

Paul C. Okonkwo [1], Rana Abdul Shakoor [1,*], Abdelbaki Benamor [2],
Adel Mohamed Amer Mohamed [3] and Mohammed Jaber F A Al-Marri [2]

[1]   Center of Advanced Materials, Qatar University, 2713 Doha, Qatar; paulokonkwo@qu.edu.qa
[2]   Gas Processing Center, Qatar University, 2713 Doha, Qatar; benamor.abdelbaki@qu.edu.qa (A.B.);
      m.almarri@qu.edu.qa (M.J.A.-M.)
[3]   Department of Metallurgical and Materials Engineering, Faculty of Petroleum and Mining Engineering,
      Suez University, 43721 Suez, Egypt; adel.mohamed25@yahoo.com
*   Correspondence: shakoor@qu.edu.qa; Tel.: +974-4403-6867

Academic Editor: Robert Tuttle
Received: 8 February 2017; Accepted: 14 March 2017; Published: 25 March 2017

Abstract: Recently, the API X100 steel has emerged as an important pipeline material for transportation of crude oil and natural gas. At the same time, the presence of significant amounts of hydrogen sulfide (H_2S) in natural gas and crude oil cause pipeline materials to corrode, which affects their integrity. In this study, the effect of H_2S concentration on the corrosion behavior of API X100 in 3.5% NaCl solution is presented. The H_2S gas was bubbled into saline solutions for different durations, and the corrosion tests were then performed using potentiodynamic polarization and electrochemical impedance spectroscopy (EIS). X-ray photoelectron spectroscopy (XPS), X-ray diffraction (XRD), atomic force microscopy (AFM), and scanning electron microscopy (SEM) techniques were used to characterize the corroded surface. The results indicate that the corrosion rate of API X100 steel decreases with increasing H_2S bubbling time due to the increase in H_2S concentration in 3.5% NaCl solutions. It is noticed that an accumulation of a critical amount of hydrogen in the metal can result in hydrogen-induced crack initiation and propagation. It was further observed that, when the stress limit of a crystalline layer is exceeded, micro-cracking of the formed protective sulfide layer (mackinawite) occurs on the API X100 steel surface, which may affect the reliability of the pipeline system.

Keywords: H_2S concentration; corrosion; sour environment; API X100 steel; sulfide layer

1. Introduction

Pipelines are one of the most convenient means of transporting petroleum and its products from one region to another. Carbon steels (C-steel) are commonly used as a material for the transportation pipelines because of their economic advantages and their ability to withstand operating pressure [1,2]. However, the steel surfaces of pipelines are constantly exposed to corrosive environments during the transportation of the petroleum and its products; hence, the integrity of the pipeline system is always found to be affected [3–5]. The sulfur content in the petroleum and its products have been shown to be instrumental in the internal corrosion of C-steel pipelines [6]. Sulfide is predominant in aqueous solutions through industrial waste and various biological processes and has a significant influence on the aqueous corrosion of steels [7]. Other parameters such as temperature, fluid, velocity, and microstructure also are influential in the corrosion rate of C-steel used in the petroleum industry [5,8–11].

Recently, attentions have been drawn to the sour corrosion of C-steel due to a poor understanding of sulfide corrosion absorption. Many researchers have shown that the formed iron sulfide on the steel surface during the sour corrosion process can be either protective or non-protective [12,13]. It has been reported that the formed corrosion product layer during the corrosion of API X52 is significantly composed of iron sulfide and various oxides, which enhance the protective properties of the film [4]. In a similar study, the corrosion of API X52 steel in H_2S revealed that the increase in the thickness of formed iron sulfide film influences the corrosion rate of C-steel [14]. It has also been reported that, when the stress limit of the formed sulfide layer exceeds, micro-cracks within the formed layers have formed, allowing a more rapid penetration of sulfide and chloride species into the deposit, resulting in an increase in the corrosion rate [15].

The properties and the composition of the steel play critical roles in the formation, nature and stability of the sulfide layer and consequently its corrosion prevention behavior [3]. A recent study [16] has shown that the chemical composition and microstructure of C-steel significantly influence its resistance to corrosion in wet H_2S environments. However, the needs to increase the strength and reduce the corrosion rate of existing C-steel pipeline materials have inspired new interest in the quest for higher-strength steels [17,18]. Despite several contributions of different authors regarding a shared understanding of the corrosion of steel materials [19,20], few studies have investigated the corrosion in a sour environment [21,22], especially at low temperatures and with higher grades of carbon steel. The application of this high strength steel as a pipeline material is aimed at enhancing the corrosion resistance of pipeline steel. Nevertheless, many of the electrochemical characteristics of these high strength low-alloy steel materials have not yet been reported in the literature [23–27]. Furthermore, the corrosion behavior and mechanism of these steels in sulfide environments have not yet been fully understood. Here, the corrosion behavior and mechanism of API X100 steel in sour 3.5% NaCl solution at different H_2S concentrations are presented using experimental tools, while surface characterization techniques are used to characterize the corroded surface.

2. Materials and Methods

2.1. Material Preparation

The pipeline steel plates supplied by Hebei Yineng Pipeline Group Co., Ltd. (Shandong, China) were cast into slabs after ladle refining and hot-rolled into 10 mm thickness plates. This process was followed by solid-solution fine-grain strengthening, precipitation, and controlled cooling to achieve the desirable microstructure. The microstructure of the as-received samples were found to be pearlitic, formed by cooling the austenite. A representative microstructure of the target steel is shown in Figure 1.

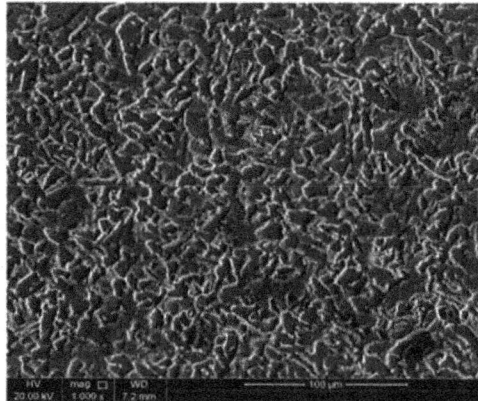

Figure 1. Scanning electron microscopy (SEM) micrograph of API X100 steel.

The chemical composition of the API X100 steel used in the present study was determined by an optical emission (OE) spectrometer, ARL 3460 (Thermo Fisher Scientific SARL, Ecublens,

Switzerland). The sample preparation procedure and the chemical composition are detailed elsewhere [28,29].

2.2. Electrochemical Measurement Setup

Different electrochemical experiments were carried out at various H_2S concentrations using an experimental setup that was described elsewhere [28]. Figure 2 shows the schematic diagram of the corrosion set-up used in this study.

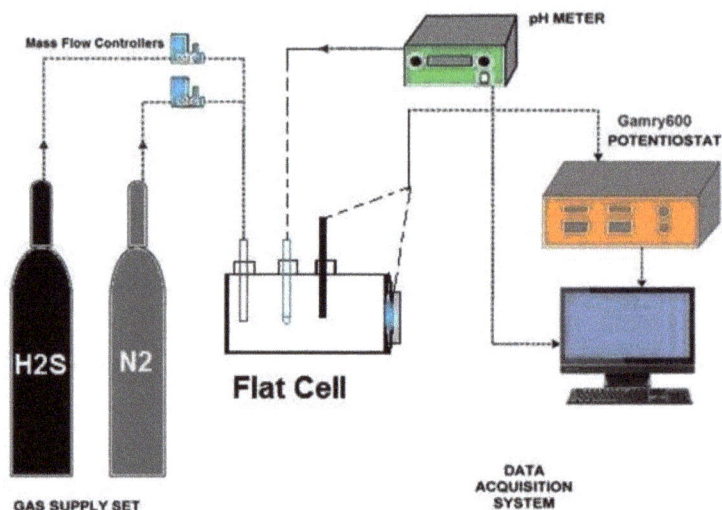

Figure 2. Schematic diagram of the corrosion test set-up employed to study the corrosion behavior of API X100 steel at room temperature.

The set-up was designed to allow and control the bubbling of both H_2S and N_2 into the saline electrolyte solution. In the experimental study, the work was designed to investigate the corrosion behavior of API X100 steel material in 3.5% NaCl solutions bubbled with H_2S gas at different durations. De-aeration was carried out using nitrogen gas for 2 h prior to bubbling H_2S gas into the saline solutions at a pressure of 0.5 MPa and regulated using a flow meter (CVG Technocrafts, Mumbai, India) for different time intervals before each test. In addition, both the pH and the H_2S concentration of the solution were measured and calculated, respectively, at the end of each test (Table 1). The actual concentration of H_2S in the solution was determined following the NACE standard [30], as shown in Table 1.

Table 1. The pH and H_2S concentration used in the tests.

H_2S Gas Bubbling Duration (h)	H_2S Concentration (ppm)	PH
0	0	7.13
1	7	6.58
3	14	6.02
6	21	4.91

For reproducibility, each test was repeated three times. The electrochemical impedance spectrometer (EIS) experiments were conducted within a frequency range of 0.1 to 100 KHz starting from the higher limit towards the lower one. The potentiodynamic polarization experiments were done using a Gamry Reference Eco potentiostat (Gamry Instruments, Warminster, PA, USA) at a scan rate of 0.167 m·Vs^{-1}. The purity of the used gas, H$_2$S, was 99.99% and was purchased from Buzwair Scientific & Technical Gases, Doha, Qatar. The 3.5% NaCl solution was prepared using deionized water with a conductivity of 18.6 μ Siemens. A JEOL JSM-7800F scanning electron microscope (SEM, Peabody, MA, USA) and atomic force microscopy (AFM) (JPK Instruments, Berlin, Germany) were used to capture micrographs to document and compare the surfaces before and after the corrosion. Energy dispersion X-ray (EDX) (AZoNano, Manchester, UK) and Kratos Axis Ultra DLD X-ray photoelectron spectroscopy (XPS) units (Kratos Analytical Ltd, Manchester, UK) were used to qualitatively and quantitatively measure the elemental composition of the formed corrosion product layers at different H$_2$S concentrations.

3. Results and Discussion

3.1. Potentiodynamic Polarization

Figure 3 presents the typical potentiodynamic polarization curve of API X100 steels in 3.5% NaCl solution and at different H$_2$S concentrations.

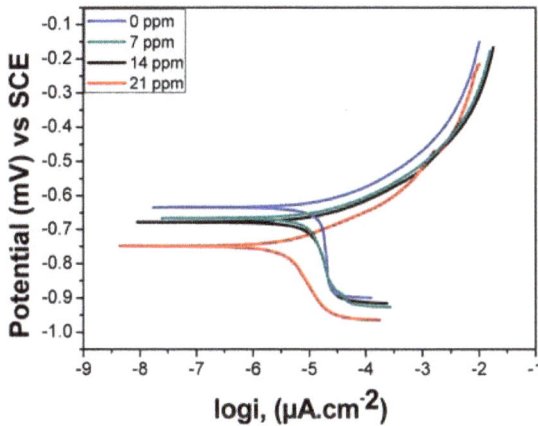

Figure 3. Potentiodynamic polarization curves of API X100 steels.

The corrosion current decreases as the anodic current and the potential were scanned from open circuit potential (OCP) in the anodic direction at different H$_2$S bubbling durations, which confirms that the protective nature of the formed sulfide film increases due to the increase in bubbling duration. The decrease in the cathodic current with increasing bubbling duration is attributed to two opposing factors: the pH and the H$_2$S gas concentrations in the solution [10]. Table 2 shows the Tafel parameters derived from the Tafel measurements shown in Figure 3.

Table 2. The Tafel analysis parameters derived from the Tafel measurement shown in Figure 3.

H$_2$S Concentration (ppm)	E_{corr} (V)	I_{corr} (μA·cm^{-2})	βc (V·decade^{-1})	βa (V·decade^{-1}) × 10^{-3}
0	−0.633 ± 0.011	52.29 ± 0.091	1.011 ± 0.012	98.92 ± 0.007
7	−0.667 ± 0.020	28.52 ± 0.054	1.018 ± 0.013	112.1 ± 0.013
14	−0.679 ± 0.015	22.15 ± 0.013	1.012 ± 0.006	116.7 ± 0.008
21	−0.748 ± 0.012	6.89 ± 0.017	535.8 ± 0.014	103.8 ± 0.015

3.2. Electrochemical Impedance Spectroscopy (EIS)

The resistance of carbon steel materials to corrosion depends significantly on the physicochemical properties of the material [31]. Figure 4 shows the measured electrochemical impedance spectra

(Nyquist and Bode formats) within a frequency range of 0.1 to 100 KHz starting from the highest to the lowest frequency limit. The corresponding fitted data (dotted line) are also within a frequency range of 0.1 to 100 KHz in 3.5% NaCl solutions saturated with H_2S gas at 0, 7, 14, and 21 ppm concentrations. All values were obtained using the Gamry Echem Analysis Software technique [32].

Nyquist and Bode with the phase angle plots obtained for API X100 steel materials tested in 3.5% NaCl solution with and without H_2S are shown in Figure 4. An increase in resistance with increasing H_2S concentration can be seen in the impedance behavior of the API X100 steel. The Nyquist plot of the API X100 steel without H_2S displays a defined single smaller semicircular shape, while the tests with H_2S showed increasing semicircular curves (Figure 4). The fitting was done using the equivalent circuit shown in Figure 5. The values of the fitted data are shown in Table 3. In the diagram, R_s represents the solution resistance; CPE in the circuit is the constant phase element; W is the mass transfer Warburg element [33]. R_{ct} stands for the charge transfer resistance, and R_{po} is the pore resistance.

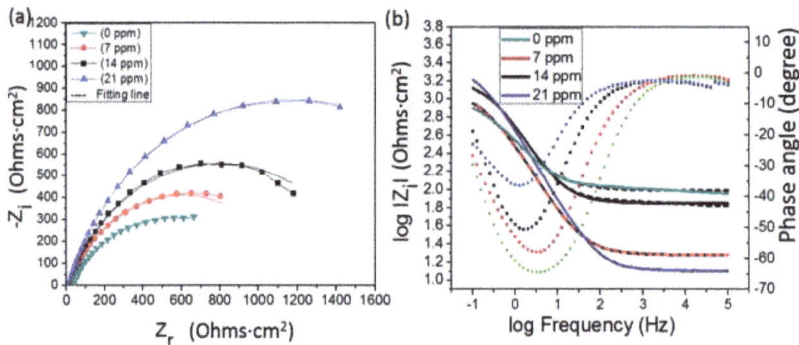

Figure 4. Measured Electrochemical Impedance Spectroscopy (EIS) data represented in (**a**) Nyquist and (**b**) Bode with phase angle represented in a dotted line format for API X100 steel at different durations in 3.5% NaCl solutions with hydrogen sulfide within a frequency range of 0.1 to 100 KHz.

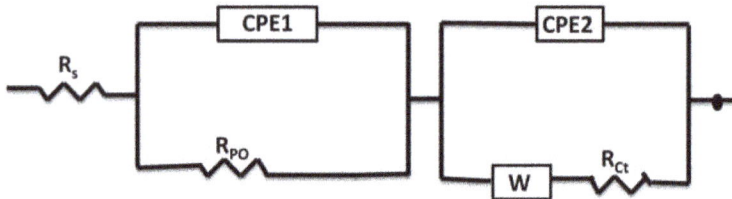

Figure 5. The equivalent electrical circuit model used to analysis the EIS data. CPE: constant phase element.

Table 3. Influence of hydrogen sulfide concentration on the corrosion of API X100 steel—data exported from fitted circuits.

H_2S Concentration (ppm)	R_s (ohm·cm²)	R_{po} (ohm·cm²)	CPE1 (Ss^{n}cm^{-2}) $\times 10^{-3}$	n_1	R_{ct} (ohm·cm²)	CPE2 (Ss^{n}cm^{-2}) $\times 10^{-4}$	n_2
0	12.73	52.95	4.52	0.735	840	5.17	0.869
7	19.05	67.13	3.31	0.718	1095	4.82	0.785
14	70.99	98.22	2.78	0.632	1440	3.09	0.807
21	98.19	106.10	1.52	0.640	1747	2.17	0.676

It can be observed that an increase in hydrogen sulfide concentration increased the charge transfer resistance (R_{ct}) of the steel [34].

The increase in R_{ct} indicates that the sulfide film is becoming more protective. Additionally, the higher the R_{ct} is, the lower the double layer capacitance will be. As can be seen in Table 3, the most protective sulfide layer was at a hydrogen sulfide concentration of 21 ppm. The corrosion resistance of the API X100 steel increases when the hydrogen sulfide concentration is increased from 0 to 21 ppm. This observation is attributed to the iron sulfide film formation and growth [13,35], which is in good agreement with the findings of Tang et al. [36], who observed similar iron sulfide film deposits on the SAE-1020 carbon steel when exposed to a sour environment. Furthermore, the increase in the hydrogen sulfide concentrations from 14 to 21 ppm showed evidence of crack development, which may be ascribed to hydrogen-induced cracking (HIC). The hydrogen-induced crack (HIC) has been reported to be one of the most significant damage modes in a sour environment [37,38]. It is believed that hydrogen atoms produced due to surface corrosion of the steel diffuse into it through microstructural defects that exist in the material [38]. The accumulation of a critical amount of hydrogen in the metal defeats results in HIC initiation and propagation, as recently reported by Kittel et al. [39].

It is demonstrated that the corrosion resistance of API X100 steel material enhances significantly with increasing sulfide concentration, and can be attributed to the formation of a protective iron sulfide layer on the API X100 steel surface when exposed to an aqueous hydrogen sulfide environment. A previous study [40] has shown that a chemical reaction occurs in a sour environment when H_2S is bubbled into the aqueous solution. However, some authors have reported the occurrence of supersaturated mackinawite when low concentrations of H_2S are combined with an increase in Fe^{2+} concentration, resulting in the growth of mackinawite on the steel surface [41–44]. Furthermore, several authors [45,46] have also reported the significant effects of H_2S concentration on the sulfide layer formation.

3.3. Surface Characterizations

The atomic force microscopy (AFM) technique was used for a better understanding of the corroded API X100 steel surface. The 3D AFM images of the corroded surfaces are presented in Figure 6a–d. It is evident that the surface roughness increases with increasing H_2S concentration. The specimens corroded at highest H_2S concentration have the highest surface roughness (Ra = 400 nm), while the specimens tested in the absence of H_2S have the lowest surface roughness (Ra = 50 nm). This behavior can be ascribed to the increase in the formation of more scale layers on the API X100 steel surfaces with increasing concentration of H_2S (Figure 6).

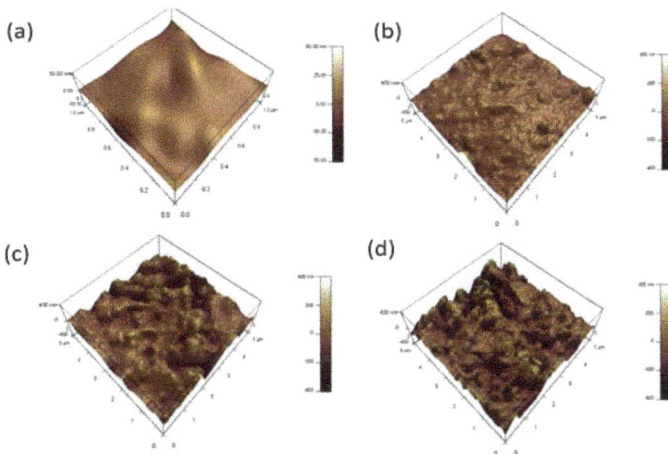

Figure 6. 3D atomic force microscopy (AFM) images of API X100 steel exposed to a sour environment at different H_2S concentrations of (**a**) 0 ppm; (**b**) 7 ppm; (**c**) 14 ppm; (**d**) 21 ppm.

Evidence of the sulfide layer formation of sulfide layer on the surface of the API X100 steel specimen when tested in an H_2S environment was confirmed by EDX analysis. The presence of sulfur (S) peak in the EDX spectra shown in Figure 7b–d clearly confirms the formation of a sulphide layer on the surface of all API X100 steel tested under H_2S environment. However, there was no evidence of "S" in the spectra taken from the test specimen performed without H_2S (Figure 7a).

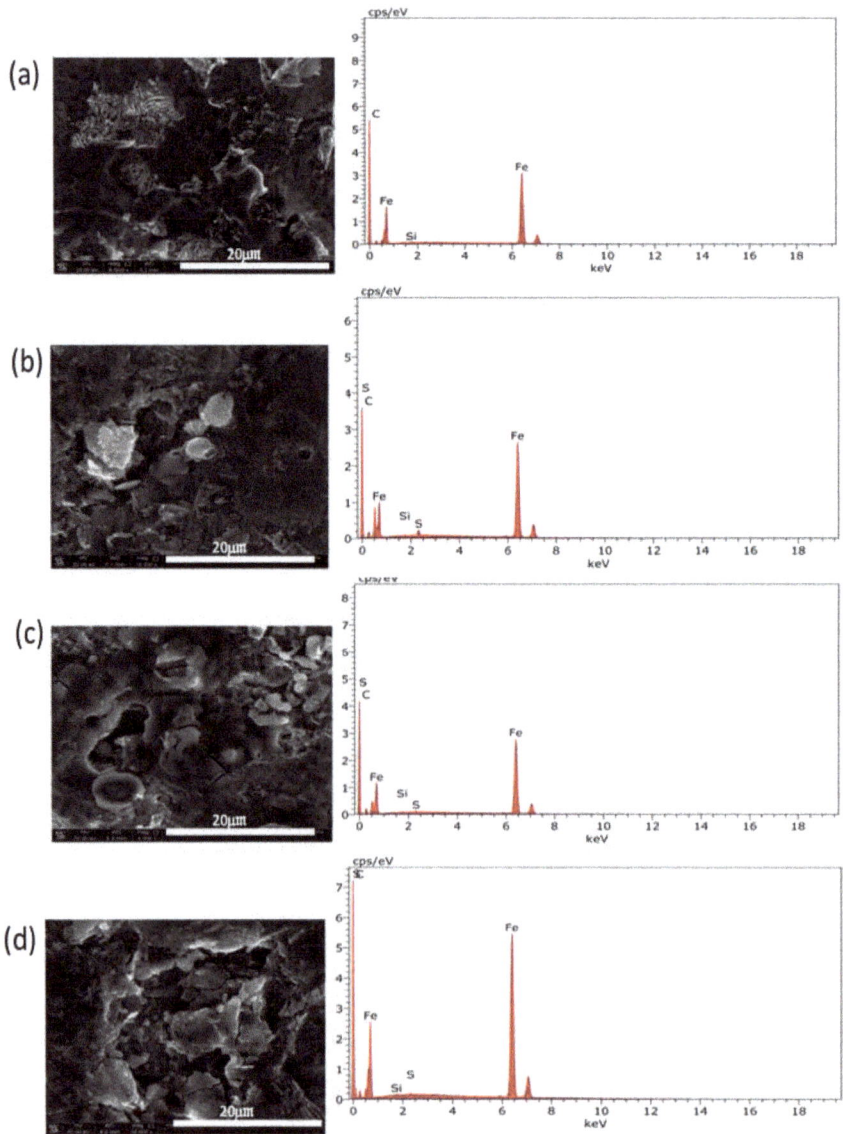

Figure 7. SEM images of API X100 steel material for tests conducted in 3.5% NaCl (**a**) without H_2S, (**b**) with 7 ppm of H_2S, (**c**) with 14 ppm H_2S, and (**d**) with 21 ppm H_2S.

Evidence of cracks can be seen on the sulfide layers at different H_2S concentrations, especially at higher concentrations (Figure 7). The phenomenon of the formation of a sulfide film and its cracking is influenced by test solution conditions [47]. It is clear from Figure 7 that deposits of iron sulfide layers have been built up on the sample surface (Figure 7). SEM micrographs shown in Figure 7 reveal that the sulfide layers appear finer and more clustered as the H_2S concentration is increased, which can be attributed to the formation of a more homogeneous protective iron sulfide layer on the steel surface. It has been reported that the continuous diffusion of hydrogen sulfide on a steel surface can lead to generate internal stresses into the formed sulfide layer [15]. Upon exceeding the stress limit of the crystalline layer, micro-cracking of the formed layers occurs as observed in Figure 7b–d. Analysis of the corroded metal surface state can be carried out using surface X-ray photoelectron spectroscopy analysis techniques. Figure 8 displays the survey XPS spectra of API X100 steel after exposure to different H_2S concentrations. The presence of iron peaks at 709.82 eV, 713.53 eV and a carbon peak at 285.5 eV can be noticed. However, the peaks at 163 and 170 eV binding energy (Figure 8b) indicate the presence of sulfur and thus deduce the possible formation of sulfide layers on the corroded API X100 steel surfaces. This is in good agreement with previous studies [48].

Figure 9 shows the XRD spectra of the corroded API X100 steel samples in saline and sour solutions. XRD peaks located at different 2θ values confirm the presence of different iron compounds. The test performed in the absence of H_2S shows a diffraction pattern typical of the peak of iron oxide (Fe_2O_3, 2θ = 65.08°) [4]. However, the presence of iron sulphide also referred to as mackinawite (FeS, 2θ = 82.43°) was observed on the surface of API X100 steel surface when exposed to different H_2S concentrations.

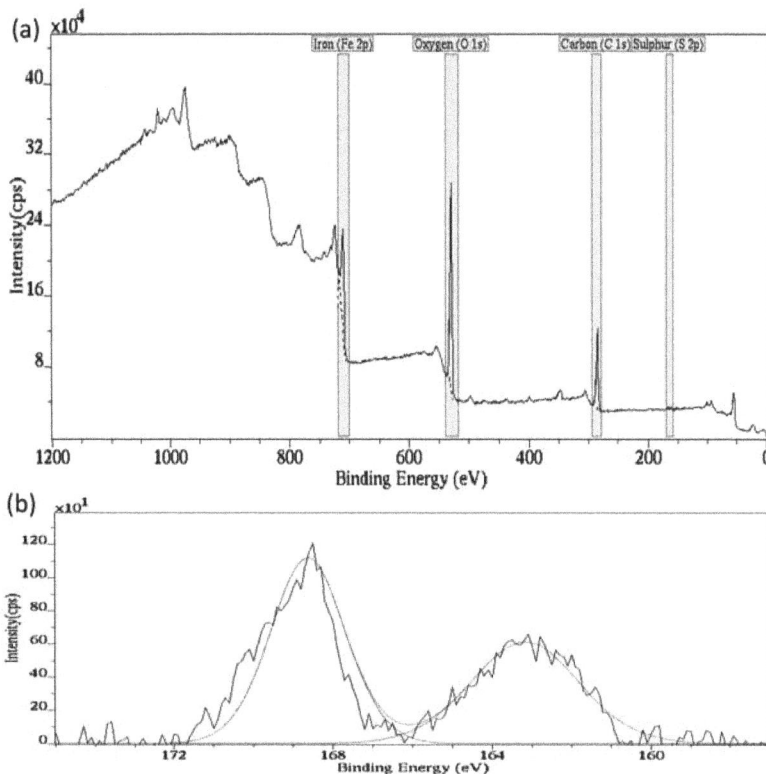

Figure 8. X-ray photoelectron spectroscopy (XPS) fitting spectra obtained from (**a**) the survey and (**b**) the sulfur spectra with a fitted line of corroded API X100 steel at 21 ppm of H_2S.

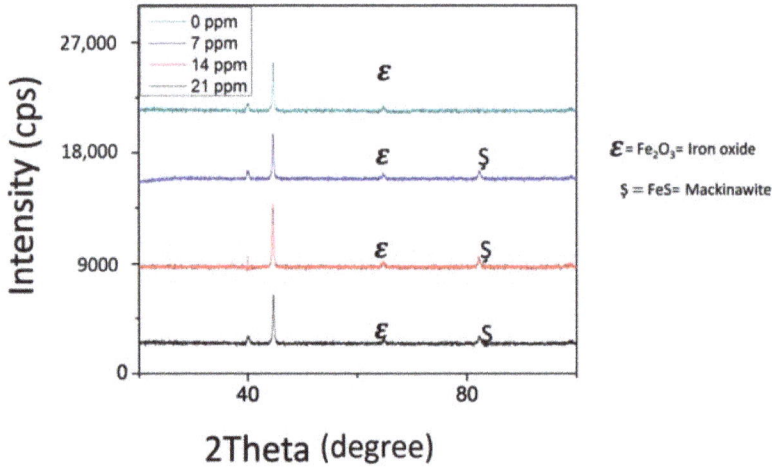

Figure 9. The XRD spectra of API X100 with the corrosion product films at different H_2S concentrations. Scanning was at $2°S^{-1}$.

Several researchers have reported that the initial formation of mackinawite layer on the steel surface before the formation of the more stable sulfide species depends on the sample material and the test conditions [49,50]. The presence of this iron sulphide (mackinawite) layer provides some protection to the surface of the material from further corrosion due to its structural orientation [51], resulting in a decrease in the corrosion rate. This protective iron sulfide layer becomes more homogeneous and dense with increasing H_2S concentration as can be observed in the SEM micrographs (Figure 7). Similar studies have shown that iron and sulfide films were observed on API X52 pipeline steel surface when exposed to sour environments [4]. However, their quantity is sensitive to the experimental procedure [4,52]. Subsequently, iron sulfide layer (mackinawite) is formed due to the presence of iron and sulfide ions in the solution. Nevertheless, an increase in the sulfide concentration as seen in this test decreases the iron concentration in the solution, resulting in the precipitation of protective iron sulfide on the steel surface [53].

4. Conclusions

In order to understand the effect of hydrogen sulfide on the corrosion behavior of API X100 steel, sets of experiments in the presence and absence of sulfide in 3.5% NaCl were performed on API X100 steel. Based on the experimental results, the following can be concluded:

1. EDX and XPS analyses confirm the presence of sulfur on the corroded API X100 steel surfaces at different H_2S concentrations.
2. A significant decrease in the corrosion rate of API X100 steel in 3.5% NaCl solutions containing various concentrations of H_2S is noticed which can be attributed to the formation of sulfide layer.
3. SEM micrographs confirm that the formed protective sulfide layer becomes more homogenous and denser as the H_2S concentration is increased.
4. The formation of more homogeneous, dense and protective iron sulfide layer (mackinawite) is responsible for the decrease in the corrosion rate with increasing concentration of H_2S.

Acknowledgments: This publication was made possible by National Priorities Research Program (NPRP) Grant 6-027-2-010 from the Qatar National Research Fund (a member of the Qatar Foundation). Statements made herein are solely the responsibility of the authors.

Author Contributions: Paul C. Okonkwo, Rana Abdul Shakoor, and Adel Mohamed Amer Mohamed conceived and designed the experiments; Paul C. Okonkwo performed the experiments; Rana Abdul Shakoor, Adel Mohamed Amer Mohamed, Abdelbaki Benamor, and Mohammed Jaber F A Al-Marri supervised experimental work and data analysis; Rana Abdul Shakoor and Adel Mohamed Amer Mohamed contributed the analysis tools and reviewed the manuscript; Paul C. Okonkwo wrote the manuscript.

Conflicts of Interest: The authors declare no conflict of interest.

References

1. Igi, S.; Sakimoto, T.; Endo, S. Effect of Internal Pressure on Tensile Strain Capacity of X80 Pipeline. *Procedia Eng.* **2011**, *10*, 1451–1456. [CrossRef]

2. Jones, N.; Birch, R.S. Influence of Internal Pressure on the Impact Behavior of Steel Pipelines. *J. Press. Vessel. Technol.* **1996**, *118*, 464–471. [CrossRef]

3. Zhao, M.-C.; Tang, B.; Shan, Y.-Y.; Yang, K. Role of microstructure on sulfide stress cracking of oil and gas pipeline steels. *Metall. Mater. Trans. A* **2003**, *34*, 1089–1096.

4. Hernández-Espejel, A.; Domínguez-Crespo, M.A.; Cabrera-Sierra, R.; Rodríguez-Meneses, C.; Arce-Estrada, E.M. Investigations of corrosion films formed on API-X52 pipeline steel in acid sour media. *Corros. Sci.* **2010**, *52*, 2258–2267. [CrossRef]

5. López, H.F.; Raghunath, R.; Albarran, J.L.; Martinez, L. Microstructural aspects of sulfide stress cracking in an API X-80 pipeline steel. *Metall. Mater. Trans. A* **1996**, *27*, 3601–3611. [CrossRef]

6. Nasirpouri, F.; Mostafaei, A.; Fathyunes, L.; Jafari, R. Assessment of localized corrosion in carbon steel tube-grade AISI 1045 used in output oil–gas separator vessel of desalination unit in oil refinery industry. *Eng. Fail. Anal.* **2014**, *40*, 75–88. [CrossRef]

7. Lens, P.N.L.; Visser, A.; Janssen, A.J.H.; Hulshoff Pol, L.W.; Lettinga, G. Biotechnological Treatment of Sulfate-Rich Wastewaters. *Crit. Rev. Environ. Sci. Technol.* **1998**, *28*, 41–88. [CrossRef]

8. Lucio-Garcia, M.A.; Gonzalez-Rodriguez, J.G.; Casales, M.; Martinez, L.; Chacon-Nava, J.G.; Neri-Flores, M.A.; Martinez-Villafañe, A. Effect of heat treatment on H_2S corrosion of a micro-alloyed C–Mn steel. *Corros. Sci.* **2009**, *51*, 2380–2386. [CrossRef]

9. Qi, Y.; Luo, H.; Zheng, S.; Chen, C.; Lv, Z.; Xiong, M. Effect of temperature on the corrosion behavior of carbon steel in hydrogen sulphide environments. *Int. J. Electrochem. Sci.* **2014**, *9*, 2101–2112.

10. Liu, M.; Wang, J.Q.; Ke, W.; Han, E.H. Corrosion Behavior of X52 Anti-H_2S Pipeline Steel Exposed to High H_2S Concentration Solutions at 90 °C. *J. Mater. Sci. Technol.* **2014**, *30*, 504–510. [CrossRef]

11. Okonkwo, P.C.; Mohamed, A.M.A. Erosion-Corrosion in Oil and Gas Industry: A Review. *Int. J. Metall. Mater. Sci. Eng.* **2014**, *4*, 7–28.

12. Yin, Z.F.; Zhao, W.Z.; Bai, Z.Q.; Feng, Y.R.; Zhou, W.J. Corrosion behavior of SM 80SS tube steel in stimulant solution containing H_2S and CO_2. *Electrochim. Acta* **2008**, *53*, 3690–3700. [CrossRef]

13. Vedage, H.; Ramanarayanan, T.A.; Mumford, J.D.; Smith, S.N. Electrochemical Growth of Iron Sulfide Films in H_2S-Saturated Chloride Media. *Corrosion* **1993**, *49*, 114–121. [CrossRef]

14. Bai, P.; Zheng, S.; Chen, C. Electrochemical characteristics of the early corrosion stages of API X52 steel exposed to H_2S environments. *Mater. Chem. Phys.* **2015**, *149–150*, 295–301. [CrossRef]

15. Schulte, M.; Schutze, M. The Role of Scale Stresses in the Sulfidation of Steels. *Oxid. Met.* **1999**, *51*, 55–77. [CrossRef]

16. Carneiro, R.A.; Ratnapuli, R.C.; de Freitas Cunha Lins, V. The influence of chemical composition and microstructure of API linepipe steels on hydrogen-induced cracking and sulfide stress corrosion cracking. *Mater. Sci. Eng. A* **2003**, *357*, 104–110. [CrossRef]

17. Xu, M.; Li, W.; Zhou, Y.; Yang, X.; Wang, Z.; Li, Z. Effect of pressure on corrosion behavior of X60, X65, X70, and X80 carbon steels in water-unsaturated supercritical CO_2 environments. *Int. J. Greenh. Gas Control.* **2016**, *51*, 357–368. [CrossRef]

18. Gadala, I.M.; Alfantazi, A. A study of X100 pipeline steel passivation in mildly alkaline bicarbonate solutions using electrochemical impedance spectroscopy under potentiodynamic conditions and Mott–Schottky. *Appl. Surf. Sci.* **2015**, *357*, 356–368. [CrossRef]

19. Eliyan, F.F.; Alfantazi, A. Mechanisms of Corrosion and Electrochemical Significance of Metallurgy and Environment with Corrosion of Iron and Steel in Bicarbonate and Carbonate Solutions—A Review. *Corrosion* **2014**, *70*, 880–898. [CrossRef]

20. Sherar, B.W.A.; Keech, P.G.; Noël, J.J.; Worthingham, R.G.; Shoesmith, D.W. Effect of Sulfide on Carbon Steel Corrosion in Anaerobic Near-Neutral pH Saline Solutions. *Corrosion* **2012**, *69*, 67–76. [CrossRef]
21. Papavinasam, S.; Doiron, A.; Revie, R.W. Model to Predict Internal Pitting Corrosion of Oil and Gas Pipelines. *Corrosion* **2010**, *66*. [CrossRef]
22. Sridhar, N.; Dunn, D.S.; Anderko, A.M.; Lencka, M.M.; Schutt, H.U. Effects of Water and Gas Compositions on the Internal Corrosion of Gas Pipelines—Modeling and Experimental Studies. *Corrosion* **2001**, *57*, 221–235. [CrossRef]
23. Han, J.; Carey, J.W.; Zhang, J. Effect of sodium chloride on corrosion of mild steel in CO_2-saturated brines. *J. Appl. Electrochem.* **2011**, *41*, 741–749. [CrossRef]
24. Shi, J.; Sun, W.; Jiang, J.; Zhang, Y. Influence of chloride concentration and pre-passivation on the pitting corrosion resistance of low-alloy reinforcing steel in simulated concrete pore solution. *Constr. Build. Mater.* **2016**, *111*, 805–813. [CrossRef]
25. Abayarathna, D.; Naraghi, A.R.; Wang, S.H. *The Effect of Surface Films on Corrosion of Carbon Steel in a CO_2–H_2S–H_2O System in Corrosion/2005*; NACE: Houston, TX, USA, 2005.
26. Valcarce, M.B.; Vázquez, M. Carbon steel passivity examined in alkaline solutions: The effect of chloride and nitrite ions. *Electrochim. Acta* **2008**, *53*, 5007–5015. [CrossRef]
27. Rao, T.S.; Sairam, T.N.; Viswanathan, B.; Nair, K.V.K. Carbon steel corrosion by iron oxidising and sulphate reducing bacteria in a freshwater cooling system. *Corros. Sci.* **2000**, *42*, 1417–1431. [CrossRef]
28. Paul, C.O.; Ahmed, E.; Mohamed, A.M.A. Effect of Temperature on the Corrosion Behavior of API X80 Steel Pipeline. *Int. J. Electrochem. Sci.* **2015**, *10*, 10246–10260.
29. Okonkwo, P.; Shakoor, R.; Zagho, M.; Mohamed, A. Erosion Behaviour of API X100 Pipeline Steel at Various Impact Angles and Particle Speeds. *Metals* **2016**, *6*, 232. [CrossRef]
30. TM0284–2003, N.S. *Evaluation of Pipeline and Pressure Vessel Steels for Resistance to Hydrogen-induced Cracking*; NACE International: Houston, TX, USA, 2003.
31. Saravanamoorthy, S.; Velmathi, S. Physiochemical interactions of chiral Schiff bases on high carbon steel surface: Corrosion inhibition in acidic media. *Prog. Org. Coat.* **2013**, *76*, 1527–1535. [CrossRef]
32. Musa, A.Y.; Kadhum, A.A.H.; Mohamad, A.B.; Rahoma, A.A.B.; Mesmari, H. Electrochemical and quantum chemical calculations on 4,4-dimethyloxazolidine-2-thione as inhibitor for mild steel corrosion in hydrochloric acid. *J. Mol. Struct.* **2010**, *969*, 233–237. [CrossRef]
33. Liu, H.J.; Xu, Q.; Yan, C.W.; Cao, Y.Z.; Qiao, Y.L. The Effect of Temperature on the Electrochemical Behavior of the V(IV)/V(V) Couple on a Graphite Electrode. *Int. J. Electrochem. Sci.* **2011**, *6*, 3483–3496.
34. Islam, M.A.; Farhat, Z.N. Characterization of the Corrosion Layer on Pipeline Steel in Sweet Environment. *J. Mater. Eng. Perform.* **2015**, *24*, 3142–3158. [CrossRef]
35. Huang, H.-H.; Tsai, W.-T.; Lee, J.-T. Electrochemical behavior of the simulated heat-affected zone of A516 carbon steel in H_2S solution. *Electrochim. Acta* **1996**, *41*, 1191–1199. [CrossRef]
36. Tang, J.; Shao, Y.; Guo, J.; Zhang, T.; Meng, G.; Wang, F. The effect of H_2S concentration on the corrosion behavior of carbon steel at 90 °C. *Corros. Sci.* **2010**, *52*, 2050–2058. [CrossRef]
37. Elboujdaini, M.; Revie, R.W. Metallurgical factors in stress corrosion cracking (SCC) and hydrogen-induced cracking (HIC). *J. Solid State Electrochem.* **2009**, *13*, 1091–1099. [CrossRef]
38. Lu, B.T.; Luo, J.L. Relationship between Yield Strength and Near-Neutral pH Stress Corrosion Cracking Resistance of Pipeline Steels—An Effect of Microstructure. *Corrosion* **2006**, *62*, 129–140. [CrossRef]
39. Kittel, J.; Smanio, V.; Fregonese, M.; Garnier, L.; Lefebvre, X. Hydrogen-induced cracking (HIC) testing of low alloy steel in sour environment: Impact of time of exposure on the extent of damage. *Corros. Sci.* **2010**, *52*, 1386–1392. [CrossRef]
40. Ma, H.; Cheng, X.; Li, G.; Chen, S.; Quan, Z.; Zhao, S.; Niu, L. The influence of hydrogen sulfide on corrosion of iron under different conditions. *Corros. Sci.* **2000**, *42*, 1669–1683. [CrossRef]
41. Buckley, A.N.; Woods, R. The surface oxidation of pyrite. *Appl. Surf. Sci.* **1987**, *27*, 437–452. [CrossRef]
42. Hamilton, I.C.; Woods, R. An investigation of surface oxidation of pyrite and pyrrhotite by linear potential sweep voltammetry. *J. Electroanal. Chem. Interfacial Electrochem.* **1981**, *118*, 327–343. [CrossRef]
43. King, R.A.; Miller, J.D.A.; Smith, J.S. Corrosion of Mild Steel by Iron Sulphides. *Br. Corros. J.* **1973**, *8*, 137–141.
44. Lee, W.; Lewandowski, Z.; Nielsen, P.H.; Hamilton, W.A. Role of sulfate-reducing bacteria in corrosion of mild steel: A review. *Biofouling* **1995**, *8*, 165–194. [CrossRef]

45. Choi, Y.-S.; Nesic, S.; Ling, S. Effect of H_2S on the CO_2 corrosion of carbon steel in acidic solutions. *Electrochimica Acta* **2011**, *56*, 1752–1760. [CrossRef]

46. Sardisco, J.B.; Pitts, R.E. Corrosion of Iron in an H_2S-CO_2-H_2O System Mechanism of Sulfide Film Formation and Kinetics of Corrosion Reaction. *Corrosion* **1965**, *21*, 245–253. [CrossRef]

47. Bonis, M.; Crolet, J.-L. Practical aspects of the influence of in situ pH on H_2S-induced cracking. *Corros. Sci.* **1987**, *27*, 1059–1070. [CrossRef]

48. Heuer, J.K.; Stubbins, J.F. An XPS characterization of $FeCO_3$ films from CO_2 corrosion. *Corros. Sci.* **1999**, *41*, 1231–1243. [CrossRef]

49. Rickard, D. Metastable Sedimentary Iron Sulfides. *Dev. Sedimentol.* **2012**, *65*, 195–231.

50. Bai, P.; Zheng, S.; Chen, C.; Zhao, H. Investigation of the iron-sulfide phase transformation in nanoscale. *Cryst. Growth Des.* **2014**, *14*, 4295–4302. [CrossRef]

51. Solehudin, A.; Nurdin, I.; Suratno, W.; Agma, M. EIS Study of Temperature and H_2S Concentration Effect on API 5LX65 Carbon Steel Corrosion in Chloride Solution. *J. Mater. Sci. Eng. A* **2011**, *1*, 496–505.

52. Lowson, R.T. Aqueous oxidation of pyrite by molecular oxygen. *Chem. Rev.* **1982**, *82*, 461–497. [CrossRef]

53. El Mendili, Y.; Abdelouas, A.; Bardeau, J.F. Impact of a sulphidogenic environment on the corrosion behavior of carbon steel at 90 °C. *RSC Adv.* **2013**, *3*, 15148–15156. [CrossRef]

metals

MDPI

Article

Characterization of the Microstructure, Mechanical Properties, and Corrosion Resistance of a Friction-Stir-Welded Joint of Hyper Duplex Stainless Steel

Jianchun Li [1,2], Xinglong Liu [1], Guoping Li [2], Peide Han [1,*] and Wei Liang [1]

[1]   College of Materials Science and Engineering, Taiyuan University of Technology, Taiyuan 030024, China;
     lijc02@tisco.com.cn (J.L.); imliuxl@163.com (X.L.); liangwky@126.com (W.L.)
[2]   Taiyuan Iron and Steel (Group) Company Ltd., Taiyuan 030003, China; ligp@tisco.com.cn
*   Correspondence: hanpeide@tyut.edu.cn; Tel.: +86-351-6018398

Academic Editor: Robert Tuttle
Received: 10 January 2017; Accepted: 5 April 2017; Published: 13 April 2017

Abstract: This study investigates the microstructure, mechanical properties, and corrosion resistance of a friction-stir-welded joint of the hyper duplex stainless steel SAF2707. Friction stir welding (FSW) is performed at a tool rotation rate of 400 rpm and a welding speed of 100 mm/min. The microstructure of the joints is examined using scanning electron microscopy and X-ray diffraction. Tensile test and fractography are subsequently employed to evaluate the mechanical properties of the joints. Results show that the grain size of the stir zone (SZ) is smaller than that of the base metal (BM). Electron back-scattered diffraction analysis reveals that fine-equiaxed grains form in the SZ because of the dynamic recrystallization during the FSW. These grains become increasingly pronounced in the austenite phase. The tensile specimens consistently fail in the BM, implying that the welded joint is an overmatch to the BM. Moreover, the welded joints consist of finer grains and thus display higher tensile strength than their BMs. Potentiodynamic polarization curves and impedance spectroscopy both demonstrate that the corrosion resistance of the SZ is superior to that of the base material.

Keywords: friction stir weld; hyper duplex stainless steel; microstructure; mechanical properties; corrosion resistance

---

1. Introduction

Given the favorable combination of mechanical and corrosion properties and excellent weldability of duplex stainless steels (DSSs), DSSs have become more attractive than austenitic and ferritic stainless steels in terms of applications [1,2]. In the process of their development, super stainless steels have been widely applied in environments characterized by relatively excessive corrosion and high pressure. However, they exhibit incapacities because of the short service life in such areas as oil exploration, which strives for deeper water over 2500 m and requires stronger corrosion resistant [3]. Thus, a new high-alloyed DSS demonstrating excellent corrosion resistance and enhanced strength is necessary. SAF2707HD is a new type of DSS in the form of a hyper stainless steel developed by Sandvik Corporation. It exhibits excellent performance in relevant test evaluations. The superior corrosion resistance of DSSs extends their application to aggressive chloride environments, such as hot tropical seawater, seabed piping systems, and deep wells [3,4]. The corrosion resistance of DSSs is determined by the fraction of α, γ, and detrimental intermetallic phases, such as σ, which often renders Cr- and Mo-depleted zones susceptible to high-rate dissolution [5–7]. Given their inherent brittleness, these undesirable intermetallic phases, which easily form during hot rolling or welding, often degrade the mechanical properties of DSSs [8,9].

In most applications, selected construction materials must undergo a welding process. However, problems can occur when welding DSSs [10]. For example, fusion welding may destroy the favorable duplex microstructure of DSSs and produce coarse ferrite grains, intergranular brittle and intergranular austenite phases in the weld metal and heat-affected zone [11]. In addition, high ferrite content and coarse grains reduce the corrosion resistance and mechanical properties of welded joints [12,13]. Moreover, σ phase in DSSs may precipitate during welding because of high Cr, Mo, and N concentrations. Studies have shown that the holding time of precipitating 1% σ phase at the most critical temperature is 134 s, whereas that for hyper DSS (HDSS) is 69 s [14,15].

Several welding methods have been used to address these problems. Among these methods, friction stir welding (FSW) has recently attracted considerable attention. FSW is a solid-state technology wherein a material is subjected to intense plastic deformation at an elevated temperature, thereby generating fine and equiaxed recrystallized grains. The main advantages of FSW are solid-state welding, absence of metal addition, and fast cooling. In addition, FSW prevents precipitation. Initially, FSW is mainly used in metals with low-melting points, such as Al and Mg alloys, because the tools used in FSW display limited capacities at high temperatures [16,17]. With the improvement in tools used in FSW, the application of FSW is extended to materials with high melting points, such as steel and titanium alloys. As a result, numerous investigations have been conducted on FSW of steels and other alloys with high melting points.

Existing investigations report that only a few studies have explored the effect of FSW on the microstructure, mechanical properties, and corrosion properties of DSSs [18–20]. Sato [18] has demonstrated that FSW elicits the desirable mechanical properties of SAF2507; these properties include defect-free weld and generation of fine and equiaxed recrystallized grains in the stir zone (SZ). Saeid [19] has investigated the mechanism of the grain refinement of DSSs by using electron backscattered diffraction (EBSD) and found that the mechanism is correlated with different mechanisms of dynamic recrystallization. Furthermore, the corrosive behavior of lean DSSs subjected to FSW was studied by Sarlak, who found that the resistance of DSSs to the local and pitting corrosion of weldments is superior to that of base metals (BMs) [20].

Studies have shown that FSW considerably affects the mechanical properties and corrosion resistance of DSSs during welding. However, to the best of our knowledge, studies on the welding of HDSS using FSW are few. Studies on HDSS welding mainly focus on the influence of shielding gases on corrosion resistance [21,22]. No systematic research has investigated the corrosion resistance and mechanical properties of HDSS weldments. Therefore, in this study, HDSS is welded using FSW to investigate the microstructure, mechanical properties, and resistance to corrosion of HDSSs that underwent FSW.

2. Materials & Experiment

SAF2707HD was prepared using vacuum induction furnace and electro-slag remelting furnace. Table 1 shows the chemical composition of SAF2707HD. The experimental steel was supplied in the form of a cold rolled and annealed (1200 °C for 1 h) plate (150 mm × 100 mm × 3 mm). The plates were ground to remove the oxide on the top and bottom surfaces to ensure the welding quality. Welding was performed using an FSW machine (China FSW Center, Beijing, China) parallel to the rolling direction at a welding speed of 100 mm/min and a tool rotation rate of 400 rpm. The welding tool was made of tungsten-rhenium (W-RE) with a shoulder diameter of 20 mm and a tapered probe. The pin was tapered from 8 mm at the shoulder to 4.5 mm with a length of 2.8 mm. Figure 1 presents the schematic of the tool used in the FSW experiment. During the welding process, the tilt angle was 0° and downward force was controlled at 20 kN. An argon atmosphere around the tool was introduced as shielding gas to prevent excessive surface oxidation. Figure 2 presents a typical surface morphology diagram of SAF2707HD butt joint subjected to FSW. In this diagram, the welding surface is smooth, and macroscopic cracks, air holes, and tunnel defects are absent along the welding line, indicating that the welding parameter can produce sufficient heat input to ensure the flow of plastic materials.

Table 1. Chemical composition of base material (wt. %).

Elements	C	N	Si	Mn	Cr	Ni	Co	Mo	Cu	Ce	Fe
Content (wt. %)	0.02	0.47	0.42	1.00	27.07	6.71	0.99	4.66	0.85	0.03	Bal.

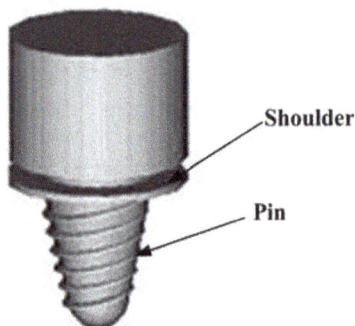

Figure 1. Schematic illustration of the welding tool used in the experiment [23].

Figure 2. Surface morphology of friction stir welding (FSW) butt joint.

Following FSW, a metallographic examination was conducted on a transverse section of the welded joint via optical microscopy (OM) (Leica, Heerbrugg, Switzerland) and scanning electron microscopy (SEM) (FEI, Hillsboro, OR, USA) equipped with an INCA energy dispersive X-ray spectrometer system (EDX) (Oxford Instruments, Oxon, UK). The specimen for OM was chemically etched with 25 g of KOH, 15 g of KSCN, and 100 mL of distilled water at a temperature of approximately 65 °C. Prior to EBSD, the specimen was electrolytically polished using 8 mL of perchloricacid, 5 mL of glycerol, 100 mL of ethanol, 10 mL of 2-butoxy ethanol, and 15 mL of distilled water at room temperature. Electrolytic polishing was performed at 30 V and at a 15 s storage time, respectively. A Philips X-Ray diffractometer (Philips, Eindhoven, Netherlands) was used to identify the phase formed on the surface of the SZ during the FSW.

Mechanical properties were investigated mainly through Vickers hardness and tensile tests. The tensile test was conducted at ambient temperature on an Instron testing machine (Instron, Boston, MA, USA). Vickers hardness test was performed using a LECO LM-700ATV hardness testing machine (LECO, St. Joseph, UT, USA). The hardness profiles were measured along and perpendicular to the mid-thickness of the joint with regular intervals of 0.5 mm at a load of 50 g for 15 s. The samples used in the tensile test were the welded joints extracted from the weldments. Figure 3 shows the sizes of tensile sample of the welded joints. The loading force was perpendicular to the direction of the welding at a constant cross head speed of 1 mm/s. The tensile section fracture was observed through SEM.

A three-electrode electrochemical cell system was introduced to assess the electrochemical corrosion of the BM and welded joints. A three-electrode system comprises a working electrode, platinum (Pt) foil, and a saturated calomel electrode (SCE). A Pt electrode and SCE work as counter and reference electrodes, respectively. All potentials quoted were related to SCE. The BM and weldment acting as working electrodes were sealed with fastening screws and then combined with a seal ring to avoid crevice corrosion. The exposed electrode area was 1 cm^2. All measurements were performed three times to ensure their accuracy and repeatability. Potentiodynamic polarization curves were obtained in 3.5 wt. % NaCl solution at 25 °C $\pm$ 1 °C at a scanning rate of 0.5 mV/s. The scanning region was from −0.2 V to a certain current at which transpassivity occurred.

Electrochemical impedance spectroscopy (EIS) (Princeton Applied Research, Oak Ridge, TN, USA) was conducted to evaluate the electrochemical behavior of the oxide film formed on the surface of the stainless steel. Prior to all the experiments, the specimens were immersed in a test solution for 30 min to reach a stationary condition. The scanning frequency region ranges from 0.1 mHz to 100,000 Hz. The measured EIS data was fitted with Zview 2.70 (Scribner Associates, Southern Pines, NC, USA).

Figure 3. (**a**) The front; (**b**) sectional views of the transverse tensile specimen.

3. Results and Discussion

3.1. Microstructural Evolution

Figure 4 shows a typical low-magnification overview of the different sections of SAF2707 welded joint subjected to FSW. The joint without internal defects exhibits a high degree of community. In the cross section (Figure 4a), the left and right sides of the joint center comprise the retreating side and advancing side of the rotational tool, respectively. A welded joint can be divided into three zones: the SZ, the thermo-mechanically affected zone (TMAZ), and the BM. Additionally, the figure shows that the SZ is characteristic of a ring vortex, which is produced through the flow of materials driven by the thread on the pin. A similar microstructural characterization is performed in related studies [24,25]. Based on the advancing and retreating sides of the working tool, the TMAZ located adjacent to the SZ can be classified into TMAZ–AS and TMAZ–RS. As verified by other authors [18–20,24], no obvious heat affected zone is observed because of the relatively low heat input..

Figure 4b shows the schematic of microstructure of the as-received BM. The BM exhibits a microstructure of austenitic islands embedded in ferrite matrix. Figure 4c–f shows the microstructures at different joint positions. As shown in Figure 4c, the SZ located in the joint center displays a stretched and two-phase microstructure. Compared with BMs, the SZ contains more elongated and finer austenitic islands. Similar to the typical microstructure of Al alloy and steel joints welded through FSW [16,17,24–27], the average grain size is significantly finer in the SZ than in the BM because the SZ experiences higher temperature and has undergone significantly plastic deformation. No significantly grain coarsening and σ phase are observed across the welded joint. During FSW, the material hardly melts because of the low peak temperature. These factors possibly account for the absence of grain coarsening and σ phase across the welded joint. Moreover, the region in the SZ adjacent to the TMAZ-AS (Figure 4d) consists of the finest equiaxed grains. As explained by Liu [26], the material lying in the advancing side of the SZ is slightly displaced forward from its original position. The material is rotated using a pin tool and deposited at roughly the same location as its initial position. The material flow of FSW suggests that the material on the advancing side of the SZ rotates and advances with the tool, and the material on the retreating side is entrained, indicating a relatively small deformation [27]. This mechanism of material flow suggests that the material in the SZ at the advancing side undergoes the most severe deformation and has a tendency to disrupt a duplex microstructure that produces the finest grains.

As shown in Figure 4e,f, the TMAZ–AS/SZ boundary is clearly sharper in the advancing side than that on the retreating side. This result is consistent with previous findings [18,24]. The difference between these boundaries can be attributed to the difference in shear stress between the advancing and retreating sides of the working tool. The shear force of the advancing side of the tool induces a larger plastic deformation than that induced in the retreating side. One study [18] has shown that the advancing side experiences the most severe deformation because of the material flow mechanism induced by the difference in plastic deformation around the pin tool during the FSW process. The TMAZ microstructure is distorted, and its grains are re-oriented perpendicular to the rotational direction of the tool. The asymmetry of FSW is evident in the TMAZ–AS and TMAZ–RS microstructures.

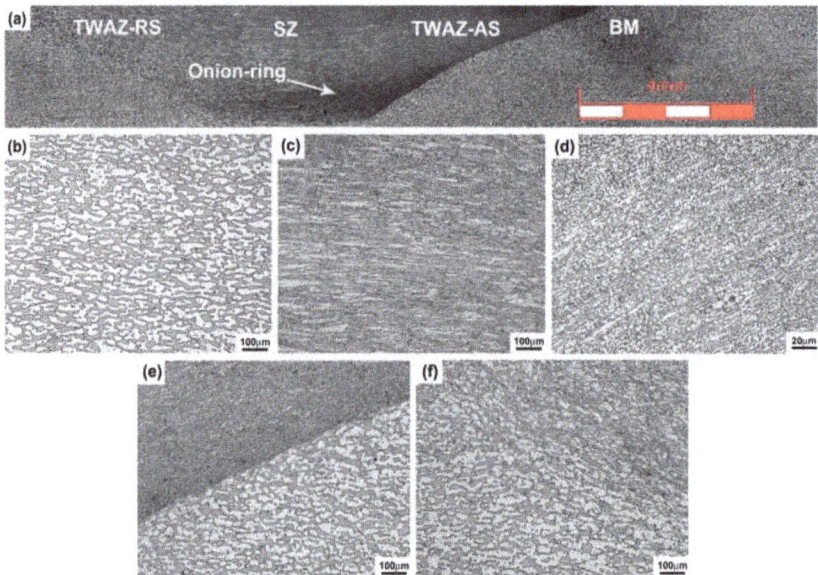

Figure 4. Optical microscopy of cross section of the joint (**a**); related microstructure (**b**) base metal (BM); (**c**) stir zone (SZ); (**d**) stir zone—advancing side (SZ-AS); (**e**) thermo-mechanically affected zone—advancing side (TMAZ-AS); (**f**) thermo-mechanically affected zone—retreating side (TMAZ-RS).

Figure 5 shows the backscatter electron (BSE) images of the SZ in different positions after electrolytic polishing. The grain size distribution is inhomogeneous along the thickness direction of the welded joint. The grain size decreases gradually from the top surface to the bottom of the welded joint. The heat during welding process is generated by the friction between the rotating tool and material interface, and the contact between the tool shoulder and the sample surface generates most of the heat; thus, the maximum temperature should be observed near the top surface. Moreover, given that the bottom of the testing plate comes in contact with the metal operating platform and that the rate of heat dissipation is faster, the cooling rate near the bottom is higher than that at the top surface. Grain size along the thickness direction may be attributed to the graded distribution of the peak temperature and to the varying cooling rates. The high peak temperature and slow cooling rate are regarded as being indicative of the remarkable growth of grains near the top surface.

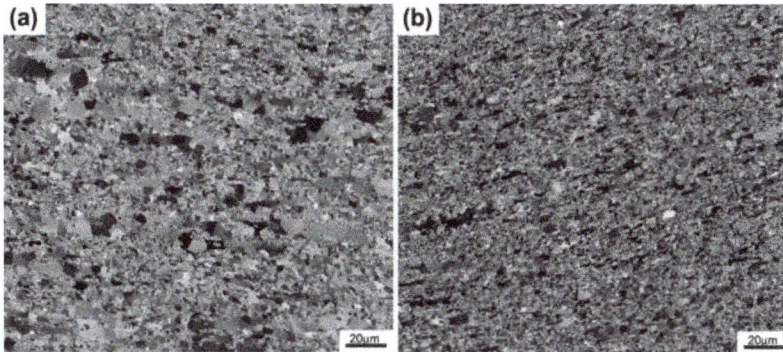

Figure 5. Backscatter electron (BSE) images of the SZ after electrolytic polishing (**a**) near the top surface; (**b**) at the bottom.

Figure 6 shows the EBSD maps of the BM and SZ. The result of the EBSD analysis on the BM reveals an austenite phase, which uses a higher number of grains (mostly twin boundaries) than the ferrite phase. The average grain sizes of the austenite and ferrite phases are 9 and 9.5 μm, respectively. The SZ displays fine and equiaxed grain microstructures. The average grain size of the austenite phase in the SZ is approximately 1.6 μm, which is smaller than that in the ferrite phase (2.2 μm). The amounts of twin grains are considerably lower than those of the BM, suggesting that the intense deformation associated with FSW disintegrates the initial annealed microstructure of BMs. The duplex phase of the SZ is finer than that of the BM. The grain size difference between the SZ and the BM, as observed in the microstructure and EBSD maps, suggests that dynamic recrystallization (DRX) occurs in the austenite and ferrite phases during FSW. As reported in related studies on the FSW, microstructures in the SZ experiences severe deformation and high temperature; thus, DRX leads to fine grains [18–20,24–26].

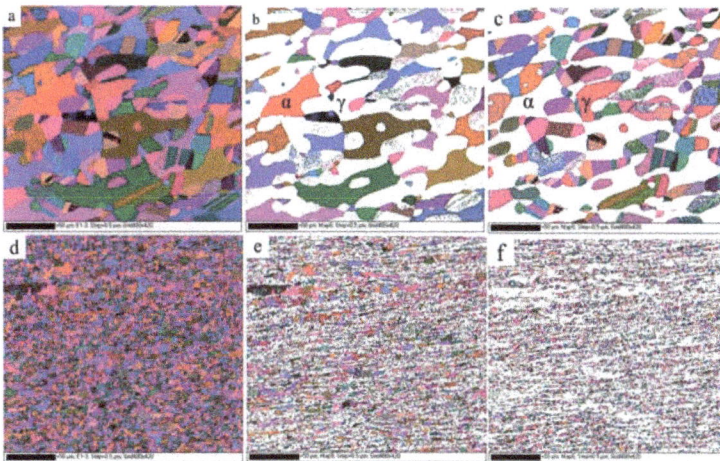

Figure 6. EBSD maps of the BM and the SZ: (**a**) map for the BM; (**b**) ferrite of the BM; (**c**) austenite of the BM; (**d**) map for the SZ; (**e**) ferrite of the SZ; (**f**) austenite of the SZ.

In the SZ, grain size in the austenite and ferrite phases slightly differ; that is, the austenite phase is slightly more pronounced than the ferrite phase. The grain size of DRX is known to be associated with the deformation and diffusion rate of atoms. During FSW, deformation mainly occurs in the ferrite phase because it is considerably weak at high temperatures. Fang et al. have demonstrated that, during the hot deformation of a new DSS containing high nitrogen content, the wedge crack is nucleated at the interface between the ferrite and austenite phases and easily propagates toward the ferrite phase; hence, the ferrite phase is softer than the austenite phase at high temperatures [28]. Consequently, great strain occurs in the ferrite phase and supplies additional energy for nucleation. Moreover, the ferrite phase has a higher diffusion rate of atoms compared with the austenite phase at high temperatures [28]. Both of them can produce a recrystallization process, which is followed by grain growth in the ferrite phase. According to earlier studies, DRX, including continuous DRX (CDRX), or discontinuous DRX (DDRX) may exist in the SZ during the FSW of DSSs [24–26,29,30]. High stacking fault energy (SFE) of the ferrite phase facilitates CDRX. By contrast, low SFE of the austenite phase facilitates DDRX.

Based on a simple mode of heat generation during FSW described by Schmidt [31], maximum heat is generated in interfaces of shoulder and plate. SAF2707HD is a high-alloyed steel, and the precipitation rate of the σ phase is generally fast. XRD was used to identify the different phases based on their locations in the XRD pattern. Additionally, XRD was used to clearly understand whether σ phase occurs on SZ. Figure 7 shows the XRD profile of the SZ surface. The ferrite and austenite phases are readily observed in the XRD pattern, and their corresponding diffraction peaks are marked. The σ phase was not formed during the FSW process, or its amount is so low that it cannot be detected via XRD. Although the introduction of high strain and DRX can promote a mutual diffusion of alloying elements [32], the time for nucleation of the σ phase is insufficient because of a short welding thermal cycle.

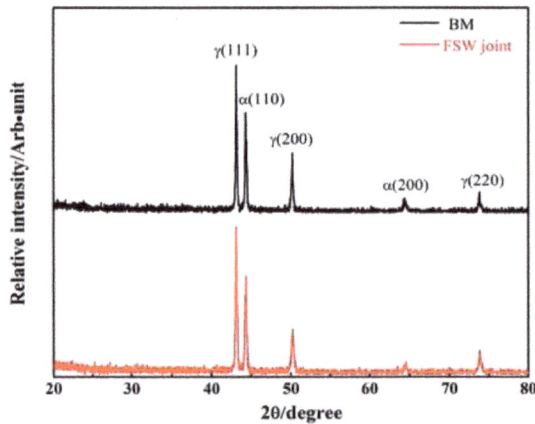

Figure 7. XRD pattern of base metal and SZ.

During solid-state transition, austenite formation in the fusion zone of DSSs depends on cooling rate. Followed by a fully ferrite phase, nucleation of the austenite phase preferentially lies in the grain boundaries of the ferrite grains during the weld cooling cycle. Previous evidence demonstrates that using high-heat inputs and slow cooling rates during the GTA process produce microstructures containing considerable amounts of various forms of the austenite phase, such as grain boundary austenite and secondary austenite [33]. In the present study, microstructural characterization shows that the intergranular austenite phase is absent in all regions across the weld. In addition, these regions contained a microstructure consisting of a ferrite matrix with austenite islands. As shown in Figure 7, the XRD spectra of the BM and the SZ consist of ferrite and austenite phases. The intensities of the α phase diffraction peak (110) and γ phase diffraction peak (111) in the SZ and the BM are nearly equal, implying that the ratio of the ferrite/austenite phase of the SZ is close to that of the BM. Overall, the ratio of the ferrite/austenite phase in the SZ is similar to that in the BM, as revealed by related studies [18,24]. The similarity of the ratios of the ferrite/austenite phase between SZ and BM indicates that the peak temperature is below the transition temperature between austenite and ferrite during FSW of DSSs.

3.2. Mechanical Characterization

Figure 8 shows the hardness values, as a function of the distance from the weld center, across the joint. The maximum hardness with minimal fluctuation lies in the SZ. Compared with the BM, the SZ shows a smaller fluctuation. BMs contain large grains, and the hardness of its ferrite and austenite phases differs. Thus, fluctuation in the BM is greater than that of the SZ. As the distance from the SZ increases, the hardness values gradually decrease. The difference in hardness between the BM to the TMAZ ranges from 303 HV to 355 HV. The main mechanism of this remarkable welding technique involves fine grains occurring in the SZ as a result of DRX without changes in proper phase balance. Moreover, as indicated by the microstructure of the SZ, the σ phase precipitation has not occurred. Hence, as explained by Hajian [34], the dominant mode toward an increase in hardness in the SZ is related to grain refinement, according to the fine-grain strengthening theory. Similarly, the values of hardness gradually decrease from the bottom to the surface of the weld joint (shown in Figure 9).

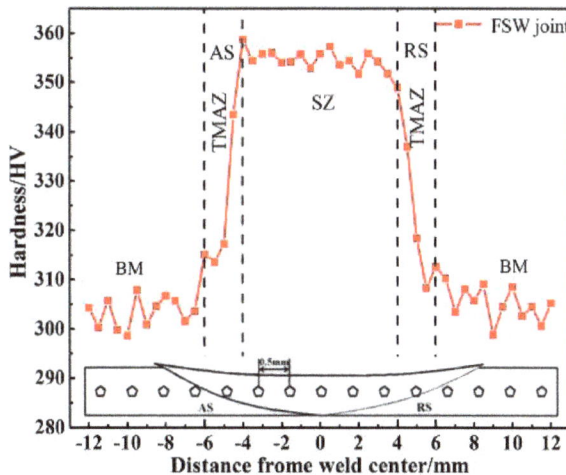

Figure 8. Hardness profile across the joint.

Figure 10a,b shows the stress–strain curves and transverse tensile properties of the BM and FSW HDSS weld joint. The experimental result shows that the 0.2% offset yield and ultimate tensile strength of the weld metal overmatch those of the BM, consistent with the Vickers hardness results. According to fine-grain strengthening theory, the elevated strength of the weldment relative to that of the BM is consistent with the results of microstructural observation. As reported by Esmailzadeh [30], owing to the presence of fine grains in SZs, the strength of welded SZ joints is more elevated than that of BMs. Fine grains accompanied by additional grain boundaries inhibit dislocation motion, thereby increasing the strength. The failure is consistently located in the BM and far from the FSW joint (Figure 11), indicating that the BM is weaker than the FSW joint. The tensile test results show that FSW produces defect-free HDSS welds. However, elongation of the FSW joint is 30%, which is lower than that of the

BM, and their difference is 6%. Li [35] explained that severe plastic deformation in the SZ reduces ductility because of high-density dislocations generated in the SZ. Plastic deformation mainly occurs on the base metal, implying that the strength of the weldments is influenced by the FSW joint.

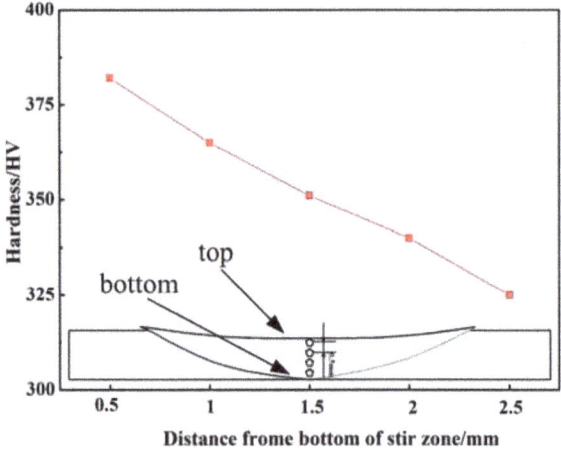

Figure 9. Hardness profile along the weld thickness.

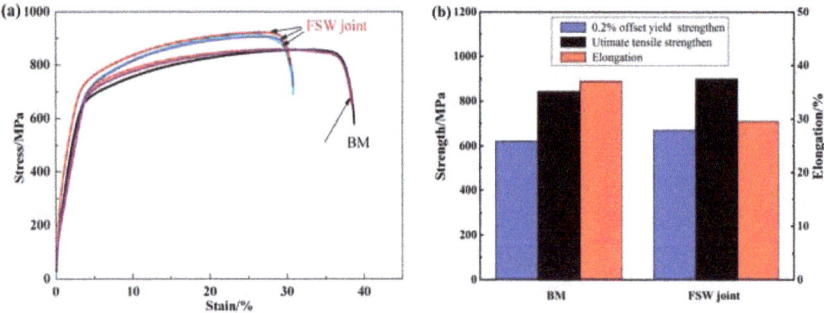

Figure 10. (**a**) Stress–strain curves of the BM and FSW joint; and (**b**) results of tensile test.

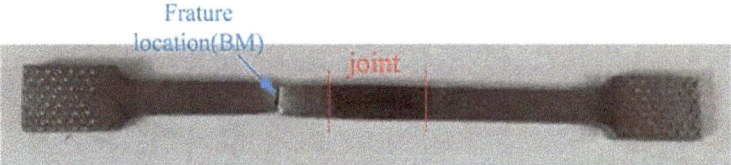

Figure 11. The macro pictures of the failure in the BM.

Figure 12a,b shows the optical micrograph of the tensile surface and SEM morphology of the weldment fracture surface. The phase appearing reddish brown in the OM is the ferrite phase (α) and the phase appearing white is the austenite phase (γ). The typical EDX results of dual phases are shown in Figure 12c,d, respectively. Crack initiation apparently occurred in the interface of the ferrite and austenite phases. The cracks propagated toward the austenite phase, which is softer than the ferrite phase at room temperature as a result of the low SFE. The SEM morphology of the weldment fracture surface is characterized by a dimple with a microvoid, a feature of a ductile fracture. However, size distribution of the dimple is inhomogeneous, which corresponds to low elongation [35].

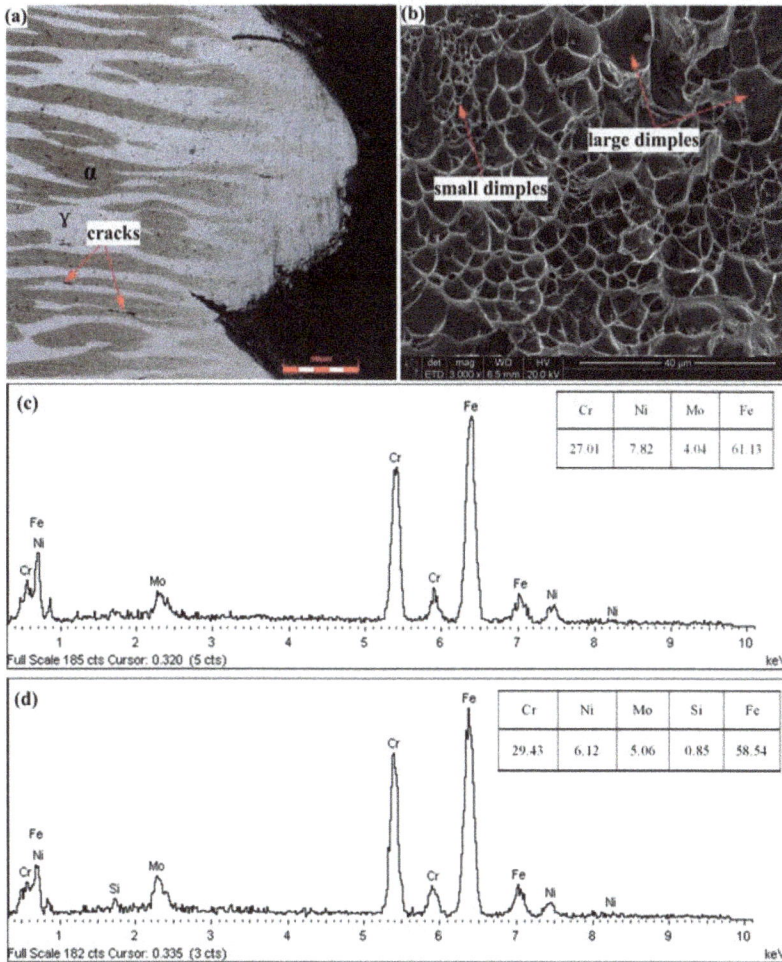

Figure 12. (**a**) Optical micrograph of the tensile surface; (**b**) SEM morphology of the weldment fracture surfaces; (**c**) EDX of the austenite phase; (**d**) EDX of the ferrite phase.

3.3. Corrosion Resistance

The effect of FSW on corrosion resistance of the SZ was investigated in a 3.5% NaCl solution at 25 °C ± 1 °C (Figure 13). The polarization curves of the BM and the SZ appear to have similar shapes and contain a wide passive region around 1.2 V. The pitting potentials of the BM and the SZ are 0.95 and 0.98 Vsce, respectively. The difference in corrosion resistance between the pitting potential (E_{pit}) and passive current density (i_{pass}) is positive. That is, the SZ displays a greater corrosion resistance than the BM. In addition, the magnitude of variation in corrosion potential (E_{corr}) corresponds to the pitting potential. Along with the mechanism of the FSW, fine grains are generated in the SZ. A large number of grain boundaries exhibit active positions susceptible to corrosion. They also supply ionic conduction channels contributing to anodic reactions. Ralston [36] has demonstrated that materials with fine grain structures are expected to be more corrosion resistant. The improved corrosion resistance can be attributed to the fine grains in the SZ; that is, an increase in the grain boundary density surface acting as diffusion paths of atoms facilitates passive film growth. Similar results have been reported [20,37]. As reported by LvJinlong [37], fine grains strengthen the stability and uniformity and improve the corrosion resistance of passive films of 2205 duplex stainless steels. Another study has demonstrated that the fine grains achieved by FSW increases the corrosion resistance of DSSs [20].

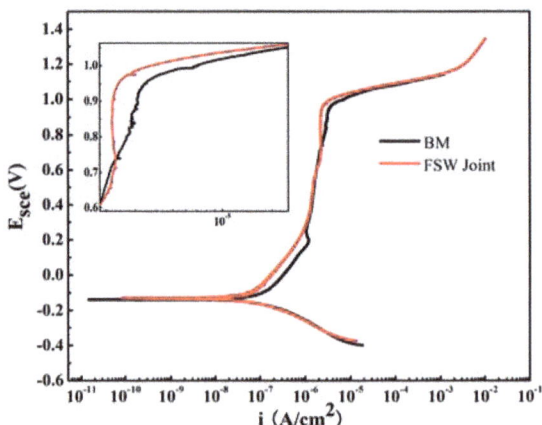

Figure 13. Potentiodynamic polarization curves of the BM andthe SZ in a 3.5 wt. %NaCl solution.

The electrochemical properties of the passive film were evaluated via EIS, a technique for accessing relaxation phenomena. Figure 14a shows the typical Nyquist plots of base and weld metals in 3.5% NaCl. Two successive depressed semicircles appear in the Nyquist plot. The diameter of the capacitive semicircle in the Nyquist diagrams is correlated with corrosion resistance. Thus, the numerical size of the charge transfer resistance of the passive film is reflected [38]. The weldment has a larger impedance diameter than the BM, indicating superior corrosion resistance of the former. Figure 14b shows the corresponding Bode plots of the BM and the SZ. Two well-defined time constants are observed in the two Bode-phase formats. This observation indicates that a passive film has formed on the 2707 HDSS surface and it consists of a bilayer structure. As reported [37,38], a passive film, including its active or passive region, cannot be ideally homogeneous.

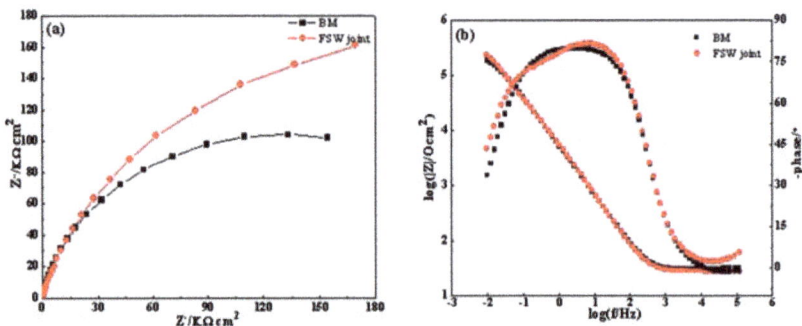

Figure 14. Nyquist (**a**); and Bode (**b**) plots for the BM and the SZ in a 3.5 wt. % NaCl solution.

An equivalent circuit (EC) has been proposed to be used to interpret the interface phenomenon between an electrode and a sample [37–39]. The maximal phase angle value in both Bode-phase plots is lower than 90°. Such behavior can be attributed to the deviation from the ideal capacitor behavior. Therefore, constant phase elements (CPEs) instead of a pure capacitance are used in EC to describe the frequency-independent phase shift betweenalternating potential and its current response. The corresponding impedance of a CPE is defined by the expression $Z_{CPE} = Z_0(jw)^n$. n, which is a CPE exponent, describes the deviation from the ideal capacitance. When $n = 1$, the CPE represents an ideal capacitance. For $0.5 < n < 1$, n decreases with the increase in capacitance dispersion. When $n = 0.5$, the CPE symbolizes a Warburg impedance characteristic of diffusion.

Figure 15 shows the EC proposed to model the electrode/sample interface. This circuit contains two time constants that match well the experimental results. Rs is the resistance of the solution. Q_1 represents the capacitive behavior of the passive film resulting from the formation of the ionic paths across the passive film, and constitutes the first time constant laid at a high frequency coupled with a resistance R_1. R_1 is the ionic diffusion resistance of the outer passive film. The second time constant is located in the low frequency region comprising Q_2 and R_2. Q_2 symbolizes the capacitive behavior of a finite-length passive film, and R_2 is the diffusion resistance. Dispersion constant n consistently lies between 0.5 and 1.

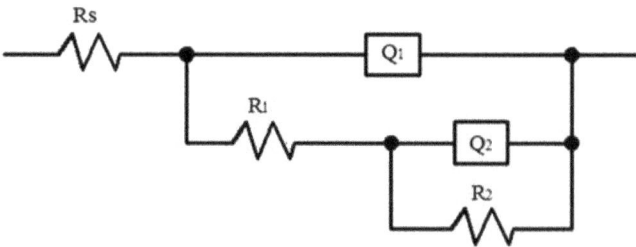

Figure 15. Equivalent circuit used for simulating bilayer oxide film on the BM andthe SZ.

Table 2 shows the corresponding fitting results. The film on the stainless steels consists of a two-layer oxide film, and these layers are denoted as R_1 and R_2 [40,41]. The quantitative values of R_1 and R_2 of the SZ are higher than those of the BM. This result suggests that the oxide film that formed on the surface of the SZ is more homogeneous and resistant than that formed on the surface of the BM. LvJinlong [37] demonstrated that the oxide film on the ultrafine grained DSS 2205 consists mainly of Cr_2O_3 and Fe_2O_3, whereas the coarse grain comprises FeO and $Cr(OH)_3$, which are less protective than Cr_2O_3 and Fe_2O_3. EIS test results confirm that the BM exhibits relatively poor corrosion performance compared with the weldment. These results are consistent with the potentiodynamic polarization curve test results. Meanwhile, BM has far lower R_1 value than the SZ, indicating that the outer passive film of the former is defective and characterized by large porosities and active regions. The corrosion resistance of passive films is mainly governed by the inner passive film, and this role has been proven by the numerical difference between R_1 and R_2. As mentioned by ČrtomirDonik [41], the dominance of the inner passive film of DSS 2205 can be attributed to the presence of chromium oxide between the surface oxide and bulk material at all temperatures.

Table 2. The EIS fitting results of the BM and SZ.

Area	R_s ($\Omega \cdot cm^2$)	Q_1 ($uF \cdot cm^{-2}$)	R_1 ($k\Omega \cdot cm^2$)	n_1	R_2 ($k\Omega \cdot cm^2$)	Q_2 ($uF \cdot cm^{-2}$)	n_2
BM	30.53	19.10	0.51	0.98	262.50	16.50	0.77
SZ	27.61	29.25	47.35	0.94	304.50	12.94	0.84

4. Conclusions

This work investigates the microstructure, mechanical properties, and corrosion resistance of the FSW joint. The main conclusions drawn based on the experimental results are as follows:

(1) A high-quality FSW joint without deleterious phases is obtained by rapidly cooling the samples. The welded joint microstructure contains fine equiaxed grains in the SZ. The grains are exceptionally pronounced in the austenite phase.

(2) Failure of the tensile specimens is consistently located in the BM zone, suggesting that the welded joint overly matches the BM. Owing to their fine grains, the welded joints display a higher tensile strength than their BMs.

(3) In the potentiodynamic polarization and impedance spectroscopy tests in NaCl solution, all SZs exhibit corrosion resistance superior to that of the BM.

Acknowledgments: This work was financially supported by The National Science Foundation of China (No. 51371123), The High-Tech Research Development Program (863 Program) of China (No. 2015AA034301), The Specialized Research Foundation of the Doctoral Program for Institution of Higher Education (Nos. 2013140211003, 20131402120004), The Natural Science Foundation of Shanxi Province (Nos. 2014011002-1, 201601D202034), Fund of the China Scholarship Council (CSC) and The Projects of International Cooperation in Shanxi (No. 201603D421026).

Author Contributions: This work was designed and mainly done by Jianchun Li, and he also took the job of the data analysis and write this paper. Corrosion resistance experiment was conducted by Xinglong Liu. Duplex stainless steel SAF2707 plate was provided by Guoping Li. Some constructive advices were obtained from Peide Han and Wei Liang.

Conflicts of Interest: The authors declare no conflict of interest.

References

1. Hayes, F.H. Phase equilibria in duplex stainless steel. *J. Less Common Met.* **1985**, *114*, 89–96. [CrossRef]
2. Weibull, I. Duplex stainless steels and their application, particularly in centrifugal separators: Part B Corrosion resistance. *Mater. Des.* **1987**, *8*, 82–88. [CrossRef]
3. Kenneth, G.; Marie-Louise, N.; Martin, H.; Eduardo, G. *Sandvik SAF 2707HD (UNS S32707)—A Hyper-Duplex Stainless Steel for Severe Chloride Containing Environments*; Internal Lecture No. S-51-63; Sandvik Materials Technology: Sandviken, Sweden, 2006.
4. Chail, G.; Kangas, P. Super and hyper duplex stainless steels: Structure, properties and application. *Procedia Struct. Integr.* **2016**, *2*, 1755–1762. [CrossRef]
5. Ha, H.; Jang, M.; Lee, T.; Moon, J. Understanding the relation between phase fraction and pitting corrosion resistance of UNS S32759 stainless steel. *Mater. Charact.* **2015**, *106*, 338–345. [CrossRef]
6. Ghosh, S.K.; Mondal, S. High temperature ageing behaviour of a duplex stainless steel. *Mater. Charact.* **2008**, *59*, 1776–1783. [CrossRef]
7. Sathirachinda, N.; Pettersson, R.; Pan, J. Depletion effects at phase boundaries in 2205 duplex stainless steel characterized with SKPFM and TEM/EDS. *Corros. Sci.* **2009**, *51*, 1850–1860. [CrossRef]
8. Deng, B.; Wang, Z.; Jiang, Y.; Sun, T.; Xu, J.; Li, J. Effect of cycles on the corrosion on the corrosion and mechanical properties of UNS S31803 duplex stainless steel. *Corros. Sci.* **2009**, *51*, 2969–2975. [CrossRef]
9. Badji, R.; Bouabdallah, M.; Bacroix, B. Phase transformation and mechanical behavior in annealed 2205 duplex stainless seel welds. *Mater. Charact.* **2008**, *59*, 447–453. [CrossRef]
10. Stenvall, P.; Holmquist, M. *Weld Properties of Sandvik SAF 2707 HD*; Sandvik materials technology: Sandviken, Sweden, 2008; Volume 11.
11. Pramanik, A.; Littlefair, G.; Basak, A.K. Weldability of duplex stainless steel. *Mater. Manuf. Process* **2015**, *30*, 1053–1068. [CrossRef]
12. Hsieh, R.I.; Liou, H.Y.; Pan, Y.T. Effects of cooling time and alloying elements on the microstructure of the Gleeble-simulated heat affected zone of 22% Cr duplex stainless steels. *J. Mater. Sci. Perform.* **2001**, *10*, 526–536. [CrossRef]

13. Pardal, J.M.; Tavares, S.S.M.; Cindra Fonseca, M.; de Souza, J.A.; Côrte, R.R.A.; de Abreu, H.F.G. Influence of the grain size on deleterious phase precipitation in superduplex stainless steel UNS S32750. *Mater. Charact.* **2009**, *60*, 165–172. [CrossRef]

14. Sieurin, H.; Sandstrom, R. Sigma phase precipitation in duplex stainless steel 2205. *Mater. Sci. Eng. A* **2007**, *444*, 271–276. [CrossRef]

15. Kim, S.M.; Kim, J.; Tae, K.; Park, K.; Lee, C.S. Effect of Ce addition on secondary phase transformation and mechanical properties of 27Cr-7Ni hyper duplex stainless steels. *Mater. Sci. Eng. A* **2013**, *573*, 27–36. [CrossRef]

16. Chen, Z.W.; Cui, S. On the forming mechanism of banded structures in aluminium alloy friction stir welds. *Scr. Mater.* **2008**, *58*, 417–420. [CrossRef]

17. Hirata, T.; Oguri, T.; Hagino, H.; Tanaka, T. Influence of friction stir welding parameters on grain size and formablity in 5083 aluminum alloy. *Mater. Sci. Eng. A* **2007**, *456*, 344–349. [CrossRef]

18. Sato, Y.S.; Nelson, T.W.; Sterling, C.J.; Steel, R.J.; Pettersson, C.O. Microstructure and mechanical properties of friction Stir Welded SAF 2507 super duplex stainless steel. *Mater. Sci. Eng. A* **2005**, *397*, 376–384. [CrossRef]

19. Saeid, T.; Abdollah-Zadeh, A.; Shibayanagi, T.; Ikeuchi, K.; Assadi, H. On the formation of grain structure during friction stir welding of duplex stainless steel. *Mater. Sci. Eng. A* **2010**, *527*, 6484–6488. [CrossRef]

20. Sarlak, H.; Atapour, M.; Esmailzadeh, M. Corrosion behavior of friction stir welded lean duplex stainless steel. *Mater. Des.* **2015**, *66*, 209–216. [CrossRef]

21. Kim, S.; Jang, S.; Lee, I.; Park, Y. Effects of solution heat-treatment and nitrogen in shielding gas on the resistance to pitting corrosion of hyper hyper duplex stainless steel welds. *Corros. Sci.* **2011**, *53*, 1939–1947. [CrossRef]

22. Jang, S.; Kim, S.; Lee, I.; Park, Y. Effect of shielding gas composition on phase transformation and mechanism of pitting corrosion of hyper duplex stainless steel welds. *Mater. Trans.* **2011**, *52*, 1228–1236. [CrossRef]

23. Thomas, W.M.; Johnson, K.I.; Wiesner, C.S. Friction stir welding–recent developments in tool and process technologies. *Adv. Eng. Mater.* **2003**, *5*, 485–490. [CrossRef]

24. Saeid, T.; Abdollah-Zadeh, A.; Assadi, H.; MalekGhaini, F. Effect of friction stir welding speed on the microstructure and mechanical properties of a duplex stainless steel. *Mater. Sci. Eng. A* **2008**, *496*, 262–268. [CrossRef]

25. Wang, D.; Ni, D.R.; Xiao, B.L.; Ma, Z.Y.; Wang, W.; Yang, K. Microstructure evolution and mechanical properties of friction stir joint of Fe-Cr-Mn-Mo-N austenite stainless steel. *Mater. Des.* **2014**, *64*, 355–359. [CrossRef]

26. Liu, F.C.; Nelson, T.W. In-situ material flow pattern around probe during friction stir welding of austenite stainless steel. *Mater. Des.* **2016**, *110*, 354–364. [CrossRef]

27. Guerra, M.; Schmidt, C.; McClure, J.C.; Murr, L.E.; Nunes, A.C. Flow patterns during friction welding. *Mater. Charact.* **2003**, *49*, 95–101. [CrossRef]

28. Fang, Y.L.; Liu, Z.Y.; Song, H.M.; Jiang, L.Z. Hot deformation behavior of a new austenite-ferrite duplex stainless containing high content of nitrogen. *Mater. Sci. Eng. A* **2009**, *526*, 128–133. [CrossRef]

29. Magnani, M.; Terada, M.; Lino, A.O.; Tallo, V.P.; Fonseca, E.B.; Santos, T.F.; Ramirez, A.J. Microstructural and electrochemical characterization of friction stir welded duplex stainless steels. *Int. J. Electrochem. Sci.* **2014**, *9*, 2966–2977.

30. Santos, T.F.; López, E.A.; Fonseca, E.B.; Ramirez, A.J. Friction stir welding of duplex and superduplex stainless steel and some aspects of microstructural characterization and mechanical performance. *Mater. Res.* **2016**, *19*, 117–131. [CrossRef]

31. Esmailzadeh, M.; Shamanian, M.; Kermanpur, A.; Saeid, T. Microstructure and mechanical properties of friction Stir Welded lean duplex stainless steel. *Mater. Sci. Eng. A* **2013**, *561*, 486–491. [CrossRef]

32. Sato, Y.S.; Harayama, N.; Kokawa, H.; Inoue, H.; Tadokoro, Y.; Tsuge, S. Evaluation of microstructure and properties in friction stir welded superaustenitic stainless steel. *Sci. Technol. Weld. Join.* **2009**, *14*, 202–209. [CrossRef]

33. Ramkumar, K.D.; Thiruvengatam, G.; Sudharsan, S.P.; Mishra, D.; Arivazhagan, N.; Sridhar, R. Characterization of weld strength and impact toughness in the multi-pass welding of super-duplex stainless steel UNS2750. *Mater. Des.* **2014**, *60*, 125–135. [CrossRef]

34. Hajian, M.; Abdollah, A.; Rezaei-Nejad, S.S.; Assaddi, H.; Hadavi, S.M.M.; Chung, K.; Shokouhimehr, M. Microstructure and mechanical properties of friction stir processed AISI 316L stainless steel. *Mater. Des.* **2015**, *67*, 82–94. [CrossRef]

35. Li, H.B.; Jiang, Z.H.; Feng, H.; Zhang, S.C.; Li, L.; Han, P.D. Microstructure, mechanical, and corrosion properties of friction Stir Welded high nitrogen nickel-free austenitic stainless steel. *Mater. Des.* **2015**, *84*, 291–299. [CrossRef]

36. Rlston, K.D.; Birbilis, N.; Davies, C.H.J. Revealing the relationship between grain size and corrosion rate of metals. *Scr. Mater.* **2010**, *63*, 1201–1204. [CrossRef]

37. Jinlong, L.; Tongxiang, L.; Chen, W.; Limin, D. Comparision of corrosion properties of passive films formed on coarse grained and ultrafined grained AISI 2205 duplex stainless steels. *J. Electroanal.* **2015**, *757*, 263–269. [CrossRef]

38. Luo, H.; Dong, C.F.; Xiao, K.; Li, X.G. Characterization of passive film on 2205 duplex stainless steel in sodium thiosulphate solution. *Appl. Surf. Sci.* **2011**, *258*, 631–639. [CrossRef]

39. Eghlimi, A.; Shamanian, M.; Raeissi, K. Effect of current type on microstructure and corrosion resistance of super duplex stainless steel claddings produced by the tungsten arc welding process. *Surf. Coat. Technol.* **2014**, *244*, 45–51. [CrossRef]

40. Jinlong, L.; Tongxiang, L.; Limin, D.; Chen, W. Influence of sensitization on microstructure and passive property of AISI 2205duplex stainless steel. *Corros. Sci.* **2016**, *104*, 144–151. [CrossRef]

41. Donik, Č.; Kocijan, A.; Grant, J.T.; Jenko, M.; Drenik, A.; Pihlar, B. XPS study of duplex stainless steel oxidized by oxygen atoms. *Corros. Sci.* **2009**, *51*, 827–832. [CrossRef]

metals

MDPI

Article

Tribological Properties of Plough Shares Made of Pearlitic and Martensitic Steels

Tomasz Stawicki [1], Beata Białobrzeska [2,*] and Piotr Kostencki [1]

[1] Department of Agrotechnical Systems Engineering, Faculty of Environmental Management and Agriculture, West Pomeranian University of Technology, Papieża Pawła VI Street 1, 71-459 Szczecin, Poland; Tomasz.Stawicki@zut.edu.pl (T.S.); Piotr.Kostencki@zut.edu.pl (P.K.)

[2] Welding and Strength of Materials, Department of Materials Science, Faculty of Mechanical Engineering, Wroclaw University of Technology, Smoluchowskiego Street 25, 50-370 Wrocław, Poland

* Correspondence: beata.bialobrzeska@pwr.edu.pl; Tel.: +48-71-320-38-45

Academic Editor: Robert Tuttle
Received: 11 March 2017; Accepted: 11 April 2017; Published: 14 April 2017

Abstract: Tribological properties of ploughshares made of pearlitic and martensitic steels were compared in field tests. Sectional ploughshares consisting of separate share-points and trapezoidal parts were subjected to examinations. Contours of the examined parts were similar, but the thickness of the parts made of pearlitic steel was 1 to 3 mm greater for the share-points and 0.5 to 2 mm greater for the trapezoidal parts. Within the tests, sandy loams, loams, and loamy sands with circa (ca.) 13% humidity were cultivated. A greater intensity of thickness reduction and mass wear of the parts made of pearlitic steel was found, which indicates a lower resistance of this steel to wear in soil. However, contour changes of the share-points and the trapezoidal parts made of pearlitic and martensitic steels were comparable, which was probably influenced by the greater thickness of the parts made of pearlitic steel. The roughness of the rake faces of the parts made of pearlitic steel was greater than that for the parts made of martensitic steel, which can be attributed to lower hardness of the former. The largest differences occurred between maximum peak heights of the roughness profile values (R_p), which indicates stronger ridging in the case of pearlitic steel. Scanning electron microscope (SEM) observations of the rake faces showed that martensitic steel was subjected to wear mostly by microcutting, but pearlitic steel was principally worn by microcutting and microploughing. During tillage, only one share-point made of pearlitic steel was broken. However, the main disadvantage of these parts was that their bending was related to the lower mechanical strength of pearlitic steel.

Keywords: plowshares; wear; durability; martensitic steel; pearlitic steel

---

1. Introduction

Tillage still belongs to popular operations of soil cultivation and this indicates a need to carry out research aimed at the improvement of the design solutions of ploughshares. Ploughshares, like many other parts working in soil, are subjected during operation to intensive abrasion wear [1–3] that results in a change in their geometry and a reduction of mass. In consequence, after reaching the so-called ultimate limit state, these components must be replaced with new ones. Thus, users of agricultural equipment expect that the manufacturers offer durable replacement parts. The economic aspect, i.e., the low price of the products, is also significant.

In spite of many years of laboratory and field research, wear of the parts operating in soil (including ploughshares) constitutes a still valid tribological and operating problem, because of the complexity of the mechanisms occurring in these processes. The research performed so far includes, among others, an evaluation of the relation between wear processes and the geometry of particles composing abrasive soil mass, as well as the physico-chemical conditions of the working environment, described by the granulometric composition of soil, its humidity and consistency, the content of skeleton particles, reactions, etc. [4–7]. As far as humidity alone is concerned, its diversified influence on the wear dynamics of ploughshares has been demonstrated. It was observed that, with increasing soil humidity, the wear rate of ploughshares is lower in loamy and clay soils, but is higher in sandy soils [5,8,9]. In soils with low humidity, the intensive wear of ploughshares was observed on their flank faces, which resulted in a relatively quick reduction in the length of their share-points and in the width of their trapezoidal parts. On the other hand, in soils with high humidity, a relatively quick thickness reduction in shares has been found [10,11]. The complexity of the wear mechanisms of the parts operating in soil is also shown by the research in that a relation between the wear rate of steel and the reaction of the working environment has been demonstrated [12,13].

Users of agricultural equipment have no effect on most of the factors related to the operating conditions of tillage tools, but are interested in a long durability of the working parts. The relatively well-known and still-in-development methods that of improving the durability of ploughshares include pad-welding with hard alloys in the areas subjected to intensive wear. The effectiveness of this solution has been confirmed in laboratories and field research many times [14–18]. Research on the anti-wear properties of steels containing micro-additions of boron (e.g., steels Hardox, Raex, SSAB Boron 33) and steels surface-hardened by diffusive boriding [19–21] have been undertaken. The possibility of using oxide ceramics and plates made of sintered carbides as abrasion-resistant superficial layers of the components operating in soil has also been considered [22–26]. It should also be mentioned that a beneficial economic effect of plough cultivation can be reached by using sectional ploughshares, currently offered by several manufacturers. Such a solution makes it possible to utilize the trapezoidal parts of shares more completely [11].

The present research was aimed at comparing tribological properties of ploughshares made of pearlitic and martensitic steels, while the starting material for the manufacture of pearlitic shares were railway rails withdrawn from use, being a cheap base material.

2. Materials and Methods

2.1. Objects of the Research

Two variants of ploughshares were used in the research, differing basically in properties of the used steels. One of the steels was characterized by a martensitic structure (shares from a well-known manufacturer of agricultural tooling), and the other steel by a pearlitic structure (shares from a manufacturer that utilizes recycled railway rails). It should be mentioned that the usefulness of the shares made of pearlitic steel is not well recognized. However, the shares made of martensitic steel, used in this research, are renowned, and this is why they were selected as reference objects. In both variants, the shares were sectioned, i.e., composed of separate share-points and trapezoidal parts.

The chemical composition of the steels used for the examined ploughshares was determined by spectral analysis using a glow discharge spectrometer GDS500A (Leco Corporation, Saint Joseph, MO, USA). The examined material was placed under atmosphere low-pressure argon. A negative potential between 800 and 1200 V was applied to the specimens. Results are given in Table 1.

Microscopic examinations were carried out on cross sections of steel specimens etched with 3% HNO_3 (according to PN-H-04512-1975) at magnifications between 100 and 500×. Observations were made using a metallographic microscope (Nikon Corporation, Tokyo, Japan) coupled with a charge-coupled device (CCD) camera. The structure of the ploughshares made of railway rails was found to be pearlitic (non-homogeneous in some places) with ferrite crystallized on grain boundaries of former austenite. Moreover, numerous inclusions were observed, probably sulfides, see Figure 1a,b. The structure of the reference shares was found to be martensitic, composed of medium-carbon and fine-acicular tempered martensite, see Figure 1c.

Table 1. The chemical composition of steels used for ploughshares.

Element	Concentration [wt %]	
	Pearlitic Steel	Martensitic Steel
C	0.809	0.362
Mn	0.884	1.270
Si	0.214	0.230
P	0.022	0.013
S	0.025	0.006
Cr	0.061	0.256
Ni	0.058	0.084
Mo	0.005	0.019
V	0.001	0.000
Cu	0.074	0.138
Al	0.016	0.031
Ti	0.002	0.043
Co	0.019	0.005
As	0.032	0.000
B	0.0004	0.0023
Pb	0.002	0.002
Zr	0.008	0.004
Fe	Rem.	Rem.

(a) (b) (c)

Figure 1. Microstructures of ploughshare materials: (**a**) pearlitic steel with ferrite on grain boundaries (indicated by arrows); (**b**) pearlitic steel with ferrite on grain boundaries, in some places non-homogeneous, with visible inclusions, probably sulfides (indicated by an arrow); (**c**) martensitic steel, the structure of fine-acicular tempered martensite.

For the steels under examination, hardness, impact strength, tensile strength, proof stress, and ultimate elongation were determined. Brinell hardness was measured according to EN ISO 6506-1:2014-12 using a Zwick/Roell tester (Zwick Roell Gruppe, Ulm, Germany), under 187.5 kG (1839 N) for 15 s. Charpy V-notch tests were carried out according to EN ISO 67 148-1:2010 at ambient temperature, on the Zwick Roell pendulum hammer RPK300 (Zwick Roell Gruppe, Ulm, Germany) on specimens with dimensions $b = 10$ mm $\times$ $h = 7.5$ mm, cut out in longitudinal and perpendicular directions in relation to the plastic working direction. Fracture surfaces of the specimens were analyzed using a scanning electron microscope JEOL JSM-5800LV (Joel Ltd., Tokyo, Japan) coupled with an X-ray analyzer, Oxford LINK ISIS-300 (Oxford Instruments, Abingdon, UK). Tensile tests were carried out according to EN ISO 6892-1:2016-09 on a tester MTS 810 (MTS Systems, Eden Prairie, MN, USA) using flat specimens 40 mm long, with a cross section with a gauge length of 5 × 10 mm. The specimens were cut-out from trapezoidal parts of the ploughshares along their length. Results of the measurements and tests are given in Table 2.

Table 2. Mechanical properties of the examined materials.

Parameter		Material			
		Pearlitic Steel		Martensitic Steel	
Hardness [HBW]	share-point	258.0	s = 4.7	476.0	s = 3.5
	trapezoidal part	265.4	s = 2.8	472.2	s = 1.8
Impact strength [J/cm^2]—relation	longitudinal	4	s = 0.3	37	s = 0.3
to plastic working direction	perpendicular	3	s = 0.3	25	s = 2.6
Tensile strength R_m [MPa]		911	s = 99	1833	s = 40
Proof stress $R_{p0.2}$ [MPa]		494	s = 70	1482	s = 35
Ultimate elongation A [%]		10	s = 1.0	10	s = 0.4

s—standard deviation.

Microscopic analysis of fracture surfaces obtained in impact tests showed 100% of the brittle zone for pearlitic steel, but significant plastic zones in the side parts for martensitic steel, see Figure 2a–d.

Specimens of pearlitic steel showed cleavage fractures with fragments of plastic fracture between the facets, with characteristic profiles of "rivers", see Figure 2e–h. In the places where cracks propagated through grain boundaries, "fan" profiles could sometimes be seen (Figure 2h). On surfaces of the facets, marks of secondary cracks also occurred, propagating at a certain angle to the main cracking surface, see Figure 2e–h.

Fractures of the specimens of martensitic steel for both longitudinal and perpendicular directions are shown in Figure 2i–l. Side zones, as well as small zones under the notch, are plastic fractures with voids of various diameters, see Figure 2i. Brittle non-metallic inclusions that initiated the fracture occur within larger voids. Pits have parabolic shapes, which evidences action of tangential forces during cracking. The central zone is the so-called quasi-cleavage fracture (Figure 2j,k). This is typical for steels with martensitic and bainitic structures. Even though these facets are similar to cleavage facets (because of occurrence of the "river" profile), it is almost impossible to identify crystallographic planes. Ridges of quasi-cleavage facets are characterized by developed topography, which also evidences intensive plastic deformation during their creation. Moreover, numerous transverse cracks and cavities left by non-metallic inclusions were observed, see Figure 2k,l.

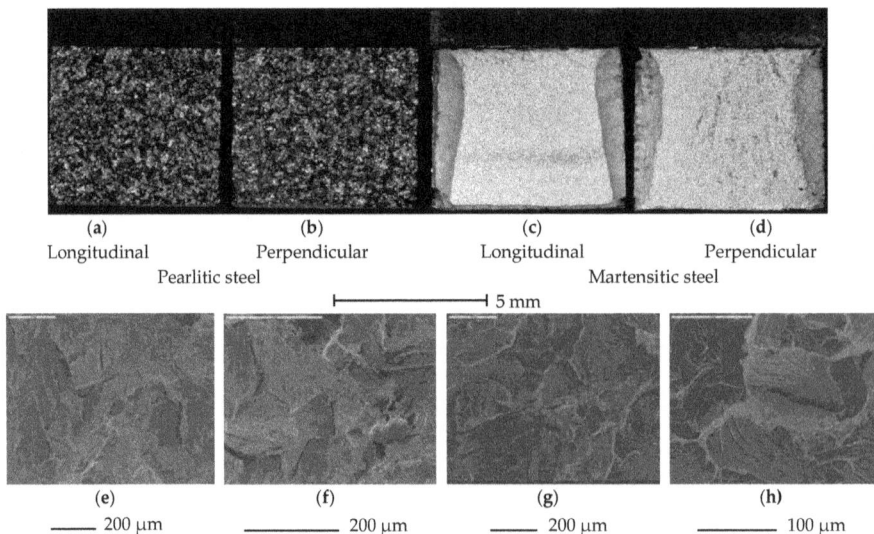

(a)	(b)	(c)	(d)
Longitudinal	Perpendicular	Longitudinal	Perpendicular
Pearlitic steel		Martensitic steel	

⊢————————⊣ 5 mm

| (e) | (f) | (g) | (h) |
| ——— 200 μm | ————— 200 μm | — 200 μm | ——— 100 μm |

Figure 2. *Cont.*

| (i) | (j) | (k) | (l) |
| ——— 100 μm | ——— 200 μm | ——— 100 μm | ——— 200 μm |

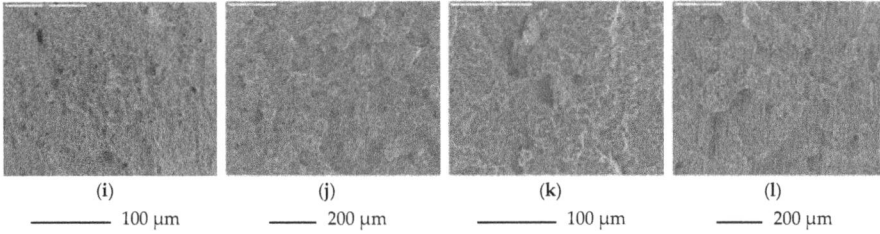

Figure 2. Fracture surfaces of pearlitic and martensitic steels after the impact test. (**a–d**) macroscopic images of fracture surfaces after the impact test; Pearlitic steel: (**e,f**) cleavage fracture with characteristic "river" profile (longitudinal), (**g,h**) cleavage fracture with characteristic "river" and "fan" profiles (perpendicular); Martensitic steel: (**i**) plastic side zone (longitudinal), (**j**) quasi-cleavage fracture with developed topography (longitudinal), (**k**) cavities after non-metallic inclusions (longitudinal), (**l**) cavities after non-metallic inclusions (perpendicular).

Characteristic dimensions of the examined ploughshares are given in Table 3. The basic difference in their geometry is thickness. The share-points and trapezoidal parts made of pearlitic steel were thicker (Table 3, dimensions W_3 and W_8) than those made of martensitic steel. This was reflected in the mass: the share-points and trapezoidal parts made of pearlitic steel were respectively ca. 17% and 7% heavier than those made of martensitic steel.

Table 3. Dimensions and masses of the examined ploughshares.

Ploughshare Material	Dimensions [mm]									
	Share-Points						Trapezoidal Parts			W_{10}
	W_1	W_2	W_3	W_4/a	W_5/a	W_6	W_7	W_8	W_9/a	
Martensitic steel	97	335	9–15 [1)]	28/3	27/3	496	149	11	42/4	650
Pearlitic steel	99	338	12–16 [1)]	28/5	25/5	480	146	12–13	30/5	643
	Mass [g]									
Martensitic steel	1846.6		$s = 22.0$				4588.3		$s = 19.2$	
Pearlitic steel	2157.4		$s = 38.0$				4907.4		$s = 52.3$	

a—blade thickness; [1)] Maximum thickness occurred at the field edge, close to half of its length; *s*—standard deviation.

2.2. Conditions of Field Testing

Field tests were carried out between the 8th and 16th of August 2016 on the fields of Agrofirma Witkowo Cooperative seated in Witkowo (geographic coordinates 53°25′ N, 15°16′ E), using a seven-furrow reversible plough EuroDiamant (Lemken GmbH & Co. KG, Alpen, Germany) with full-moldboard bodies. The plough bodies were equipped with automatic, spring overload protection. For each material variant, 14 share-points and 7 trapezoidal parts were prepared. The parts made of pearlitic steel were installed on one side of the plough, and those made of martensitic steel on the other side. Thanks to the application of a reversible plough, the shares worked in corresponding soil and service conditions, in spite of some variability of these conditions during the tests. This way, it was possible to conclude about usability of the shares on the grounds of their wear. During the examinations, the areas cultivated by the share-points and trapezoidal parts were respectively 8.43–12.29 and 20.93 ha per body. After the tillage of that area was complete, part of the shares reached ultimate wear limits, and the other shares still had a small material "margin" for further operation.

Parameters describing soil and service conditions of ploughshare operation are given in Table 4. Percentages of individual soils present in the tillable layer were established on the grounds of soil-agricultural maps elaborated for the cultivated area. (These maps were elaborated in the middle of the 20th century, when somewhat different principles of soil classification were applied; nevertheless, it is possible to assess the percentages of individual soils with quite good approximation using the U.S. Department of Agriculture (USDA) criteria. Percentages of skeleton particles were determined by sieving soil samples taken from the tillable layer through a 2 mm sieve. The percentage of fine stones (3 to 14 cm) was determined by selecting such stones from 1 m^2 of the tillable layer. The number of large stones was determined on the grounds of the number of activations of the mechanisms protecting the plough bodies against overload on a determined field area. Volumetric density and humidity of soil was determined using 100 cm^3 Kopecky's cylinders by a drying and weighing method (according to PN-R-55003:1990). Consistency was determined with a spring meter using a cone diameter of 16.6 mm with an apex angle of 30°. Shearing stresses were measured with a vane tester Geonor H-60 (Geonor Inc., Augusta, ME, USA), equipped with a cross 20 mm wide and 40 mm high. The width and depth of the tillage was determined according to the guidelines in PN-90/R-55021. Tillage speed was determined by measuring the cultivator travel time with a stop-watch on the distance of 50 m.

Table 4. Operating conditions of ploughshares.

Quantity	Determined Average Value		Soil Layer
Percentages of soil grades in tillable layer of the cultivated area [%]	sandy loams and loams	63	Tillable layer
	loamy sands	34	
	sands	3	
Percentage of skeleton particles (fraction 2 to 30 mm) [%]	2.4; s = 1.0		
Stoniness	fine stones (3–14 cm): 15 t/ha; s = 8 (168,000 pcs./ha; s = 80,000) large stones: 28 pcs./ha; s = 31		
Reaction [pH$_{KCL}$]	6.07–6.33		
Humus content [%]	1.78; s = 0.25		
Actual humidity [wt %]	13.5; s = 1.8		0–15 cm
	13.1; s = 1.3		15–30 cm
Volumetric density [g/cm^3]	1.45; s = 0.06		0–15 cm
	1.49; s = 0.13		15–30 cm
Consistency [kPa]	789; s = 166		0–15 cm
	1866; s = 496		15–30 cm
Shearing stresses [kPa]	47.0; s = 10.3		0–15 cm
	65.2; s = 19.2		15–30 cm

Table 4. *Cont.*

Quantity	Determined Average Value	Soil Layer
Working width [m]	0.45; s = 0.04	
Working depth [cm]	20.5; s = 3.5	
Speed [m/s]	2.78; s = 0.25	

Besides percentages of soil grades, average values for individual quantities were calculated on the grounds of the data obtained in six individual days of testing; *s*—standard deviation.

In order to make the granularity of the cultivated soils more clear, exemplary percentages of individual granulometric fractions in the tillable layer are given in Table 5. Granularity was measured by the Casagrande's method modified by Prószyński.

Table 5. Exemplary percentages of granulometric fractions in the cultivated soil.

Sample No.	Percentages of A Granulometric Fraction [%]								Granulometric Group
	Sand				Silt			Clay	
	very Coarse $1.0 < d \leq 2.0$	Coarse $0.5 < d \leq 1.0$	Medium $0.25 < d \leq 0.5$	Fine $0.10 < d \leq 0.25$	very Fine $0.05 < d \leq 0.10$	Coarse $0.02 < d \leq 0.05$	Fine $0.002 < d \leq 0.02$	$d \leq 0.002$	
1	1.7	5.8	11.7	34.2	20.1	10.8	10.8	4.9	FSL
2	1.8	6.5	12.3	30.9	19.0	12.8	10.8	5.9	FSL
3	1.7	6.3	10.9	22.4	23.4	10.8	14.7	9.8	SL
4	1.9	5.4	11.1	34.4	19.7	8.8	12.8	5.9	FSL
5	1.1	3.6	6.9	15.2	22.7	24.8	15.8	9.9	L
6	1.4	4.5	9.3	36.4	23.8	6.9	11.8	5.9	FSL

d—size of soil grain [mm]; FSL—fine sandy loam; SL—sandy loam; L—loam.

2.3. Evaluation of Wear of the Examined Parts

To evaluate intensity of wear of the examined parts, the following parameters were used: unit mass wear, unit contour change, and unit thickness reduction, determined on the grounds of absolute wear of the parts after cultivation of a determined area. Unit mass wear was determined as the ratio of mass loss of the part to the tilled area attributed to this part (understood as the area tilled to the moment of replacing the part, divided by the number of bodies, i.e., 14 pcs.). The parameters related to the geometry changes of the parts were determined for selected measurement places, as shown in Figure 3. The unit contour change of a part (in mm/ha) was determined as the ratio of the absolute contour change of the part in a given measuring line to the area cultivated by this part. A unit thickness reduction (also in mm/ha) was determined as the ratio of the absolute thickness reduction of the part in a given measuring point to the area cultivated by this part.

Contours of the parts were measured with a slide caliper, and thickness was measured with a micrometer caliper. For contour measurements, a specially prepared "measuring table" was used, making measurements of the distances between bases and edges of the parts possible. Differences between the measurements of the new parts and of the used ones formed the contour change in the given measuring line. For thickness measurements, a template was used, making repeated measurements at the same point possible.

The width of wear bands formed on the flank faces of the parts as a result of the abrasive action of soil from the furrow bottom was also measured. Measurements of this parameter were taken at Lines L_2 and L_3 at the share-points and Lines L_4, L_5 and L_6 at the trapezoidal parts, see Figure 3. The measurements were made with a slide caliper in the direction perpendicular to the blade lines of the parts.

Measuring places for contour changes: measuring lines from L_1 to L_6.

Measuring places for thickness changes: measuring points from g_1 to g_{13}.

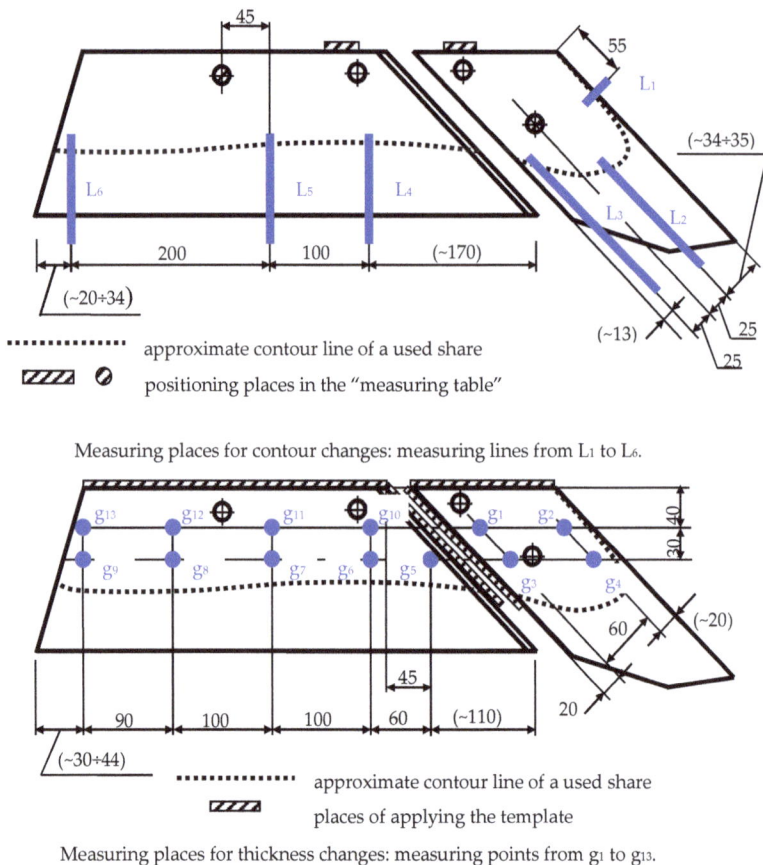

Figure 3. Measuring places for geometry changes of the examined shares, all results are in mm.

2.4. Evaluation of Surface Condition of the Parts after Operation in Soil

The microgeometry of the rake faces of the parts was measured with a profilographometer, Hommel Tester T1000 (Jenoptic AG, Jena, Germany). The measurements were taken on measuring lengths 4.8 mm (according to EN ISO 4288:2011), repeated four times for each part. The roughness of the surface was described by the arithmetical mean deviation of the roughness profile R_a, the total height of the roughness profile R_t, the maximum valley depth of the roughness profile R_v, and the maximum peak height of the roughness profile R_p. Observation of the rake faces of the examined parts was also carried out using a scanning electron microscope, JEOL JSM-5800LV (Joel Ltd., Tokyo, Japan), coupled with an X-ray microanalyzer, Oxford LINK ISIS-300 (Oxford Instruments, Abingdon, UK).

2.5. Statistical Analysis of Measurement Results

Measurement results concerning unit wear, contour change, and thickness reduction were complemented with statistical analysis in order to verify whether the share-points and trapezoidal parts made of comparable steels are characterized by significantly different intensities of wear. Analysis was performed using a Student's t-test and the method of uncorrelated variables. The accepted significance level α was 0.05. A similar analysis was performed for roughness parameters.

3. Results

3.1. Wear of the Examined Parts

During tillage, one of the parts was broken: a share-point made of pearlitic steel broke across, in the line running through the lower assembly hole, see Figure 4. After the tests, almost all share-points made of pearlitic steel were somewhat bent, see Figure 5. Looking in the direction of their movement in soil, the share-points were bent "forward" or "backward". The bends were small and did not exclude the share-points from further operation. It is probable that the "backward" bends were caused by collisions of the shares with stones occurring in soil. The "forward" bends could also result from collisions with stones (at the reaction force directed upwards) or by hitting against the ground on the headlands. A consequence of "forward" bending of the share–points was the necessity to replace them somewhat earlier, because the lower fitting bolts and nuts were uncovered and subjected to abrasive wear. Bending of the trapezoidal parts (Figure 5) was observed on five examined shares but, principally, at the end of their service life, when their thickness was significantly reduced as a result of abrasive wear. Therefore, in the test conditions, bending of the parts made of pearlitic steel was a typical phenomenon.

Considering the above, wear demonstrated by unit mass wear, unit thickness reduction, and unit contour change of the share-points and trapezoidal parts made of pearlitic steel was evaluated in all the parts, both those bent and not bent.

Figure 4. Broken share-point part made of pearlitic steel.

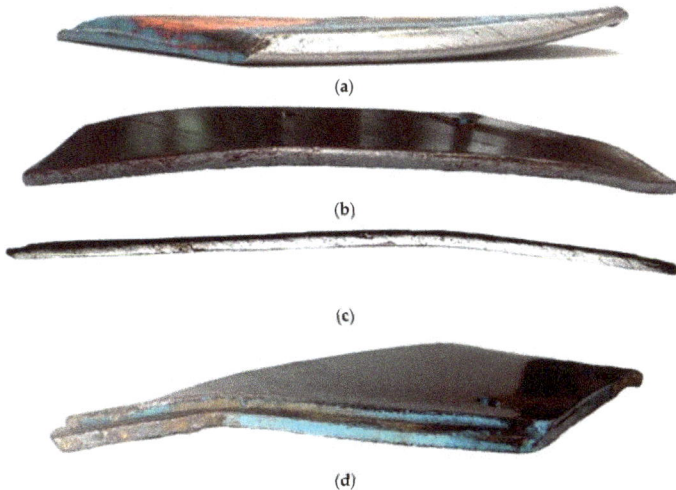

(a)

(b)

(c)

(d)

Figure 5. Bending of share-point and trapezoidal parts made of pearlitic steel: (**a**) a share-point part bent "backward"; (**b**,**c**) longitudinal bend of a trapezoidal part; (**d**) transverse bend of a trapezoidal part. The presented parts are relatively strongly bent.

Figure 6 shows values of the unit mass wear of the examined shares. Significant differences between these values for the two used steels (at intensity levels of respectively $p = 0.00025$ and 0.007 for the share-points and trapezoidal parts; statistically significant differences are marked by asterisks in the figures). In both share-points and trapezoidal parts made of pearlitic steel, more intensive wear was observed compared to the parts made of martensitic steel: ca. 1.42 times more for the share-points and ca. 1.23 times more for the trapezoidal parts.

The trapezoidal parts were characterized by a greater unit mass wear than the share-points: ca. 1.2 times greater for the parts made of pearlitic steel and ca. 1.4 times greater for the parts made of martensitic steel. This can be explained by a ca. 2.4 times larger rake face area in the trapezoidal parts, while it is obvious that material loss occurred also from the sides of the flank faces and field edges of the ploughshares.

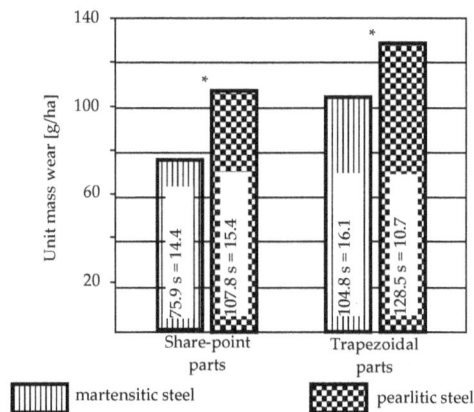

Figure 6. Unit mass wear of share-points and trapezoidal parts of the examined ploughshares (s—standard deviation, *—statistically significant differences).

The measurement results of the unit thickness reduction of the share-points and trapezoidal parts of the ploughshares are shown in Figure 7. At each measuring point, the values found in the parts made of pearlitic steel were higher than those found in the parts made of martensitic steel: from 0.02 to 0.15 mm/ha for the share-points and from 0.04 to 0.11 mm/ha for the trapezoidal parts. In the share-points, statistically different values were found in three measuring points: g_1 ($p = 0.001$), g_2 ($p = 0.002$), and g_4 ($p = 0.003$), see Figure 3. In the trapezoidal parts, statistically different values of unit thickness reduction were found in all the measuring points ($p = 0.0000005–0.0042$).

For unit thickness reduction, linear regression analysis was performed in order to find a correlation between the values for the parts made of pearlitic and of martensitic steel. Obvious boundary conditions were accepted for this regression: zero thickness reduction of the parts made of martensitic steel corresponds to zero thickness reduction of the parts made of pearlitic steel. Results of the analysis are shown in Figure 8.

A very strong correlation was found between the unit thickness reduction of the parts made of pearlitic and that made of martensitic steel, indicating a more intensive wear of pearlitic parts: ca. 1.4 times for the share-points and 1.8 times for the trapezoidal parts.

Unit thickness reduction of the parts was greater in the places located in the lower measuring line (measuring points from g_3 to g_9). The share-points were subjected to the most intensive wear in the area adjacent to the field edge and the nearest to the blade (measuring point g_4), which can be related to the highest pressure exerted by the soil on that surface fragment. In the case of the trapezoidal parts, these areas were extreme measuring points located at the lower measuring line (measuring points g_5 and g_9).

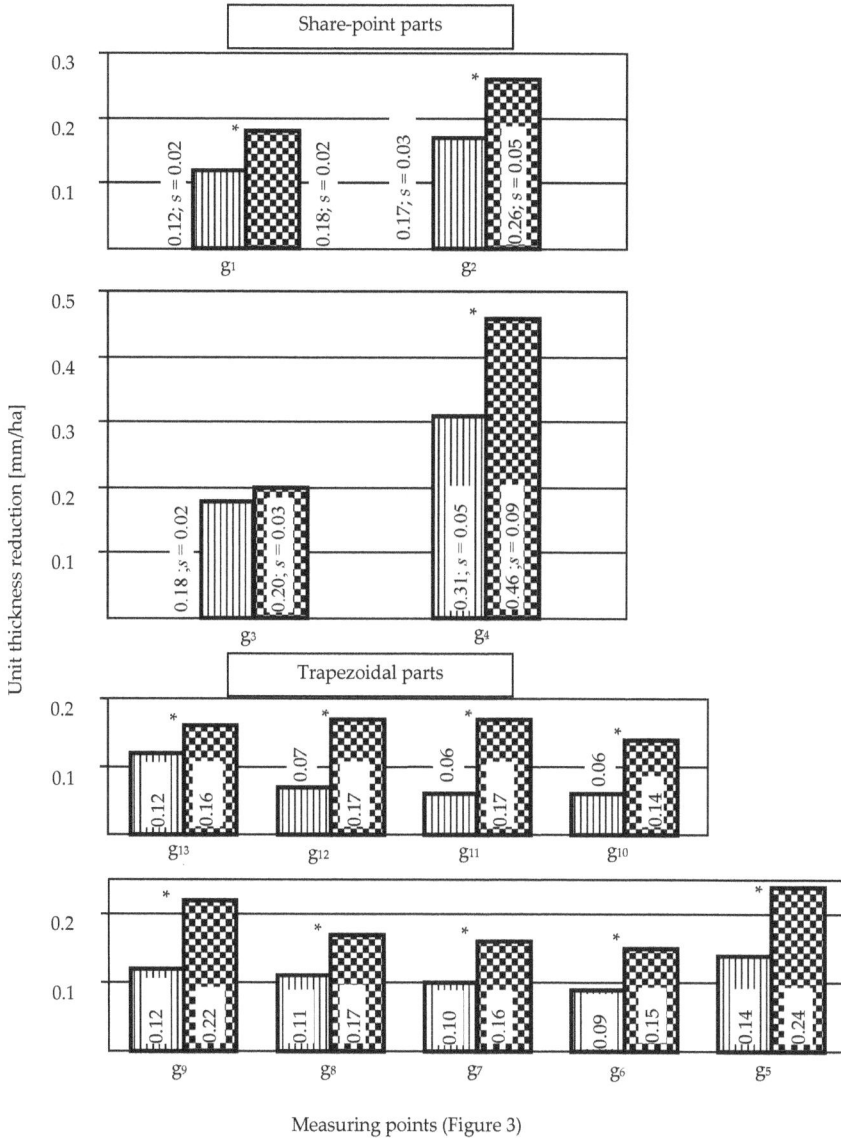

Figure 7. Unit thickness reduction of share-point parts and trapezoidal parts of the examined ploughshares-standard deviation for trapezoidal parts: s = 0.01–0.03 mm/ha. (*—statistically significant differences).

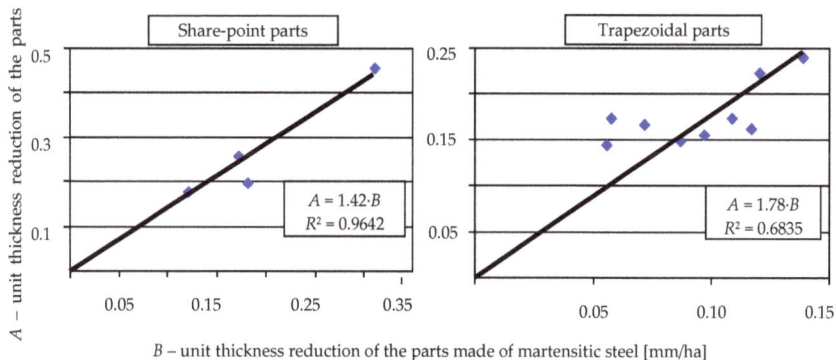

Figure 8. Relation between unit thickness reduction of the parts made of pearlitic and of martensitic steel (R^2—coefficient of determination).

Figure 9 shows the values of the unit contour change of the examined parts. At most of the measuring lines, no statistically significant differences were found for the parts made of pearlitic and of martensitic steel. Such a significant difference occurred only at Line L_3, used in the share-points.

For the share-points, unit contour changes were much greater than those for the trapezoidal parts, of course in addition to the wear at Measuring Line L_1. A slight tendency for the wear intensity of the trapezoidal parts to increase with increasing distance from the share-points was also observed (increasing parameter values in Lines L_4, L_5, and L_6).

On the grounds of unit contour change of the elements, it is possible to evaluate their durability. It was found that absolute length reduction of the share-points measured at Line L_2 (corresponding to limit wear condition) was 96.6 mm, and the width reduction of the trapezoidal parts measured at Line L_4 was 63.7 mm. Utilization of the parts beyond the accepted limit wear condition in Lines L_2 and L_4 (Figure 3) would result in abrasion of the bolts and nuts located under the parts (Figure 10) which, with regard to operation of the plough, should be considered as unacceptable. At durability evaluation, the simplifying assumption was accepted: the intensity of contour changes of the parts at Measuring Lines L_2 and L_4 does not depend on the cultivated area.

Since no statistically significant differences were found between unit contour changes at Lines L_2 and L_4, the averaged value of this parameter (determined for the parts made of both martensitic and pearlitic steel) was accepted in the calculations. The so evaluated durability was ca. 11.5 ha for the share-points and 26.8 ha for the trapezoidal parts. Thus, in the testing conditions, the share-points showed a ca. 2.3-fold lower durability compared to that of the trapezoidal parts. It should be noted that the evaluated durability of the share-points made of pearlitic steel concerns the parts that would not be bent or would be bent "backwards".

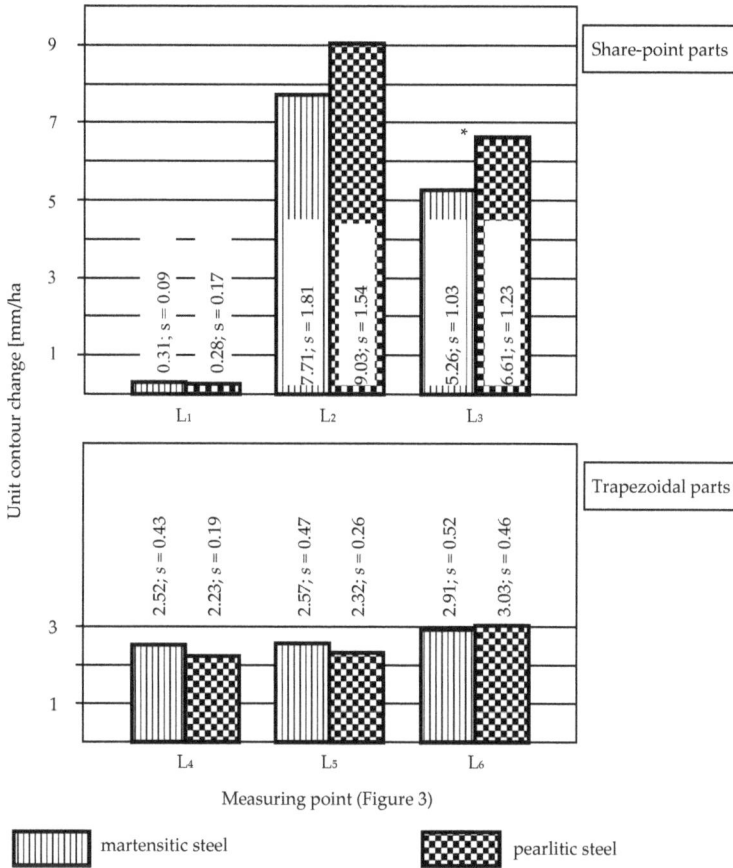

Figure 9. Unit contour change of share-points and trapezoidal parts of the examined ploughshares (*s*—standard deviation, *—statistically significant differences).

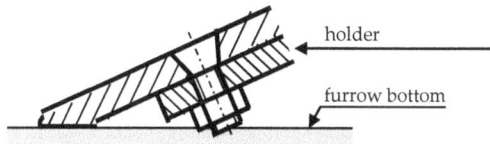

Figure 10. Limit wear condition of ploughshares resulting from the contour change.

Bending of the share-points made of pearlitic steel resulted in various widths of the wear band formed on the flank faces of the parts, see Figures 11 and 12. This parameter is important insofar as, at the wide wear band, difficulties can happen in sinking the plough or in maintaining its set working depth. Bending the share-points "forward" resulted in a ca. 1.6–1.2-fold wider wear band at Measuring Lines L_2 and L_3, respectively.

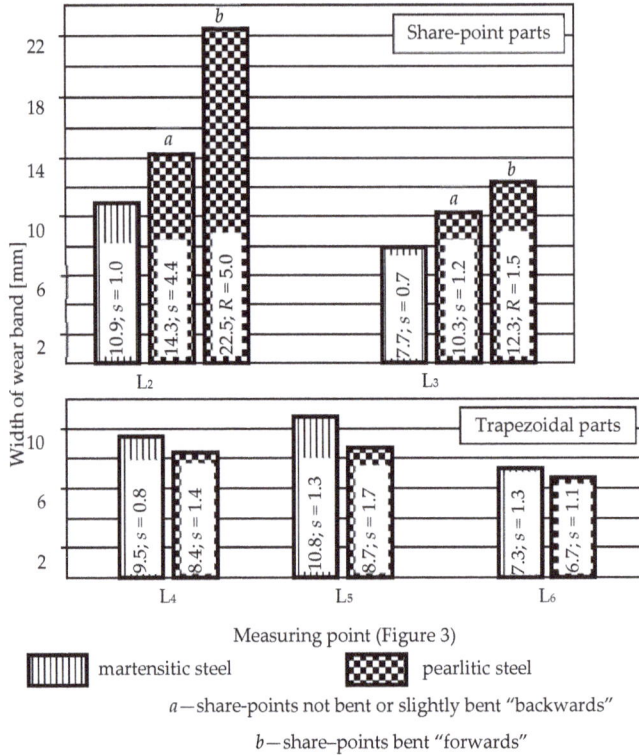

Figure 11. Width of the wear band created on the flank faces of the share-points and trapezoidal parts after the tillage test: 8.43–12.29 and 20.93 ha/body, respectively. *s*—standard deviation, *R*—range. No significance test was performed for this parameter.

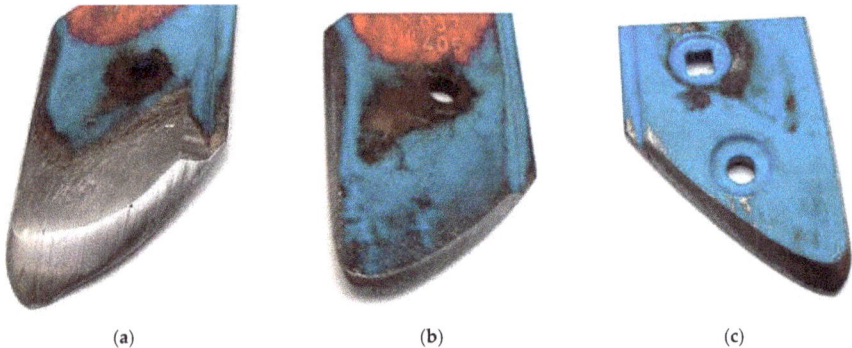

Figure 12. Wear band on the flank faces of the share-points: (**a**) pearlitic steel, "forward" bent; (**b**) pearlitic steel, slightly "backward" bent; (**c**) martensitic steel.

The width of the wear bands in the share-points made of pearlitic steel, which were not bent "forward", was ca. 1.3 times greater than that in the parts made of martensitic steel. However, the width of the wear bands in the trapezoidal parts made of pearlitic steel was ca. 1.1–1.2 times smaller (Figure 11), which can be related to their clearly greater thickness reduction, see Figure 5.

In some share-points made of pearlitic steel, a bending of their tips on a small length and a curling of the blade edges were observed, see Figure 13. These phenomena contributed to more irregular thickness reduction of the parts. Bending the share-points "forward" facilitated the a.m. processes because of the smaller thickness of the parts in their blade areas.

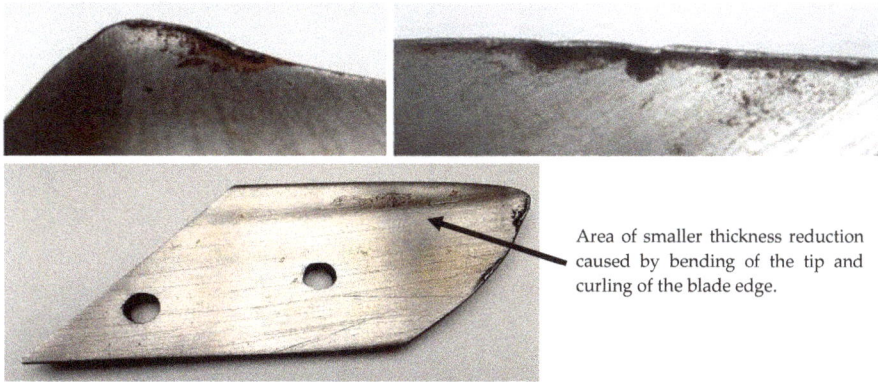

Area of smaller thickness reduction caused by bending of the tip and curling of the blade edge.

Figure 13. Exemplary bending of the tip and curling of the blade edge in the share-point made of pearlitic steel.

3.2. Condition of the Parts after Operation in Soil

Values of roughness parameters of the rake faces of the examined parts are given in Figure 14, and SEM images of these surfaces are shown in Figure 15.

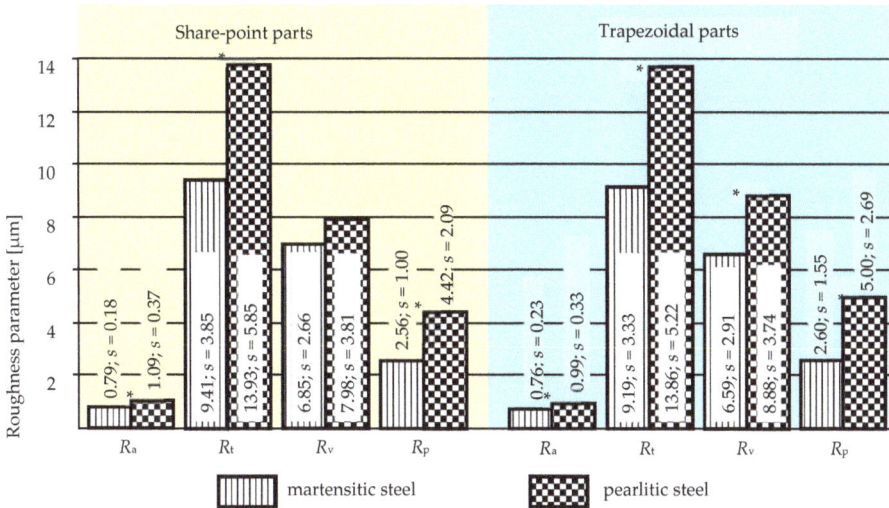

Figure 14. The roughness of the rake faces of the examined ploughshares: R_a—arithmetical mean deviation of the roughness profile, R_t—total height of the roughness profile, R_v—maximum valley depth of the roughness profile, R_p—maximum peak height of the roughness profile (*s*—standard deviation, *—statistically significant differences).

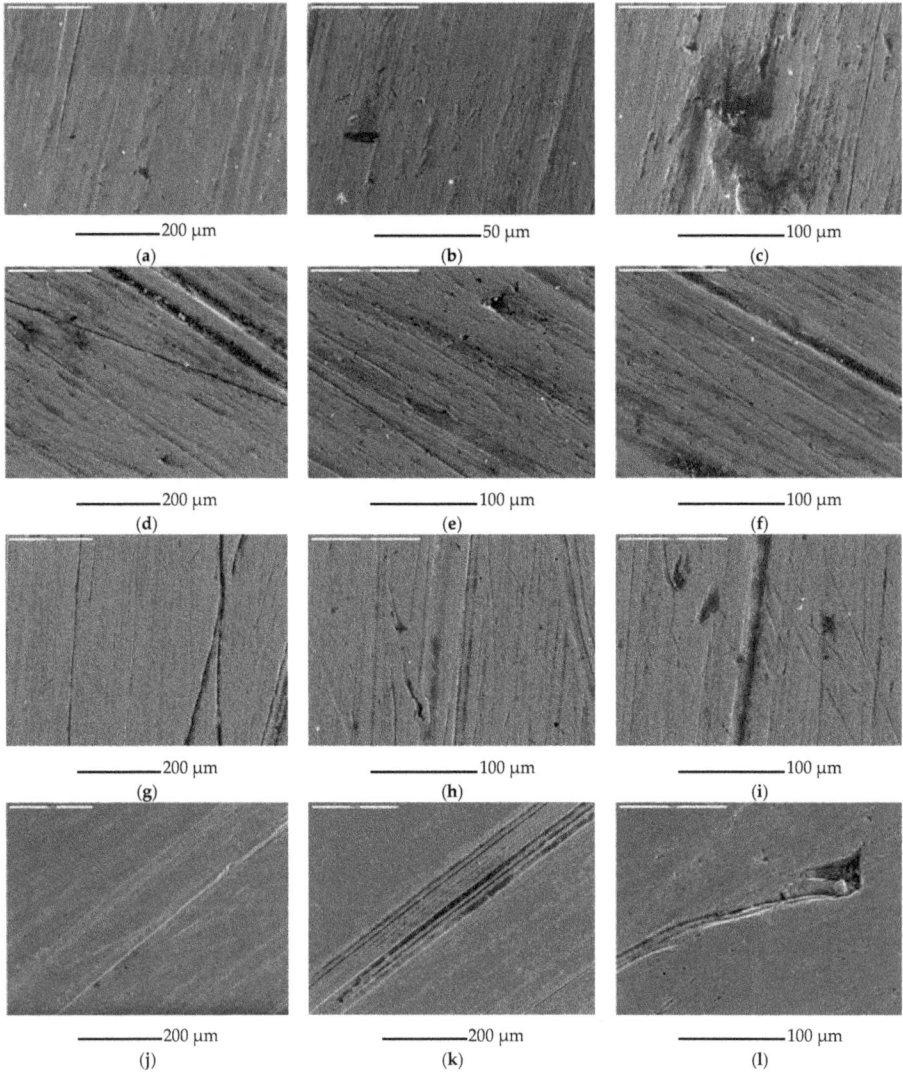

Figure 15. Rake faces of the examined parts (SEM). Rake face surface of a share-point made of pearlitic steel: (**a**) scratches and grooves, (**b**) pit hole, (**c**) fragments of embedded abrasive particle; Surface of a trapezoidal part made of pearlitic steel: (**d**) scratches and grooves, (**e**) fragments of embedded abrasive particle, (**f**) groove and plastic deformation; Surface of a share-point made of martensitic steel: (**g**) scratches and grooves, (**h**) holes formed during a particle impact, (**i**) groove and holes formed during a particle impact and embedded abrasive particle; Surface of a trapezoidal part made of martensitic steel: (**j**) scratches and grooves, (**k**) grooves, (**l**) groove formed during a particle impact and embedded abrasive particle.

Statistically significant differences were found in the share-points for the parameters R_a ($p = 0.0003$), R_t ($p = 0.0006$), and R_p ($p = 0.0001$), and in the trapezoidal parts—for the parameters R_a ($p = 0.0048$), R_t ($p = 0.0003$), R_v ($p = 0.0130$), and R_p ($p = 0.0002$).

The rake faces of the share-points and trapezoidal parts made of pearlitic steel were characterized by a greater roughness than corresponding surfaces of the parts made of martensitic steel. The average roughness R_a was respectively ca. 1.4 and 1.3 times greater for the share-points and trapezoidal parts, the total height of the roughness profile R_t was ca. 1.5 times greater for both share-points and trapezoidal parts, and the maximum peak height of the roughness profile R_p was respectively as much as ca. 1.7 and 1.9 times greater.

Analysis of the wear surface of a share-point made of pearlitic steel shows rough surface topography due to microcutting and microploughing caused by abrasive particles. The scratches and grooves are wide, deep, and randomly orientated, see Figure 15a. There are also visible scratches and scrapes randomly orientated. Moreover, some traces of spalling and pitting, originating in the places where larger areas of material were removed (Figure 15b), were observed at the surfaces. An examination of some pit holes showed the plastic deformation and displacement of material from the impact sites to the crater rims. These phenomena caused formation of grooves with pronounced lips at their rear ends. Abrasive particles could also cause surface fatigue. In these lips, detachment could occur due to these particles. In some pit holes, fragments of embedded abrasive particles are visible (Figure 15c). It can be seen on a surface of a trapezoidal part made of pearlitic steel that the proportion of microploughing increases (Figure 15d–f). The other observation on the surfaces of a share-point (Figure 15g–i) and a trapezoidal part (Figure 15j–l) made of martensitic steel is that scratches and grooves are arranged parallel to the movement direction of abrasive particles along the surface. The grooves are relatively narrow and shallow, but some areas are visible, where scratches and grooves are more concentrated (Figure 15k). Some shallow, randomly orientated scratches can be observed. Grooves formed by moving soil particles and plastically deformed material are visible on the crater rims, see Figure 15i,l.

4. Discussion

A striking result of the research is that values of the unit contour change of the parts made of pearlitic and of martensitic steel are comparable, in the context of the demonstrated differences in mass wear and the intensity of their thickness reduction, see Figures 4 and 5. In most of the measuring places, unit thickness reduction was found to be significantly greater for the ploughshares made of pearlitic steel; significantly higher also were the values of unit mass wear of these parts. However, such a difference did not occur in the case of the parameter of unit contour change. As was already mentioned, the thickness of the parts made of pearlitic steel was greater than the thickness of those made of martensitic steel—by 1 to 3 mm in the share-points and by 0.5 to 2 mm in the trapezoidal parts (Table 3). This probably contributed to the smaller unit contour change of the parts made of pearlitic steel and thus to their longer durability, which is determined by contour change. In consequence, the increased thickness of the parts made of pearlitic steel resulted in a comparable durability of the parts made of pearlitic and martensitic steels. It should be added that pearlitic steel is characterized by a relatively greater resistance to abrasive wear because of the content of a hard cementite phase. This feature certainly affected the obtained results of the tests [27,28].

Some literature data (obtained in laboratory conditions) indicate a more intensive wear of steel working in soil along with its lower hardness, while the intensity of wear depends on the chemical composition of steel and on working conditions [29]. Therefore, the high wear intensity of pearlitic steel with lower hardness in comparison to martensitic steel corresponds—to some extent—with the general regularity in the literature.

During the tests, one share-point part made of pearlitic steel was broken in spite of an >8-fold lower impact strength of pearlitic steel in relation to that of martensitic steel (Table 2), which indicates that these parts can break. However, a significant weak point of both the share-points and the trapezoidal parts made of pearlitic steel is their bending during tillage, which did not occur in the case of the parts made of martensitic steel. This was favored by the low mechanical strength of pearlitic steel (R_m = 911 MPa, $R_{p0.2}$ = 494 MPa, respectively, 2- and 3-fold lower than that of martensitic steel, see

Table 2). Bending was facilitated by a reduced cross section of the parts and was caused by thickness reduction during operation (Figure 5). The changed geometry of the parts, caused by their bending, could—to some extent—influence the processes of undercutting and crushing the soil, as well as the working resistance of the plough. As was mentioned before, the "forward" bending of the share-points resulted also in their somewhat earlier replacement due to uncovering the fitting bolts and to the creation of a wide wear band. It seems that a possible way of strengthening the parts could be further increasing their thickness, but it is uncertain whether that would be an effective solution with so clearly a lower strength of pearlitic steel. At the same time, this way is restricted by requirements of the plough design. In addition, a wide wear band is created on the flank faces of the thicker parts, which can result in a longer distance of the plough sinking on the headlands and in periodic reduction of the set tillage depth.

Larger roughness found on the rake faces of the parts made of pearlitic steel (Figure 14) can be related to lower mechanical properties of this steel and, especially, with ca. 1.8 times lower hardness than that of martensitic steel, see Table 2. Higher roughness values indicate a more intensive course of wear of pearlitic steel. The maximum peak height of the roughness profile values (R_p) on the share-points and trapezoidal parts made of pearlitic steel were respectively ca. 1.7 and 1.9 times greater than those for the parts made of martensitic steel. Thus, in the course of the wear process of pearlitic steel, the material was subjected to more intensive plastic deformation, characteristic of ridging, which was confirmed by SEM observations and former literature data [30–33].

It is also interesting that values of the parameters R_a, R_t, R_v, and R_p determined for the share-points and trapezoidal parts made of the same material slightly differ from each other, see Figure 14. Generally, the values found on the trapezoidal parts were a bit lower, but the differences did not exceed 14% of the parameter value. In this case, the performed roughness measurements did not show any statistically significant differences. This advocates a similar interaction of elementary wear processes occurring on rake faces of the share-points and trapezoidal parts. Therefore, considering a more intensive unit thickness reduction of the share-points in comparison to that of trapezoidal parts, it can be assumed that the more intensive wear of the share-points resulted from a greater intensity of the occurring wear processes.

It should be added that, for a user of agricultural equipment, the purchase cost of the applied replacement parts is an important factor in addition to their durability. The railway rails withdrawn from use are a cheap raw material that can be subjected to material recycling. One of the forms of such action is manufacturing ploughshares from recycled rails. In this case, the low price of raw material significantly influences the prices of ploughshares that are in Poland—ca. 2 times cheaper than those produced by renowned manufacturers of agricultural equipment.

5. Conclusions

(1) In the tillage tests carried out, the intensity of thickness reduction of ploughshares was greater for the parts made of pearlitic steel than that for the parts made of martensitic steel: ca. 1.4 times greater for share-points and ca. 1.8 times greater for trapezoidal parts. Unit mass wear was also more intensive for the parts made of pearlitic steel: 1.4 times greater for share-points and 1.2 times greater for trapezoidal parts. This indicates that pearlitic steel, compared to martensitic steel, has a lower resistance to wear in soil.

(2) After operation in soil, the rake faces of the parts made of pearlitic steel showed greater roughness than the parts made of martensitic steel, which can be attributed to pearlitic steel's lower hardness. This indicates a more intensive run of wear phenomena in pearlitic steel. The maximum peak height of the roughness profile values (R_p) were as much as respectively 1.7 and 1.9 times greater for the share-points and trapezoidal parts made of pearlitic steel in comparison to those made of martensitic steel. Therefore, in the parts made of pearlitic steel, plastic deformation of the surface layer material destroyed by soil particles was significantly greater.

(3) In spite of the 8-fold lower impact strength of pearlitic steel compared to that of martensitic steel, only one part was broken (the plough with overload protection of the bodies was used). However, most of the share-points and trapezoidal parts made of pearlitic steel were somewhat bent, which was favored by the successive reduction of the cross-section area of the parts as a result of thickness reduction. Nevertheless, in this case, the parts need not be replaced immediately. Earlier replacement was necessary in the case of "forward" bending of a share-point because the fitting bolts became uncovered. Wide wear bands were created on the flank faces of the so shaped share-points. One compensation for the described disadvantage of the parts made of pearlitic steel is their lower price.

(4) With pearlitic steel's greater susceptibility to wear in soil, the parts made of this steel were characterized by comparable values of unit contour change in relation to the parts made of martensitic steel. This fact was probably influenced by the greater thickness of the parts made of pearlitic steel (increased by 1 to 3 mm in share-points and by 0.5 to 2 mm in trapezoidal parts). Therefore, it is probable that, if the pearlitic share-points were not subjected to bending, their durability would be comparable to that of the parts made of martensitic steel.

(5) The bending of the ploughshares made of pearlitic steel is their definite disadvantage, since it influences to some extent interaction between the parts and soil. The bending strength of the parts could be increased by the parts' increased thickness, although this action is restricted by specific design solutions of the plough. It should be also emphasized that wide wear bands are formed on thicker parts, which can result in a lower agricultural quality of tillage.

Author Contributions: Tomasz Stawicki contributed reagents/materials/analysis tools, conceived and designed the experiments, analysed data, wrote the paper; Beata Białobrzeska contributed reagents/materials/analysis tools, conceived and designed the experiments, analysed data, wrote the paper; Piotr Kostencki contributed reagents/materials/analysis tools, conceived and designed the experiments, analysed data, wrote the paper.

Conflicts of Interest: The authors declare no conflict of interest.

References

1. Owsiak, Z.; Zużycie lemieszy pługów Cz., I. Charakter zużycia i stan graniczny lemieszy pługów. *Rocz. Nauk. Rol.* **1988**, *77*, 69–75.

2. Kushwaha, R.L.; Shi, J. Investigation of wear of agricultural tillage tools. *Lubr. Eng.* **1989**, *47*, 219–222.

3. Müller, M.; Hrabě, P. Overlay materials used for increasing lifetime of machine parts working under conditions of intensive abrasion. *Res. Agric. Eng.* **2013**, *59*, 16–22.

4. Hamblin, M.G.; Stachowiak, G.W. Description of abrasive particle shape and its relation to two-body abrasive wear. *Tribol. Trans.* **1996**, *39*, 803–810. [CrossRef]

5. Natsis, A.; Papadakis, G.; Pitsilis, J. The Influence of soil type, soil water and share sharpness of a mouldboard plough on energy consumption, rate of work and tillage quality. *J. Agric. Eng. Res.* **1999**, *72*, 171–176. [CrossRef]

6. Natsis, A.; Petropoulos, G.; Pandazaras, C. Influence of local soil conditions on mouldboard ploughshare abrasive wear. *Tribol. Int.* **2008**, *41*, 151–157. [CrossRef]

7. Napiórkowski, J. Zużyciowe oddziaływanie gleby na elementy robocze narzędzi rolniczych. *Inżynieria Rol.* **2005**, *9*, 3–171.

8. Yu, H.J.; Bhole, S.D. Development of prototype abrasive wear tester for tillage tool material. *Tribol. Int.* **1990**, *23*, 309–316. [CrossRef]

9. Miller, E.A. Wear in tillage tools. In *Wear Control Handbook*; Peterson, M. D., Winer, W.O., Eds.; ASME: New York, NY, USA, 1984; pp. 987–998.

10. Kostencki, P. Geometria zużycia lemieszy płużnych użytkowanych w glebach piaszczystych. *Probl. Inżynierii Rol.* **2007**, *3*, 49–64.

11. Kostencki, P.; Nowowiejski, R. Wytrzymałość ścierna wybranych lemieszy płużnych podczas uprawy pyłu zwykłego o dwóch stanach nawilgocenia. *Tribology* **2006**, *2*, 123–142.

12. Napiórkowski, J. Wpływ odczynu gleby na intensywność zużycia elementów roboczych. *Tribology* **1997**, 793–801.

13. Stabryła, J. Research on the degradation process of agricultural tools in soil. *Probl. Eksploat.* **2007**, *4*, 223–232.

14. Napiórkowski, J.; Michalski, R. Zwiększanie trwałości elementów roboczych w glebie metodami napawania. In Proceedings of the III Ogólnopolska Konferencja Naukowo-Techniczna, Jachranka, Poland, 24–26 May 1995; pp. II-27–II-37.

15. Bayhan, Y. Reduction of wear via hardfacing of chisel ploughshare. *Tribol. Int.* **2006**, *39*, 570–574. [CrossRef]

16. Hrabě, P.; Müller, M. Research of overlays influence on ploughshare lifetime. *Res. Agric. Eng.* **2013**, *59*, 147–152.

17. Novák, P.; Müller, M.; Hrabě, P. Research of a material and structural solution in the area of conventional soil processing. *Agron. Res.* **2014**, *12*, 143–150.

18. Horvat, Z.; Filipovic, D.; Kosutic, S.; Emert, R. Reduction of mouldboard plough share wear by a combination technique of hardfacing. *Tribol. Int.* **2008**, *41*, 778–782. [CrossRef]

19. Białobrzeska, B.; Kostencki, P. Abrasive wear characteristics of selected low-alloy boron steels as measured in both field experiments and laboratory tests. *Wear* **2015**, *328*, 149–159. [CrossRef]

20. Er, U.; Par, B. Wear of plowshare components in SAE 950C steel surface hardened by powder boriding. *Wear* **2006**, *261*, 251–255. [CrossRef]

21. Tian, Z.; Sun, W.; Shang, M.; Jiang, X.; Han, W.; Li, L. Application of boronizing technology on ploughshares and study on the abrasive wear characteristics under low stress of boronized layer. In Proceedings of the International Symposium on Agricultural Engineering (89-ISAE), Beijing, China, 12–15 September 1989; International Academic Publishers: Beijing, China, 1989; pp. 248–249.

22. Foley, A.G.; Lawton, P.J.; Barker, A.W.; McLees, V.A. The use of alumina ceramic to reduce wear of soil-engaging components. *J. Agric. Eng. Res.* **1984**, *30*, 37–46. [CrossRef]

23. Napiórkowski, J.; Ligier, K. Wear testing of α-Al$_2$O$_3$ oxide ceramic in a diverse abrasive soil mass. *Tribology* **2014**, *1*, 63–74.

24. Napiórkowski, J.; Ligier, K.; Pękalski, G. Tribological properties of cemented carbides in abrasive soil mass). *Tribology* **2014**, *2*, 123–134.

25. Müller, M.; Chotěborský, R.; Valášek, P.; Hloch, S. Unusual possibility of wear resistance increase research in the sphere of soil cultivation. *Teh. Vjes.* **2013**, *20*, 641–646.

26. Kostencki, P.; Stawicki, T.; Białobrzeska, B. Durability and wear geometry of subsoiler shanks provided with sintered carbide plates. *Tribol. Int.* **2016**, *104*, 19–35. [CrossRef]

27. Perez-Unzueta, A.J.; Beynon, J.H. Microstructure and wear resistance of pearlitic rail steels. *Wear* **1993**, *162*, 173–182. [CrossRef]

28. Viafara, C.C.; Castro, M.I.; Velez, J.M.; Toro, A. Unlubricated sliding wear of pearlitic and bainitic steels. *Wear* **2005**, *259*, 405–411. [CrossRef]

29. Severnev, M.M. *Wear of Agricultural Machine Parts*; USDA/Amerind Publishing CO. Pyt. Ltd.: New Delhi, India; Washington, DC, USA, 1984.

30. Kazemipour, M.; Shokrollahi, H.; Sharafi, S. The influence of the matrix microstructure on abrasive wear resistance of heat-treated Fe–32Cr–4.5C wt % hardfacing alloy. *Tribol. Lett.* **2010**, *39*, 181–192. [CrossRef]

31. Hokkirigawa, K.; Kato, K. An experimental and theoretical investigation of ploughing, cutting and wedge formation during abrasive wear. *Tribol. Int.* **1988**, *21*, 51–57. [CrossRef]

32. Sapate, S.G.; Selokar, A.; Garg, N. Experimental investigation of hardfaced martensitic steel under slurry abrasion conditions. *Mater. Des.* **2010**, *31*, 4001–4006. [CrossRef]

33. Turenne, S.; Lavallee, F.; Masounave, J. Matrix microstructure effect on the abrasion wear resistance of high–chromium white cast iron. *J. Mater. Sci.* **1989**, *24*, 3021–3028. [CrossRef]

![metals logo] *metals*

MDPI

Article

Critical Condition of Dynamic Recrystallization in 35CrMo Steel

Yuanchun Huang [1,2,3,*]**, Sanxing Wang** [1]**, Zhengbing Xiao** [1,2] **and Hui Liu** [2]

1   College of Mechanical and Electrical Engineering, Central South University, Changsha 410083, China;
    wangsanxing@csu.edu.cn (S.W.); xiaozb@csu.edu.cn (Z.X.)
2   Light Alloy Research Institute, Central South University, Changsha 410012, China; liuhui2015@csu.edu.cn
3   State Key Laboratory of High Performance Complex Manufacturing, Central South University,
    Changsha 410083, China
*   Correspondence: science@csu.edu.cn; Tel.: +86-731-8887-6315

Academic Editor: Robert Tuttle
Received: 15 April 2017; Accepted: 4 May 2017; Published: 9 May 2017

Abstract: The compression deformation behaviors of 35CrMo steel at different conditions was studied by using Gleeble-3810 thermo-simulation machine under large strain. The results indicate that the flow stress curves of 35CrMo steel is affected by strain rate and deformation temperature, showing the characteristics of dynamic recovery (DRV) and dynamic recrystallization (DRX), which is the main softening mechanism of 35CrMo steel. The activation energy (Q) and Zener–Hollomon parameter (Z parameter) expression for thermal deformation of this steel was calculated by linear regression. The inflection point on the curve of strain hardening rate and flow stress (θ-σ curve) corresponds to the beginning of DRX, and the critical strain of DRX increases with the decrease of deformation temperature and the increase of strain rate. Based on the inflection point criterion, the constitutive equation of the critical strain of DRX of 35CrMo steel was established: $\varepsilon_c = 0.000232Z^{0.1673}$, which reflects the variation of the critical strain of DRX with the Z parameter. In addition, through metallographic observation, the rationality of the inflection point criterion in determining the critical strain of DRX of 35CrMo steel was verified, and the DRX state diagram was established.

Keywords: 35CrMo steel; dynamic recrystallization; work hardening; critical condition

1. Introduction

Microstructure control of metals and alloys during thermal processing is an important key because it allows control of the final microstructure of the alloy as well as the required mechanical properties. One of the most important mechanisms of microstructure control in thermal deformation is dynamic recrystallization (DRX), which occurs in several metals and alloys [1–3]. The traditional view is that the emergence of the peak stress of the rheological curve represents the occurrence of DRX, but for some materials with strain hardening properties the rheological curve does not show significant peak characteristics, even if the presence of DRX is confirmed by metallographic observation. Although the critical strain of DRX can be determined by the metallographic method, this technique requires many samples before and after the critical stress. In addition, the phase change during the cooling process from the thermal deformation temperature also changes the deformation structure, which makes the metallographic analysis difficult. Therefore, it is very important to find a method to obtain the critical strain of DRX without using metallographic analysis. Poliak and Jonas [4–6] believed that once the DRX occurs, regardless of whether there is a stress peak on the stress–strain curve, the inflection point is shown on the θ-σ curve, and the critical condition is considered to be the maximum on the $d\theta/d\sigma$-σ curve and the inflection point of the θ-σ curve corresponding. Najafizadeh and Mirzadeh [7,8] proposed to describe the θ-σ curve with a cubic polynomial, which makes it easy and quantifiable to determine the inflection point. Many scholars [9–11] applied this theory and method, in the metals and alloys to determine the DRX critical strain to achieve better results.

Due to its good wear, impact, corrosion, and fatigue resistances—and its price being much lower than that of chromium-nickel alloy steel with the same mechanical properties—35CrMo steel is widely used in processing of mechanical products, vehicle assembly, mining, and other industries. It is heavily used in the manufacturing of various small and medium-sized parts such as driveshafts, crankshafts, and fasteners [12–15]. In 2003, Zhang [12] on the ordinary casting 35CrMo steel hot compression test, and build the corresponding dynamic recrystallization model. In 2010, SV Sajadi et al. [13] studied the dynamic recovery (DRV) behavior of 34CrMo4 (domestic 35CrMo grade) steel at 900–1100 °C/0.001–0.1 s^{-1}, and constructed the rheological stress of steel under the deformation parameters prediction equation, and achieved good application effect. Xu et al. [14] studied the effect of rheological forming parameters on the rheological resistance of 34CrMo4 steel, and constructed the high-temperature rheological constitutive equation of steel based on macroscopic phenomenology. Furthermore, Xiao et al. [15] found that the DRX behavior of 35CrMo steel can not only refine the grain in the process of thermal deformation, but also can effectively inhibit the crack. Unfortunately, the DRX critical condition of 35CrMo steel is not given.

In this paper, the reason of the inflection point on the θ-σ curve of 35CrMo steel is analyzed by work hardening theory, and the softening mechanism of the stress–strain curves was studied. The DRX inflection point criterion, focused on the critical strain calculation for the initiation of the DRX of 35CrMo steel under various deformation conditions and verifies its rationality in combination with microstructure analysis.

2. Experimental Materials and Procedures

35CrMo steel is widely used as the bearing part and rotary shaft material of the marine positioning system. The chemical composition is shown in Table 1, and the original microstructure grains are coarse and irregular, as shown in Figure 1. Before the experiment, the steel was processed into a cylindrical sample of 10 mm and a height of 12 mm by wire cutting. The graphite sheet is padded between the ends of the specimen and the contact surface of the indenter and the lubricant is applied at both ends of the sample to keep the temperature and deformation more uniform during heating and compression.

The thermal simulation compression test was carried out on a Gleeble-3810 thermo-simulation machine (Central South University, Changsha, China) in the temperatures range of 850–1150 °C at an interval of 100 °C and at strain rates ranging from 0.01 to 10 s^{-1}. All specimens were heated up to 1200 °C at 10 °C/s, soaked for 120 s to eliminate the thermal gradient, and then cooled to the deformation temperature at 10 °C/s. Before the compression tests, samples were maintained at deformation temperature for 120 s. The reduction in height was 60% (true strain 0.85) at the end of the compression tests, and each sample after compression was quenched in water immediately to retain the deformed microstructure. In order to observe the DRX under different strains, four sets of experiments were carried out at the deformation temperature of 1050 °C and the strain rate of 0.1 s^{-1}, and the strain was the critical strain (0.115), the maximum strain (0.22), the softening strain (0.31) and the steady strain (0.62), respectively. The sample was subjected to grinding and polishing, and was etched at 60 to 80 °C in a self-formulated etchant (2.5 g of picric acid, 50 mL of distilled water and 1 mL of hydrochloric acid) bath for 4 to 10 min, with an Olympus DSX500 optical microscope (Olympus Corporation, Tokyo, Japan) to observe the microstructure.

Table 1. Chemical composite of 35CrMo steel in mass fraction.

Element	C	Si	Mn	Cr	Mo	S	P	Al	Ni	Fe
Content (wt. %)	0.34	0.21	0.56	0.95	0.19	0.0051	0.019	0.0032	0.05	Bal.

Figure 1. Original microstructure.

3. Results and Discussion

3.1. Work Hardening Phenomenon

The flow stress of metals is affected by the change of internal dislocation density. According to the theory of work hardening [16–20], the work hardening rate with stress changes can be divided into five stages: I—easy slip phase, II—linear hardening stage, III—DRV hardening stage, IV—large strain hardening stage, and V—DRX softening stage, as shown in Figure 2.

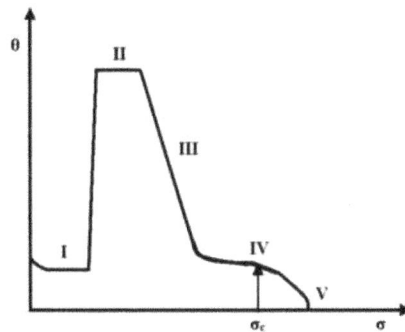

Figure 2. Sketch map of strain hardening rate θ and stress σ.

DRX occurs in the transition zone of IV stage to the V stage. With the increase in strain accumulated to a certain amount, the material hardening will enter the IV stage. The dislocation cell wall in the cellular structure absorbs the dislocation of the active dislocation and increases the dislocation density of the cell wall, forming the subgrain boundary, and the cellular structure becomes the substructure. Sub-structure formation, due to its strength on the material has a certain contribution, leading to a decline in the work hardening rate of slowing down, however, without a point of inflection on the θ-σ curve. Obtaining this point of inflection requires introduction of a stage V which can only be brought about by a change in microstructural mechanisms. Gottstein [17] proposes to associate this change with the mobility of sub-boundaries.

When the boundaries move they incorporate and annihilate the dislocations of the swept cell interior. This introduces a new dynamic recovery mechanism and, compared to stage IV behavior, speeds up attainment of a steady-state with a dynamically stable and equiaxed sub grain structure. Also, it generates a stage V of the hardening behavior as a rapid transition to steady-state [17,21]. In this stage, the deformation to a certain extent, the deformation of the dislocation density within the organization reached a certain critical value, microstructural heterogeneity or boundary kinetics may play a role in rendering microstructure behavior stable or unstable at the critical point. Consequently, there will be a point of inflection at the transition from stages IV to V, as shown in Figure 2. The corresponding stress at the inflection point is the DRX critical stress σ_c, and the lnθ-lnσ curve and the lnθ-ϵ curve show inflection point characteristics [5,7], Thus, by analyzing the strain hardening rate θ and the inflection point on the flow stress σ curve or the maximum value on the dθ/dσ-σ curve, the critical condition of DRX can be obtained.

3.2. Characteristics of Stress–Strain Curves

Flow stress and dislocation density are closely linked [22,23]. The austenite phase in 35CrMo steel has low stacking fault energy, and its recovery process is slow. The DRV cannot be synchronized to offset the accumulation of dislocation during hot deformation, so DRX occurs under a certain critical deformation condition. During the DRX process, a large number of dislocations are eliminated by the large angle interface of the recrystallization core, which also acts as a softening effect on the alloy during hot working. Therefore, DRV and DRX reduce the dislocation density of the steel in the hot working process, which is an effective softening mechanism in the hot deformation.

Figure 3 shows the typical flow curves of 35CrMo steel under different deformation conditions, with similar rheological characteristics: at the beginning of deformation, flow stress increases rapidly due to the strong effect of work hardening. Subsequently, the dynamic softening effect increases with the increase of the strain, resulting in a slow increase in flow stress, which remains stable after a peak or a slight decrease, showing DRV and DRX characteristics, respectively. At higher strain rate (10 s^{-1}) and lower temperatures (850 °C), the flow stress is almost constant after reaching a peak (peak point is not obvious), so the test steel did not occur DRX before the peak stress, the corresponding softening mechanism for DRV. On the contrary, at lower strain rates (0.01 s^{-1}) and higher temperature (1050 °C), DRX has occurred before peak stress appears (peak point is obvious) when the strain exceeds critical strain. The driving force of DRX is the removal of dislocations [24]. The increase in work hardening and dislocation density leads to critical microstructural changes that are conditional for new grain nucleation and new high-angle boundary growth. The lower the strain rate and the higher the deformation temperature, the more obvious the effect of DRX on the stress–strain curves [25], so the higher the plasticity.

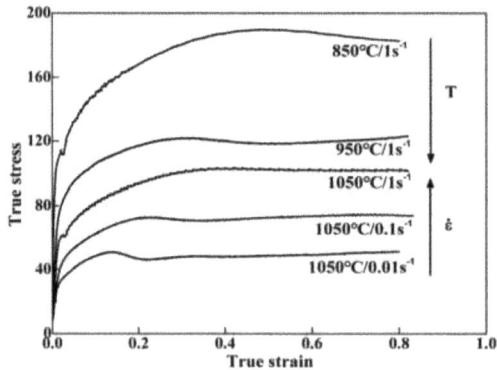

Figure 3. True stress–strain curves of 35CrMo steel at different deformation conditions.

3.3. The Activation Energy of 35CrMo Steel

Arrhenius type equation is widely used to obtain the activation energy of the metals and alloys, and the influence of strain rate and temperature on the deformation behavior can be expressed by the Zener–Hollomon parameter in the exponential equation. It can be expressed as follows [26–29]:

$$\dot{\varepsilon} = AF(\sigma)\exp[-Q/(RT)] \tag{1}$$

$$Z = \dot{\varepsilon}\exp(\frac{Q}{RT}) \tag{2}$$

Here,

$$\begin{cases} F(\sigma) = \sigma^{N}, & (\alpha\sigma < 0.8) \\ F(\sigma) = \exp(\beta\sigma), & (\alpha\sigma > 1.2) \\ F(\sigma) = [\sinh(\alpha\sigma)]^{n}, & \text{For all } \sigma \end{cases} \tag{3}$$

where $\dot{\varepsilon}$ is the strain rate (s^{-1}), A, α, β, N, and n, are material constants, and $\alpha = \beta/N$, Q is the apparent activation energy for deformation, R is the gas constant (8.3145 J·mol^{-1}·K^{-1}), T is the absolute temperature (K), σ is the true stress (MPa).

Substituting Equation (3) into Equation (1) and taking the logarithm on both sides, gives

$$\begin{cases} \ln\dot{\varepsilon} = \ln A + N\ln\sigma - \frac{Q}{RT} & (\alpha\sigma < 0.8) \\ \ln\dot{\varepsilon}\ln A + \beta\sigma - \frac{Q}{RT} & (\alpha\sigma > 1.2) \\ \ln\dot{\varepsilon} = \ln A + n\ln[\sinh(\alpha\sigma)] - \frac{Q}{RT} & \text{For all } \sigma \end{cases} \tag{4}$$

In the case of lower stress (i.e., $\alpha\sigma < 0.8$), the first equation can be used to calculate the value of N, under higher stress conditions (i.e., $\alpha\sigma > 1.2$), using the second equation to calculate β value. Figure 4a,b shows the plots of $\ln\sigma - \ln\dot{\varepsilon}$ and $\sigma_p - \ln\dot{\varepsilon}$ when $\varepsilon = 0.1$, $\sigma = \sigma_p$, respectively, the linear regression was used to fit the slope, and the average slope was $N = 9.473307$, $\beta = 0.067817$. Then $\alpha = \beta/N = 7.159 \times 10^{-3}$.

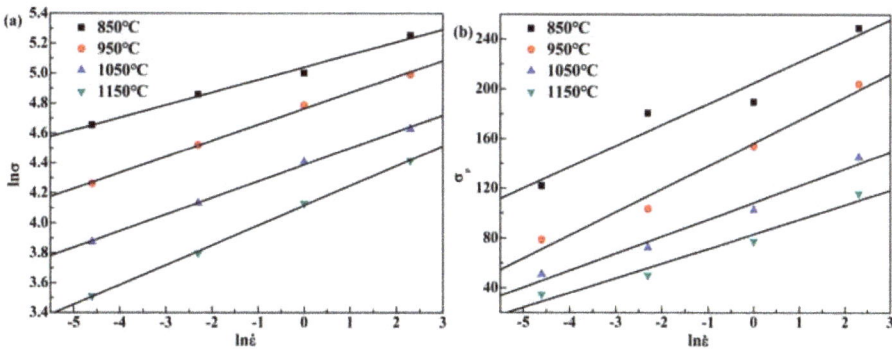

Figure 4. (a) $\ln\sigma - \ln\dot{\varepsilon}$ plots; and (b) $\sigma_p - \ln\dot{\varepsilon}$ plots under different deformation conditions.

Substituting the value of α into Equation (4), the plots of $\ln[\sinh(\alpha\sigma)] - \ln\dot{\varepsilon}$ can be obtained as shown in Figure 5a, and the value of n can be computed as 8.074677 by linear fitting method.

For a given strain rate $\dot{\varepsilon}$, the activation energy Q can be obtained from the following equation,

$$Q = Rn\left[\frac{\partial\ln\sinh(\alpha\sigma)}{\partial\ln(1/T)}\right] \tag{5}$$

The plots of $\ln[\sinh(\alpha\sigma)] - 1000/T$ as shown in Figure 5b, and the average value of slope is 6.084558. With the combination of R and n, the activation energy $Q = 408.498$ kJ/mol. Then the Z parameter can be expressed as,

$$Z = \dot{\varepsilon}\exp\left(\frac{408.498 \times 10^3}{RT}\right) \tag{6}$$

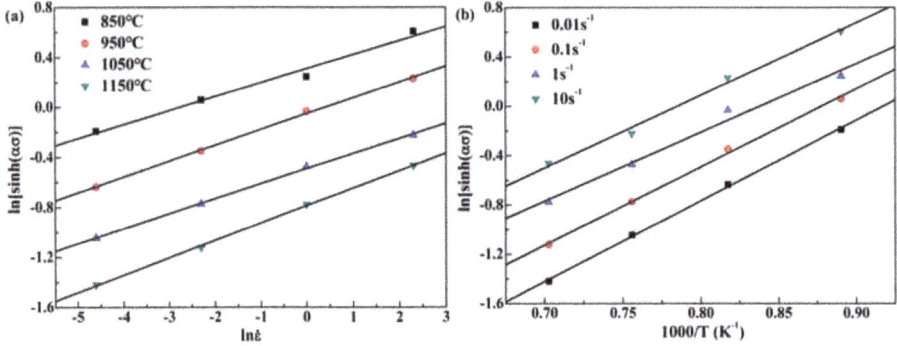

Figure 5. (a) $\ln[\sinh(\alpha\sigma)] - \ln\dot{\varepsilon}$ plots under different temperature; and (b) $\ln[\sinh(\alpha\sigma)] - 1000/T$ plots at different strain rates.

3.4. Critical Stress for Initiation of Dynamic Recrystallization (DRX)

Taking the deformation conditions $1050\,^\circ$C/0.1 s^{-1} as an example, due to the rheological curve obtained by the experiment is not a smooth curve in the mathematical sense, the differential curve can cause the irregularities of the differential curve and affect the analysis of the data. In this case, the stress–strain curve (0 to peak strain) is first fitted with nine polynomial functions (Equation (7)), and then the differential operation is performed. This will avoid the effect of data irregularities on the differential operation, as shown in Figure 6.

$$\sigma = 8.56328 \times 10^{10}\varepsilon^9 - 8.00356 \times 10^{10}\varepsilon^8 + 3.22096 \times 10^{10}\varepsilon^7 - 7.3067 \times 10^9\varepsilon^6 + 1.02741 \times 10^9\varepsilon^5$$
$$-9.27781 \times 10^7\varepsilon^4 + 5.39153 \times 10^6\varepsilon^3 - 1.96558 \times 10^5\varepsilon^2 + 4.386 \times 10^3\varepsilon + 1.93894 \tag{7}$$

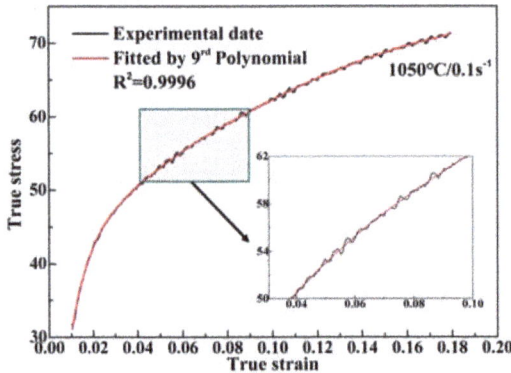

Figure 6. Contrast between experimental curve and fitted curve deformed at temperature of 1050 °C and strain rate of 0.1 s^{-1}.

The stress σ in Figure 6 was derived from 0 to the peak stress σ_p, and the relationship between the strain hardening rate θ and the true stress σ as shown in Figure 7a is obtained. The curve in the figure was fitted with cubic polynomial to get θ-σ equation:

$$\theta = -3.86 \times 10^{-2}\sigma^3 + 0.75 \times 10\sigma^2 - 4.88 \times 10^2\sigma + 1.06 \times 10^4 \tag{8}$$

Again, the dθ/dσ-σ curve was derived and plotted, as shown in Figure 7b. The corresponding stress σ is the critical strain ε_c when $d^2\theta/d\sigma^2 = 0$. Then the critical stress σ_c of 35CrMo steel is obtained as follows:

$$\sigma_c = -\frac{7.5}{-0.0386 \times 3} = 64.77 (\text{MPa}) \tag{9}$$

The corresponding critical strain ε_c is 0.096.

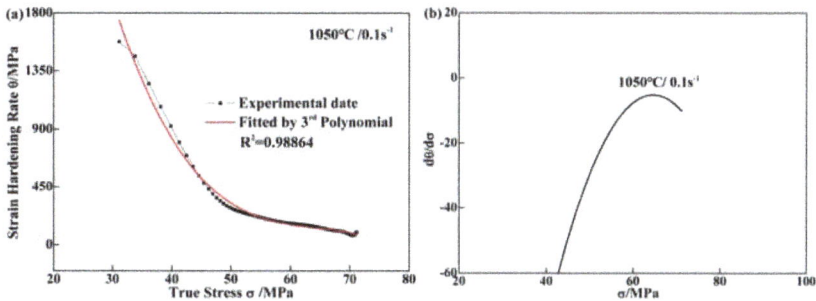

Figure 7. Curves of θ versus σ (**a**) and dθ/dσ versus σ (**b**) deformed at 1050 °C with 0.1 s^{-1}.

The critical stress and critical strain of DRX of 35CrMo steel at different temperatures and strain rates can be obtained from Figures 8 and 9. It is shown that the inflection point at the critical condition of DRX does not appear at low temperature (850 °C). With the increase of temperature, and at lower strain rate—i.e., 950 °C/0.01 s^{-1}, 1050 °C/(0.01, 0.1, 1 s^{-1})—the existence of inflection point was observed on the θ-σ curves. After the temperature reached 1150 °C, DRX occurred all conditions before the peak stress appeared. This proves that the high temperature and low strain rate is favorable for the formation of DRX of 35CrMo steel. Table 2 shows the value of the inflection point on the θ-σ curves' corresponding critical stress.

Table 2. Parameters fitted for the strain hardening rate as a function of flow stress.

T (°C)	Strain Rate(s^{-1})	θ/σ Relation	σ_c (MPa)	σ_p (MPa)
950	0.01	$\theta = -1.72 \times 10^{-1}\sigma^3 + 3.34 \times 10\sigma^2 - 2.15 \times 10^3\sigma + 4.60 \times 10^4$	64.73	78.87
	0.01	$\theta = -1.81 \times 10^{-1}\sigma^3 + 2.33 \times 10\sigma^2 - 1.01 \times 10^3\sigma + 1.49 \times 10^4$	43.03	50.86
1050	0.1	$\theta = -3.86 \times 10^{-2}\sigma^3 + 0.75 \times 10\sigma^2 - 4.88 \times 10^2\sigma + 1.06 \times 10^4$	64.77	72.97
	1	$\theta = -6.04 \times 10^{-2}\sigma^3 + 1.49 \times 10\sigma^2 - 1.23 \times 10^3\sigma + 3.41 \times 10^4$	82.21	102.75
	0.01	$\theta = -2.78 \times 10^{-1}\sigma^3 + 2.37 \times 10\sigma^2 - 6.87 \times 10^2\sigma + 6.87 \times 10^3$	28.45	34.67
1150	0.1	$\theta = -9.06 \times 10^{-2}\sigma^3 + 1.16 \times 10\sigma^2 - 6.00 \times 10^2\sigma + 7.38 \times 10^3$	42.77	50.27
	1	$\theta = -1.63 \times 10^{-2}\sigma^3 + 0.31 \times 10\sigma^2 - 1.99 \times 10^2\sigma + 4.63 \times 10^3$	62.40	77.59
	10	$\theta = -2.80 \times 10^{-2}\sigma^3 + 0.24 \times 10\sigma^2 - 2.33 \times 10^2\sigma + 7.93 \times 10^3$	97.64	115.86

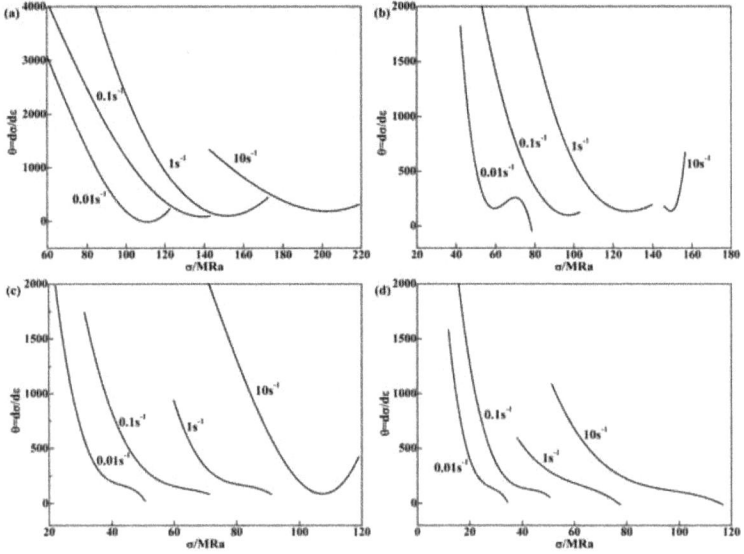

Figure 8. Curves of θ versus σ for 35CrMo steel under different deformation temperatures: (**a**) 850 °C, (**b**) 950 °C, (**c**) 1050 °C, (**d**) 1150 °C.

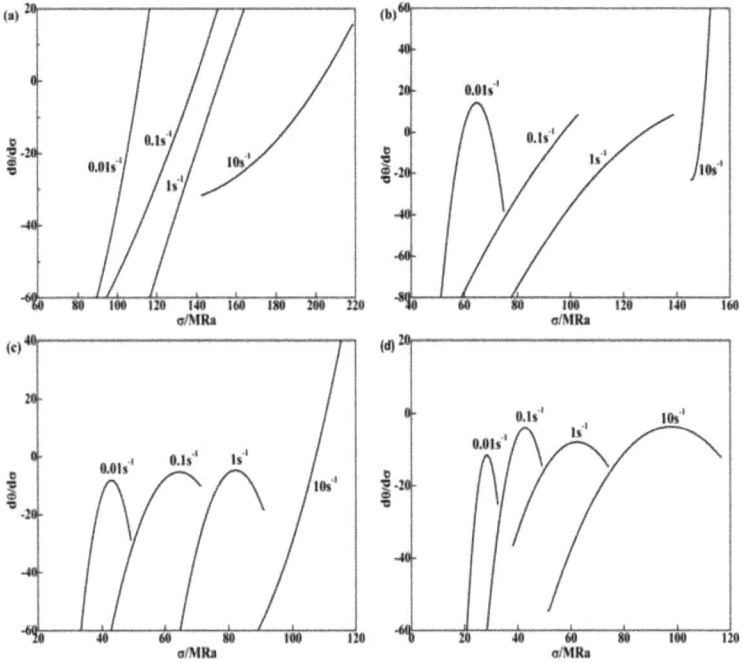

Figure 9. Curves of dθ/dσ versus σ for 35CrMo steel under different deformation temperatures: (**a**) 850 °C, (**b**) 950 °C, (**c**) 1050 °C, (**d**) 1150 °C.

Figure 10 shows the curves of critical stress (σ_c) versus temperature (T) and critical strain (ε_c) versus T for 35CrMo steel under different deformation conditions. It is shown that the critical stress and critical strain increases with temperature decrease and strain rate increase, showing negative temperature sensitivity and positive strain rate sensitivity. This is because, at the same deformation temperature, the greater the strain rate, the shorter the deformation time and decreased dislocation density, and the critical strain of the recrystallization increases. As the temperature increases, the internal drive energy becomes larger and the dislocation motion occurs more easily, increasing the likelihood of DRX, so that the critical strain of DRX becomes smaller.

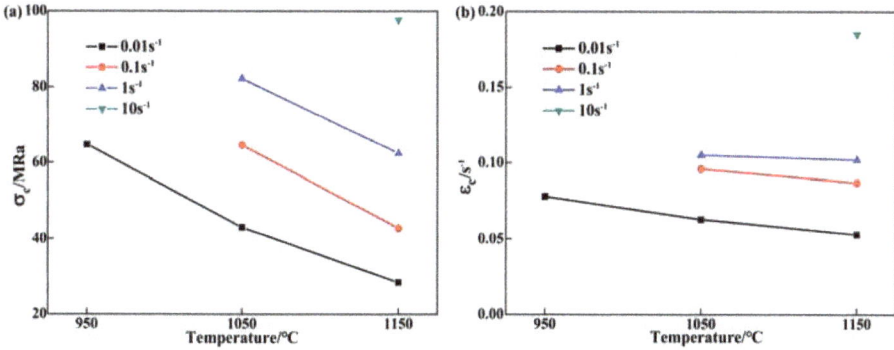

Figure 10. Curves of (**a**) σ_c versus T and (**b**) ε_c versus T for 35CrMo steel under different deformation conditions.

In general, the critical strain model of DRX can be expressed by the following equation [11,24]:

$$\varepsilon_c = a_1 \varepsilon_p \tag{10}$$

$$\varepsilon_p = a Z^c \tag{11}$$

As shown in Figure 11, the plots of ε_c-ε_p and $\ln\varepsilon_p$-$\ln Z$ can determine the unknown parameters a and a_1 of the Equations (10) and (11), i.e., $a = 0.31912$, $a_1 = 7.28 \times 10^{-4}$, $c = 0.1673$ are obtained from linear fit. The critical strain and the peak strain satisfy the relationship: $\varepsilon_c = 0.31912\,\varepsilon_p$, and the mathematical model of peak strain is: $\varepsilon_p = 7.28 \times 10^{-4} Z^{0.17001}$. Combined with the above two formulas, the critical strain of DRX constitutive model is obtained: $\varepsilon_c = 2.32 \times 10^{-4} Z^{0.1673}$.

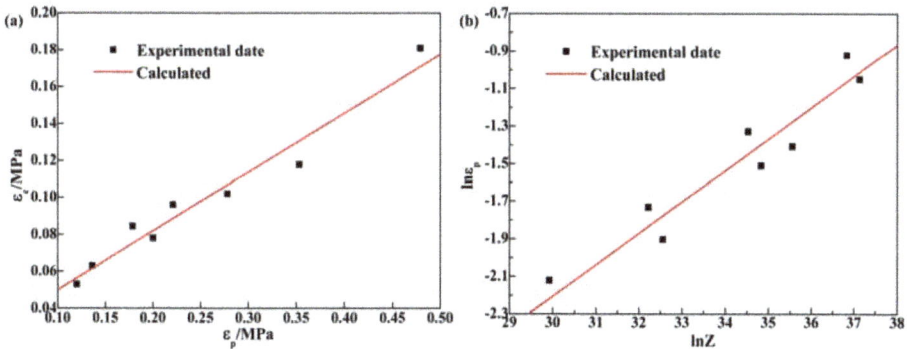

Figure 11. The plots of (**a**) ε_c-ε_p; (**b**) $\ln\varepsilon_p$-$\ln Z$.

3.5. Microstructure Evolution and DRX State Diagram

Figure 12 shows the relationship between θ and ε deformed at 1050 °C with 0.1 s^{-1}. The corresponding strain A, B, C, and D in the curve are the critical strain (ε_c), the peak strain (ε_p), the strain at the maximum softening rate (ε^*) and the strain at steady-state flow stress (ε_s). Figure 13 shows the microstructure corresponding to four points.

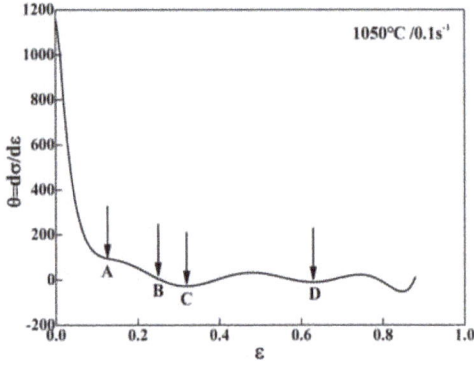

Figure 12. Relationship between θ and σ deformed at 1050 °C with 0.1 s^{-1}.

In the early stage of deformation, the austenite grains of test steels are crushed and elongated, and the grain boundaries are locally migrated. When the strain reaches the critical strain ε_c, a small amount of fresh equiaxed dynamic recrystallized grains are deformed (Figure 13a). When the strain reaches the critical strain ε_c, and the work hardening rate is 0 at this moment, the experimental steel part of the DRX, deformed grains and coexistence of equiaxed recrystallized grains (Figure 13b), and then the DRX softening will be greater than the work hardening effect. When the strain reaches the maximum softening rate at strain ε^*, the work hardening is minimized, the DRX is basically completed, and the microstructure is transformed into inhomogeneous recrystallized grains (Figure 13c). With the further increase of the deformation, the softening effect of recrystallization tends to balance the work hardening produced by the increase of the dislocation density, the rheological curve enters the steady state stage, where the microstructure is fully DRX and grows up to equiaxed austenite grains (Figure 13d).

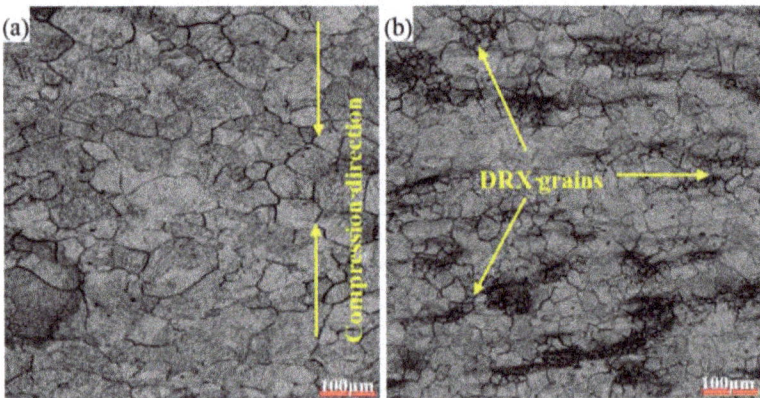

Figure 13. *Cont.*

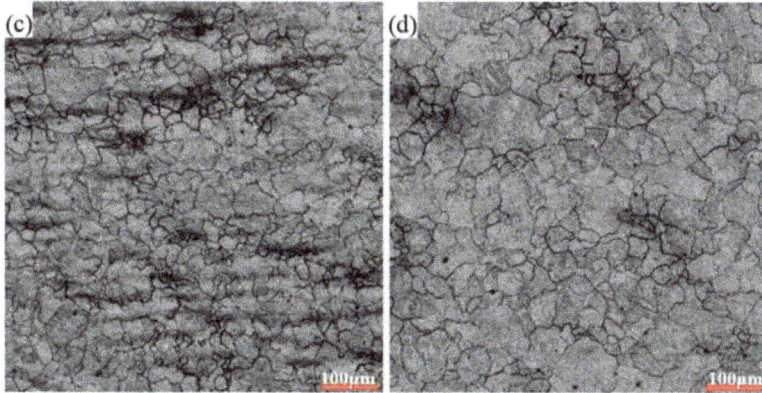

Figure 13. Microstructure corresponding to point A, B, C and D in Figure 9 (**a**) A; (**b**) B; (**c**) C; (**d**) D. DRX: dynamic recrystallization.

The DRX state diagram of 35CrMo steel is drawn from the experimental data. As shown in Figure 14, where zone A is the work hardening zone, zone B is the partial DRX zone, zone C is the full DRX zone. The influence of strain and Z parameters on the recrystallization of 35CrMo steel was quantitatively described by the DRX state diagram. It can be seen from the DRX state diagram that with increase of Z parameters—i.e., with decreasing deformation temperature and increase of the strain rate, the critical strain of the material gradually increases—that is to say, DRX becomes more difficult.

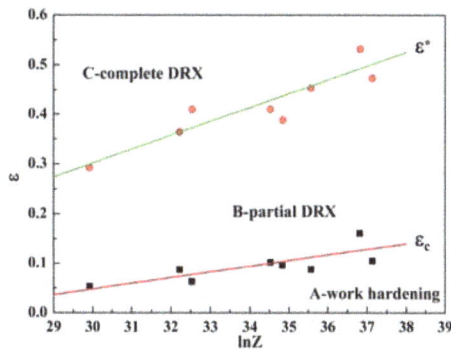

Figure 14. State diagram of DRX for 35CrMo steel under hot temperature deformation.

4. Conclusions

(1) In the work hardening process, the dislocation density accumulated to a certain critical value, microstructural heterogeneity or boundary kinetics may play a role to render a microstructure that behaves stable or unstable at the critical point. Consequently, there will be a point of inflection which can be expressed on the θ-σ curve at the transition from large strain hardening stage to DRX softening stage.

(2) DRV and DRX mechanisms are the softening mechanisms of 35CrMo steel during hot deformation, whose flow curves exhibit DRV and DRX types and followed by a steady-state flow. When the DRX occurs in the 35CrMo steel, the θ-σ curve has an inflection point, and the maximum value appears on the dθ/dσ-σ curve. The critical strain of the DRX of 35CrMo steel can be determined by the inflection point criterion.

(3) The experimental results show that the activation energy of 35CrMo steel in this paper is $Q = 408.498$ kJ/mol, and the expression of Z parameter is $Z = \dot{\varepsilon} \exp\left(\frac{408.498 \times 10^3}{RT}\right)$.

(4) The critical stress σ_c and the critical strain ε_c of 35CrMo steel increase with the decrease of the deformation temperature and the increase of the strain rate. The critical strain and the peak strain satisfy the relationship: $\varepsilon_c = 0.31912\varepsilon_p$ and the critical strain model is $\varepsilon_c = 2.32 \times 10^{-4} Z^{0.1673}$. With the increase of Z parameters—i.e., with the decrease of deformation temperature and the increase of strain rate—the DRX of 35CrMo steel becomes more difficult, which can be reflected by the state diagram of DRX for 35CrMo steel.

Acknowledgments: The authors are grateful for the financial support from the National Program on Key Basic Research Project of China (No.2014CB046702).

Author Contributions: Yuanchun Huang and Sanxing Wang conceived and designed the experiments; Zhengbing Xiao and Hui Liu performed the experiments; Yuanchun Huang contributed reagents/materials/analysis tools; and Sanxing Wang wrote the paper.

Conflicts of Interest: The authors declare no conflict of interest.

References

1. Galiyev, A.; Kaibyshev, R.; Gottstein, G. Correlation of plastic deformation and dynamic recrystallization in magnesium alloy ZK60. *Acta Mater.* **2001**, *49*, 1199–1207. [CrossRef]
2. Mejía, I.; Bedolla-Jacuinde, A.; Maldonado, C.; Cabrera, J.M. Determination of the critical conditions for the initiation of dynamic recrystallization in boron microalloyed steels. *Mater. Sci. Eng. A* **2011**, *528*, 4133–4140. [CrossRef]
3. Takaki, T.; Yoshimoto, C.; Yamanaka, A.; Tomita, Y. Multiscale modeling of hot-working with dynamic recrystallization by coupling microstructure evolution and macroscopic mechanical behavior. *Int. J. Plast.* **2014**, *52*, 105–116. [CrossRef]
4. Poliak, E.I.; Jonas, J.J. A one-parameter approach to determining the critical conditions for the initiation of dynamic recrystallization. *Acta Mater.* **1996**, *44*, 127–136. [CrossRef]
5. Poliak, E.I.; Jonas, J.J. Initiation of Dynamic Recrystallization in Constant Strain Rate Hot Deformation. *ISIJ Int.* **2003**, *43*, 684–691. [CrossRef]
6. Poliak, E.I.; Jonas, J.J. Critical Strain for Dynamic Recrystallization in Variable Strain Rate Hot Deformation. *ISIJ Int.* **2003**, *43*, 692–700. [CrossRef]
7. Najafizadeh, A.; Jonas, J.J. Predicting the Critical Stress for Initiation of Dynamic Recrystallization. *ISIJ Int.* **2006**, *46*, 1679–1684. [CrossRef]
8. Mirzadeh, H.; Najafizadeh, A. Prediction of the critical conditions for initiation of dynamic recrystallization. *Mater. Des.* **2010**, *31*, 1174–1179. [CrossRef]
9. Xu, Y.; Tang, D.; Song, Y.; Pan, X. Dynamic recrystallization kinetics model of X70 pipeline steel. *Mater. Des.* **2012**, *39*, 168–174. [CrossRef]
10. Li, L.; Ye, B.; Liu, S.; Hu, S.; Li, B. Inverse analysis of the stress–strain curve to determine the materials models of work hardening and dynamic recovery. *Mater. Sci. Eng. A* **2015**, *636*, 243–248. [CrossRef]
11. Zhang, P.; Yi, C.; Chen, G.; Qin, H.; Wang, C. Constitutive model based on dynamic recrystallization behavior during thermal deformation of a nickel-based superalloy. *Metals* **2016**, *6*. [CrossRef]
12. Zhang, B.; Zhang, H.B.; Ruan, X.Y.; Zhang, Y. Hot deformation behavior and dynamic recrystallization model of 35CrMo steel. *Acta Metall. Sin.* **2003**, *16*, 183–191.
13. Far, S.V.S.; Ketabchi, M.; Nourani, M.R. Hot Deformation Characteristics of 34CrMo4 Steel. *J. Iron Steel Res.* **2010**, *17*, 65–69.
14. Xu, W.; Zou, M.; Zhang, L. Constitutive analysis to predict the hot deformation behavior of 34CrMo4 steel with an optimum solution method for stress multiplier. *Int. J Press. Vessel. Pip.* **2014**, *123*, 70–76. [CrossRef]
15. Xiao, Z.; Huang, Y.; Liu, H.; Wang, S. Hot Tensile and Fracture Behavior of 35CrMo Steel at Elevated Temperature and Strain Rate. *Metals* **2016**, *6*, 210. [CrossRef]
16. Estrin, Y. Dislocation theory based constitutive modelling: Foundations and applications. *J. Mater. Process. Technol.* **1998**, *80–81*, 33–39. [CrossRef]

17. Gottstein, G.; Frommert, M.; Goerdeler, M.; Schäfer, N. Prediction of the critical conditions for dynamic recrystallization in the austenitic steel 800H. *Mater. Sci. Eng. A* **2004**, *387*, 604–608. [CrossRef]

18. Pantleon, W. Stage IV work-hardening related to disorientations in dislocation structures. *Mater. Sci. Eng. A* **2004**, *387–389*, 257–261. [CrossRef]

19. Alexandrov, I.V.; Valiev, R.Z. Nanostructures from severe plastic deformation and mechanisms of large-strain work hardening. *Nanostruct. Mater.* **1999**, *12*, 709–712. [CrossRef]

20. Sevillano, J.G.; van Houtte, P.; Aernoudt, E. Large strain work hardening and textures. *Prog. Mater. Sci.* **1980**, *25*, 69–134. [CrossRef]

21. Gottstein, G.; Brünger, E.; Frommert, M.; Goerdeler, M.; Zeng, M. Prediction of the critical conditions for dynamic recrystallization in metals. *Z. Metallkunde* **2003**, *94*, 628–635. [CrossRef]

22. Edington, J.W.; Smallman, R.E. The relationship between flow stress and dislocation density in deformed vanadium. *Acta Metall.* **1964**, *12*, 1313–1328. [CrossRef]

23. Zhang, L.; Sekido, N.; Ohmura, T. Real time correlation between flow stress and dislocation density in steel during deformation. *Mater. Sci. Eng. A* **2014**, *611*, 188–193. [CrossRef]

24. Solhjoo, S. Determination of critical strain for initiation of dynamic recrystallization. *Mater. Des.* **2010**, *31*, 1360–1364. [CrossRef]

25. Sun, W.P.; Hawbolt, E.B. Comparison between Static and Metadynamic Recrystallization. An Application to the Hot Rolling of Steels. *ISIJ Int.* **1997**, *37*, 1000–1009. [CrossRef]

26. Sellars, C.M. The kinetics of softening processes during hot working of austenite. *Czechoslov. J. Phys.* **1985**, *35*, 239–248. [CrossRef]

27. Meysami, M.; Mousavi, S.A.A.A. Study on the behavior of medium carbon vanadium microalloyed steel by hot compression test. *Mater. Sci. Eng. A* **2011**, *528*, 3049–3055. [CrossRef]

28. Sellars, C.M.; Mctegart, W.J. On the mechanism of hot deformation. *Acta Metall.* **1966**, *14*, 1136–1138. [CrossRef]

29. Zener, C.; Hollomon, J.H. Effect of Strain Rate Upon Plastic Flow of Steel. *J. Appl. Phys.* **1944**, *15*, 22–32. [CrossRef]

![metals logo] *metals*

MDPI

Article

Effect of Current on Structure and Macrosegregation in Dual Alloy Ingot Processed by Electroslag Remelting

Yu Liu [1,2], Zhao Zhang [1,2], Guangqiang Li [1,2,3,*], Qiang Wang [1,2], Li Wang [1,2] and Baokuan Li [4]

[1]   The State Key Laboratory of Refractories and Metallurgy, Wuhan University of Science and Technology, Wuhan 430081, China; liuyuwust@yeah.net (Y.L.); zhangzhaowust@163.com (Z.Z.); wangqiangwust@wust.edu.cn (Q.W.); wustwangli@163.com (L.W.)
[2]   Key Laboratory for Ferrous Metallurgy and Resources Utilization of Ministry of Education, Wuhan University of Science and Technology, Wuhan 430081, China
[3]   Collaborative Innovation Center of Steel Technology, University of Science and Technology Beijing, Beijing 100083, China
[4]   School of Metallurgy, Northeastern University, Shenyang 110004, China; libk@smm.neu.edu.cn
*   Correspondence: liguangqiang@wust.edu.cn; Tel./Fax: +86-27-6886-2665

Academic Editor: Robert Tuttle
Received: 6 April 2017; Accepted: 13 May 2017; Published: 24 May 2017

Abstract: Macrosegregation is a very common problem for the quality control of all cast ingots. The effect of current on the structure and macrosegregation in dual alloy ingot processed by electroslag remelting (ESR) was investigated experimentally with various analytical methods. In this study, the electrode consisted of NiCrMoV alloy bar (upper part) and CrMoV alloy (lower part) with a diameter of 55 mm, was remelted in a laboratory-scale ESR furnace with the slag containing 30 mass pct alumina and 70 mass pct calcium fluoride under an open air atmosphere. The results show that the macrostructures of three ingots processed by electroslagremelting with different currents are nearly similar. The thin equiaxed grains region and the columnar grains region are formed under the ingot surface, the latter region is the dominant part of the ingot. The typical columnar structure shows no discontinuity among the NiCrMoV alloy zone, the CrMoV alloy zone, and the transition zone in three ingots. With the increase of the current, the grain growth angle increases due to the deeper molten metal pool. The secondary dendrite arm spacing (SDAS) firstly decreases, then increases. The SDAS is dominated by the combined effect of the local solidification rate and the width of mushy region. With the current increasing from 1500 A to 1800 A and 2100 A, the width of the transition zone decreases from 147 mm to 115 mm and 102 mm. The macrosegregation becomes more severe due to the fiercer flows forced by the Lorentz force and the thermal buoyancy force. The cooling rate firstly increases, then decreases, due to the effect of the flows between the mushy region and metal pool and the temperature gradient at the mushy zone of the solidification front. With a current of 1800 A, the SDAS is the smallest and cooling rate is the fastest, indicating that less dendrite segregation and finer precipitates exist in the ingot. Under the comprehensive consideration, the dual alloy ingot processed by the ESR with a current of 1800 A is the best because it has the smallest SDAS, the appropriate grain growth angle, moderate macrosegregation and thickness of the transition zone.

Keywords: current; structure; macrosegregation; electroslag remelting; dual alloy; transition zone

1. Introduction

With the wide application of the single-cylinder steam turbine using combined cycle, the power generation efficiency has been greatly improved. The rotor made of traditional bolted high/intermediate pressure-low pressure shaft has been unable to meet the increasing demand for power generation efficiency.

Compared with the conventional bolted shaft parts, the dual alloy single shaft significantly improves power generation efficiency. Such a dual alloy shaft can be manufactured by welding, however, it requires a long production cycle [1]. Numerous creep tests indicate that elevated temperature and high pressure have a fatal impact on the life of the material due to the degradation of material structure during service under harsh conditions [2–4]. In order to improve the dual alloy single shaft quality and yield, ESR can be applied to produce dual alloy shaft. During the ESR manufacture process, two pre-melted rods containing different alloy compositions are joined into a single electrode by welding, and then the single electrode is remelted by ESR technology [1].

ESR is a combined process for steel melting, refining and casting. The molten slag is added into the water-cooled copper mold, and then the electrode is inserted into the molten slag. The alternating current passing through the loop creates Joule heating in the highly resistive molten slag. The electrode is heated to melt in a slag pool, and the molten metal moves down along the melting front to form a metal droplet at the electrode tip. The metal droplets sink through the molten slag to form a molten metal pool in the mold [5]. Inclusions and harmful elements are largely removed during this process. The interaction between the alternating current and the self-induced magnetic field produces a Lorentz force [6]. With the heat transferring to the mold, the molten metal is solidified to form a structure compact ingot. The cooling conditions of ESR also give a directional solidification in the ingot [7], which improves the ability of the rotor to resist high temperature creep and fatigue due to the elimination of the transverse grain boundaries. The final properties and quality of the ESR ingot heavily rely on the structure forming during the solidification process.

During the production of the ESR dual alloy ingot, there is a transition zone (TZ) with a large composition variation in the ingot [8]. In order to obtain a high performance ESR ingot, it is indispensable to maintain a successive structure and to achieve a narrow chemical transition zone (TZ) because a discrete structure and twisted chemical transition zone might increase the risk of running the rotor at elevated temperature due to thermal expansion mismatch [9]. In our previous study [8], two structures in the solidified ingot are observed—one is a quite narrow, fine, equiaxed grains region at the edge of the ingot, and the other is a columnar grains region inside the ingot, playing a dominant role. The secondary dendrite arm spacing (SDAS) and the grain growth direction are the most important micro- and macrostructure factors for ingot performance [10]. The ESR ingot with a small grain growth angle (the angle between grain growth direction and axis of ingot) can exhibit improved hot forging performance [11]. The secondary dendrite arm spacing has a significant influence on dendrite segregation. The larger the secondary dendrite arm spacing is, the severer dendrite segregation is [12]. Enormous mathematical simulation and experiments [10–14] have proved that the solidification conditions have a vital effect on the structure. The solidification conditions are dominated by fluid flow and heat transfer, which has been extensively studied by researchers [15–18]. Furthermore, the macrosegregation also occurs in ingot and is mainly attributed to the uneven distribution of the solute between the solid and liquid phases during the ESR process [5]. Some researchers used mathematical models to study the element redistribution in ingots [19,20]. The solute transport is mainly attributed to the fluid flows, which is influenced by the joint effect of the thermal buoyancy, the solutal buoyancy, and the Lorentz force during the ESR process [1,21,22]. The current is a significant parameter in the ESR process, which can greatly affect the electromagnetic fields, temperature field and the metal pool shape. Medina studied the influence of voltage and melting current on crystal orientation in ingots, as well as melting rate and ingot surface quality [23]. However, the effect of the current on the structural evolution and macrosegregation in dual ingot processed by ESR has not been studied by the experimental method. Therefore, it is essential to systematically study the effect of current on the structure and macrosegregation in dual ingot.

Because of these factors, the authors were motivated to experimentally explore the underlying effect of current on the structure and macrosegregation in dual ingot. The subtle changes of the structure and macrosegregation in different zones (CrMoV zone, the transition zone (TZ), and NiCrMoV zone) of ESR ingot were determined. This work is designed to provide fundamental knowledge of the evolution of the structure and of macrosegregation in the dual alloy ingot that is processed by ESR.

2. Experimental Section

2.1. Experimental Apparatus and Method

A laboratory-scale ESR furnace (Herz, Shanghai, China) was employed to remelt consumable electrode under an open air atmosphere. The inner diameter and the height of water-cooled copper mold were 120 and 600 mm, respectively. The electrode consisted of two pieces of pre-melted bars (Diameter: 55 mm), and was joined by welding. The upper part was a NiCrMoV alloy bar (Elec. NiCrMoV), and the lower part was a CrMoV alloy bar (Elec. CrMoV). The chemical composition of the electrode determined by the ICP-AES (Tailun, Shanghai, China), and the carbon and sulfur analyzer (Jinbo, Wuxi, China) is listed in Table 1. The slag was comprised of 30 mass pct alumina and 70 mass pct calcium fluoride. The thickness and weight of the slag layer were about 60 mm and 2.3 kg, respectively. Three laboratory experiments were performed, the alternating current of which were 1500 A, 1800 A and 2100 A, respectively, which was generated by a transformer. The voltage was about 30–45 V and the frequency of the alternating current was constant at 50 Hz. The electrode immersion depth was about 10 mm and was controlled by an electrode lifting device powered by an electromotor. The remelting time was recorded by a stopwatch to calculate the time-average melt rate (kg/h). W3Re/W25Re thermocouple was used to measure the slag temperature in the experiment.

Table 1. Chemical composition of consumable electrode used in present experiment (wt. %).

Electrode	C	Mn	Si	P	S	Cr	Ni	Mo	Nb	Al	Ti	T.[O]
Elec. NiCrMoV	0.106	1.67	0.37	0.018	0.039	16.28	7.45	0.117	0.039	0.009	0.012	0.0156
Elec. CrMoV	0.074	3.94	0.40	0.020	0.011	12.25	5.85	0.140	0.032	0.014	0.008	0.0121

2.2. Specimen Preparation and Analyzing Methods

After the ESR process, three dual alloy ingots were obtained and the weight of each ingot was 35.5 kg. Each ingot was evenly divided into two halves along the length direction by wire-electrode cutting. Steel filings were obtained along the longitudinal centerline of the section every 20 mm by drilling, and along the transverse radius every 15 mm for Cr and C analysis, as shown in Figure 1. The Cr mass fraction of each ingot was measured by ICP-AES, and the C mass fraction was analyzed by the carbon and sulfur analyzer. Three slices were taken from the upper (NiCrMoV), middle (TZ), and the lower parts (CrMoV) of each ingot, and three 6 mm × 6 mm × 6 mm specimens were then sampled from three slices, respectively (Figure 1). The NiCrMoV zone, the TZ (transition zone), and the CrMoV zone were distinguished according to the composition profile along the ingot axial. Another part of the ingot was prepared for macrostructure observation, which was ground, polished, and finally etched via the aqua regia for a certain time. Figure 2 demonstrates the macrostructure of three dual alloy ingots. The nine specimens sampled from three ingots for metallographic observation were etched at 75 °C in a picric acid solution, then assessed by optical microscopy (OM, Carl Zeiss, Jena, Germany)

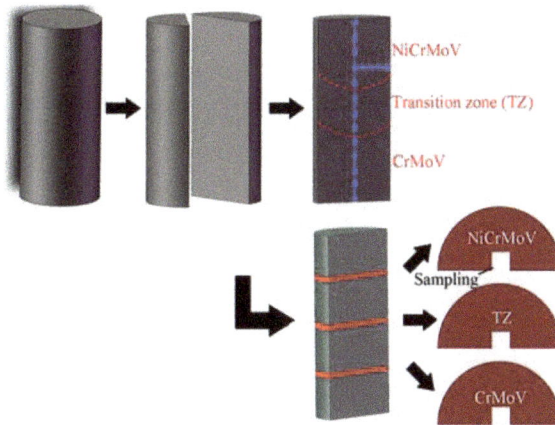

Figure 1. Schematic drawing of the dissection of the ESR dual alloy ingot.

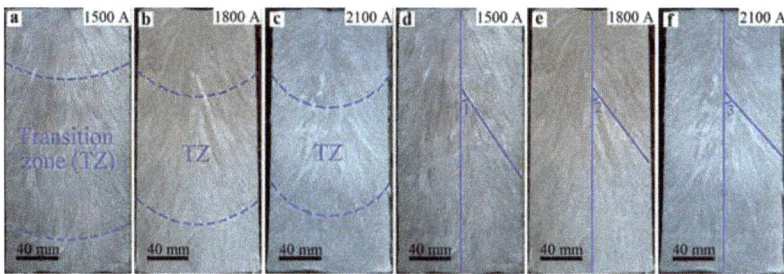

Figure 2. The macrostructures of the ingots processed by electroslag remelting with different currents of (**a,d**) 1500 A, (**b,e**) 1800 A and (**c,f**) 2100 A. The zone between two dotted lines in (**a–c**) is the transition zone.

3. Results and Discussion

3.1. Macro- and Microstructure Characterization of Three Ingots Made via ESR with Different Currents

Figure 2 shows the macrostructures of three ingots processes by electroslag remelting with different currents. The zone between two dotted lines in Figure 2a–c is the transition zone (TZ). The typical columnar structure indicates that no discontinuity among the NiCrMoV zone, the CrMoV zone, and the transition zone in three ingots occurs (Figure 2a–c). The macrostructures of three ingots are nearly similar—a thin equiaxed grains region is located under the ingot surface, and the columnar grains region exists inside the ingot, composing the dominant part of ingot. At the beginning of the ESR process, the heat loss to the base plate is dominant. Columnar crystals nucleate at the bottom of the ingot and grow up vertically (Figure 2d–f). With the solidification front advancing, the solidified part acts as a heat choke, and the heat loss to the mold wall increases and cooling intensity decreases gradually due to less heat transferring to the base plate. As a result, the inclined columnar crystals nucleate in the vicinity of the lateral wall and grow to hinder the vertical crystals (Figure 2d–f). The inclined columnar crystals form an inverted chevron structure with a certain angle. The grain growth angle of inclined columnar crystals in three ingots with different currents of 1500 A, 1800 A and 2100 A is 35.1°, 38.6° and 43.7°, respectively (from Angle 1 to Angle 2 and Angle 3 in Figure 2d–f). It is well known that the grain growth direction is perpendicular to the solidification front of the molten metal pool. Figure 3 shows the average melting rate (kg/h) and average slag temperature with different currents, indicating that the melting rate and slag temperature increase with the increase of the current. In general, with the current increasing, the melting rate increases and the molten metal pool becomes deeper due to more Joule heat produced by the larger current passing through the liquid slag [1,5]. A deeper molten metal pool results in a larger grain growth angle in the test with a larger current.

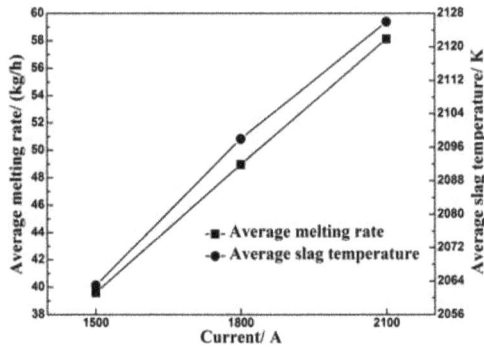

Figure 3. Average melting rate and average slag temperature with different currents.

The variation in optical microstructures of three ingots processed by electroslag remelting with different currents is shown in Figure 4. The dendritic structure is formed throughout the cross section, which gradually becomes coarsened from the CrMoV zone to TZ and NiCrMoV zone along the direction from bottom to the top of each ingot because the cooling intensity decreases gradually from the bottom to top of the ingot. When the current increases from 1500 A to 1800 A and 2100 A, the dendritic structure firstly becomes fine, then coarsened. In order to quantitatively analyze the difference of dendritic structure, the average secondary dendrite arm-spacing (SDAS) at NiCrMoV zone and TZ of ingots was measured, as shown in Figure 5. The average SDAS is the average of four SDAS's at three different points. It should be noted that the SDAS of CrMoV zone was not measured because the crystal structure in CrMoV zone is mainly equiaxed crystal. It can be seen from Figure 5 that SDAS of NiCrMoV zone varies from 61.5 μm to 49.6 μm and 65.3 μm with the current increasing from 1500 A to 1800 A and 2100 A, which varies from 49.2 μm to 40.3 μm and 49.7 μm in TZ.

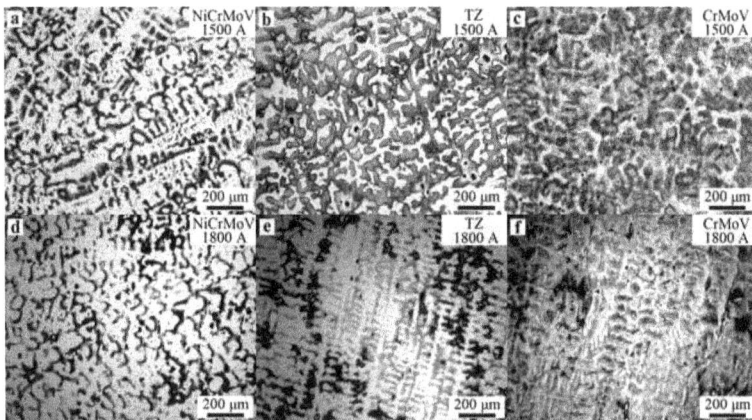

Figure 4. *Cont.*

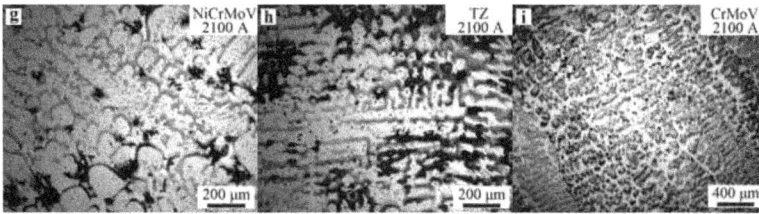

Figure 4. Optical micrographs of the (**a,d,g**) NiCrMoV zone, (**b,e,h**) TZ and (**c,f,i**) CrMoV zone in the ingots processed by electroslag remelting with different currents of (**a–c**) 1500 A, (**d–f**) 1800 A and (**g–i**) 2100 A.

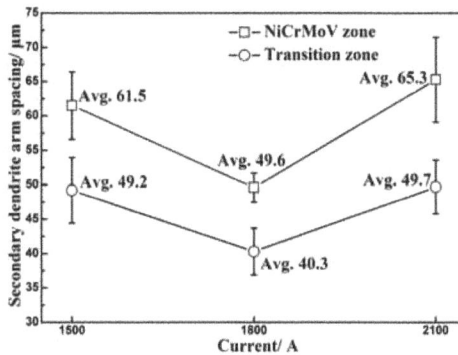

Figure 5. Secondary dendritic arm spacing at different positions (NiCrMoV zone and transition zone) of the ingots processed by electroslag remelting with different currents.

The SDAS are dominated by the local solidification time (*LST*/min). *LST* represents the time that an alloy stays in the solid-liquid two-phase zone (mushy zone). With the increase of *LST*, the SDAS increases. The *LST* can be calculated using following equation [5,24]:

$$LST = Xr/v \tag{1}$$

It can be inferred from Equation (1) that *LST* is dominated by the combined effect of *Xr* (the width of mushy zone/mm) and *v* (the local solidification rate/mm/min). As shown in Figure 6, the local solidification rate (*v*) is perpendicular to the tangent of the solidus curve and has an Angle θ (grain growth angle) to the axis of the ingot. The remelting rate (v_M/mm/min) is parallel to the axis, which represents rising velocity of solidus. The local solidification rate (*v*) can be calculated using the remelting rate (v_M) as this geometric relation [24]:

$$v = v_M \times \cos \theta \tag{2}$$

The average remelting rate (v_M) can be obtained by the remelting time recorded by the stopwatch. The local solidification rate calculated by Equation (2) and average remelting rate are shown in Figure 7. It indicates that the remelting rate increases evenly with the increase of current. However, correspondingly, the magnitude of increase in local solidification rate (*v*) decreases due to the increase of the grain growth angle (Angle θ). With the increase of the current, the temperature gradient at the solidification front decreases and the width of the mushy zone increases (*Xr*) [5]. *v* increases more rapidly than *Xr* while the current increases from 1500 A to 1800 A, resulting in the lower *LST* and smaller SDAS. When the current increases from 1800 A to 2100 A, the magnitude of the increase in *v* is smaller than that in *Xr*. As a result, the *LST* and SDAS increase. It has been confirmed that with the SDAS decreasing, the less dendrite segregation occurs and the precipitates becomes finer [5,8,12]. The SDAS is the smallest in the ingot with 1800 A (Figure 5), implying that less dendrite segregation and finer precipitates exist in the ingot with a current of 1800 A.

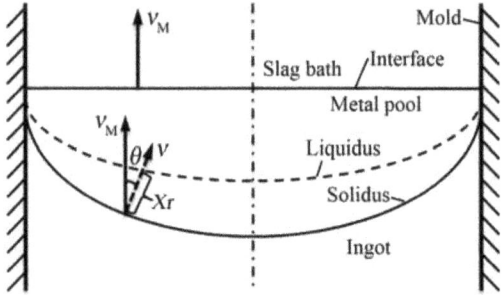

Figure 6. Schematic of solidification interface and the relationship between remelting rate and cooling rate.

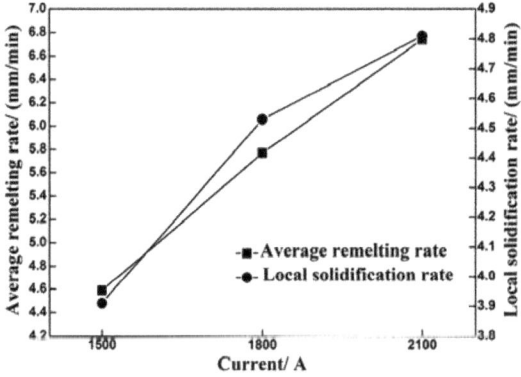

Figure 7. Average remelting rate and local solidification rate with different currents.

The cooling rate at the solidification front ($C_R/°C/s$) can be calculated by Equation (3) as follows [25]:

$$\lambda_S(\mu m) = (169.1 - 720.9 \bullet [\%C]) \bullet C_R^{-0.4935} \ (0 < [\%C] \leq 0.15) \tag{3}$$

where λ_S is the SDAS (μm) and [%C] is the carbon mass fraction, which is presented in Table 2. According to Equation (3), the cooling rate of NiCrMoV zone and TZ was calculated, as shown in Figure 8. The cooling rate first increases, then decreases with the current increasing from 1500 A to 1800 A and 2100 A (Figure 8). The cooling rate is closely related to the flows of the molten metal pool, which will be discussed in detail in Section 3.2.

Table 2. Carbon mass fraction of NiCrMoV zone and transition zone (TZ) in the ingots processed by the ESR with different currents (wt. %).

ESR Ingots	[%C]		
	1500 A	1800 A	2100 A
NiCrMoV	0.108	0.103	0.107
TZ	0.091	0.093	0.094

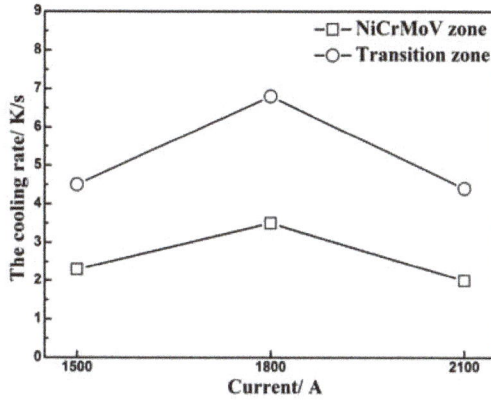

Figure 8. The cooling rates at different positions (NiCrMoV zone and transition zone) of the ingots processed by electroslag remelting with different currents.

3.2. Macrosegregation of Three Ingots Made via ESR with Different Currents

Figure 9 shows the Cr concentration distributions along the longitudinal centerline and the transverse radius in ingots processed by electroslag remelting with different currents. The Cr content increases markedly from the nominal concentration of the CrMoV alloy to that of the NiCrMoV alloy in the TZ of the ingot (Figure 9a). It can be seen from Figure 9a that Cr content along the longitudinal centerline with a current of 1500 A is the first one passing the nominal concentration of the CrMoV alloy, while is the last one exceeding the average concentration of the NiCrMoV alloy. The width of this zone with a composition fluctuation is namely the thickness of the TZ. The thickness of the TZ is about 147 mm, 115 mm and 102 mm with the current increasing from 1500 A to 1800 A and 2100 A, respectively. With the current increasing, the remelting rate (Figure 7) increases [1,25], implying a faster rising velocity of solidus. The TZ of the electrode can be melted, and then solidified into the ingot in shorter time and distance. As a result, the thickness of the TZ decreases with the increase of the current.

Along the transverse radius, the Cr content is higher at the center of all ingots than the edge (Figure 9b). In addition, the concentration gradient between the center and the edge increases with the increase of the current, implying that the severer macrosegregation occurs. It is well known that the macrosegregation is dominated by the solute transport. The flows in the electroslag remelting process have a significant influence on the solute transport [1,19].

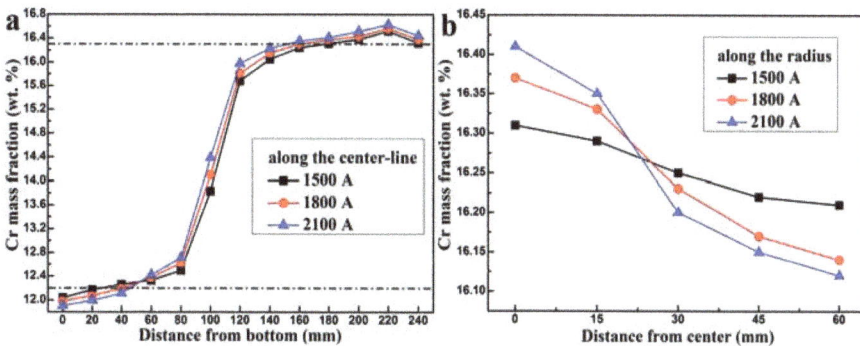

Figure 9. Cr concentration distributions along (**a**) the longitudinal centerline and (**b**) the transverse radius in ingots processed by electroslag remelting with different currents.

Figure 10 shows the illustration of the flows of the slag bath and metal pool in the ESR process. The metal at the tip of the electrode is remelted by the Joule heating created by the interaction between current and slag to form the metal droplets, which sink through the molten slag to form a crescent shaped molten metal pool that is deep in the center of the ingot and gradually becomes shallow outward in the direction of the radius. It should be noted that the effect of the droplets on the flows in the metal pool was not considered because the momentum carried by small droplets during the small scale ESR process is small. Thermal buoyancy force and Lorentz force are the main driving forces for the flows in the metal pool [26,27]. The molten metal close to the mold wall is cooled by the mold. The hot metal floats up and the cool metal sinks down, and a clockwise circulation is formed near the mold wall. At the solidification front, there is also a large temperature gradient. The cool metal with a higher density will move down along the oblique solidification front and wash out the solidifying mushy zone. The cooling intensity weakens around the base of the molten metal pool. The hot metal rises toward the slag-metal interface and then returns to the mold wall, which also forms a clockwise circular flow at the inclined solidification front. In addition, according to the Faraday's law of electromagnetic induction, a clockwise circular magnetic field (looking down from the top) would be induced by the downward current. The interaction between the clockwise circular magnetic field and the downward current creates an inward Lorentz force, which also pushes the metal from the edge to the middle.

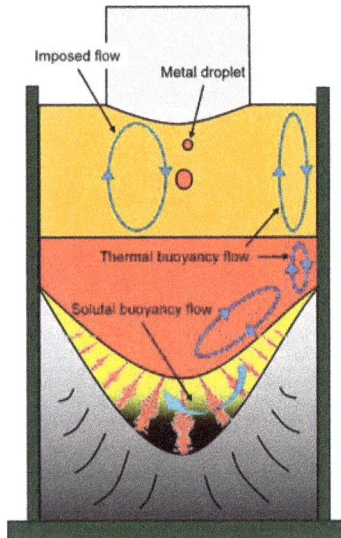

Figure 10. Illustration of the flows of the slag bath and metal pool in the ESR process.

The solute element Cr becomes enriched in the mushy zone due to the partition ratio ($k_{Cr} = 0.76$) [28]. Furthermore, the density of Cr ($\rho_{Cr} = 6900$ kg/m^3) is lower than that of iron ($\rho_{Fe} = 7500$ kg/m^3) [29]. The Cr would be enriched in the molten metal pool due to the buoyancy force, resulting in the so-called gravity segregation. At the solidification front, the solute-poor metal displaces the solute-rich metal through washing out the mushy region due to the clockwise circular flow. Inward Lorentz force also pushes the metal from the edge to the bottom center of metal pool. As a result, the Cr accumulates at the bottom center of the pool and the concentration decreases from the middle to the edge.

With the increase of the current, the slag bath temperature becomes higher (Figure 3) and the flows become faster. The reinforced heat transfer increases the melting rate of the electrode (Figure 3).

The metal droplets formed at the tip of the electrode become bigger, which brings more heat to the molten metal pool. The flows in the metal pool also become more intense. Furthermore, the inward Lorentz force also increases with the increase of the current. So, the macrosegregation becomes severer with the increase of the current (Figure 9b).

As mentioned above, the cooling rate at the solidification front first increases, then decreases with the current increasing from 1500 A to 1800 A and 2100 A (Figure 8). The cooling rate is dominated by the flows between the mushy region and metal pool and the temperature gradient at mushy zone of the solidification front (Figure 10). With the current increasing, the temperature of the metal pool increases and the Lorentz force enhances the flows more fiercely, whereas, the temperature gradient at the solidification front decreases (the width of mushy zone Xr increases) [5]. The enhanced flows accelerate the heat transfer, but the reduced temperature gradient weakens the heat transfer at the solidification front. With the current increasing from 1500 A to 1800 A, the enhanced flows play a dominant role in heat transfer and the cooling rate increases at the solidification front. When the current increases form 1800 A to 2100 A, the heat transfer fades due to the decreasing temperature gradient at the mushy zone of the solidification front and the cooling rate decreases.

4. Conclusions

Three heats with different currents were designed to investigate the effect of the current on structure and macrosegregation in dual alloy ingot processed by electroslag remelting, the following conclusions can be reached.

(1) The macrostructures of three ingots are nearly similar. The thin equiaxed grains region is situated under the ingot surface, and the columnar grains region lies inside the ingot. The typical columnar structure shows no discontinuity among the CrMoV zone, the transition zone and NiCrMoV zone in three ingots. With the increase of the current, the grain growth angle (the angle between grain growth direction and axis of ingot) increases due to the deeper molten metal pool.

(2) The SDAS firstly decreases, then increases with the increase of the current. The SDAS is dominated by the combined effect of the local solidification rate and thickness of the mushy region. With a current of 1800 A, the SDAS is the smallest and the cooling rate is the fastest, indicating that less dendrite segregation and finer precipitates exist in the ingot.

(3) With the increase of the current, the thickness of the transition zone decreases. The macrosegregation becomes severer due to the fiercer flows forced by the Lorentz force and the thermal buoyancy force. The cooling rate first increases, then decreases, which is dominated by a combined effect of the flows between the mushy region and metal pool and the temperature gradient at the mushy zone of the solidification front.

(4) Under the comprehensive consideration, the dual alloy ingot processed by the ESR with a current of 1800 A is the best due to the smallest SDAS, the appropriate grain growth angle, moderate macrosegregation and thickness of the transition zone. The present work clarifies the effect of the current on the structure and macrosegregation in the dual alloy ingot processed by the ESR, providing a reference for the parameter election of the manufacture of the dual alloy rotor to be used in steam turbines using a combined cycle.

Acknowledgments: The authors gratefully acknowledge the support from the National Natural Science Foundation of China (Grant No. 51210007) and the Key Program of Joint Funds of the National Natural Science Foundation of China and the Government of Liaoning Province (Grant No. U1508214).

Author Contributions: Guangqiang Li and Baokuan Li conceived and designed the experiments; Qiang Wang, Yu Liu, Zhao Zhang and Li Wang performed the experiments; Qiang Wang and Yu Liu analyzed the data; Baokuan Li contributed materials and tools; Yu Liu wrote the paper.

Abbreviations

The following abbreviations are used in this manuscript:

ESR	Electroslag remelting
TZ	Transition Zone
ICP-AES	Inductively Coupled Plasma-Atomic Emission Spectroscopy
OM	Optical Microscopy
SDAS	Secondary Dendrite Arm-Spacing
LST	Local Solidification Time

References

1. Wang, Q.; Yan, H.; Ren, N.; Li, B. Effect of current on solute transport in electroslag remelting dual alloy ingot. *Appl. Therm. Eng.* **2016**, *101*, 546–567. [CrossRef]
2. Zieliński, A.; Golański, G.; Sroka, M. Comparing the methods in determining residual life on the basis of creep tests of low-alloy Cr-Mo-V cast steels operated beyond the design service life. *Int. J. Press. Vessels Pip.* **2017**, *152*, 1–6. [CrossRef]
3. Sroka, M.; Zieliński, A.; Dziuba-Kałuża, M. Assessment of the Residual Life of Steam Pipeline Material beyond the Computational Working Time. *Metals* **2017**, *7*, 82. [CrossRef]
4. Zieliński, A.; Golański, G.; Sroka, M. Estimation of long-term creep strength in austenitic power plant steels. *Mater. Sci. Technol.* **2016**, *32*, 780–785. [CrossRef]
5. Li, Z.B. *Electroslag Metallurgy Theory and Practice*; Metallurgical Industry Press: Beijing, China, 2010; pp. 17–18. (In Chinese)
6. Wang, Q.; Li, B. Numerical investigation on the effect of fill ratio on macrosegregation in electroslag remelting ingot. *Appl. Therm. Eng.* **2015**, *91*, 116–125. [CrossRef]
7. Jiang, Z.H. *Physical Chemistry and Transport Phenomena for Electroslag Metallurgy*; Northeastern University Press: Shenyang, China, 2000; pp. 143–146. (In Chinese)
8. Liu, Y.; Zhang, Z.; Li, G.; Wang, Q; Wang, L.; Li, B. The structural evolution and segregation in a dual alloy ingot processed by electroslag remelting. *Metals* **2016**, *6*, 325. [CrossRef]
9. Kajikawa, K.; Ganesh, S.; Kimura, K.; Kudo, H.; Nakamura, T.; Tanaka, Y.; Schwant, R.; Gatazka, F.; Yang, L. Forging for advanced trubine applications: Development of multiple alloy rotor forging for turbine application. *Ironmak. Steelmak.* **2007**, *34*, 216–220. [CrossRef]
10. Rao, L.; Zhao, J.H.; Zhao, Z.X.; Ding, G.; Geng, M.P. Macro-and microstructure evolution of 5CrNiMo steel ingots during electroslag remelting process. *J. Iron Steel Res. Int.* **2014**, *21*, 644–652.
11. Chang, L.Z.; Shi, X.F.; Yang, H.S.; Li, Z.B. Effect of low-frequency AC power supply during electroslag remelting on qualities of alloy steel. *J. Iron Steel Res. Int.* **2009**, *16*, 7–11. [CrossRef]
12. Liu, Z.; Li, J.; Fu, H. Dendritic arm spacing and microsegregation of directionally solidified superalloy DZ22 at various solidification rates. *Acta Metall. Sin.* **1995**, *31*, 329–332. (In Chinese).
13. Li, B.; Wang, Q.; Wang, F.; Chen, M. A coupled cellular automaton-finite-element mathematical model for the multiscale phenomena of electroslag remelting H13 die steel ingot. *JOM* **2014**, *66*, 1153–1165. [CrossRef]
14. Ma, D.; Zhou, J.; Chen, Z.; Zhang, Z.; Chen, Q.; Li, D. Influence of thermal homogenization treatment on structure and impact toughness of H13 ESR steel. *J. Iron Steel Res. Int.* **2009**, *16*, 56–60. [CrossRef]
15. Mitchell, A. Solidification in remelting processes. *Mater. Sci. Eng. A* **2005**, *413*, 10–18. [CrossRef]
16. Hernandez-Morales, B.; Mitchell, A. Review of mathematical models of fluid flow, heat transfer, and mass transfer in electroslag remelting process. *Ironmak. Steelmak.* **1999**, *26*, 423–438. [CrossRef]
17. Mitchell, A.; Joshi, S.; Cameron, J. Electrode temperature gradients in the electroslag process. *Metall. Trans.* **1971**, *2*, 561–567. [CrossRef]
18. Mitchell, A.; Joshi, S. The thermal characteristics of the electroslag process. *Metall. Trans.* **1973**, *4*, 631–642. [CrossRef]
19. Fezi, K.; Yanke, J.; Krane, M.J.M. Macrosegregation during electroslag remelting of alloy 625. *Metall. Mater. Trans. B* **2015**, *46*, 766–779. [CrossRef]
20. Wang, Q.; Wang, F.; Li, B.; Tsukihashi, F. A three-dimensional comprehensive model for prediction of macrosegregation in electroslag remelting ingot. *ISIJ Int.* **2015**, *55*, 1010–1016. [CrossRef]

21. Jardy, A.; Ablitzer, D.; Wadier, J.F. Magnetohydronamic and thermal behavior of electroslag remelting slags. *Metall. Mater. Trans. B* **1991**, *22*, 111–120. [CrossRef]

22. Kharicha, A.; Ludwig, A.; Wu, M. On melting of electrodes during electro-slag remelting. *ISIJ Int.* **2014**, *54*, 1621–1628. [CrossRef]

23. Medina, S.F.; Andres, M.P. Electrical Field in the Resistivity Medium (Slag) of the ESR Process: Influence on Ingot Production and Quality. *Ironmak. Steelmak.* **1987**, *14*, 110–121.

24. Zhong, Y.; Qiang, L.; Fang, Y.; Wang, H.; Peng, M.; Dong, L.; Zheng, T.; Lei, Z.; Ren, W.; Ren, Z. Effect of transverse static magnetic field on microstructure and properties of GCr15 bearing steel in electroslag continuous casting process. *Mater. Sci. Eng. A* **2016**, *660*, 118–126. [CrossRef]

25. Won, Y.M.; Thomas, B.G. Simple model of microsegregation during solidification of steels. *Metall. Mater. Trans. A* **2001**, *32*, 1755–1767. [CrossRef]

26. Dong, J.; Cui, J.; Zeng, X.; Ding, W. Effect of low-frequency electromagnetic field on microstructures and macrosegregation of Φ270 mm DC ingots of an Al-Zn-Mg-Cu-Zr alloy. *Mater. Lett.* **2005**, *59*, 1502–1506. [CrossRef]

27. Zhang, B.; Cui, J.; Lu, G. Effect of low-frequency magnetic field on macrosegregation of continuous casting aluminum alloys. *Mater. Lett.* **2003**, *57*, 1707–1711. [CrossRef]

28. Schneider, M.C.; Beckermann, C. Formation of macrosegregation by multicomponent thermosolutal convection during the solidification of steel. *Metall. Mater. Trans. A* **1995**, *26*, 2373–2388. [CrossRef]

29. Weber, V.; Jardy, A.; Dussoubs, B.; Ablitzer, D.; Ryberon, S.; Schmitt, V.; Hans, S.; Poisson, H. A comprehensive model of the electroslag remelting process: Description and validation. *Metall. Mater. Trans. B* **2009**, *40*, 271–280. [CrossRef]

![metals logo] *metals*
MDPI

Article

Studying Mechanical Properties and Micro Deformation of Ultrafine-Grained Structures in Austenitic Stainless Steel

Na Gong [1], Hui-Bin Wu [1,2,*], Zhi-Chen Yu [1], Gang Niu [2] and Da Zhang [1]

[1] Institute of Engineering Technology, University of Science and Technology Beijing, Beijing 100083, China; gongnana.cheng@gmail.com (N.G.); yuzhichen@163.com (Z.-C.Y.); claycn@outlook.com (D.Z.)
[2] Collaborative Innovation Center of Steel Technology, University of Science and Technology Beijing, Beijing 100083, China; ustbniug@163.com
*   Correspondence: wuhb@ustb.edu.cn; Tel.: +86-010-6233-2617

Academic Editor: Robert Tuttle
Received: 17 April 2017; Accepted: 19 May 2017; Published: 24 May 2017

Abstract: Eighty percent heavy cold thickness reduction and reversion transformation in the temperature range 700–950 °C for 60 s were performed to obtain the reverted ultrafine-grained (UFG) structure in 304 austenitic stainless steel. Through mechanical property experiments and transmission electron microscopy (TEM) of micro deformation of the UFG austenite structure, the tensile fractographs showed that for specimens annealed at 700–950 °C, the most frequent dimple sizes were approximately 0.1–0.3 μm and 1–1.5 μm. With the increase in annealing temperature, the dimple size distribution of nano-sized grains turned to micron-size. TEM micro deformation experiments showed that specimens annealed at 700 °C tended to crack quickly. In the grain annealed at 870 °C, partial dislocations were irregularly separated in the crystal or piled up normal to the grain boundaries; stacking faults were blocked by grain boundaries of small grains; twins held back the glide of the dislocations. In the grain annealed at 950 °C, the deformation twins were perpendicular to ε martensite. Fine grain was considered a strengthening phase in the UFG structure and difficult to break.

Keywords: 304 austenitic stainless steel; UFG; mechanical experiment; micro deformation; fractograph; dislocations

1. Introduction

Austenitic stainless steel (ASS) with its enhanced yield strength, high work hardening property, excellent weldability, and improved corrosion resistance has been put to use in many fields such as engineering applications and in everyday utensils [1–3]. Recently, the combination of high strength and high ductility in ultrafine-grained (UFG) structured ASS were achieved via operating severe deformation and reversion annealing treatment [4,5]. There are numerous recent papers discussing the deformation mechanisms in UFG steels from the microscopic or macroscopic view using conventional tensile testing and nanoindentation with TEM examination [4–6]. It is thought that deformation mechanisms in nanostructured metals can be different from those in coarse grained structures. It is suggested that partial dislocation generated at grain boundaries may be the main activity in UFG materials [7,8]. Allain et al. [9] has reported that the mechanical ε martensitic transformation only occurs if the stacking fault energy (SFE) is lower than 18 mJ/m^2, that mechanical twinning occurs at SFE roughly in the range 12–35 mJ/m^2 [10], and glide of dislocation would occur when SFE exceeds 45 mJ/m^2. For 304 austenitic stainless steel under uniaxial tension, two transformation mechanisms were proposed according to the SFE [11–14]: (a) stress-induced-transformation $\gamma \rightarrow \varepsilon \rightarrow \alpha'$ (<18 mJ/m^2) and (b) strain-induced-transformation, $\gamma \rightarrow$ deformation twinning $\rightarrow \alpha'$ (12–35 mJ/m^2).

For materials with low SFE, the plastic deformation mode may change from dislocation slip to deformation twinning, which is important for material strengthening. It is reported that ASS commonly exhibits ductile failure controlled by dislocation flow or their mutual interactions. Meanwhile, the mechanism of ductile failure is well developed. Three parts of the mechanism are void nucleation, growth, and coalescence to form a crack, ending with fracture [15,16]. The ductile fracture of material consists of void nucleation and growth, which are governed by the motion of dislocations. It was realized that further information on the behavior of dislocations in the material might be obtained from a study of their motion. It has been possible to study the motion of dislocations in 304 ASS [14,17,18]. In the present investigation, a careful analysis of UFG of the fracture feature morphologies and the relation to the mechanical properties has been made in the ASS. TEM micro deformation tensile test of ASS with UFG structure deformation experiments were conducted to discuss the state of deformation mechanisms and fracture mechanisms.

2. Material and Experimental Methods

The 304 ASS used in this study had chemical composition (weight percent) Fe–0.04C–0.16Si–1.52Mn–17.8Cr–8.1Ni–0.005P–0.005S. Several specimens with of dimension 7.9 mm × 80 mm × 600 mm were machined for subsequent solution treatment and thermomechanical processes. The plates were solution-treated at 1050 °C for 12 min. The solution-treated specimens were cold rolled to 80% reduction in thickness and were subjected to reversion transformation at temperatures of 700–950 °C for 60 s to obtain a reverted UFG austenite structure.

Phase characterization was conducted during cold rolling and annealing by electron backscatter diffraction (EBSD, ZEISS ULTRA 55, Carl Zeiss, Germany) and X-ray diffractometry (Rigaku DMAX-RB with Cu-Kα radiation, Rigaku, Tokyo, Japan) (XRD). Before EBSD and XRD, the specimens were prepared by electropolishing at 15 V for 30 s to remove deformation-induced martensite on the surface; the electrolyte contained 20 vol % perchloric acid and 80 vol % ethanol.

Tensile tests were carried out at room temperature using the CMT5605 tensile machine (SANS Testing Machine Co., Ltd., Shenzhen, China) and Vickers micro-hardness values were measured on an HV-1000 micro-Vickers durometer (Shanghai optical instrument factory, Shanghai, China)to obtain the mechanical properties of the UFG structure. After sample fracture, scanning electron microscopy (SEM, FEI Quanta-450; FEI Corporation, Hillsboro, OR, USA) was used to obtain SEM fractographs. Image processing technique was employed on the digital fractographs to describe the two-dimensional dimple features on the fracture surfaces using Image-Pro Plus (Version 6.0, Media Cybernetics, Inc., Rockville, MD, USA, 2006).

Three tensile specimens for micro deformation TEM (transmission electron microscopy) observation were cut from the UFG austenite. Subsequently, the foils for the tensile observation by TEM were thinned until perforation by twin-jet electropolishing apparatus in a solution of 10 vol % of perchloric acid and 90 vol % of ethanol at a voltage of 36 V and a temperature of 0 °C. Tensile micro deformation was carried out in a JEM2100 TEM (JEOL Ltd., Tokyo, Japan). TEM was used to observe the inner feature of the UFG structure until cracks appeared in the thin area.

3. Results and Discussion

3.1. Microstructure Characterizations of the 304 ASS with UFG Structure

The SFE of the ASS is the key factor determining whether twinning, martensite transformation, or dislocation glide will dominate the deformation process. Schramm and Reed [19] proposed Equation (1) for calculating SFE. The SFE (in mJ/m^2) of 304 ASS can be calculated by

$$SFE = -53 + 6.2 \times C_{Ni} + 0.7 \times C_{Cr} + 3.2 \times C_{Mn} + 9.3 \times C_{Mo} \tag{1}$$

where C_{Ni}, C_{Cr}, C_{Mn} and C_{Mo} are the content values (in wt %) of Ni, Cr, Mn, and Mo, respectively. From the chemical composition, SFE of the sample used in this study was evaluated to be ~14.5 mJ/m^2.

X-ray patterns of the experimental alloy in the solution-treated condition after 80% cold rolling and upon reversion annealing at various temperatures are depicted in Figure 1. The solution-treated specimen exhibited entirely austenite peaks. After 80% cold rolling, the microstructure of the 304 ASS changed to mainly martensite due to the heavy cold reduction, as shown by XRD pattern B in Figure 1. Upon reversion annealing, the martensite reverted to austenite. The intensity of the austenite peaks increased with increasing temperature in the range 700–950 °C, which indicates an increase in the volume fraction of austenite with respect to martensite. Upon annealing at 700 °C, the microstructure mostly consisted of austenite (accounting for 65%) along with a small amount of retained martensite (accounting for 35%). On the other hand, increasing the temperature to 820 °C resulted in reversion of almost all the martensite to austenite, whose volume fraction was 99.5% (calculated from Figure 1).

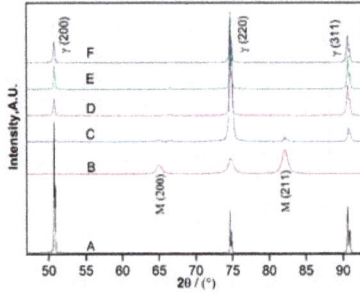

Figure 1. X-ray diffractometry (XRD) patterns for each stage of thermomechanical treatment: (A) after solution annealing; (B) after 80% cold rolling, and reversion annealing at (C) 700 °C; (D) 820 °C; (E) 870 °C; and (F) 950 °C for 60 s.

EBSD showed that for the specimen solution-treated at 1050 °C for 12 min, the grain size was approximately 20–40 μm, as shown in Figure 2a. For the specimens annealed at 820 °C (shown in Figure 2b), the average grain size was about 500 nm, while the specimens annealed at 870 °C (shown in Figure 2c) had average grain size of about 2 μm. With annealing temperature increasing to 950 °C (shown in Figure 2d), the average grain size was about 5 μm. However, the grain size of the sample annealed at 700 °C for 60 s was too small to be discerned by EBSD. Therefore, in this work, TEM observations were performed on the sample annealed at 700 °C for 60 s to measure the grain size, as shown in Figure 3. The statistical result indicates that the grain size was about 150 nm, with 35% martensite mixed with austenite.

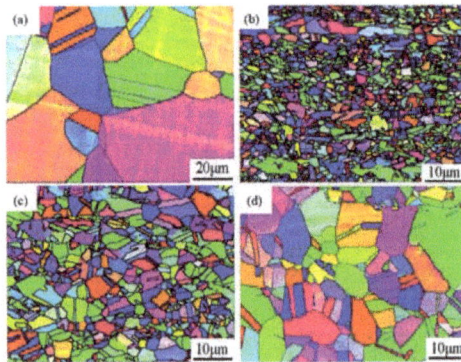

Figure 2. Electron backscatter diffraction (EBSD) micrograph for each stage of treatment: (**a**) after solution treatment; and reversion annealing at (**b**) 820 °C; (**c**) 870 °C; and (**d**) 950 °C for 60 s.

Figure 3. Transmission electron microscopy (TEM) micrographs showing the morphology of grains in 304 austenitic stainless steel (ASS) after annealing at 700 °C for 60 s.

3.2. Mechanical Experiments for the 304 ASS with UFG Structure

Tensile tests were carried out at room temperature. The tensile fractographs of four specimens are shown in Figure 4a–d. Image processing (IP) was used to characterize the two-dimensional geometry of dimples to obtain the dimple diameter and its distribution on the fracture surfaces, as shown in Figure 4e–h. Figure 5 shows the dimple size distribution after annealing at different temperatures. As shown in Figure 5, the most frequent dimple sizes were approximately 0.1–0.3 μm and 1–1.5 μm. With the increase of annealing temperature, the nano-sized dimple size distribution changed to micron-sized.

The curves of strength properties and ductility properties with average dimple diameter at different annealing temperatures are shown in Figures 6 and 7, respectively. With increase in annealing temperature, the strengths decreased, while the elongation and the average size of the dimple increased. In comparison with the specimen annealed at 700 °C, the other three UFG specimens (annealed at 820, 870, and 950 °C) exhibited significantly higher ductility and lower strength. The yield strength of the specimen annealed at 700 °C was 1028 ± 14 MPa, which is approximately 3 times higher than that of the specimen annealed at 950 °C and the ultimate tensile strength was 1157 ± 20 MPa. Meanwhile, the gap between the yield strength and tensile strength increased with increasing annealing temperature. The elongation-to-failure were 8.2 ± 0.3%, 50.3 ± 0.2%, 53.7 ± 0.9%, and 62.5 ± 0.7% for the specimens annealed at 700, 820, 870 and 950 °C, respectively. Specimens with UFG structure annealed at 820, 870 and 950 °C led to higher uniform elongation during the tensile test.

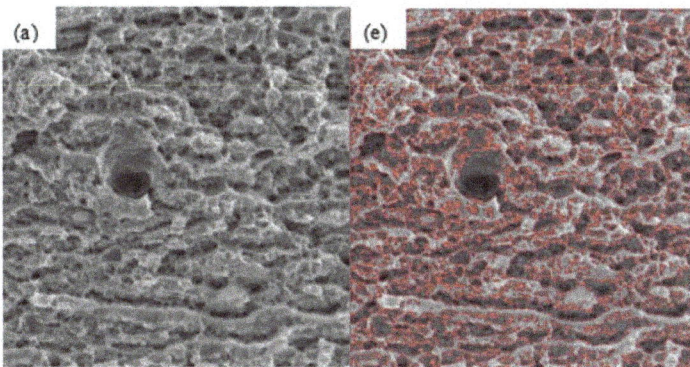

Figure 4. *Cont.*

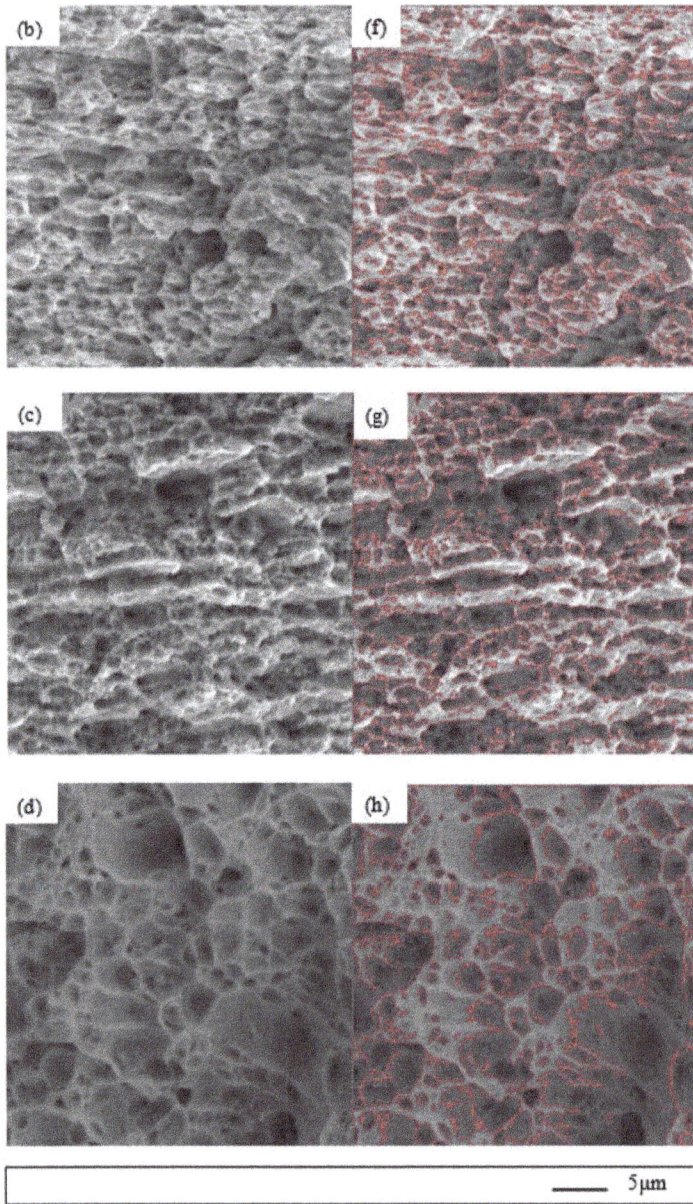

Figure 4. Scanning electron microscopy (SEM) fractographs of 304 ASS annealed at (**a**) 700 °C; (**b**) 820 °C; (**c**) 870 °C; and (**d**) 950 °C for 60 s; and the corresponding void networks obtained by image processing on the SEM images: (**e**) 700 °C; (**f**) 820 °C; (**g**) 870 °C; and (**h**) 950 °C.

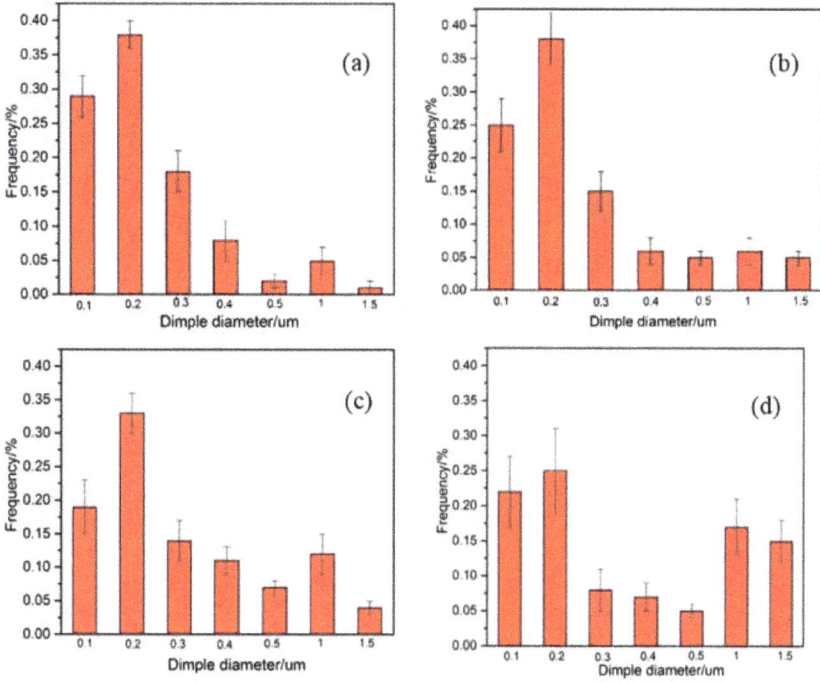

Figure 5. Dimple size distribution after annealing at (**a**) 700 °C, (**b**) 820 °C, (**c**) 870 °C, and (**d**) 950 °C for 60 s.

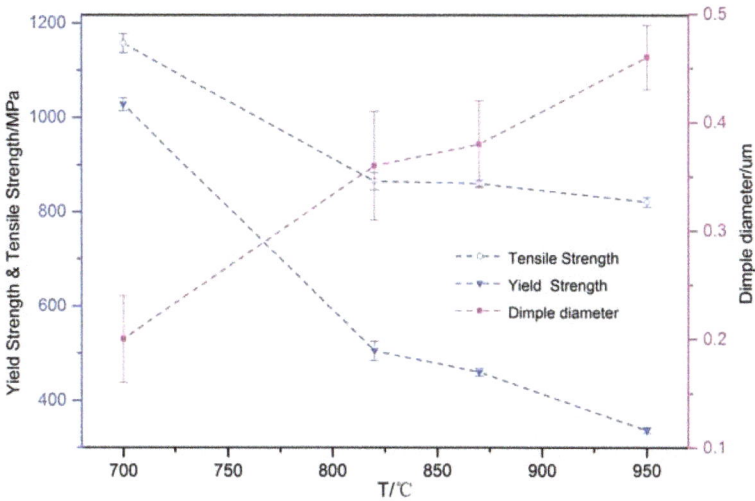

Figure 6. Strength properties with average dimple diameter at different annealing temperatures.

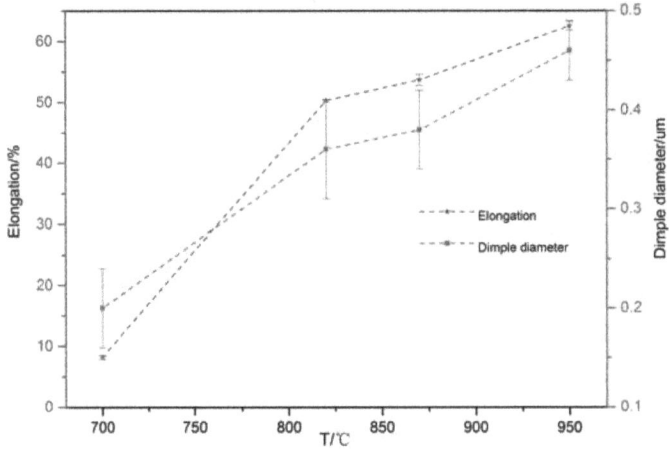

Figure 7. Ductility properties with average dimple diameter at different annealing temperatures.

3.3. Micro Deformation Experiments for the 304 ASS with UFG Structure

3.3.1. Deformation Mechanisms for Specimen Annealed at 700 °C

Three UFG foils for the micro deformation were studied out via TEM. The stress direction in the tensile test is shown in Figures 8, 10 and 13. It was found that the specimen annealed at 700 °C during the micro deformation experiment tended to crack quickly. This may be attributed to the combined effect of the low ductility of the UFG and the existence of martensite. From the micro deformation observation shown in Figure 8, it was found that some pile-ups of closely dense dislocations ended in cracks and the dislocation pile-ups disappeared after the crack. Further, the dislocation pile-ups only took place on either side of the crack. In other words, the stress concentrated on the tip of the crack with inc mentals-183011 reasing stress during the tensile test. No dislocation pile ups appeared in the other grains. Further straining could not produce additional interactions on the grains since cracks had generated elsewhere, which relieved the stress on the dislocation pile-ups. Figure 9 shows the phase map after micro deformation tensile test of the specimen annealed at 700 °C for 60 s, taken 5–10 μm away from the crack. The blue color represents the austenite phase and the red color represents the martensite phase. From the micro deformation of the foils, it was found that part of the strain induced martensite transformation from austenite.

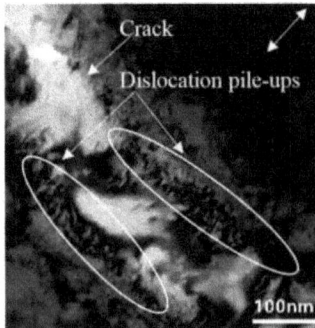

Figure 8. Schematic illustration of the crack propagation after the micro deformation tensile test, showing specimens annealed at 700 °C for 60 s (bidirectional arrow: stress direction).

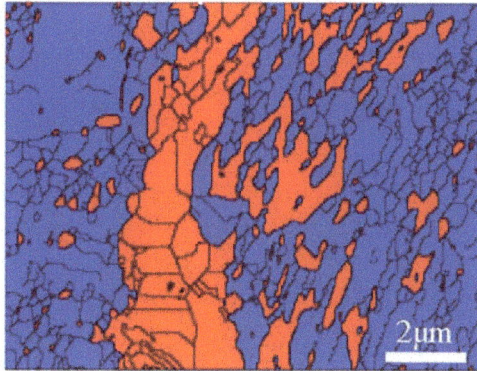

Figure 9. Phase map of the specimen annealed at 700 °C for 60 s after micro deformation tensile test, 5–10 μm away from the crack (blue-austenite, red-martensite).

3.3.2. Deformation Mechanisms for Specimen Annealed at 870 °C

Because of the low SFE of the 304 ASS in this study, partial dislocations that act as glide dislocations were more widely separated than in high SFE materials, as shown in Figure 10a. A few isolated dislocations, pile-ups, or irregular networks appeared in some areas (see the label in Figure 10a). Most of the irregular objects appeared to be partial dislocations in the pile-ups. This means that it was easier to activate a partial dislocation than a full dislocation in the UFG microstructure with average grain size 2 μm. The space of the dislocations in the pile-ups did not vary regularly in the manner expected from the calculations of Eshelby et al. [20], but were much more irregular. Figure 10b marks a set of partial dislocations arranged in a row and breaking up to some distance. The picture shows that the piled-up structure (arranged at grain boundaries) that was used to illuminate the relation between flow stress and grain size is largely correct [21]. It was confirmed that these dislocations are parallel to $(111)_\gamma$ slip planes, hence the planes that contain these dislocations must be $(111)_\gamma$ slip planes for all orientations observed [22]. In some areas, multiple striped bands appeared, as shown in Figure 10c. For the steel with average grain size of 2 μm, it was found that numerous dislocations were blocked by grain boundaries of small grains. It was thought that these bands are due to stacking faults produced by the movement of partial dislocations. In contrast to the motion of dislocations in aluminum, cross slipping has not been observed within the individual grains [12]. This can be attributed to the large width of stacking faults, speculated from the low SFE and the subsequent difficulty of developing constrictions. In addition, as shown in Figure 10d, twins with high-angle boundary (see black arrow in Figure 10d) retarded the glide of the dislocations could therefore enhance the strain-hardening rate, as shown in Figure 11 [6]. These structures are typical in low SFE materials. According to the reference [23], it is indicated that the transformation mechanism was via $\gamma \rightarrow$ deformation twinning $\rightarrow \alpha'$.

A phase map of the specimen annealed at 870 °C for 60 s after the micro deformation tensile test, 5–10 μm away from the crack is shown in Figure 12. The blue color represents the austenite phase, the red color represents the martensite phase, and the yellow color represents the ε martensite phase. It can be observed that ε martensite was included in α' martensite, which indicated that the transformation mechanism was $\gamma \rightarrow \varepsilon \rightarrow \alpha'$. As mentioned above, a conclusion can be reached that α' martensite might nucleate by deformation twins, ε martensite, or their intersections [23].

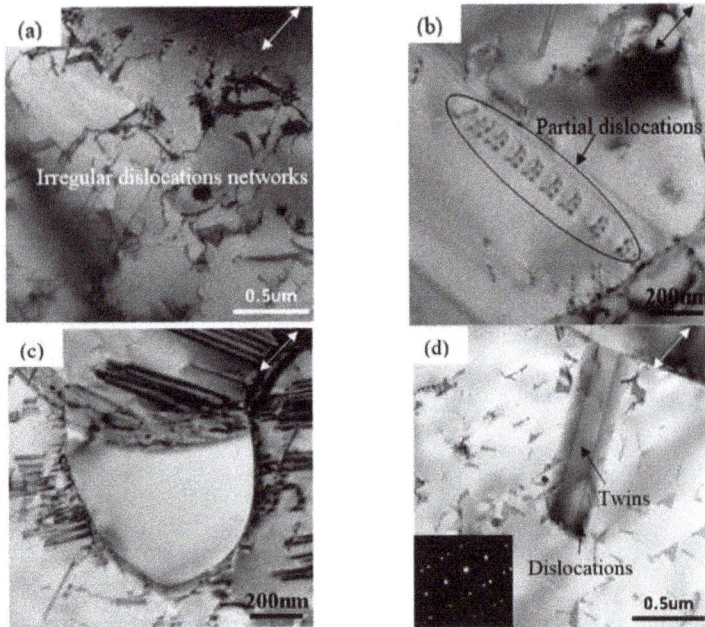

Figure 10. TEM micrographs of specimen annealed at 870 °C after micro deformation tensile test, showing (**a**) irregular distribution of partial dislocations; (**b**) partial dislocations arranged in a row; (**c**) large stacking faults around ultrafine-grained (UFG) boundary; and (**d**) twins (bidirectional arrow: stress direction).

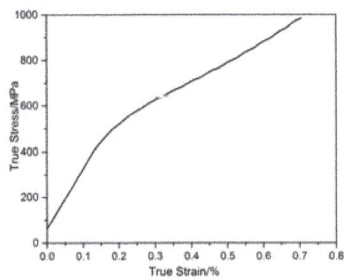

Figure 11. True stress-strain tensile data for the specimen annealed at 870 °C for 60 s.

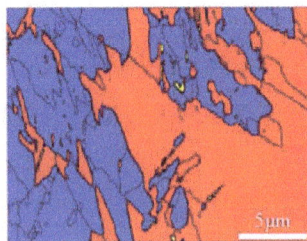

Figure 12. Phase map of the specimen annealed at 870 °C for 60 s after micro deformation tensile test, 5–10 μm away from the crack (blue-austenite, red-martensite, yellow-ε martensite).

3.3.3. Deformation Mechanism for Specimen Annealed at 950 °C

ε martensite was found as marked in Figure 13a. Meanwhile, the deformation twins clearly appeared in the deformed austenite microstructure (circled in Figure 13a). In addition, the directions of all deformation twins were nearly perpendicular to that of ε martensite, which is consistent with the results of Li et al. [14], who reported that deformation twins grow along the direction perpendicular to ε martensite with increasing deformation. In fcc metals, stacking faults and deformation twins can be dissociated from a screw dislocation or a 60° dislocation [13]. Stacking faults overlapped within the $(111)_\gamma$ plane of the fcc crystal, which promoted the generation of deformation twins [24]. As the $(111)_\gamma$ plane slipped due to the increase of deformation, the $(10\bar{1}0)_\varepsilon$ planes were perpendicular to the $(111)_\gamma$ planes [25]. Hence, the observation on deformation twins in the direction vertical to ε martensite is correct. According to calculation of the SFE of 304 ASS, the energy is in the range 12–35 mJ/m^2.

Dense dislocation arrangements accumulated inside grains of size 2 μm, while thin dislocation was observed inside grains of size 500 nm, as shown in Figure 13b. Apparently, this differs from the specimen annealed at 870 °C in terms of the transformation of UFG during the tensile test. It is indicated that the coarse grain generates dislocation pile-ups more easily than the fine grain, which is considered a strengthening phase in the UFG structure and difficult to break.

Phase map of the specimen annealed at 950 °C for 60 s after the micro deformation tensile test, 10–15 μm away from the crack is shown in Figure 14. The blue color represents the austenite phase, the red color represents the martensite phase, and the yellow color represents the ε martensite phase. The ε martensite appeared as a band included in α′ martensite, which indicated that the transformation mechanism was $\gamma \rightarrow \varepsilon \rightarrow \alpha'$, namely the stress-induced-transformation.

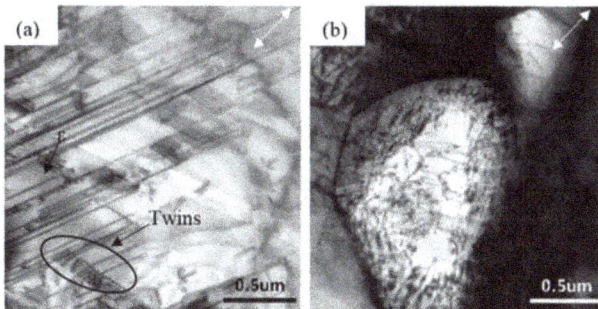

Figure 13. TEM micrographs of specimen annealed at 950 °C after micro deformation tensile test showing (**a**) ε and twins observed in coarse grains; (**b**) dislocations in UFG structure (bidirectional arrow: stress direction).

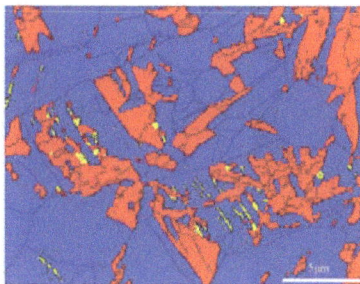

Figure 14. Phase map of the specimen annealed at 950 °C for 60 s after micro deformation tensile test, 10–15 μm away from the crack (blue-austenite, red-martensite, yellow-ε martensite).

4. Conclusions

The tensile fractographs of the obtained UFG structure show that the most frequent dimple sizes were approximately 0.1–0.3 μm and 1–1.5 μm. With increasing annealing temperature, nano-sized grains grew to micron-sized grains and the strengths decreased, while the elongation and average size of the dimples increased.

For the 304 ASS with different grain sizes during the micro deformation experiment, (a) the specimen annealed at 700 °C tended to crack quickly. This may be attributed to the combined effect of the low ductility of the UFG and the existence of martensite. (b) In the sample annealed at 870 °C with average grain size 2 μm, partial dislocations were widely separated in the irregular crystals or piled up normal to the grain boundaries; stacking faults were blocked by grain boundaries of small grains; twins prevented gliding of the dislocations. It can be concluded that α' martensite might nucleate through deformation twins, ε martensite or their intersections. (c) The deformation twins of the sample annealed at 950 °C with average grain size 5 μm were perpendicular to the ε martensite. Fine grain was considered as the strengthening phase in the UFG structure and was difficult to break.

Acknowledgments: This work was financially supported by the National Natural Science Foundation of China (Grant No. 51474031).

Author Contributions: Na Gong and Hui-Bin Wu conceived and designed the experiments; Na Gong and Zhi-Chen Yu performed the experiments; Na Gong and Gang Niu analyzed the data; Na Gong and Zhi-Chen Yu were responsible for language modification; Da Zhang contributed reagents/materials/analysis tools; Na Gong wrote the paper.

Conflicts of Interest: The authors declare no conflict of interest.

References

1. Huang, G.L.; Matlock, D.K.; Krauss, G. Martensite formation, strain rate sensitivity, and deformation behavior of type 304 stainless steel sheet. *Metall. Trans. A* **1989**, *20*, 1239–1246. [CrossRef]
2. Hedayati, A.; Najafizadeh, A.; Kermanpur, A.; Forouzan, F. The effect of cold rolling regime on microstructure and mechanical properties of AISI 304L stainless steel. *J. Mater. Process. Technol.* **2010**, *210*, 1017–1022. [CrossRef]
3. Choi, J.; Jin, W. Strain induced martensite formation and its effect on strain hardening behavior in the cold drawn 304 austenitic stainless steels. *Scr. Mater.* **1997**, *36*, 99–104. [CrossRef]
4. Misra, R.D.K.; Kumar, B.R.; Somani, M.; Karjalainen, P. Deformation processes during tensile straining of ultrafine/nanograined structures formed by reversion in metastable austenitic steels. *Scr. Mater.* **2008**, *59*, 79–82. [CrossRef]
5. Challa, V.S.A.; Wan, X.L.; Somani, M.C.; Karjalainen, L.P.; Misra, R.D.K. Significance of interplay between austenite stability and deformation mechanisms in governing three-stage work hardening behavior of phase-reversion induced nanograined/ultrafine-grained (NG/UFG) stainless steels with high strength-high ductility combination. *Scr. Mater.* **2014**, *86*, 60–63.
6. Misra, R.D.K.; Zhang, Z.; Jia, Z.; Somani, M.C.; Karjalainen, L.P. Probing deformation processes in near-defect free volume in high strength–high ductility nanograined/ultrafine-grained (NG/UFG) metastable austenitic stainless steels. *Scr. Mater.* **2010**, *63*, 1057–1060. [CrossRef]
7. Shen, F.; Zhou, J.; Liu, Y.; Zhu, R.; Zhang, S.; Wang, Y. Deformation twinning mechanism and its effects on the mechanical behaviors of ultrafine grained and nanocrystalline copper. *Comp. Mater. Sci.* **2010**, *49*, 226–235. [CrossRef]
8. Capolungo, L.; Cherkaoui, M.; Qu, J. On the elastic-viscoplastic behavior of nanocrystalline materials. *Int. J. Plast.* **2007**, *23*, 561–591. [CrossRef]
9. Allain, S.; Chateau, J.P.; Bouaziz, O. A physical model of the twinning-induced plasticity effect in a high manganese austenitic steel. *Mater. Sci. Eng. A* **2004**, *387*, 143–147. [CrossRef]
10. Cohen, G.B.O. A general mechanism of martensitic nucleation: Part I. General concepts and the FCC → HCP transformation. *Metall. Mater. Trans. A* **1976**, *7*, 1897–1904.
11. Mangonon, P.L.; Thomas, G. The martensite phases in 304 stainless steel. *Metall. Mater. Trans. B* **1970**, *1*, 1577–1586. [CrossRef]

12. Liao, X.Z.; Zhou, F.; Lavernia, E.J.; Srinivasan, S.G.; Baskes, M.I.; He, D.W.; Zhu, Y.T. Deformation mechanism in nanocrystalline Al: Partial dislocation slip. *Appl. Phys. Lett.* **2003**, *83*, 632–634. [CrossRef]

13. Shen, Y.F.; Li, X.X.; Sun, X.; Wang, Y.D.; Zuo, L. Twinning and martensite in a 304 austenitic stainless steel. *Mater. Sci. Eng. A* **2012**, *552*, 514–522. [CrossRef]

14. Li, X.; Ding, W.; Cao, J.; Ye, L.; Chen, J. In situ tem observation on martensitic transformation during tensile deformation of sus304 metastable austenitic stainless steel. *Acta Metall. Sin.* **2015**, *28*, 302–306. [CrossRef]

15. Gurson, A.L. Continuum theory of ductile rupture by void nucleation and growth: Part I—Yield criteria and flow rules for porous ductile media. *J. Eng. Mater. Technol.* **1977**, *99*, 2–15. [CrossRef]

16. Bandstra, J.P.; Koss, D.A.; Geltmacher, A.; Matic, P.; Everett, R.K. Modeling void coalescence during ductile fracture of a steel. *Mater. Sci. Eng. A* **2004**, *366*, 269–281. [CrossRef]

17. Lee, T.C.; Robertson, I.M.; Birnbaum, H.K. TEM in situ deformation study of the interaction of lattice dislocations with grain boundaries in metals. *Philos. Mag. A* **1990**, *62*, 131–153. [CrossRef]

18. Altenberger, I.; Stach, E.A.; Liu, G.; Nalla, R.K.; Ritchie, R.O. An in situ transmission electron microscope study of the thermal stability of near-surface microstructures induced by deep rolling and laser-shock peening. *Scr. Mater.* **2003**, *48*, 1593–1598. [CrossRef]

19. Schramm, R.E.; Reed, R.P. Stacking fault energies of seven commercial austenitic stainless steels. *Metall. Mater. Trans. A* **1975**, *6*, 1345–1351. [CrossRef]

20. Eshelby, J.D.; Frank, F.C.; Nabarro, F.R.N. XLI. The equilibrium of linear arrays of dislocations. *Lond. Edinb. Dublin Philos. Mag. J. Sci.* **1951**, *42*, 351–364. [CrossRef]

21. Petch, N.J. The fracture of metals. *Prog. Met. Phys.* **1954**, *5*, 1–52. [CrossRef]

22. Whelan, M.J.; Hirsch, P.B.; Horne, R.W.; Bollmann, W. Dislocations and stacking faults in stainless steel. *Proc. R. Soc. Lond. Ser. A* **1957**, *240*, 524–538. [CrossRef]

23. Olson, G.B.; Cohen, M. A mechanism for the strain-induced nucleation of martensitic transformations. *J. Less Common Met.* **1972**, *28*, 107–118. [CrossRef]

24. Li, X.; Chen, J.; Ye, L.; Ding, W.; Song, P. Influence of strain rate on tensile characteristics of SUS304 metastable austenitic stainless steel. *Acta Metall. Sin.* **2013**, *26*, 657–662. [CrossRef]

25. Tao, K.; Wall, J.J.; Li, H.; Brown, D.W.; Vogel, S.C.; Choo, H. In situ neutron diffraction study of grain-orientation-dependent phase transformation in 304L stainless steel at a cryogenic temperature. *J. Appl. Phys.* **2006**, *100*, 123515. [CrossRef]

![metals logo] *metals* MDPI

Article

The Effect of Niobium on the Changing Behavior of Non-Metallic Inclusions in Solid Alloys Deoxidized with Mn and Si during Heat Treatment at 1473 K

Chengsong Liu [1], Xiaoqin Liu [1,*], Shufeng Yang [2], Jingshe Li [2], Hongwei Ni [1] and Fei Ye [1]

[1] The State Key Laboratory of Refractories and Metallurgy, Wuhan University of Science and Technology, Wuhan 430081, China; liuchengsong@wust.edu.cn (C.L.); nihongwei@wust.edu.cn (H.N.); yefeishangnan@163.com (F.Y.)

[2] School of Metallurgical and Ecological Engineering, University of Science and Technology Beijing, Beijing 100083, China; yangshufeng@ustb.edu.cn (S.Y.); lijingshe@ustb.edu.cn (J.L.)

* Correspondence: liuxiaoqin@wust.edu.cn; Tel.: +86-27-6886-2811

Received: 4 April 2017; Accepted: 13 June 2017; Published: 16 June 2017

Abstract: To clarify the effect of niobium (Nb) on the changing behavior of oxide inclusions in alloys containing different concentrations of Mn, Si, and Nb, heat treatment experiments at 1473 K were conducted and changes in the morphology, size, quantity, and composition of these inclusions were investigated. The stability of the oxide inclusions in both molten and solid Fe-Mn-Si-Nb alloys was also estimated by thermodynamic calculation using available data. Results showed that the change in the composition of the oxide inclusions owing to heat treatment depended on the concentrations of Nb and Si in the alloy. $MnO-SiO_2$-type oxide inclusions gradually transformed into $MnO-Nb_2O_5$-type or $MnO-SiO_2$- & $MnO-Nb_2O_5$-type inclusions in low-Si and high-Nb alloys after heating for 60 min. However, the shape of the inclusions did not change clearly. It was indicated that, during the heat treatment at 1473 K, an interface chemical reaction between the Fe-Mn-Si-Nb alloys and the $MnO-SiO_2$-type oxide inclusions occurred according to the experimental and calculation results.

Keywords: niobium; non-metallic inclusion; heat treatment; interfacial reaction; modification

1. Introduction

By oxide metallurgy technology, fine and dispersed non-metallic inclusions in steel could not only distribute around austenite boundaries to restrain grain growth but also act as heterogeneous nucleus to promote acicular ferrite, which are able to greatly improve the mechanical properties of steel [1–4]. Heat treatment is a new approach to precisely controlling and optimizing the physicochemical characteristics of non-metallic inclusions in steel and greatly expanding the application of oxide metallurgy technology in steel production, which is attracting more and more attention. Shibata et al. [5] reported that, in the Fe-Cr alloy containing 10 mass % Cr, $MnO-SiO_2$-type inclusions transformed into $MnO-Cr_2O_3$-type inclusions with a low Si content (<0.1 mass %) after heat treatment, while at high Si content (>0.3 mass %), the $MnO-SiO_2$-type inclusion was stable. Choi et al. [6] hypothesized that, in an as-cast Fe-0.028 mass % Ti alloy, Ti–O inclusions with a small amount of Fe gradually changed to Ti–Fe–O after heating at 1473 K for 180 min. The fraction of fine and coarse inclusions increased and decreased, respectively. Shao et al. [7] confirmed that shape variation of slender MnS was greatly influenced by the heating rate. As heating rate rose from 0.5 to 2 K/s, the amount of split MnS decreased; while the heating rate exceeded 6 K/s, the slender MnS remained unchanged. Liu et al. [8] studied solid-state reactions between an Fe-Al-Ca alloy and an Al_2O_3-CaO-FeO oxide during heat treatment at 1473 K and found that some Al_2O_3 particles and $CaO·Al_2O_3$ branch inclusions precipitated as reaction products in the alloy near the alloy-oxide interface, which caused the Al content in the alloy to decrease. They also [9] investigated an interfacial reaction mechanism between $CaO-SiO_2-Al_2O_3-MgO-MnO$ oxides and a Fe-Mn-Si alloy during heating. It was proved that the overall MnO and SiO_2 content of the oxide decreased and increased, respectively, after the "solid–liquid" reaction between the oxide and alloy at 1473 K.

As an important alloying element with a broad application prospect, the niobium (Nb) added into the alloy is beneficial for the precipitation of inclusion particles such as Nb(C, N), which prevents the growth of austenite grain by the pinning effect at the grain boundaries during heat treatment [10,11]. By combining with solute Nb in alloy, Nb(C, N) particles can also inhibit the recrystallization of deformed austenite through the drag effect and promote fine acicular-ferrite. Moreover, Nb contributes to the dispersive distribution of inclusions and a flexible adjustment of the toughness of the alloy in a broad range by controlling the induced precipitation and cooling rate. Therefore, Nb in the alloy can not only enhance the strength of the alloy but also improve the toughness, the high temperature oxidation resistance, and the corrosion resistance of the alloy and reduce the brittle transition temperature of the alloy to obtain a better welding and forming performance. Nevertheless, other than the precipitation of particles such as Nb(C, N), heat treatment also results in changes in composition, morphology, size, and the physicochemical characteristics of other inclusions with the influence of elemental Nb, which directly affects the quality and mechanical properties of the alloy.

In the present study, experiments were designed to reveal the effect of Nb on the changing behavior of non-metallic inclusions in the Fe-Mn-Si-Nb alloy including morphology, types, compositions, and quantities during heat treatment at 1473 K. The stability of the oxide inclusions in both molten and solid Fe-Mn-Si-Nb alloys was also discussed in light of thermodynamic calculations.

2. Experimental Methods

2.1. Materials

Initial compositions of the alloys with different concentrations of Mn, Si, and Nb are summarized in Table 1. A 6 kg Fe-Mn-Si ingot was prepared by melting electrolytic iron with ferromanganese and silicon powder in an arc melting furnace. Then, 200 g of the premelted Fe-Mn-Si alloy was taken from the position between the center and the edge of the circular cross section in the middle part of the ingot and then re-melted with the addition of high-grade ferroniobium in an MgO crucible (outer diameter 48 mm, inner diameter 38 mm, height 115 mm) at 1873 K for 15 min, under an air atmosphere in an electric resistance pipe furnace. After that, the crucible was taken out from the pipe and quenched by immersing the crucible into water. Due to the high cooling rate and small size of the alloy specimen, concentrations of Mn, Si, and Nb were firstly confirmed to be homogeneous by using an Electron Probe Microscopic Analyzer (EPMA) (JEOL, Tokyo, Japan) whose primary importance is the ability to acquire precise, quantitative elemental analyses at very small "spot" sizes (as little as 1–2 microns), primarily by wavelength-dispersive spectroscopy (WDS), and then measured and verified by Inductively Coupled Plasma Optical Emission Spectrometer (ICP-OES) (Thermo Fisher Scientific, Waltham, MA, USA).

Table 1. Initial compositions of the alloy deoxidized with Mn and Si for heat treatment at 1473 K.

	Sample		Fe	Mn	Si	Nb
			Mass %			
1	High Si	High Nb	97.57	0.63	1.09	0.61
2	Medium Si	High Nb	98.21	0.67	0.37	0.65
3	Low Si	High Nb	98.60	0.61	0.025	0.67
4	High Si	Medium Nb	98.07	0.68	1.12	0.13
5	Medium Si	Medium Nb	98.72	0.72	0.41	0.15
6	Low Si	Medium Nb	99.11	0.65	0.028	0.16
7	High Si	Low Nb	98.13	0.67	1.06	0.04
8	Medium Si	Low Nb	98.84	0.72	0.39	0.05
9	Low Si	Low Nb	99.20	0.73	0.023	0.05

2.2. Heat Treatment

The quenched alloy specimens were machined into cylindrical pieces (ϕ 10 mm × 20 mm) and hanged in the reaction pipe of an electric furnace by Mo wire. Then, specimens were heated at 1473 K for 30 min and 60 min, and the atmosphere was replaced by high purity Ar gas (99.9 mass %, flow rate: about 300 cm^3/min). After above heat treatment, the specimens were immediately taken out of the reaction pipe and quenched by immersion into water. The temperature curve of heat treatment experiment is given in Figure 1. Schematic diagram of the furnace equipped with specimen quenching under a controlled Ar gas atmosphere for heat treatment is shown in Figure 2.

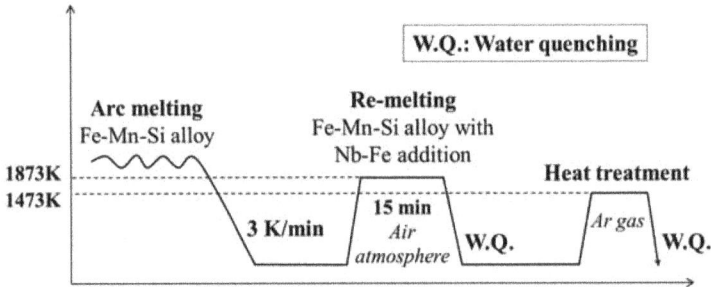

Figure 1. Temperature curve of melting and heat treatment experiments for Fe-Mn-Si-Nb specimens.

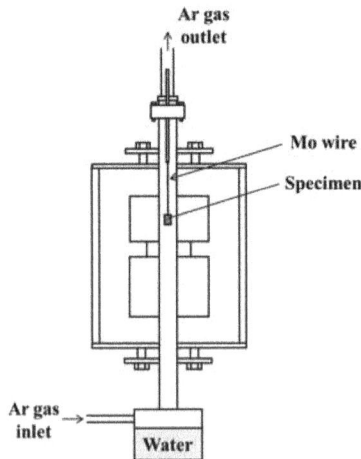

Figure 2. Schematic diagram of the furnace equipped with specimen quenching under controlled Ar gas atmosphere for heat treatment at 1473 K.

2.3. EPMA Analysis

As cast and heated specimens were embedded in the polyester resin, and the cross section of each specimen was ground with SiC sandpapers and polished with diamond polishing paste to prepare metallographic samples. Characteristics including morphology, types, quantities, and compositions of over 50 inclusions in each metallographic sample before and after heat treatment for 30 min and 60 min were observed and analyzed by an Electron Probe Microscopic Analyzer (EPMA), and the compositions of Fe, Mn, Si, and Nb were determined.

3. Results

3.1. Influence of Heat Treatment at 1473 K on the Morphology and Compositions of Oxide Inclusions

Figure 3 shows the morphology and compositions of typical oxide inclusions in Alloy Sample 2 (Mn: 0.67 mass %, Si: 0.37 mass %, Nb: 0.65 mass %). In an as-cast sample, nearly spherically shaped inclusions were observed. Their typical chemical composition, measured by using the EPMA, was 51 mol % MnO-5 mol % SiO_2-44 mol % Nb_2O_5. After the heat treatment at 1473 K for 30 and 60 min, the shape of inclusions did not clearly change. The chemical compositions of these inclusions were 48 mol % MnO-2 mol % SiO_2-50 mol % Nb_2O_5 and 37 mol % MnO-1 mol % SiO_2-62 mol % Nb_2O_5, respectively. This meant that only MnO-Nb_2O_5-type inclusions were observed in the case where Nb content in the alloy sample was higher than 0.6 mass %.

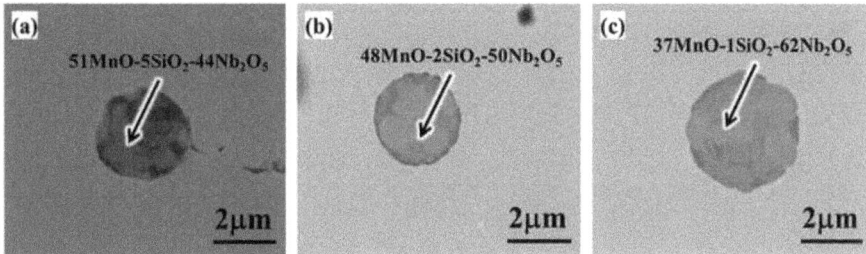

Figure 3. Morphology and compositions of typical oxide inclusions in Alloy Sample 2: (**a**) as cast; (**b**) heat treatment for 30 min; (**c**) heat treatment for 60 min.

Figure 4 shows the morphology and compositions of typical oxide inclusions in Alloy Sample 4 (Mn: 0.68 mass %, Si: 1.12 mass %, Nb: 0.13 mass %). The spherically shaped MnO-SiO_2-type inclusion was also found in an as-cast sample. Its typical chemical composition was 52 mol % MnO-46 mol % SiO_2-2 mol % Nb_2O_5. After heating at 1473 K for 30 min, although the shape of inclusions still did not clearly change, many inclusions had two phases in an inclusion, as Figure 4b shows the typical morphology of inclusions in Alloy Sample 4. One is MnO-SiO_2-type inclusion containing approximately 11 mol % Nb_2O_5. The other is Mn-Si-Nb inclusion, whose average composition is 41 mol % MnO-24 mol % SiO_2-34 mol % Nb_2O_5. As the heating time increased to 60 min, the Nb_2O_5 content of both phases in an inclusion increased, as shown in Figure 4c. Their typical chemical compositions were 61 mol % MnO-23 mol % SiO_2-16 mol % Nb_2O_5 and 49 mol % MnO-7 mol % SiO_2-44 mol % Nb_2O_5, respectively. This phenomenon indicated that the chemical composition of the oxide inclusions changed by heat treatment in the cases of medium Nb content and high Si content.

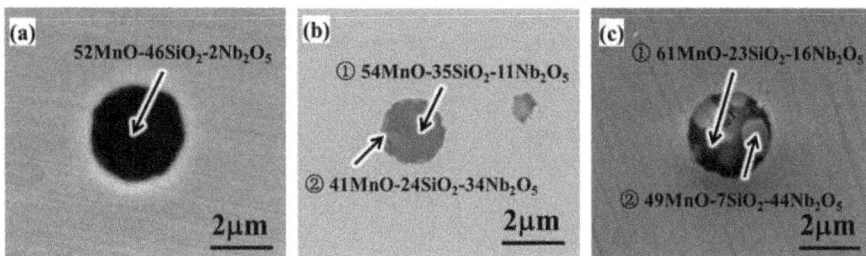

Figure 4. Morphology and compositions of typical oxide inclusions in Alloy Sample 4: (**a**) as cast; (**b**) heat treatment for 30 min; (**c**) heat treatment for 60 min.

The morphology and compositions of typical oxide inclusions observed in Alloy Sample 8 (Mn: 0.72 mass %, Si: 0.39 mass %, Nb: 0.05 mass %) were shown in Figure 5. The main shape and composition of inclusions found in the as-cast sample and their changing behavior with heating were similar to those observed in the case of Alloy Sample 4. Many observed inclusions in Alloy Sample 8 after heating also had two phases as shown in Figure 5b,c, and the Nb_2O_5 content of both phases in an inclusion appears to be relatively lower, dependent on the concentrations of Nb and Si in the alloy. After heat treatment at 1473 K for 30 min, typical chemical compositions of the two phases in an inclusion were 47 mol % MnO-45 mol % SiO_2-8 mol % Nb_2O_5 and 54 mol % MnO-23 mol % SiO_2-23 mol % Nb_2O_5, respectively. After heat treatment for 60 min, the concentrations of Nb_2O_5 in the two phases in an inclusion increased to 12 mol % and 31 mol %, while SiO_2 concentrations decreased to 37 mol % and 22 mol %, respectively.

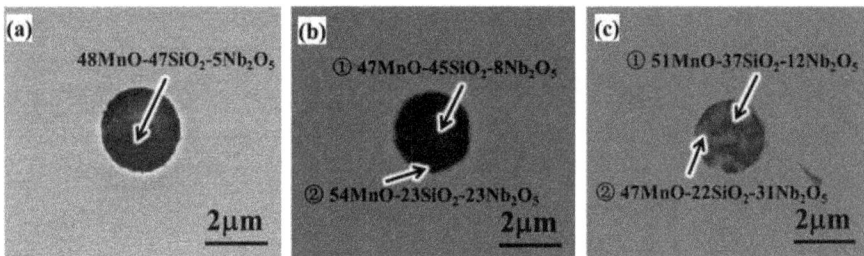

Figure 5. Morphology and compositions of typical oxide inclusions in Alloy Sample 8: (**a**) as cast; (**b**) heat treatment for 30 min; (**c**) heat treatment for 60 min.

3.2. Influence of Nb, Si and Mn Contents of the Alloy on Change in Type and Quantity of Oxide Inclusions

The dependence of chemical compositions of stable oxide inclusions on the concentrations of Mn, Si, and Nb in different alloy samples before and after heating at 1473 K for 60 min is shown in Table 2. In this table, alloy samples containing MnO-SiO_2-type and MnO-Nb_2O_5-type inclusions are denoted by white circles and black circles, respectively. Moreover, the alloy sample containing both types of inclusions is denoted by crosses. Only MnO-Nb_2O_5-type inclusions formed in Alloy Samples 1–3 containing about 0.65 mass % Nb before and after heating at 1473 K for 60 min, which was very stable although Si content varied from 0.025 mass % to 1.10 mass %. As for the chemical compositions of oxide inclusions in Alloy Samples 4–6 containing about 0.15 mass % Nb, with the decrease in Si content in the as-cast sample, MnO-SiO_2-type inclusions gradually disappeared and MnO-SiO_2- & MnO-Nb_2O_5-type and MnO-Nb_2O_5-type inclusions formed. After the heat treatment at 1473 K, original MnO-SiO_2-type and MnO-SiO_2- & MnO-Nb_2O_5-type inclusions in the as-cast samples changed to MnO-SiO_2- & MnO-Nb_2O_5-type and MnO-Nb_2O_5-type inclusions, respectively. When the Nb content decreased to 0.05 mass % in Alloy Samples 7–9, inclusions in the as-cast samples were mainly MnO-SiO_2-type or MnO-SiO_2- & MnO-Nb_2O_5-type before heat treatment depending on the concentration of Si in the alloy. After the heat treatment, some of the original MnO-SiO_2-type inclusions transformed into two-phase MnO-SiO_2- & MnO-Nb_2O_5-type inclusions, while some of the original two-phase MnO-SiO_2- & MnO-Nb_2O_5-type inclusions transformed into MnO-Nb_2O_5-type inclusions, although the transformation was incomplete. The results strongly indicate that the relatively high Si content and low Nb content retard the conversion from a MnO-SiO_2-type to a MnO-Nb_2O_5-type inclusion. This retardation occurs as long as non-equilibrium conditions persist between the alloy and the MnO–SiO_2-type oxide inclusion.

Figure 6 shows the statistical analysis of the density of oxide inclusions changing from MnO-SiO_2-type to MnO-SiO_2- & MnO-Nb_2O_5-type and MnO–Nb_2O_5-type in Alloy Samples 2, 4, and 8. At least 120 random inclusions in each sample including MnO-SiO_2-type, MnO-SiO_2- & MnO-Nb_2O_5-type and MnO-Nb_2O_5-type were measured and analyzed by EPMA.

Table 2. Dependence of chemical compositions of stable oxide inclusions on the concentrations of Mn, Si, and Nb in different alloy samples before and after heating at 1473 K for 60 min.

Sample	No.	Heating Time	As Cast	60 min
	1	High Si (1.10 mass %)	●	●
Fe-0.65 mass % Nb	2	Medium Si (0.40 mass %)	●	●
	3	Low Si (0.025 mass %)	●	●
	4	High Si (1.10 mass %)	○×	×
Fe-0.15 mass % Nb	5	Medium Si (0.40 mass %)	×	×●
	6	Low Si (0.025 mass %)	●	●
	7	High Si (1.10 mass %)	○	○×
Fe-0.05 mass % Nb	8	Medium Si (0.40 mass %)	○	○×
	9	Low Si (0.025 mass %)	×	×●

Note: ○—MnO-SiO$_2$-type; ×—MnO-SiO$_2$- & MnO-Nb$_2$O$_5$-type; ●—MnO-Nb$_2$O$_5$-type.

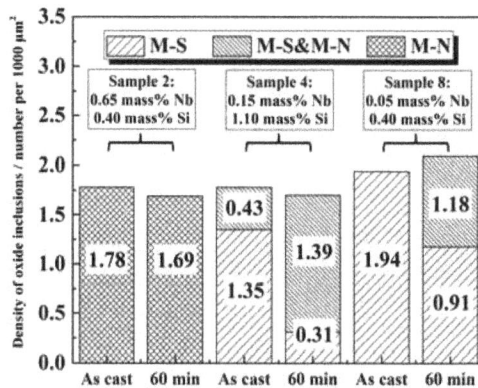

Figure 6. Statistical analysis of the quantity of oxide inclusions changing from MnO-SiO$_2$-type to MnO-SiO$_2$- & MnO-Nb$_2$O$_5$-type and MnO-Nb$_2$O$_5$-type in Alloy Samples 2, 4, and 8.

It is clear that all inclusions observed in Alloy Sample 2 were MnO-Nb$_2$O$_5$-type before and after the heating at 1473 K. For Alloy Sample 4, there were 1.35 MnO-SiO$_2$-type inclusions and 0.43 MnO-SiO$_2$- & MnO-Nb$_2$O$_5$-type inclusions within the area of 1000 μm^2 in the as-cast sample, while after the heat treatment, the density of MnO-SiO$_2$-type inclusions and MnO-SiO$_2$- & MnO-Nb$_2$O$_5$-type inclusions increased to 1.39 and decreased to 0.31 in the area of 1000 μm^2, respectively. In Alloy Sample 8, the original density of MnO-SiO$_2$-type inclusions in the as cast was 1.94 per 1000 μm^2, and then about half of them changed to MnO-SiO$_2$- & MnO-Nb$_2$O$_5$-type inclusions after the heating at 1473 K for 60 min. By comparing the statistical results of the density of oxide inclusions in Alloy Samples 2, 4, and 8 before and after heat treatment, it was proved that there was no new formation of MnO-Nb$_2$O$_5$-type and MnO-SiO$_2$- & MnO-Nb$_2$O$_5$-type inclusions during heating besides the transformation from the MnO-SiO$_2$-type inclusions. It was also clearly found that the relatively high Nb content and low Si content promoted the transformation from MnO-SiO$_2$-type inclusion to two-phase MnO-SiO$_2$- & MnO-Nb$_2$O$_5$-type and MnO-Nb$_2$O$_5$-type inclusions.

4. Discussion

4.1. Mechanism of the Interface Reaction between the Alloy and M-S-Type Inclusion

Elemental Mn, Si, and Nb in the alloy could react with dissolved oxygen to form complex oxide inclusions, due to exposure to air atmosphere when preparing the alloy samples. It is generally known that both manganese and niobium are transition metals and could take several different valences in oxides including Mn^{2+}, Mn^{3+}, Mn^{4+}, Nb^{2+}, Nb^{3+}, Nb^{4+}, and Nb^{5+}. Under the oxygen partial pressure of steelmaking, Mn^{2+} is dominant among other different valence states in its oxide form of MnO [12], and Nb seems to be stable as Nb^{5+} in solid oxide under an oxidizing atmosphere (i.e., Nb_2O_5) [13]. In addition, compared with SiO, SiO_2 is much more stable during the oxidation of silicon in the alloy. Therefore, during the re-melting of Fe-Mn-Si-Nb alloy at 1873 K under air atmosphere, stable simple oxides of Mn, Si, and Nb were MnO, SiO_2, and Nb_2O_5, respectively. Moreover, according to the stable components in the ternary system Nb_2O_5-MnO-SiO_2 [14], MnO-SiO_2-type, MnO-Nb_2O_5-type, and MnO-SiO_2- & MnO-Nb_2O_5-type oxides could be acquired, except for SiO_2-Nb_2O_5-type oxide.

In this study, influenced by the chemical compositions of the alloys, the changing behavior of inclusion from original MnO-SiO_2-type to MnO-SiO_2- & MnO-Nb_2O_5-type and MnO-Nb_2O_5-type was confirmed after heating at 1473 K for 60 min. It was indicated that there are interfacial reactions that occur between the alloy and the MnO-SiO_2-type inclusions in some alloy samples. Inclusions equilibrated with a molten alloy at 1873 K are no longer stable in a solid-state alloy. A schematic of the interface reaction mechanism is shown in Figure 7. As the heat treatment time increased, the Nb in the alloy gradually diffused and reacted with the inclusions, thereby resulting in the formation of MnO-SiO_2- & MnO-Nb_2O_5-type and MnO-Nb_2O_5-type inclusions, which could be expressed as Equations (1)–(3).

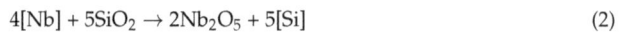

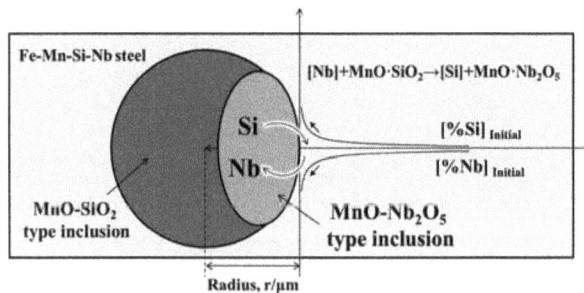

$$MnO\text{-}SiO_2 \rightarrow MnO + SiO_2 \tag{1}$$

$$4[Nb] + 5SiO_2 \rightarrow 2Nb_2O_5 + 5[Si] \tag{2}$$

$$MnO + Nb_2O_5 \rightarrow MnO\text{-}Nb_2O_5 \tag{3}$$

Figure 7. Schematic of interface reaction mechanism between the Fe-Mn-Si-Nb alloy and MnO-SiO_2-type inclusions.

The transformation from original MnO-SiO_2-type to MnO-SiO_2- & MnO-Nb_2O_5-type and MnO-Nb_2O_5-type is beneficial to reducing cracks in alloys during rolling due to the decrease in melting point and hardness [14]. In addition, as long as the mechanism of the interface reaction between the alloy and M-S-type inclusion is clarified, it is probable to control and optimize the physiochemical characteristics of the inclusions in the niobium alloy and obtain fine and dispersed non-metallic inclusions by an appropriate heat treatment processes.

4.2. The Equilibrium Relation between the Alloy and the Oxide Inclusions

The chemical compositions of stable oxide inclusions in the alloy samples containing different concentrations of Mn, Si, and Nb, at 1873 K and 1473 K, are estimated. This estimation is performed using Wagner's model for a multicomponent solution system [15], the regular solution model [16], and correlative thermodynamic data shown in Table 3. Oxygen activities for the formation of SiO_2, and MnO-SiO_2-type and MnO-Nb_2O_5-type inclusions, are then compared, based on the assumption that the most stable oxide inclusion is obtained at the lowest oxygen activity. In addition, it was observed that a large proportion of final stable oxide inclusions were approximate $MnO \cdot SiO_2$ and $MnO \cdot Nb_2O_5$ formation based on EPMA analysis. Therefore, equilibrium calculation was conducted by assuming the formation of pure $MnO \cdot SiO_2$ and $MnO \cdot Nb_2O_5$. The interaction coefficient of each element in the alloys, which is represented as e_i^j, is listed in Table 4. These thermodynamic data (see Tables 3 and 4 [17]) are also employed for approximate calculations of the equilibrium conditions between the alloy sample and oxide inclusion at 1473 K, although these data are inadequate for extrapolation to this temperature. The activities of MnO and SiO_2 in manganese silicate were also estimated and verified using the experimental results from [17]. At 1473 K, $MnO \cdot SiO_2$ and $MnO \cdot Nb_2O_5$ exist in the solid state, and their activities are assumed to be unity.

Table 3. Basic thermodynamic data for the equilibrium calculation.

No.	Reaction	$\Delta G^\ominus$ (J/mol)	Reference
1	$Nb(s) = [Nb]$	$-134,260 + 33.05T$	[18]
2	$4Nb(s) + 5O_2(g) = 2Nb_2O_5(s)$	$-3,770,150 + 834.95T$	[19]
3	$Mn(l) + 1/2O_2(g) = MnO(s)$	$-399,000 + 82.4T$	[19]
4	$O_2(g) = [O]$	$-234,304 - 5.78T$	[20]
5	$Mn(l) = [Mn]$	$4083.6 - 38.16T$	[20]
6	$[Si] + 2[O] = SiO_2(s)$	$-576,440 + 218.2T$	[21]
7	$MnO(s) + SiO_2(s) = MnSiO_3(s)$	$-27,960 + 2.42T$	[21]
8	$MnO(s) + Nb_2O_5(s) = MnNb_2O_6(s)$	$86,940 - 49.6T$	[22]

Note: T—Temperature, K.

Table 4. Interaction coefficient of each element in the Fe-Mn-Si-Nb alloy (e_i^j).

j ＼ i	Mn	Si	Nb	O
Mn	0	$-1838/T + 0.964$	$413/T - 0.217$	-0.083
Si	-0.0146	0.0103	0	-0.119
Nb	0.0093	-0.01	0	$-19,970/T + 9.950$
O	-0.021	-0.066	$-3440/T + 1.717$	$-1750/T + 0.76$

Note: T—Temperature, K.

The equilibrium oxygen activities of stable oxide inclusions in alloy samples with different Nb and Si concentrations at 1873 K and 1473 K are shown in Figures 8–10. At both temperatures, $MnO \cdot Nb_2O_5$-type inclusion in 0.65 mass % Nb alloy (see Figure 8) are considerably more stable than the $MnO \cdot SiO_2$-type inclusion and the equilibrium oxygen activity associated with the $MnO \cdot Nb_2O_5$-type oxide inclusion is always lower than that of the $MnO \cdot SiO_2$-type inclusion, indicating that $MnO \cdot Nb_2O_5$ forms prior to $MnO \cdot SiO_2$. At 1873 K, as the Nb content in the alloy decreased to 0.15 mass %, some $MnO \cdot SiO_2$-type and $MnO \cdot SiO_2$- & $MnO \cdot Nb_2O_5$-type oxide inclusions were generated. A critical Si concentration of 0.78 mass % for 0.15 mass % Nb alloy and 0.20 mass % for 0.05 mass % Nb alloy for the transformation between $MnO \cdot SiO_2$-type and $MnO \cdot Nb_2O_5$-type oxide inclusion was obtained according to the thermodynamic calculation, as shown in Figures 9a and 10a. At 1473 K (see Figures 9b and 10b), this transformation occurs at a critical Si content of 2.5 mass % for 0.15 mass % Nb alloy and 0.8 mass % for 0.05 mass % Nb alloy. This calculation results support that, at an Si content of 0.4 mass % and an Nb content of 0.65 mass % (Sample 2), the stable $MnO \cdot Nb_2O_5$-type oxide inclusion is retained during heat treatment at 1473 K, and the transformation from $MnO \cdot SiO_2$-type to $MnO \cdot Nb_2O_5$-type oxide inclusion in Alloy Samples 4 and 8 was promoted after heating. These calculation results agreed, in general, with the experimental results and revealed the mechanism of interface reaction between the Fe-Mn-Si-Nb alloy and $MnO \cdot SiO_2$ oxide inclusion. More importantly, these results contribute to the prediction of stable oxide formation in alloy (before and after heat treatment), based on the Si, Mn, and Nb concentrations of the alloy.

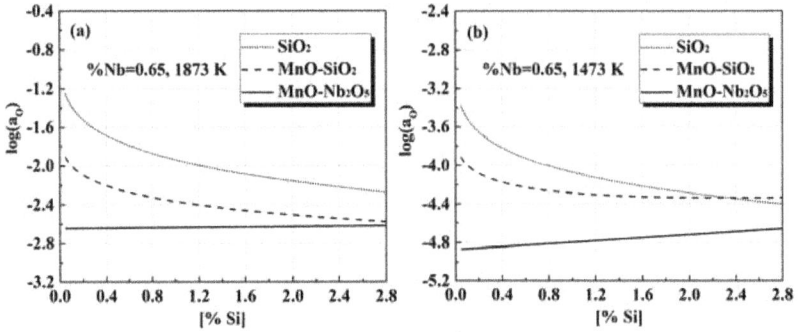

Figure 8. Equilibrium oxygen activities of stable oxide inclusions calculated at 1873 K (**a**) and 1473 K (**b**) corresponding to the alloy with 0.65 mass % Nb.

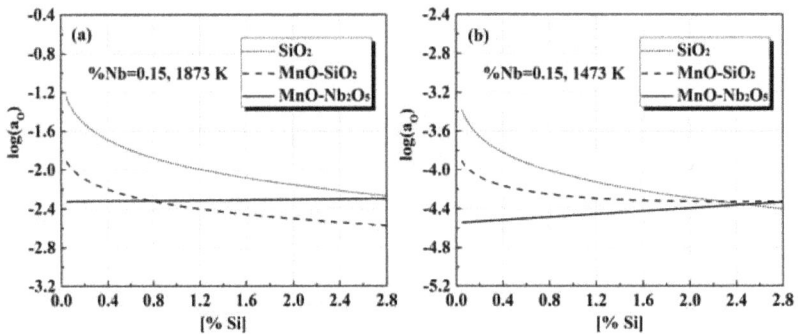

Figure 9. Equilibrium oxygen activities of stable oxide inclusions calculated at 1873 K (**a**) and 1473 K (**b**) corresponding to the alloy with 0.15 mass % Nb.

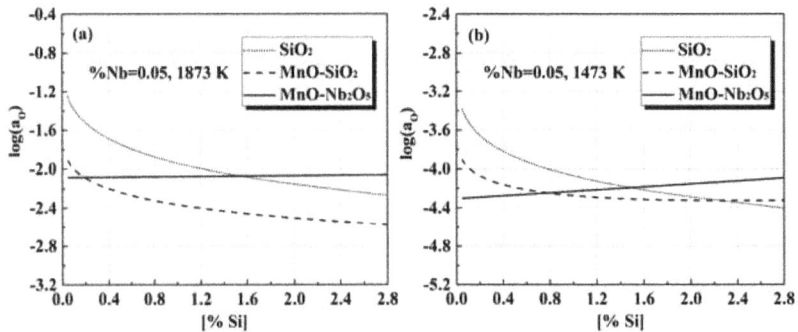

Figure 10. Equilibrium oxygen activities of stable oxide inclusions calculated at 1873 K (**a**) and 1473 K (**b**) corresponding to the alloy with 0.05 mass % Nb.

5. Conclusions

During the heating at 1473 K, the concentrations of Nb and Si in the alloys are critical for controlling the changing behavior of oxide inclusions. Stable oxide inclusions transformed from $MnO\text{-}SiO_2$-type to $MnO\text{-}Nb_2O_5$-type or $MnO\text{-}SiO_2$- & $MnO\text{-}Nb_2O_5$-type at low concentrations of Si and high concentration of Nb. It was indicated that an interface chemical reaction occurred between the Fe-Mn-Si-Nb alloy matrix and the MnO-SiO$_2$-type oxide inclusion. Estimation on the stable oxide inclusions in the alloys with different Nb concentrations by thermodynamic calculation at 1873 K and 1473 K basically matched the experimental results, thereby confirming the mechanism of the interface reaction. More significantly, the calculation results contribute to the prediction of heating-induced formation of a stable oxide, depending on the concentrations of Si, Mn, and Nb in the alloys.

Acknowledgments: This work was supported by the National Natural Science Foundation of China (Nos. 51604201 and 51574020) and the China Postdoctoral Science Foundation (No. 2016M602377).

Author Contributions: Chengsong Liu and Xiaoqin Liu conceived and designed the experiments; Chengsong Liu, Xiaoqin Liu and Fei Ye performed the experiments; Chengsong Liu, Xiaoqin Liu, Shufeng Yang, and Hongwei Ni analyzed the data; Jingshe Li and Hongwei Ni contributed reagents/materials/analysis tools; Chengsong Liu wrote the paper.

Conflicts of Interest: The authors declare no conflict of interest. The funding sponsors had no role in the design of the study; in the collection, analyses, or interpretation of data; in the writing of the manuscript; or in the decision to publish the results.

References

1. Li, Y.; Wan, X.; Cheng, L.; Wu, K. Effect of oxides on nucleation of ferrite: First principle modelling and experimental approach. *Mater. Sci. Technol.* **2016**, *32*, 88–93. [CrossRef]
2. Yuan, Q.; Xu, G.; Zhou, M.; He, B.; Hu, H. The effect of p on the microstructure and melting temperature of Fe$_2$SiO$_4$ in silicon-containing steels investigated by in situ observation. *Metals* **2017**, *7*, 37. [CrossRef]
3. Wakoh, M. Control of the size and the composition of oxide inclusions for oxides metallurgy. *Tetsu-to-Hagane* **2009**, *95*, 713–720.
4. Gao, X.; Yang, S.; Li, J.; Yang, Y.; Chattopadhyay, K.; Mclean, A. Effects of MgO nanoparticle additions on the structure and mechanical properties of continuously cast steel billets. *Metall. Mater. Trans. A* **2016**, *47B*, 461–470. [CrossRef]
5. Shibata, H.; Kimura, K.; Tanaka, T.; Kitamura, S. Mechanism of change in chemical composition of oxide inclusions in Fe-Cr Alloys deoxidized with Mn and Si by heat treatment at 1473 K. *ISIJ Int.* **2011**, *51*, 1944–1950. [CrossRef]
6. Choi, W.; Matsuura, H.; Tsukihashi, F. Changing behavior of non-metallic inclusions in solid iron deoxidized by Al-Ti addition during heating at 1473 K. *ISIJ Int.* **2011**, *51*, 1951–1956. [CrossRef]
7. Shao, X.; Wang, X.; Jiang, M.; Wang, W.; Huang, F. Effect of heat treatment conditions on shape control of large-sized elongated MnS inclusions in resulfurized free-cutting steels. *ISIJ Int.* **2011**, *51*, 1995–2001. [CrossRef]
8. Liu, C.; Yang, S.; Li, J.; Ni, H.; Zhang, X. Solid-state reaction between Fe-Al-Ca alloy and Al$_2$O$_3$-CaO-FeO oxide during heat treatment at 1473 K (1200 °C). *Metall. Mater. Trans. B* **2017**, *48B*, 1348–1357. [CrossRef]
9. Liu, C.; Ni, H.; Yang, S.; Li, J.; Ye, F. Interfacial reaction mechanism between multi-component oxides and solid alloys deoxidised by Mn and Si during heat treatment. *Ironmak. Steelmak.* **2017**, *44*. [CrossRef]
10. Koldaev, A.; D'yakonov, D.; Zaitsev, A.; Arutyunyan, N. Kinetics of the formation of nanosize niobium carbonitride precipitates in low-alloy structural steels. *Metallurgist* **2017**, *60*, 1032–1037. [CrossRef]
11. Liu, S.; Challa, V.; Natarajan, V.; Misra, R.; Sidorenko, D.; Mulholland, M.; Manohar, M.; Hartmann, J. Significant influence of carbon and niobium on the precipitation behavior and microstructural evolution and their consequent impact on mechanical properties in microalloyed steels. *Mater. Sci. Eng. A Struct.* **2017**, *683*, 70–82. [CrossRef]
12. Kang, Y.; Jung, I. Thermodynamic Modelling of pyrometallurgical oxide systems containing Mn oxides. In Proceedings of the VIII International Conference on Molten Slags, Fluxes and Salts, Santiago, Chile, 18–21 January 2009; Sánchez, M., Parra, P., Riveros, G., Díaz, C., Eds.; Quebecor World: Montreal, Chile, 2009.
13. Zhao, L. XPS study of the thermal stability of niobium surface oxides. *Acta Phys. Chim. Sin.* **1988**, *4*, 558–560.
14. Han, H.; Lin, Q.; Wei, S. Binary system 4MnO-Nb$_2$O$_5$-SiO$_2$—One of the binary sections in the ternary system MnO-Nb$_2$O$_5$-SiO$_2$. *Chin. J. Met. Sci. Technol.* **1990**, *6*, 136–138.
15. Takada, J.; Yamamoto, S.; Kikuchi, S.; Adachi, M. Determination of diffusion coefficient of oxygen in γ-iron from measurements of internal oxidation in Fe-Al alloys. *Metall. Trans. A* **1986**, *17A*, 221–229. [CrossRef]

16. Ban-Ya, S. Mathematical expression of slag-metal reactions in steelmaking process by quadratic formalism based on the regular solution model. *ISIJ Int.* **1993**, *33*, 2–11. [CrossRef]

17. Hino, M.; Ito, K. *Thermodynamic Data for Steelmaking*, 1st ed.; Tohoku University Press: Sendai, Japan, 2010; pp. 167–170.

18. Zhang, S.; Wei, S.; Wang, R.; Li, L. Electrochemical study of activity of Nb_2O_5 and MnO in Nb_2O_5-MnO-SiO_2 system. *Acta Metall. Sin.* **1990**, *26*, B11–B15.

19. Ban-Ya, S.; Tatsuhiko, E.; Mitsuo, S. *Physical Chemistry of Metals*, 1st ed.; Maruzen Press: Tokyo, Japan, 1996; pp. 204–205.

20. Turkdogan, E.T. *Physical Chemistry of High Temperature Technology*, 1st ed.; Academic Press: New York, NY, USA, 1980; p. 81.

21. Shibata, H.; Tanaka, T.; Kimura, K.; Kitamura, S. Composition change in oxide inclusions of stainless steel by heat treatment. *Ironmak. Steelmak.* **2010**, *37*, 522–528. [CrossRef]

22. Han, H.; Lin, Q.; Wei, S. Binary system MnO-Nb_2O_5. *Chin. J. Met. Sci. Technol.* **1990**, *6*, 98–102.

![metals logo] *metals*

MDPI

Article

Influence of Thickness and Chemical Composition of Hot-Rolled Bands on the Final Microstructure and Magnetic Properties of Non-Oriented Electrical Steel Sheets Subjected to Two Different Decarburizing Atmospheres

Nephtali Calvillo [1,2], Ma. de Jesús Soria [2], Armando Salinas [3], Emmanuel J. Gutiérrez [4,*], Iván A. Reyes [4] and Francisco R. Carrillo [2]

1   Altos Hornos de México S. A. B. de C. V., Prolongación Juárez, s/n, La Loma, Monclova 25770, Mexico; ncalvillor@gan.com.mx
2   Facultad de Metalurgia, Universidad Autónoma de Coahuila, Carr. 57, km. 5, Monclova 25710, Mexico; ma.soria@uadec.edu.mx (M.d.J.S.); raul.carrillo@uadec.edu.mx (F.R.C.)
3   Centro de Investigación y de Estudios Avanzados del Instituto Politécnico Nacional, Unidad Saltillo, Av. Industria Metalúrgica, 1062, Parque Industrial Saltillo-Ramos Arizpe, Ramos Arizpe 25900, Mexico; armando.salinas@cinvestav.edu.mx
4   Catedrático CONACYT-Instituto de Metalurgia de la Universidad Autónoma de San Luis Potosí, Av. Sierra Leona, 550, Lomas 2da Secc., San Luis Potosí 78210, Mexico; iareyesdo@conacyt.mx
*   Correspondence: ejgutierrezca@conacyt.mx; Tel.: +52-444-826-1450

Received: 5 May 2017; Accepted: 13 June 2017; Published: 21 June 2017

Abstract: During electrical steel processing, there are usually small variations in both chemical composition and thickness in the hot-rolled material that may lead to different magnetic properties for the same steel grade. Therefore, it is of great importance to know the effects of such variations on the final microstructure and magnetic properties of these steels. In the present investigation, samples of a specific grade of a commercial hot-rolled grain non-oriented (GNO) electrical steel were taken from different steel batches to investigate the effects of thickness and chemical composition (C, Sn, Mn and Ti) in the hot-rolled material on the final microstructure and magnetic properties (core losses and magnetic permeability) resulting from two different decarburizing annealing cycles. Hot-rolled samples were processed by cold rolling, intermediate annealing, temper-rolling and final decarburization annealing using the same processing parameters. The experimental results show that the minimum core losses and maximum magnetic permeability are obtained with the thinnest steel thickness and the largest grain size. Increasing Sb and Mn contents, and reducing the C and Ti concentrations also improve the magnetic behavior of these steels. It was also found the effect of grain size on the magnetic behavior is more significant than the one of crystallographic texture.

Keywords: non-oriented electrical steels; chemical composition; thickness; microstructure; crystallographic texture; core losses; magnetic permeability

1. Introduction

Nowadays the use of energy deals with some challenges in terms of available resources, environmental and economic impacts [1,2]. Electrical steels play an important role in the energy system including its generation, transmission, distribution and consumption.

Grain non-oriented (GNO) electrical steels are widely used in electrical equipment from the simplest domestic appliances to hybrid and pure electric vehicles [3–5]. For these applications, low core losses and high permeability are required [1–7]; these magnetic properties are strongly influenced by microstructural parameters such as residual stresses, inclusions, grain size and crystallographic texture [8–12]. Processing conditions, chemical composition and final thickness also play a crucial role in determining the magnetic behavior of these steels [13,14]. Therefore, during fabrication of GNO electrical steels, all of the above mentioned parameters must be considered to optimize the magnetic quality of these steels [15–17].

There are many previous studies regarding "the individual effects" of C, Ti, Sb, Mn and thickness on the magnetic properties of non-oriented electrical steels [18–22]. Carbon has a significant negative effect on the magnetic properties of electrical steels even when it is present in small amounts [18]. Carbon remains dissolved in Fe in the solid state even after slow cooling, and has a detrimental effect on the magnetic properties. When carbon precipitates as pearlite or free cementite its effect on the magnetic behavior is less significant but negative [18].

Darja et al. [19] studied the effects of Ti on the magnetic properties of semi-processed non-oriented electrical steel sheets. It was observed that core loss increase as the titanium content was increased. SEM analysis revealed that deterioration of the magnetic behavior was caused by the "pinning effect" of complex oxycarbonitrides, complex TiC and complex Ti(C, N).

Li et al. [20] reported that the magnetic behavior of non-oriented electrical steels can be enhanced by a texture improvement due to Sb additions. It was found that the addition of antimony inhibits the development of {111} texture and increases the intensity of Goss and {100} texture.

The influence of hot-band grain size and additions of Al and Mn on the magnetic properties of non-oriented electrical steels with 3% Si was investigated by Cardoso [21]. It was observed that the addition of manganese resulted in larger recrystallized grains after cold rolling and subsequent final annealing. Coarse grains in the hot-band and addition of Mn led to a Goss orientation component after final annealing, which resulted in an increase of the magnetic permeability.

Hunday [22] reported the influence of chemistry (variations in Si, Al, Mn, P) and hot rolling conditions (finishing temperature, the finish mill entry temperature and the transfer bar thickness) on the final magnetic properties of grain non-oriented electrical steels. The hot-rolled material was cold-rolled to a final thickness of 0.65 mm. The effects of final thickness and the amounts of Sb and Ti were not considered by Hunday [22].

Si and Al are intentionally added to these steels to improve their magnetic behavior [23]. When Si content in steel is varied, crystalline anisotropy constant K_1, electrical resistivity ρ, and saturation induction B_s change according to the following equations [12]:

$$K_1 = 5.2 - 0.5Si\% \ (10^4 \ J/m^3) \tag{1}$$

$$\rho = 12 + 11Si\% \ (\mu\Omega\cdot cm) \tag{2}$$

$$B_s = 2.16 - 0.048Si\% \tag{3}$$

The effect of Al content on these constants is similar to that of Si content. Therefore, depending on the steel grade and its application, the amount of Si + Al can vary from low [24] to high concentrations [25].

Conventionally, GNO electrical steels are processed from hot-rolling by cold rolling, continuous or batch annealing and temper rolling. Finally, laminations are subjected to a long term decarburization annealing. This later is carried out to remove residual stresses, promote grain growth and develop the optimum grain size and crystallographic texture by which the magnetic behavior of these materials can be enhanced [15–17,26].

During fabrication of these steels, there are usually small variations in both chemical composition and thickness that may lead to different microstructural characteristics, and thus can produce different magnetic properties even for the same steel grade. Therefore, it is of great importance to know the effects of such variations on the microstructure and magnetic properties of these steels.

In the present investigation, samples of a specific grade of a commercial hot-rolled low Si + Al GNO electrical steel, were taken from different steel batches to investigate the effects of thickness and chemical composition (C, Sn, Mn and Ti) on the final microstructure (grain size, crystallographic texture) and magnetic properties (core losses, magnetic permeability) of steel sheets subjected to two decarburization annealing cycles.

2. Materials and Methods

Samples of commercial hot-rolled GNO electrical steel coils were obtained from a local steelmaker (Altos Hornos de México S. A. B. de C. V., Monclova, México) considering different steel batches of a specific steel grade. Those samples that presented variations in chemical composition and thickness were selected to evaluate their effects on the magnetic properties of these steels. Although grain size and the amount of secondary phases could also vary from batch to batch. It was found experimentally that they were very similar and for this reason, they were not taken into account in the present research.

Table 1 shows the nominal chemical composition of the hot-rolled GNO electrical steel grade investigated in the present work. The experimental composition is not reported due to confidentially and privacy policies of the steel manufacturer. Chemical composition of both hot-rolled strips and annealed steel sheets samples was determined in an Espectrolab spectrometer (SPECTRO Analytical Instruments GmbH, Kleve, Germany) by optical emission spectrometry based on ASTM E403. The variations in the elements of interest in the hot-rolled bands obtained by this technique were: C, 0.030–0.035%; Sb, 0.021–0.028%; Mn, 0.49–0.62% and Ti, 0.004–0.006%. As can be observed in Table 1, the nominal chemical composition of the investigated steel grade does not consider the presence of Ti and Sb and therefore, it is of great importance to know their effects on the magnetic properties of these materials. The effect the Mn concentration is also important, since the amount of this alloying element exceeds the maximum permissible limit in this steel grade.

Table 1. Nominal chemical composition of the hot-rolled GNO electrical steel grade used in this work (wt %).

C	Si	Al	S	Mn	P	Cu	Cr	Ni	Mo
0.035	0.4	0.33	0.0033	0.55	0.013	0.34	0.023	0.033	0.019

Figure 1 shows the methodology used in the present work. The as-received hot-rolled strips, which had a thickness variability of 2.15 ± 0.15 mm, were pickled in a 4% HCL solution and cold-rolled by a thickness reduction of 70%. Cold-rolled samples were subjected to a continuous annealing at 820 °C and subsequent temper-rolling with an additional thickness reduction of 6% achieving a final thickness between 0.53 and 0.58 mm.

Longitudinal and transverse samples with respect to the rolling direction, 3 cm wide × 30 cm long, were cut and subjected to a final decarburization annealing at $T = 790$ °C for 1.5 h. Heat treatments were carried out in a laboratory KF-240-S box-type furnace (Linn High Therm GmbH, Hirschbach, Germany) under wet atmospheres. Two different gas ratios and dew points were considered to promote steel decarburization: 85% N_2-15% H_2, 21 °C and 80% N_2-20% H_2, 25 °C (Figures 1 and 2).

Texture analysis at various processing steps was carried out by orientation imaging microscopy (OIM) using a scanning electron microscope Philips XL30 (Company/Philips Electron Optics, Eindhoven, The Netherlands) equipped with a TSL-OIM system. Sample preparation for OIM included grinding to half thickness and the analyzed area for each texture map was 400 μm × 400 μm. The orientation distribution function (ODF) was calculated by the harmonic method ($L = 16$) using the TSL OIM software (Version 3.5, EDAX, Mahwah, NJ, USA, 2007).

Magnetic properties, core losses and permeability, were measured using a standardized Epstein Frame at a magnetic induction of 1.5 T and a frequency of 60 Hz according to ASTM A343. Sixteen annealed Epstein-type samples (3 cm width × 30 cm length) were used for each measurement. Microstructural evolution before and after the decarburization annealing was followed by optical microscopy Olympus GX51 (Olympus, Tokyo, Japan). Grain size was determined using comparison charts according to ASTM E112.

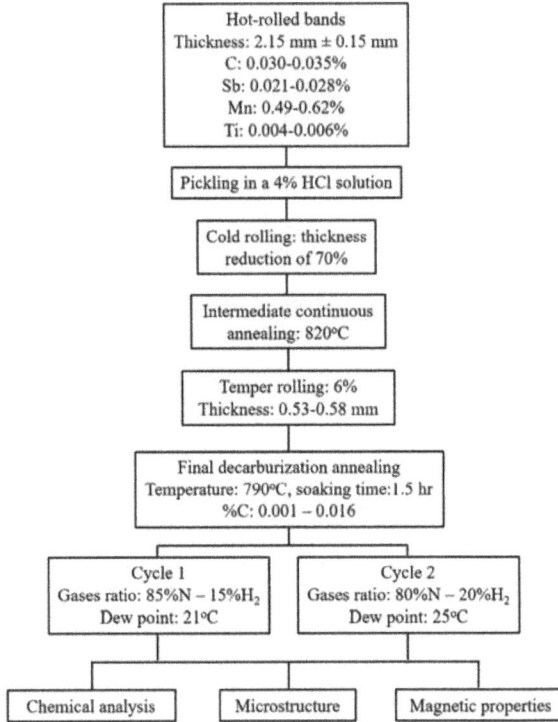

Figure 1. Methodology used in this work to evaluate the effect of chemical composition and thickness of hot-rolled bands on the final magnetic properties of grain non-oriented (GNO) electrical steels.

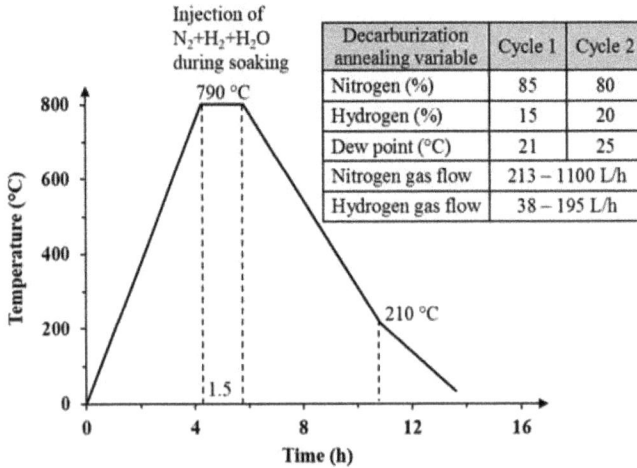

Figure 2. Schematic representation of the decarburizing thermal cycle and annealing conditions.

3. Results and Discussion

3.1. Effect of Carbon Concentration on Magnetic Properties

Figures 3 and 4 show the average values of carbon concentration, average core losses and magnetic permeability obtained from each decarburizing cycle. Important to mention is that samples subjected to Cycle 2 presented in these Figures had a thickness between 0.53–0.55 mm and those subjected to Cycle 1 had a thickness between 0.551–0.58 mm. As can be seen in Figure 3, the average value of core losses of samples subjected to both Cycle 1 and 2 is below the maximum permissible limit, which means that they both meet the core losses requirements for this steel grade. In contrast, the average magnetic permeability is above the minimum permissible value only in samples subjected to Cycle 2 (Figure 4), which means that the average value of samples subjected to Cycle 1 does not meet the requirements of this steel grade.

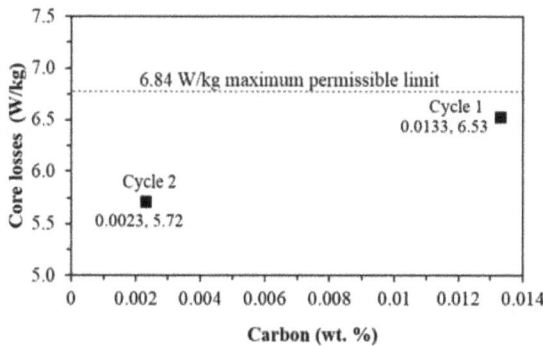

Figure 3. Effect of C content on core losses.

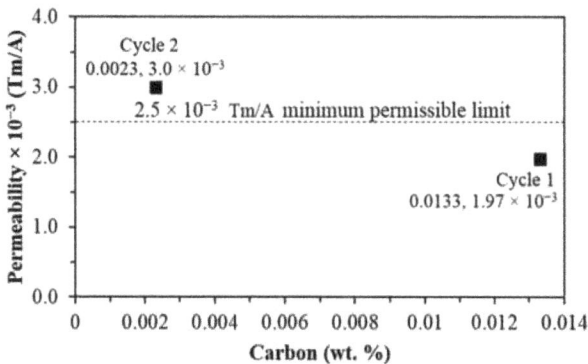

Figure 4. Effect of C concentration on magnetic permeability.

Although annealing temperature ($T = 790\ ^\circ$C) and soaking time ($t = 1.5$ h) were the same in both decarburizing cycles, there is a significant difference in the final C content, which is attributed to both thickness and changes in the decarburizing annealing conditions. Carbon content decreases from about 0.0325 wt % (average) to around 0.0133 wt % (average) when using Cycle 1 (85% N_2-15% H_2 gases ratio, and dew point of 21 $^\circ$C), and it is reduced up to 0.0023 wt % C (average) when samples are subjected to Cycle 2 (80% N_2-20% H_2 gases ratio, and dew point of 25 $^\circ$C). In addition, there exists a strong correlation between the resulting carbon concentration and core losses. The higher the carbon contents, the higher the core losses obtained. For example, core losses of 6.53 and 5.72 W/kg (average) are obtained with carbon concentrations of about 0.013% and 0.0023% C, respectively (Figure 3).

The maximum permissible core losses for the investigated GNO electrical steel grade are 6.84 W/kg, according to standard specifications. This value is indicated in Figure 3 by the dotted line for comparison with the results obtained in the present work in heat treated samples. As can be observed, the average core losses of samples subjected to Cycle 1 is very close to the maximum permissible limit, but core losses of samples annealed according to Cycle 2 are satisfactorily below the limit (Figure 3).

Magnetic permeability also changes significantly with carbon concentration, but in this case, this property increases as the amount of carbon decreases (Figure 4). Magnetic permeability varies from 1.97×10^{-3} to 3.00×10^{-3} Tm/A in samples subjected to Cycle 1 (0.013% C) and Cycle 2 (0.0023% C), respectively. Important to mention is that the minimum permissible permeability in the investigated electrical steel grade is 2.51×10^{-3} Tm/A, according to standard specifications. This value is shown in Figure 4 by the dotted line for comparison with the permeability obtained in samples subjected to the two decarburization annealing. As can be seen in this figure, while permeability of samples processed by Cycle 1 (0.013% C) is significantly lower than the minimum allowed value, the one resulting from Cycle 2 (0.0023% C) is considerably higher (Figure 4).

These results demonstrate that magnetic properties of these steels are enhanced by the reduction of C, which is favored if dew point is increased from 21 to 25 °C, and gases ratio is changed from of 85% N_2-15% H_2 to 80% N_2-20% H_2 (Figures 3 and 4).

Some researchers investigate the effect of dew point on the efficiency of decarburization in non-oriented electrical steels [27]. They found that an increase in the dew point result is a faster steel decarburization which is consistent with the results obtained in the present work. The higher decarburization rate was associated with a lower oxidation of steel with the increase of dew point. Apparently, the conditions set for Cycle 1 cause higher oxidation of steel and make carbon removal from steel difficult resulting a lower decarburization rate.

Figure 5 shows the variation of the magnetic properties, core losses and magnetic permeability, as a function of the applied decarburizing annealing cycle. Important to mention is that samples subjected to Cycle 2 presented in this Figure had a thickness between 0.53–0.55 mm and carbon concentrations less than 0.003%, while samples subjected to Cycle 1 had a thickness between 0.551–0.58 mm and carbon concentrations between 0.006% and 0.015%. It is clear that samples with lower C content result in higher permeability and lower core losses being more significant when carbon content is lower than 30 ppm (Figure 5).

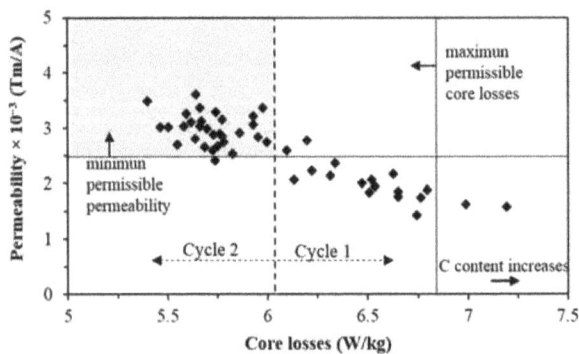

Figure 5. Relationship between magnetic permeability and core losses of decarburized GNO electrical steel.

Figure 6 presents a contour graph showing the effect of carbon content on the core losses and permeability of samples subjected to Cycle 1 and Cycle 2. Worthy of mention is that the thickness of samples selected for this Figure varied from 0.53 to 0.55 mm in those subjected to Cycle 2 and from 0.551 to 0.58 mm in samples subjected to Cycle 1. The conclusion that can be drawn from this Figure is that the combination of a lower carbon concentration and a thinner thickness result in optimum magnetic properties: lower core losses and higher permeability.

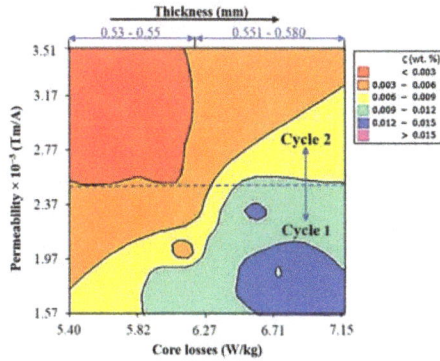

Figure 6. Contour plot showing the effect of C on magnetic properties.

The effect of carbon on the magnetic behavior of electrical steels has been extensively investigated. This alloying element reacts with other elements and forms carbides, which affect significantly the wall domain motion during magnetization of steel [8,28–30]. For this reason, samples with lower carbon concentrations result in better magnetic properties.

It can be observed in Figure 6 that when carbon content is about 0.003%, the obtained core losses vary between 5.30 and 6.07 W/kg, and permeability varies from 2.51×10^{-3} to 3.52×10^{-3} Tm/A. In contrast, with a carbon concentration of about 0.012%, core losses vary from 6.40 to 7.51 W/kg, while permeability changes in the range of 1.63×10^{-3}–2.01×10^{-3} Tm/A, respectively. The variations in the magnetic properties for a given carbon range are related to the combined effects of grain size, thickness, texture and changes in the Ti, Sb and Mn concentrations, which will be discussed in next sections.

3.2. Influence of Steel Thickness on Core Losses

Figure 7 illustrates core losses of samples subjected to Cycle 2, which showed the lowest values, the thickness in these samples varied from 0.53 to 0.58 mm. As can be seen, core losses increase as the steel thickness is increased. The total core losses are considered as the sum of hysteresis losses (W_h) and eddy current losses (W_e) [12,15,28,31]. These later, are represented by the following equation [31]:

$$W_e = k\frac{(tfB)^2}{\rho} \tag{4}$$

where k is a constant, t is the sheet thickness, f is the frequency, B is the magnetic flux density, and ρ is the resistivity.

According to Equation (4), reducing the steel thickness t causes a decrease of eddy current losses and contributes to minimize the total core losses, which is in agreement with the results obtained in this work. Eddy current losses are determined by flux per lamination and resistance of the lamination and they are, therefore, dependent on lamination thickness [14,31,32]. If steel thickness is reduced, eddy current losses and consequently total core losses are also decreased as observed in Figure 7.

The combined effects of carbon content and thickness on the magnetic properties of samples subjected to Cycle 1 and Cycle 2 are illustrated in Figure 8. As can be seen, higher thickness and higher carbon concentrations result in higher core losses. It can be concluded then, according to this figure, that to achieve the best performance in terms of magnetic properties, electrical steels must be processed with the minimum possible thickness and the lowest carbon concentration.

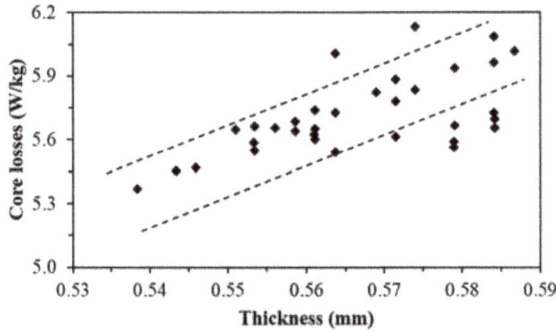

Figure 7. Effect of thickness on core losses.

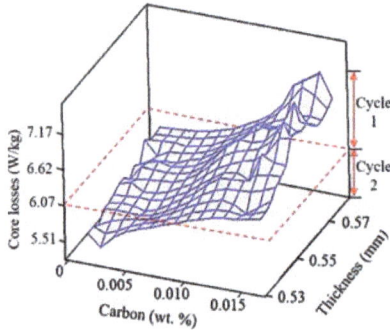

Figure 8. Effects of carbon and steel thickness on core losses.

3.3. Effect of Ti, Sb and Mn Concentrations on Core Losses

In order to evaluate the effects of Ti, Sb and Mn on the magnetic properties of annealed samples, only those samples with lower carbon concentration and smaller thickness were considered. This selection was made considering the results presented in previous sections. Figure 9 shows the variation of core losses as a function of Ti, Sn and Mn concentrations in samples with a thickness between 0.53 and 0.55 mm and carbon concentrations lower than 0.003%, which correspond to samples subjected to Cycle 2.

As can be observed, these three elements have a significant influence on the magnetic properties. An increment in the amount of Ti causes an increase of core losses, however, this property is reduced by increasing the content of Sb and Mn resulting in an enhancement of the magnetic quality (Figure 9).

Alloying elements such as Si, Al, Sn, Sb, Mn and Ti (among others), increase the resistivity of steels, ρ, and therefore, according to Equation (4), could lead to a reduction of eddy current losses and consequently to total core losses [11,33,34]. For this reason, these elements are added to satisfy the required magnetic properties of a specific electrical steel grade.

Although Ti increases the steel resistivity by which the magnetic behavior could be enhanced, the results of the present investigation show that Ti has a detrimental effect. It has been reported that Ti has a strong affinity for carbon and nitrogen and forms very stable carbonitrides. Titanium carbonitrides affect the development of favorable textures, and wall domain motion during magnetization of steel sheets, which in turn has a detrimental effect on magnetic properties [34].

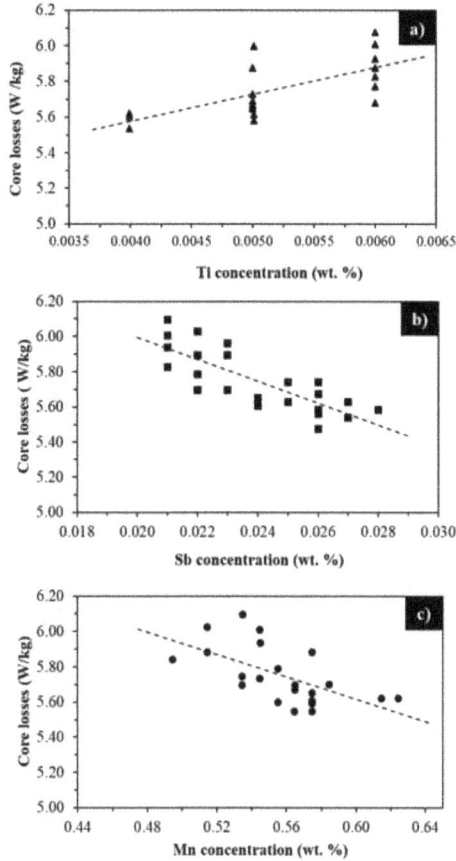

Figure 9. Effect of alloying elements on core losses: (**a**) Ti, (**b**) Sb and (**c**) Mn.

3.4. Effect of Grain Size on Magnetic Properties

Figure 10 shows a comparison between the microstructure of the temper-rolled steel before and after the decarburization annealing. As can be seen, a grain size ASTM 8 is obtained prior to decarburization (Figure 10a). However, after decarburization annealing (Cycle 2) grain size increases changing from ASTM 8 to ASTM 3 (Figure 10b). This result confirms that, the appropriate combination of small plastic deformation during temper-rolling and optimum decarburizing annealing conditions (gases ratio and dew point) favor grain growth during decarburization annealing. This result indicates that magnetic properties of the experimental electrical steels can also be enhanced by increasing grain size. Grain boundaries act as barriers for the domain wall motion during the magnetization process of steel [12]. Therefore, increasing grain size causes a reduction of the amount of obstacles to the magnetization process resulting in an enhancement of the magnetic behavior of these steels. It is well known that when grain size increases, hysteresis losses decrease, while eddy current losses increase. Therefore, there exists an optimum grain size that minimizes the total core losses (hysteresis losses + eddy current losses) [12,15,28,31].

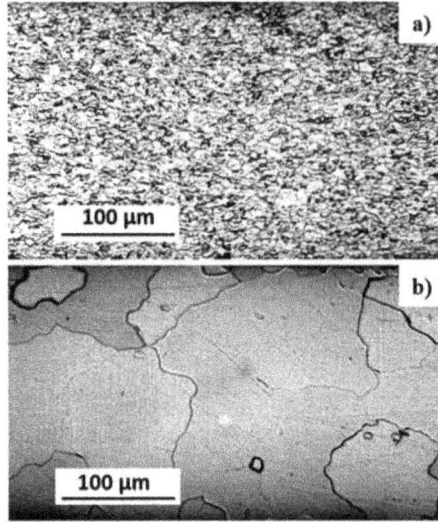

Figure 10. Microstructure of GNO electrical steel: (**a**) before and (**b**) after the decarburization annealing.

Core loss (at peak induction of 1.5 T, and at frequency of 60 Hz, sinusoidal) is considered an important parameter when deciding on a suitable steel for a particular application. Some associations provide some standards to characterize soft magnetic materials under sinusoidal supplies like the Epstein frame method [35,36] used in the present work. The Epstein frame containing the test specimens (longitudinal and transverse) to be measured constitutes a transformer for which the total losses are measured by the wattmeter method. The total core losses can be obtained by the following equation [36]:

$$P_t = P_h + P_p \left[\frac{F}{1.111} \right]^2 \tag{5}$$

where P_t = measured specific total losses, in watts per kilogram, P_h = apparent hysteresis losses, in watts per kilogram, P_p = apparent eddy current losses, in watt per kilogram and P_t = measured form factor [36].

It has also been reported that knowledge of the linear losses (longitudinal and transverse) is sufficient for designers to adequately estimate the behavior in a given machine design [8].

On the other hand, a number of theories describe the excess losses as the third component of the total core losses of fully finished, non-oriented electrical steel [37]. This third component was added as a result of the deviation between the measured and the calculated values [37]. Many authors have studied the mechanisms of their formation and there are many contradictory theories that can be found in the literature. Whereas Bertotti ascribes the excess losses to domain-wall processes, there are several authors who correlate the excess losses with the hysteresis losses and describe them as frequency-dependent hysteresis losses [37].

Although it is important to be able to actually measure excess loss as demonstrated in an extensively research literature regarding this topic, some problems abound [8]. For instance, a pure constant amplitude B vector cannot be created or rotated. This can be imposed by computer methods and powerful drive amplifiers, but in real machines such a regime never identically applies [8]. Losses are noted to differ if vector rotation is clockwise or anticlockwise, due probably to unevenness in grain texture. The directions of applied magnetizing field and attained magnetization vectors do not coincide and angular separation varies as the B vector moves [8].

According to the information above mentioned the excess losses, according to Bertotti's analysis, were not considered.

3.5. Effect of Decarburization Annealing on Crystallographic Texture

Analyses by electron backscatter diffraction (EBSD) were carried out to investigate if there was an effect of the decarburizing Cycle on the final texture. To this end the samples used for these analyses were those in the temper-rolled condition, and after being subjected to Cycle 1 and Cycle 2, which had carbon contents of 0.035%, 0.013% and 0.002%, respectively. These samples also had the lowest Ti concentration and the highest Sb and Mn contents. In the case of samples subjected to decarburization annealing they had the same thickness.

Figure 11 shows the orientation maps obtained in samples without decarburization (temper-rolled condition) and with decarburization annealing according to Cycle 1 and Cycle 2. Inverse pole figure (IPF) orientation maps use a basic RGB (red, green and blue) coloring scheme, fit to an inverse pole figure. For cubic phases, full red, green, and blue are assigned to grains whose <100>, <110> or <111> axes, respectively, are parallel to the projection direction of the IPF. Intermediate orientations are colored by an RGB mixture of the primary components. It is noteworthy that IPF has its own limitations; most notable is the coloring of pixels only by the projection-parallel crystallographic axis, independent of rotation about that axis [38,39]. Thus, grains with identical axes parallel to a specified IPF projection direction will have the same color in the IPF-based scheme, but may be in significantly different orientations. For example, two grains with <100> parallel to the surface normal are both colored red, but possess 30° of relative rotation about that axis. IPF-based orientation maps are most useful for displaying materials with strong fiber-textures and for understanding preferred orientations parallel to a sample direction of interest [38,39].

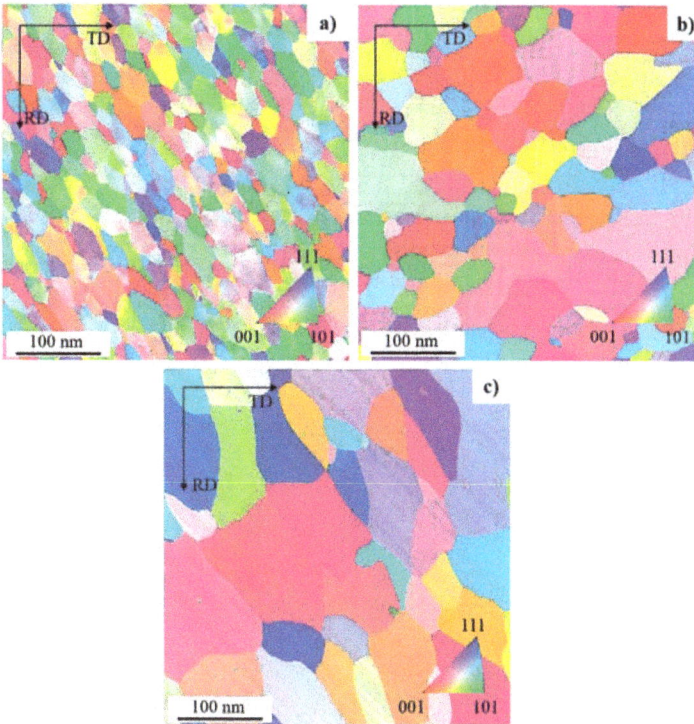

Figure 11. Electron backscatter diffraction (EBSD) color coded maps of the inverse pole figure [001] of GNO electrical steels obtained after: (**a**) temper-rolling, and decarburization annealing (**b**) Cycle 1 and (**c**) Cycle 2.

According to this Figure, it can be observed that samples exhibit significant differences in crystallographic texture developed, which means that decarburization annealing can modify this parameter.

On the other hand, ODF's are 3D representations of "Euler space", with the three Euler angles (φ_1, ϕ, φ_2) that describe the orientation of a crystal forming the axes. A crystal's orientation is represented as a point within that space. The ODF may more clearly reveal component and fiber textures. The best description of a crystallographic texture in polycrystalline materials where a crystal with a volume V_1 has an orientation g_1, a crystal with volume V_2 has an orientation g_2 and so on, can be then described quantitatively by the orientation distribution function (ODF) [38,39]. This function can be represented in the Euler space in three dimensions, however, it is generally interpreted in two dimensions maintaining constant one of the three angles (φ_1, ϕ, φ_2) [38,39].

Figure 12 shows the $\varphi_2 = 45°$ section of the ODF of temper-rolled samples (Figure 12a) and decarburized samples (Figure 12b). In this figure are also shown the ideal texture components of BCC materials [40]. The main texture components that relate to good magnetic quality in annealed non-oriented electrical steels are the cube {001}<001> and rotated cube {001}<011> [28]. In contrast, the so-called γ-fiber texture, represented by {111}<uvw>, is very detrimental for electrical steels because the lowest permeability axis, <111>, is parallel to the normal direction of the sheet and should therefore be avoided [28]. The texture components resulting from temper rolling are: $(115)[0\bar{5}1]$, $(225)[\bar{2}32]$, $(773)[1\bar{1}0]$, $(110)[3\bar{3}2]$ and $(110)[4\bar{4}7]$, represented in Figure 12a by letters A, B, C, D and E, respectively. The most important observation is the presence of orientations around the $(115)[0\bar{5}1]$ texture component near to the ideal cube texture, and the absence of texture components belonging to γ-fiber texture (Figure 12a).

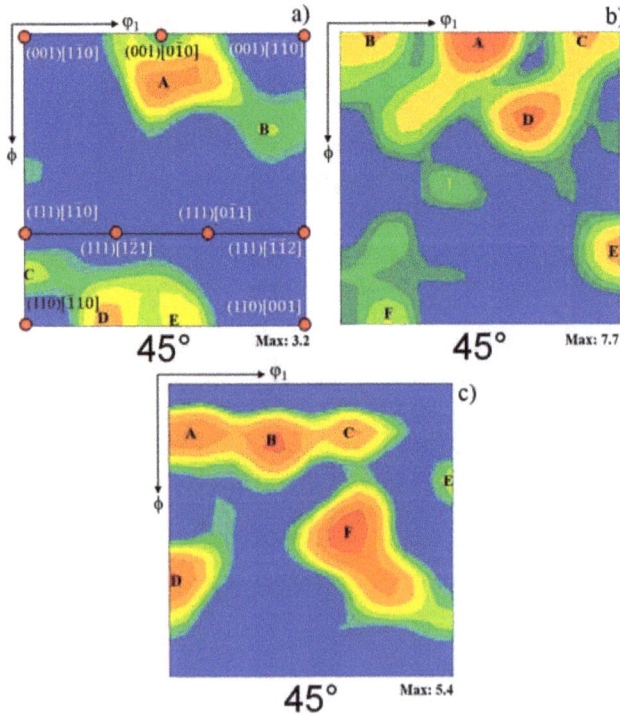

Figure 12. $\varphi_2 = 45°$ section of the orientation distribution function of GNO electrical steels after: (**a**) temper-rolling, and decarburization annealing (**b**) Cycle 1 and (**c**) Cycle 2.

The texture developed in samples subjected to Cycle 1 is characterized by components $(001)\,[0\bar{1}0]$, $(001)\,[2\bar{3}0]$, $(001)\,[\bar{2}30]$, $(113)\,[\bar{1}\bar{5}2]$, $(553)\,[\bar{3}\bar{3}10]$ and $(110)\,[8\bar{8}3]$, repsented by letters A, B, C, D, E and F in Figure 12b, respectively. The components $(001)\,[0\bar{1}0]$, $(001)\,[2\bar{3}0]$ and $(001)\,[\bar{2}30]$ which have the highest intensity, belong to the so called θ-fiber which is the more beneficial texture to optimize the magnetic behavior of these steels.

In the case of samples subjected to Cycle 2 the main texture components developed are: $(115)\,[1\bar{1}0]$, $(114)\,[4\bar{1}63]$, $(115)\,[\bar{6}\,\bar{29}\,7]$, $(554)\,[1\bar{1}0]$, $(225)\,[\bar{5}\bar{5}4]$, $(334)\,[\bar{1}\,\bar{23}\,18]$, represented in Figure 12c by letters A, B, C, D, E and F, respectively. This annealing Cycle promotes the development of orientations near to the ideal $(111)\,[1\bar{1}0]$ texture component, which as mentioned above belongs to the most detrimental γ-fiber texture. Cube and rotated cube textures, which are the most beneficial textures for optimizing the magnetic behavior of these steels are not observed after decarburization annealing conducted according to Cycle 2.

As can be seen in Figure 11, the size and volume fraction of grains with their <111> axes parallel to the <001> sample direction is small (Figure 11a). After decarburization annealing of Cycle 1 the size and the amount of these grains increase (Figure 11b) resulting in an increase of the intensity of grains with orientations along the θ-fiber texture (Figure 12a,b). In contrast, samples subjected to Cycle 2 show larger grain size (Figure 11c), a reduction of the volume fraction of grains with orientations belonging to the θ-fiber texture and an increase of the volume fraction of grains with orientations near to the ideal γ-fiber texture.

It is clear that the reduction in carbon concentration during decarburization annealing favors grain growth. The samples with higher carbon content exhibit lower size (Figure 11a), while samples with lower carbon content have a larger grain size (Figure 11b,c). In addition, these results suggest that the decarburizing Cycle not only affects the final grain size, but also the final texture.

If grain size is favored by carbon removal, thus the results obtained suggest that the conditions established in Cycle 2 favored carbon removal and consequently grain growth.

Some researchers investigate the effect of dew point on the efficiency of decarburization [27]. They found that an increase in the dew point result is a faster steel decarburization which is consistent with the results obtained in the present work. They relate the higher was associated to a lower oxidation of steel with the increase of dew point [27]. Apparently, the conditions set for Cycle 1 cause higher oxidation of steel, which retards carbon removal and results in grains with smaller size (compare Figure 11b,c).

Variations in the resulting texture can be explained as a function of the decarburization rate. Samples subjected to Cycle 1 which result in a slower carbon removal present higher intensities of components belonging to θ-fiber. The presence of these components in non-oriented electrical steels subjected to low plastic deformations and subsequent thermal treatment has been explained in terms of the mechanism of strain induced grain boundary migration (SIBM) [41]. Temper-rolling involves stored energy accumulation, which varies with the orientation of the rolling plane according to the sequence $E_{(110)} > E_{(111)} > E_{(100)}$ [41].

Apparently, grains with their planes (001) parallel to the steel surface growth during heat treatment due to their low stored energy according to the ratio before mentioned. The increase of components near to the γ-fiber texture in samples subjected to Cycle 2, could be probably related to the higher mobility of grains <111>//ND [41].

Although tempered-rolled samples and samples subjected to Cycle 1 exhibited a better texture to optimize the magnetic properties of the experimental steels (considering the higher density of texture components belonging to the θ-fiber texture and the absence of the so-called γ-fiber texture), samples subjected to Cycle 2 resulted in better magnetic properties.

Therefore, it can be concluded that the effect of grain size on the magnetic behavior is more significant than the one of texture. It has been reported that grain bourdaries act as barriers to wall-motion.

According to Figures 3–6 and Figure 8, magnetic properties are enhanced by the reduction of C concentration. Additionally, the microstructures present in Figure 10 show evidence that decarburization favors grain growth. Therefore, it can be concluded that for the experimental steel, the combined effects of C concentration and grain size are more significant than texture.

On the other hand, it has been reported that additions of Sb favor the development of (100) and (110) texture components at the expense of (111) components resulting in an improvement of the magnetic quality of these steels [20,42]. Manganese is supposed to retards the development of grains with (111) components and favors the development of grains having (100), (200) and (110) texture components [21,43], leading to an improvement of the magnetic properties of these steels.

As can be observed in Figure 12a, samples with temper rolling exhibit a set of orientation components near to the ideal cube texture, $(001)[0\bar{1}0]$, and rotated Goss $(110)[\bar{1}10]$ texture components which is consistent with the additions of Sb in the experimental steel, however, after decarburization a completely different behavior is observed suggesting that apart from chemical composition, decarburization conditions also play an important role in determining the final crystallographic textures.

4. Conclusions

From the results obtained in this investigation, it can be concluded that:

(1) Magnetic properties of the experimental steels are enhanced by the reduction of C, which is favored if dew point is increased from 21 to 25 °C, and gases ratio is changed from of 85% N_2-15% H_2 to 80% N_2-20% H_2.

(2) Grain size is also dependent on decarburizing annealing conditions, it changes from ASTM 8 in the temper rolled condition to about 4 and 3 when applying Cycle 1 (21 °C, 85% N_2-15% H_2) and Cycle 2 (25 °C, 80% N_2-20% H_2), respectively.

(3) Crystallographic texture of temper-rolled samples is characterized by the presence of components near to the ideal cube texture, and the absence of components belonging to γ-fiber texture. Cycle 1 promotes the development of components $(001)[0\bar{1}0]$, $(001)[2\bar{3}0]$ and $(001)[\bar{2}\bar{3}0]$ which belong to the so called θ-fiber texture, while Cycle 2 favors the development of orientations near to the ideal $(111)[1\bar{1}0]$ texture component, but in this case cube and rotated cube textures, were not developed.

(4) Thinner thicknesses, higher concentrations of Sb and Mn, and lower C and Ti contents lead to lower core losses and higher permeability enhancing the magnetic behavior of the experimental GNO electrical steels. Therefore, the best magnetic properties were obtained when applying Cycle 2 independently of the crystallographic texture developed.

(5) Additions of Mn and Sb increase the resistivity reducing the total core losses. Although Ti was expected to reduce core losses since it also increases the electrical resistivity, an opposite behavior was observed. Apparently, the strong affinity of Ti for carbon and nitrogen promotes the formation of stable carbonitrides which affect the wall domain motion during magnetization causing a detrimental effect on magnetic properties.

Acknowledgments: The authors of the present investigation appreciate all the facilities to carry out the experimental work at Altos Hornos de México S. A. B. de C. V. (AHMSA), Center for Research and Advances Studies of The National Politechnique Institute (CINVESTAV-Saltillo) and at The Institute of Metallurgy of The Autonomous University of San Luis Potosi (IM-UASLP).

Author Contributions: Nepthali Calvillo designed and performed the experiments and characterization of microstructure by optical microscopy; Ma. de Jesús Soria and Francisco R. Carrillo discuss the results and wrote the first draft of the manuscript; Armando Salinas realized the texture measurements by electron backscatter diffraction; Emmanuel J. Gutiérrez and Ivan A. Reyes measure the magnetic properties, review and correct the manuscript. All authors revised and approved the final version of the manuscript.

Conflicts of Interest: The authors declare no conflict of interest.

References

1. Moses, A.J. Energy efficient electrical steels: magnetic performance prediction and optimization. *Scr. Mater.* **2012**, *67*, 560–565. [CrossRef]
2. Gautam, J. Control of Surface Graded Transformation Textures in Steels for Magnetic Flux Carrying Applications. Ph.D. Thesis, Delft University of Technology, Delft, The Netherlands, March 2011.
3. Fischer, O.; Schneider, J. Influence of deformation process on the improvement of non-oriented electrical steel. *J. Magn. Magn. Mater.* **2003**, *254*, 302–306. [CrossRef]
4. Jacobs, S.; Hectors, D.; Henrotte, F.; Hafner, M.; Herranz, M.; Hameyer, K.; Goes, P. Magnetic material optimization for hybrid vehicle PMSM drive. *World Electr. Veh. J.* **2009**, *3*, 1–9.
5. Sidor, Y.; Kovac, F. Microstructural aspects of grain growth kinetics in non-oriented electrical steels. *Mater. Charact.* **2005**, *55*, 1–11. [CrossRef]
6. Bacaltchuk, C.M.B.; Castello, G.A.; Ebrahimi, M. Effect of magnetic field applied during secondary annealing on texture and grain size of silicon steel. *Scr. Mater.* **2003**, *48*, 1343–1347. [CrossRef]
7. Park, J.T.; Szpunar, J.A. Effect of initial grain size on texture evolution and magnetic properties in nonoriented electrical steels. *J. Magn. Magn. Mater.* **2009**, *321*, 1928–1932. [CrossRef]
8. Beckley, P. *Electrical Steels for Rotating Machines*, 1st ed.; The Institution of Engineering and Technology: London, UK, 2002; pp. 69–72.
9. Chang, S.K.; Huang, W.Y. Texture effect on magnetic properties by alloying specific elements in non-grain oriented silicon steels. *ISJ Int.* **2005**, *45*, 918–922. [CrossRef]
10. Tao, L.H.; Schneider, J.; Long, L.H.; Sun, Y.; Gao, F.; Hu, L.H.; Yu, S.H.; Li, L.; Qiao, G.D.; Yu, L.Z.; et al. Fabrication of high permeability non-oriented electrical steels by increasing <001> recrystallization texture using compacted strip casting processes. *J. Magn. Magn. Mater.* **2015**, *374*, 577–586.
11. Landgraf, F.J.G.; Emura, M.; Teixeira, J.C.; de Campos, M.F. Effect of grain size, deformation, aging and anisotropy on hysteresis loss of electrical steels. *J. Mag. Mag. Mater.* **2000**, *215*, 97–99. [CrossRef]
12. Matsumura, K.; Fukuda, B. Recent developments of non-oriented electrical steel sheets. *IEEE Trans. Magn.* **1984**, *5*, 1533–1537. [CrossRef]
13. Huňady, J.; Černík, M.; Hilinski, E.J.; Predmerský, M.; Magurova, A. Influence of chemistry and hot rolling conditions on high permeability non-grain oriented silicon steel. *J. Magn. Magn. Mater.* **2006**, *304*, 620–623. [CrossRef]
14. Tanaka, I.; Yashiki, H. Magnetic properties and recrystallization texture of phosphorus-added non-oriented electrical steel sheets. *J. Magn. Magn. Mater.* **2006**, *304*, 611–613. [CrossRef]
15. Salinas, J.; Salinas, A.; Gutiérrez, E.J.; Deaquino, L. Effect of processing conditions on the final microstructure and magnetic properties in non-oriented electrical steels. *J. Magn. Magn. Mater.* **2016**, *406*, 159–165. [CrossRef]
16. Chang, S.K. Texture effects on magnetic properties in high-alloyed non-oriented electrical steels. *Met. Sci. Heat Treat.* **2007**, *49*, 569–573. [CrossRef]
17. Gutiérrez, E.; Salinas, A.; Deaquino, L.; Márquez, F. High temperature oxidation and its effects on microstructural changes of hot-rolled low carbon non-oriented electrical steels during air annealing. *Oxid. Met.* **2015**, *83*, 237–252. [CrossRef]
18. Ramadam, R.; Ibrahim, S.A.; Farag, M.; Elzatahry, A.A.; Es-Saheb, M.H. Processing optimization and characterization of magnetic non-oriented electrical silicon steel. *Int. J. Electrochem. Sci.* **2012**, *7*, 3242–3251.
19. Darja, S.; Monika, J.; Jalklic, A.; Cop, A. Correlation of titanium content and core loss in non-oriented electrical steel sheets. *Metalurgija* **2010**, *49*, 37–40.
20. Li, N.; Xiang, L.; Zhao, P. Effect of antimony on the structure, texture and magnetic properties of high efficiency non-oriented electrical steel. *Adv. Mater. Res.* **2013**, *602*, 435–440. [CrossRef]
21. Cardoso, R.F.D.A.; Brandao, L.; Cunha, M.A.D. Influence of grain size and additions of Al and Mn on the magnetic properties of non-oriented electrical steels with 3 wt % Si. *Mater. Res.* **2008**, *11*, 51–55. [CrossRef]
22. Hunday, J.; Cernik, M.; Hilinski, E.; Predmersky, M.; Magurova, A. Influence of chemistry and hot rolling conditions on high permeability non-grain oriented silicon steel. *Metals* **2005**, *15*, 17–23.
23. Darja, P. Non-oriented electrical steel sheets. *Mater. Technol.* **2010**, *44*, 317–325.
24. Chaudhury, A. Low silicon non-grain-oriented electrical steel: Linking magnetic properties with metallurgical factors. *J. Magn. Magn. Mater.* **2007**, *313*, 21–28. [CrossRef]

25. Lee, H.-Y.; Hsiao, I.-C.; Tsai, M.-C. Texture Improvement of 3% Si non-oriented electrical steel. *China Steel Tech. Rep.* **2015**, *28*, 1–5.

26. Marra, K.M.; Alvarenga, E.D.A.; Buono, V.T.L. Decarburization kinetics during annealing of a semi-processed electrical steel. *ISIJ Int.* **2004**, *44*, 618–622. [CrossRef]

27. Darja, S.P.; Monika, J.; Hans, J.G.; Vasilij, P. Decarburization of non-oriented electrical steel sheets doped with selenium. *Mater. Thehnol.* **2001**, *35*, 337–341.

28. Gutiérrez, E.; Salinas, A. Effect of annealing prior to cold rolling on magnetic and mechanical properties of low carbon non-oriented electrical steels. *J. Magn. Magn. Mater.* **2011**, *323*, 2524–2530. [CrossRef]

29. Jenkins, K.; Lindenmo, M. Precipitates in electrical steels. *J. Magn. Magn. Mater.* **2008**, *320*, 2423–2429. [CrossRef]

30. PremKumar, R.; Samajdar, I.; Viswanathan, N.N.; Singal, V.; Seshadri, V. Relative effect(s) of texture and grain size on magnetic properties in a low silicon non-grain oriented electrical steel. *J. Magn. Magn. Mater.* **2003**, *264*, 75–85. [CrossRef]

31. Honda, A.; Obata, Y.; Okamura, S. History and development of non-oriented electrical steel in Kawasaki steel. *Kawasaki Steel Tech. Rep.* **1998**, *39*, 13–19.

32. De Campos, M.; Yonamine, T.; Fukuhara, M.; Landgraf, F.J.G.; Achete, C.A.; Missell, F.P. Effect of frequency on the iron losses of 0.5% and 1.5% Si nonoriented electrical steels. *IEEE Trans. Magn.* **2006**, *42*, 2812–2814. [CrossRef]

33. Barros, J.; Ros-Yanez, T.; Vandenbossche, L.; Dupré, L.; Melkebeek, J.; Houbaert, Y. The effect of Si and Al concentration gradients on the mechanical and magnetic properties of electrical steel. *J. Magn. Magn. Mater.* **2005**, *290*, 1457–1460. [CrossRef]

34. Steniner, D.; Jenko, M.; Jaklič, A.; Čop, A. The correlation between the content of titanium in steel and electromagnetic properties of oriented electrical steel. *Metalurgija* **2010**, *49*, 37–40.

35. Lotten, T.M.; Pragasen, P. Core losses in motor laminations exposed to high-frequency or nonsinusoidal excitation. *IEEE Trans. Ind. Appl.* **2004**, *40*, 1325–1332.

36. Euraopean coal and steel community. *Methods for Determining the Magnetic Properties of Magnetic Steel Sheet and Strip with 25 cm Epstein Frame*; European Committee for Standardization: Luxembourg, Luxembourg, 1975.

37. Gasper, N.; Janko, K.; Ales, N.; Darja, S.P. Correlation between the excess losses and the relative permeability in fully finished non-oriented electrical steels. *Mater. Tehnol.* **2014**, *48*, 997–1001.

38. Engler, O.; Randle, V. *Introduction to Texture Analysis: Macrotexture, Microtexture, and Orientation Mapping*, 1st ed.; CRC Press: Boca Raton, FL, USA, 2009.

39. Bunge, H.J. *Texture Analysis in Materials Science: Mathematical Methods*, 1st ed.; Butterworth-Heinemann: Berlin, Germany, 2013.

40. Castro Cerda, F.M.; Kestens, L.A.; Monsalve, A.; Petrov, R.H. The effect of ultrafast heating in cold-rolled low carbon steel: Recrystallization and texture evolution. *Metals* **2016**, *6*, 288. [CrossRef]

41. Kovac, F.; Stoyka, V.; Petryshynets, I. Strain induced grain growth in non-oriented electrical steel. *J. Magn. Magn. Mater.* **2008**, *320*, 627–630. [CrossRef]

42. Chang, S.K. Magnetic anisotropies and textures in high-alloyed non-oriented electrical steels. *ISIJ Int.* **2007**, *47*, 466–471. [CrossRef]

43. Nakayama, T.; Honjou, N.; Minaga, T.; Yashiki, H. Effects of manganese and sulfur contents and slab reheating temperatures on the magnetic properties of non-oriented semi-processed electrical steel sheets. *J. Magn. Magn. Mater.* **2001**, *234*, 55–61. [CrossRef]

metals

MDPI

Article

Bainitic Transformation and Properties of Low Carbon Carbide-Free Bainitic Steels with Cr Addition

Mingxing Zhou, Guang Xu *, Junyu Tian, Haijiang Hu and Qing Yuan

The State Key Laboratory of Refractories and Metallurgy, Key Laboratory for Ferrous Metallurgy and Resources Utilization of Ministry of Education, Wuhan University of Science and Technology, Wuhan 430081, China; kdmingxing@163.com (M.Z.); 13164178028@163.com (J.T.); hhjsunny@sina.com (H.H.); 15994235997@163.com (Q.Y.)
* Correspondence: xuguang@wust.edu.cn; Tel.: +86-027-6886-2813

Received: 28 June 2017; Accepted: 6 July 2017; Published: 10 July 2017

Abstract: Two low carbon carbide-free bainitic steels (with and without Cr addition) were designed, and each steel was treated by two kinds of heat treatment procedure (austempering and continuous cooling). The effects of Cr addition on bainitic transformation, microstructure, and properties of low carbon bainitic steels were investigated by dilatometry, metallography, X-ray diffraction, and a tensile test. The results show that Cr addition hinders the isothermal bainitic transformation, and this effect is more significant at higher transformation temperatures. In addition, Cr addition increases the tensile strength and elongation simultaneously for austempering treatment at a lower temperature. However, when the austempering temperature is higher, the strength increases and the elongation obviously decreases by Cr addition, resulting in the decrease in the product of tensile strength and elongation. Meanwhile, the austempering temperature should be lower in Cr-added steel than that in Cr-free steel in order to obtain better comprehensive properties. Moreover, for the continuous cooling treatment in the present study, the product of tensile strength and elongation significantly decreases with Cr addition due to more amounts of martensite.

Keywords: chromium; bainitic transformation; microstructure; mechanical properties; retained austenite

1. Introduction

Silicon (Si), manganese (Mn), molybdenum (Mo), and chromium (Cr) are important alloying elements in advanced high strength steels (AHSS), such as dual phases (DP) steel, quenching and partitioning (Q&P) steel, quenching-partitioning-tempering (QPT) steel, and carbide-free bainitic steel [1–5]. The amount of these elements in AHSS influences not only the transformation behavior, but also the microstructure and properties of the steel. Therefore, the optimization of the chemical composition in AHSS is always an interesting subject [6–8].

There is a continuous interest in the effects of Cr in AHSS. Han et al. [9] reported that DP steel with more Cr presents better elongation and a lower yield ratio. Li et al. [10] investigated a Cr-bearing low carbon steel treated by a Q&P process. A lath martensite, retained austenite, and lower bainite triplex microstructure was obtained, in which the tensile strength ranges from 1200 MPa to 1300 MPa and the total elongation ranges from 10% to 15%. Jirková et al. [11] found that the yield strength and elongation increase with Cr content in a Q&P steel. Ou et al. [6] investigated the microstructure and properties of QPT steels with Cr addition from a 1.35 wt. % to a 1.65 wt. %. They found that with increasing Cr content, the yield strength decreases, and the elongation first increases and subsequently decreases. Optimal microstructures and properties of steels were obtained at 1.45 wt. % and 1.55 wt. % Cr, respectively, in their study. In addition, Cr is often added in bainitic steel to increase its hardenability, so that a relatively smaller cooling rate can be used to avoid high temperature ferritic transformation and bainite can be obtained more easily [12–16]. However, this does not mean that Cr promotes bainitic transformation itself, because the increase in hardenability is mainly caused by the inhabitation of Cr on ferritic transformation. So far, only a few studies have discussed the effect of Cr on bainitic transformation. Bracke and Xu [17] stated that Cr reduces the kinetics of lower bainitic transformation in a continuous cooling process. But the effect of Cr on isothermal bainitic transformation, which is an important process to produce advanced high strength bainitic steel, is rarely reported. Therefore, it is necessary to further investigate the effects of Cr on bainitic transformation, microstructure, and the properties of bainitic steels.

In the authors' previous study, the effect of Cr on isothermal bainitic transformation was investigated [18]. However, high temperature ferritic transformation occured before isothermal bainitic transformation due to a lower cooling rate and only one heat treatment procedure was used. In the present study, two kinds of heat treatment procedure are designed, i.e., isothermal bainitic transformation (including three transformation temperatures) and continuous transformation. The effects of Cr on bainitic transformation, microstructure, and the mechanical properties of low carbon bainitic steels were investigated. The present study is more comprehensive, and some new findings different from the authors' previous study are obtained. The results are useful to the optimization of the chemical composition of low carbon bainitic steels.

2. Materials and Methods

2.1. Materials

Two low carbon bainitic steels with different Cr contents were designed and their compositions are given in Table 1. The steels were refined and cast in the form of 50 kg ingots using a laboratory-scale vacuum furnace followed by hot-rolling and then air-cooling to room temperature.

Table 1. Chemical compositions of two steels.

Steels	C	Si	Mn	Cr	Mo	N	P	S
Cr-free	0.218	1.831	2.021	/	0.227	<0.003	<0.006	<0.003
Cr-added	0.221	1.792	1.983	1.002	0.229	<0.003	<0.006	<0.003

2.2. Thermal Simulation Experiments

Thermal simulation experiments were conducted on a Gleeble 3500 simulator (DSI, New York, NY, USA). Cylindrical specimens with a diameter of 6 mm and a length of 70 mm were used. As shown in Figure 1, two kinds of heat treatment were designed for the two steels, i.e., austempering and continuous cooling. During austempering treatment, the specimens were heated to 1000 °C to obtain full austenite structure, followed by fast cooling to the isothermal bainitic transformation temperatures (400 °C, 430 °C, and 450 °C) at 30 °C/s. The bainite start temperature (B_S) and martensite start temperature (M_S) for Cr-free steel are calculated to be 524 °C and 376 °C, respectively, using the MUCG 83 program developed by Bhadeshia at Cambridge University, and they are 501 °C and 353 °C, respectively, for Cr-added steel, so that the austempering temperatures are set as 400, 430, and 450 °C. Meanwhile, 1000 °C is higher than the Ac_3 point, and the cooling rate of 30 °C/s is fast enough to avoid ferritic and pearlitic transformations (Section 3.1.1). In addition, during continuous cooling treatment, the specimens were first heated to 1000 °C and then held for 900 s, followed by cooling to room temperature at a rate of 0.5 °C/s. The dilatations along the radial direction were measured for all of the specimens during the entire experimental process.

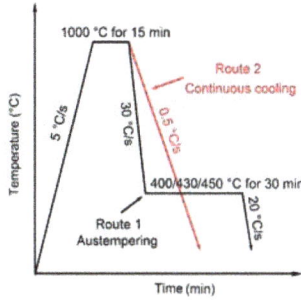

Figure 1. Experimental procedures.

2.3. Microstructure Examinations and Tensile Tests

After the thermal simulation experiments, the microstructures of the simulated specimens were examined by a Nova 400 Nano scanning electron microscope (SEM) (FEI, Hillsboro, OR, USA). In order to determine the volume fraction of retained austenite (RA), X-ray diffraction (XRD) experiments were carried out on a BRUKER D8 ADVANCE diffractometer (Bruker, Karlsruhe, Germany), using unfiltered Cu Ka radiation and operating at 40 kV and 40 mA. In addition, tensile tests were carried out on a UTM-4503 electronic universal tensile tester (SUNS, Shenzhen, China) with a cross-head speed of 1 mm/min at room temperature. Four tensile tests were repeated for each heat treatment procedure, and the corresponding average values were calculated. Then, the tensile fractures of the specimens were observed by SEM.

3. Results and Discussion

3.1. Isothermal Transformation

3.1.1. Dilatation

Figure 2 shows the dilatation versus temperature during the entire austempering treatments (400 °C) for the Cr-free and Cr-added steels. The measured Ac_1 and Ac_3 temperatures for the Cr-free steel are about 777 °C and 906 °C, respectively, and those for Cr-added steel are about 775 °C and 905 °C, respectively. Full austenite microstructures can be obtained at 1000 °C in the two steels, because the austenization temperature (1000 °C) is higher than Ac_3. Chromium addition has little effect on the Ac_1 and Ac_3 temperatures. In addition, the dilatation versus temperature curves are straight lines during the cooling process from 1000 °C to 400 °C at 30 °C/s, indicating that austenite does not decompose before the isothermal holding at 400 °C. Moreover, the apparent increase in the dilatation at 400 °C is caused by the bainitic transformation during isothermal holding.

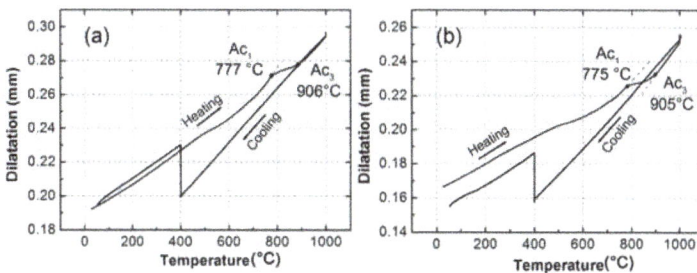

Figure 2. The dilatation versus temperature during the entire austempering treatment (400 °C) for the Cr-free and Cr-added steels: (**a**) Cr-free steel and (**b**) Cr-added steel.

The dilatation (representing the transformation amount) versus time during the isothermal holding at 400 °C, 430 °C, and 450 °C for the two steels is presented in Figure 3a. Chromium hinders the bainitic transformation kinetics, and decreases the final amount of bainitic transformation at all three temperatures. It is interesting to find that the hindrance of Cr on bainitic transformation is more obvious at higher temperatures (430 °C and 450 °C) compared with that at a lower temperature (400 °C). The time-temperature-transformation (TTT) curves of the two steels in Figure 3b, which are calculated by software JMatPro, also demonstrate that the kinetics of bainitic transformation are hindered by Cr addition. The time required to initiate bainitic transformation (2% relative transformation fraction) is listed in Table 2. The experimentally measured values are compared with the calculated values (JMatPro). The comparison shows that the calculated values are larger than the measured values, so that the kinetics of bainitic transformation calculated by JMatPro are slower compared to the experimentally measured one.

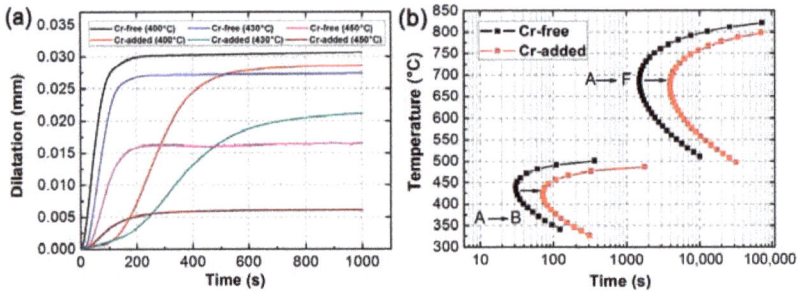

Figure 3. (**a**) The dilatation versus time during the isothermal holding at 400 °C, 430 °C, and 450 °C for the two steels; (**b**) Calculated time-temperature-transformation (TTT) curves of the Cr-free and Cr-added steels showing the start of transformations from austenite A to ferrite F and from austenite to bainite B. It shows that the kinetics of bainitic transformation are hindered by Cr addition.

Table 2. The time needed to initiate bainitic transformation (2% relative transformation fraction).

Steels	Transformation Temperature (°C)	Experimentally Measured Data (s)	Calculated Data (s)
Cr-free	400	6.0	39.4
Cr-added	400	19.4	79.8
Cr-free	430	17.9	31.0
Cr-added	430	45.0	74.3
Cr-free	450	15.3	35.8
Cr-added	450	30.1	129.8

According to the displacive approach of the bainitic transformation proposed by Bhadeshia [19,20], bainitic transformation is expected to occur when:

$$\Delta G_m < G_N \tag{1}$$

$$\Delta G^{\gamma \to \alpha} < -G_{SB} \tag{2}$$

where $\Delta G^{\gamma \to \alpha}$ is the driving force from γ to α, G_{SB} is the stored energy of bainite (about 400 J/mol), ΔG_m is the maximum driving force for the nucleation of bainitic ferrite with paraequilibrium carbon partitioning, and G_N is the universal nucleation function. Equation (1) ensures that there is a detectable nucleation rate, and Equation (2) ensures that a diffusionless growth of bainite can occur [19,20]. The values of ΔG_m and $\Delta G^{\gamma \to \alpha}$ versus temperature in Cr-free and Cr-added steels are calculated using the MUCG 83 program, and the results are shown in Figure 4. ΔG_m and $\Delta G^{\gamma \to \alpha}$ decrease with Cr addition, so that it is more difficult to meet the requirements of bainitic transformation (Equations (1) and (2)) in Cr-added steel. As a result, bainitic transformation is hindered by Cr addition. In addition, it is shown in Figure 4 that the $\Delta G^{\gamma \rightarrow \alpha}$ decreases from -1012 J/mol to -900 J/mol by Cr addition at 400 °C, and it decreases from -864 J/mol to -753 J/mol at 430 °C. The decreased driving forces account for 11% and 13% at 400 °C and 430 °C, respectively, indicating that the decreased driving force accounts for a larger proportion at a higher transformation temperature. A similar result can be obtained for ΔG_m. Therefore, the hindrance of Cr on bainitic transformation is more obvious at higher transformation temperatures.

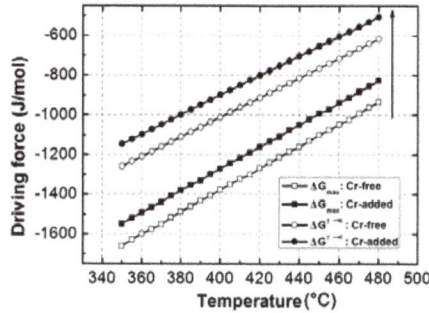

Figure 4. The values of ΔG_m and $\Delta G^{\gamma \rightarrow \alpha}$ versus temperature in Cr-free and Cr-added steels.

3.1.2. Microstructures

The typical microstructures of the Cr-free and Cr-added steels treated by austempering are shown in Figures 5 and 6. The microstructures contain lath-like bainite ferrite (BF), RA, and martensite (M). A selected region is magnified and presented in Figure 5c,d. It shows that ultra-fine BF is separated by film RA. No carbides are observed due to high silicon content [20–22]. When the transformation temperature is 400 °C (Figure 5), the amount and morphology of bainite in the two steels show no significant difference. However, the amount of bainite obviously decreases and the amount of martensite obviously increases with Cr content at 430 °C and 450 °C (Figure 6). This is consistent with the dilatation results (Figure 3).

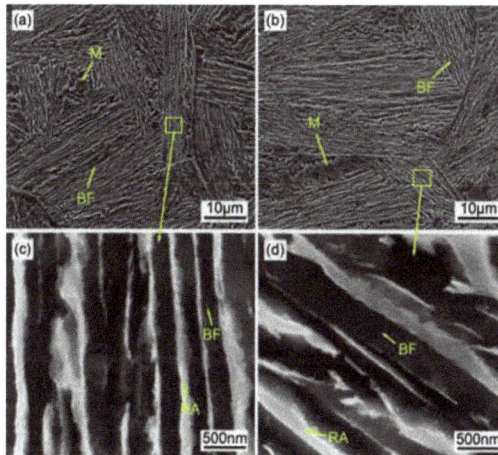

Figure 5. The typical microstructures of the Cr-free and Cr-added steels treated by austempering at 400 °C: (**a**,**c**) Cr-free steel; (**b**,**d**) Cr-added steel. M, martensite; BF, bainite ferrite; RA, retained austenite.

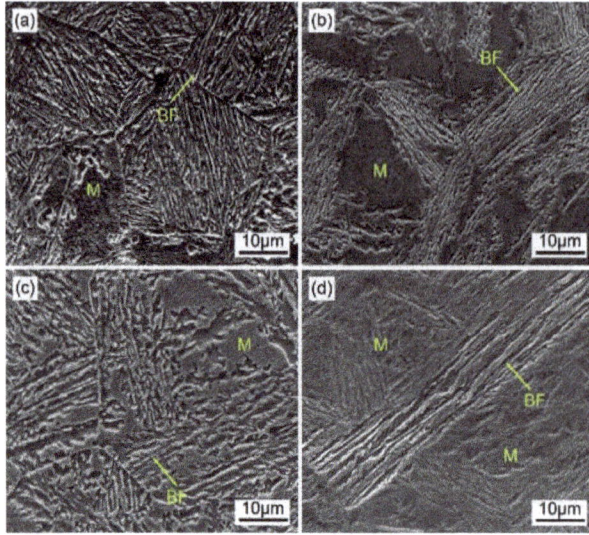

Figure 6. The typical microstructures of the Cr-free and Cr-added steels treated by austempering at 430 °C and 450 °C: (**a**) Cr-free steel, 430 °C; (**b**) Cr-added steel, 430 °C; (**c**) Cr-free steel, 450 °C; (**d**) Cr-added steel, 450 °C.

In addition, the volume fractions of RA (V_γ) in two the steels are calculated according to the integrated intensities of (200)α, (211)α, (200)γ, and (220)γ diffraction peaks based on the following equation [23,24].

$$V_i = \frac{1}{1 + G(I_\alpha / I_\gamma)} \tag{3}$$

where V_i is the volume fraction of austenite for each peak, I_α and I_γ are the corresponding integrated intensities of ferrite and austenite, and the G value is chosen as follows, 2.5 for $I_\alpha(200)/I_\gamma(200)$, 1.38 for $I_\alpha(200)/I_\gamma(220)$, 1.19 for $I_\alpha(211)/I_\gamma(200)$, and 0.06 for $I_\alpha(211)/I_\gamma(220)$ [23,24]. The results show that the volume fraction of RA increases from 6.3 vol. % to 10.1 vol. % with the addition of Cr for austempering at 400 °C, and it slightly increases from 4.0 vol. % to 4.8 vol. % for 430 °C austempering. This is because the stability of austenite is enhanced by Cr addition. In addition, the carbon content in RA (C_γ) is calculated according to the (200)γ and (220)γ diffraction peaks using the method in reference [24] and the results are given in Table 3. The carbon content in RA decreases with Cr addition, because the amount of bainitic transformation, which is accompanied by the partitioning of carbon from bainite ferrite to RA, is smaller in Cr-added steel.

Table 3. The tensile results and RA characteristics of two tested steels treated by austempering. YS, yield strength; TS, tensile strength; TE, total elongation; PSE, product of strength and elongation.

Steel	YS (MPa)	TS (MPa)	TE (%)	PSE (GPa%)	V_γ (vol. %)	C_γ (wt. %)
Cr-free (400 °C)	772 ± 16	977 ± 36	15.8 ± 0.3	15.4 ± 0.79	6.3 ± 1.1	1.18 ± 0.11
Cr-added(400 °C)	649 ± 16	1083 ± 16	18.3 ± 0.4	19.8 ± 0.52	10.1 ± 0.7	1.14 ± 0.08
Cr-free (430 °C)	620 ± 15	986 ± 13	18.5 ± 0.3	18.2 ± 0.55	4.0 ± 0.2	1.13 ± 0.06
Cr-added (430 °C)	786 ± 20	1185 ± 26	11.7 ± 0.9	13.9 ± 0.98	4.8 ± 0.3	0.99 ± 0.03
Cr-free (450 °C)	795 ± 17	1051 ± 19	13.6 ± 0.6	14.3 ± 0.42	3.6 ± 0.4	0.96 ± 0.05
Cr-added (450 °C)	1050 ± 21	1263 ± 23	5.7 ± 0.3	7.2 ± 0.58	2.8 ± 0.2	0.71 ± 0.03

3.1.3. Mechanical Properties

The tensile results of Cr-free and Cr-added steels treated by austempering are given in Table 3. The typical comparison of the engineering strain-stress curves between Cr-free and Cr-added steels treated by austempering is shown in Figure 7. It shows that when the transformation temperature is lower (400 °C), Cr-added steel shows lower yield strength (YS) compared with Cr-free steel due to more amounts of softer phase (austenite) in Cr-added steel. However, the tensile strength (TS) and the total elongation (TE) increase with Cr addition, resulting in the increase in the product of strength and elongation (PSE, TS × TE). One reason for the increase in tensile strength is that Cr acts as the role of solution strengthening. In addition, it is known that RA transforms to martensite during a tensile test, and thus improves the strength and plasticity simultaneously, which is generally termed as the transformation induced plasticity (TRIP) effect [25,26]. There is more RA in Cr-added steel treated by austempering at 400 °C, so that the strength and plasticity of the steel are higher. Moreover, when the transformation temperatures are higher (430 °C and 450 °C), the yield strength and the tensile strength increase, whereas the total elongation significantly decreases with Cr addition. The increase in strength is mainly caused by the obvious increase in martensite amount (Figure 6) and the solution strengthening of Cr. However, more amounts of martensite decreases the plasticity obviously. As a result, the PSE decreases. Therefore, at a lower transformation temperature, Cr addition increases the comprehensive property of the steel, whereas it decreases the comprehensive property of the steel at higher transformation temperatures. Moreover, in the present study, Cr-free steel shows a better comprehensive property at 430 °C, whereas Cr-added steel shows a better comprehensive property at 400 °C. Therefore, the austempering temperature should be lower for Cr-added steel than for Cr-free steel in order to obtain a better comprehensive property.

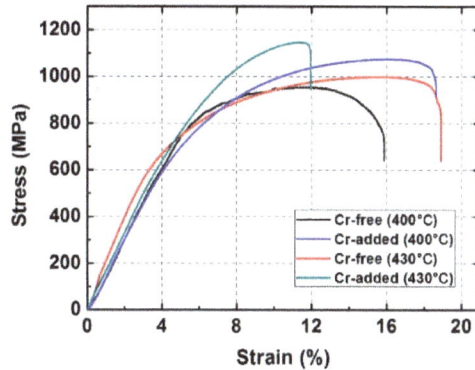

Figure 7. Typical comparison of the engineering strain-stress curves between the Cr-free and Cr-added steels treated by austempering.

Figure 8 shows the typical morphologies of tensile fractures in the two steels treated by austempering at 400 °C. Many dimples are observed in the two steels, indicating that the fracture is ductile. The fracture morphologies in the two steels show no significant difference.

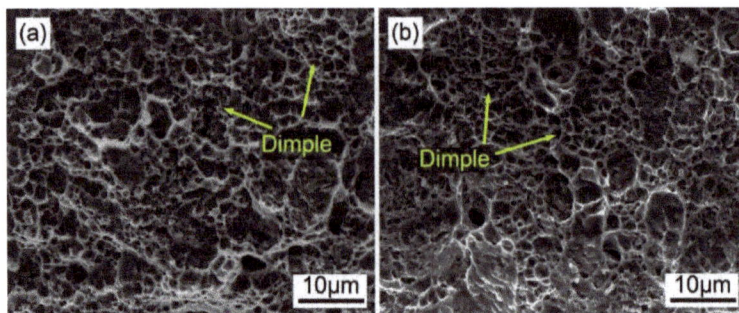

Figure 8. Morphologies of tensile fractures in two steels treated by austempering at 400 °C: (**a**) Cr-free steel; (**b**) Cr-added steel.

Carbide-free bainitic steel with ultra-fine bainitic ferrite and film-like RA is developed by Caballero et al. [12–15]. Cr is added in this advanced bainitic steel. They found that when Mn is replaced by Cr, the strength of the steel with 0.2 wt. % C decreases significantly during a continuous cooling process [15]. However, isothermal treatment is not studied in their studies. Sugimoto and his co-workers [27,28] investigated the effects of alloying elements (Cr, Mo, Ni, etc.) on the microstructure, the retained austenite, and the mechanical properties of bainitic steels treated by austempering. Their results showed that compared to the Cr-free steel, Cr addition increases the comprehensive property (PSE) of the steels at all studied austempering temperatures. The authors' previous study [18] also reported that Cr addition improves the comprehensive property of the steel austempered at 350 °C. However, the present study finds that at a lower transformation temperature (400 °C), Cr addition increases the comprehensive property of the steel, whereas Cr addition decreases the comprehensive property of the steel at higher transformation temperatures (430 °C and 450 °C). The difference between the previous result [18,27,28] and the present result may be because the austempering temperatures used in previous studies are relatively lower. In addition, the effect of Cr on bainitic transformation was not studied in references [27,28]. The authors' previous study [18] reported that Cr addition increases the amount of isothermal bainitic transformation at 350 °C, because high temperature ferritic transformation occurs in Cr-free steel and consumes some untransformed austenite before austempering, whereas no ferritic transformation occurs in Cr-added steel. In the present study, there is no ferritic transformation before isothermal bainitic transformation, and it is found that Cr addition decreases the bainite amount due to a decreased chemical driving force for bainitic transformation, and this effect is more significant at higher transformation temperatures.

3.2. Continuous Transformation

3.2.1. Dilatation

The dilatation versus temperature during cooling process for the two steels treated by continuous cooling is presented in Figure 9a. Ferrite transformation occurs at a relatively high temperature due to a slow cooling rate (0.5 °C/s). The calculated M_S temperatures for Cr-free and Cr-added steels are 376 and 353 °C, respectively, so that the region between a and b in the Cr-free steel and the region between c and d in the Cr-added steel contain not only bainitic transformation, but also martensitic transformation. The transformation start temperature decreases from 537 °C (point a) to 450 °C (point c) with Cr addition. In addition, D_1 and D_2 (Figure 9a) are dilatations caused by bainitic and martensitic transformations. D_2 is larger than D_1, indicating that the volume fractions of bainitic and martensitic transformations increase with Cr addition. This may be because the bainitic and martensitic transformations in Cr-added steel occur in lower temperature regions, so that the undercooling of the transformation is larger. In addition, the continuous cooling transformation (CCT) curves for the two steels are calculated using the software JMatPro (Figure 9b). It is obvious that the temperatures of ferritic, pearlitic, and bainitic transformations all decrease with Cr addition, which is consistent with the experimental results. As shown in Figure 9b, the calculated bainite transformation start temperatures for Cr-free and Cr-added steels are 482 °C and 451 °C, respectively, for a cooling rate of 0.5 °C/s. The calculated value (451 °C) is almost the same as the measured value (450 °C, Figure 9a) for Cr-added steel, whereas the calculated value (482 °C) is obviously lower than the measured value (537 °C, Figure 9a) for Cr-free steel.

In addition, the relative change of transformation fraction versus temperature during the cooling process from point a to b in Cr-free steel, and from point c to d in Cr-added steel, are calculated using the lever rule and the results are shown in Figure 10b. An example is given in Figure 9a. At a certain temperature, the relative volume fraction of transformation is determined as ON/MN [29,30]. In general, the transformation kinetics are hindered by Cr addition (Figure 9b). This is because the lower transformation temperature region makes the diffusion of carbon slow down during the nucleation process of bainitic transformation [20]. In addition, the slope of the transformed fraction curves (df/dT) is shown in Figure 10c. The slope represents the transformation rate. There is a peak value in each slope curve, which corresponds to the maximum transformation rate. First, the maximum transformation rate appears later in Cr-added steel, indicating that the transformation is delayed by Cr addition. Second, the maximum transformation rate in Cr-added steel is larger than that in Cr-free steel. The faster transformation rate may be caused by martensite transformation.

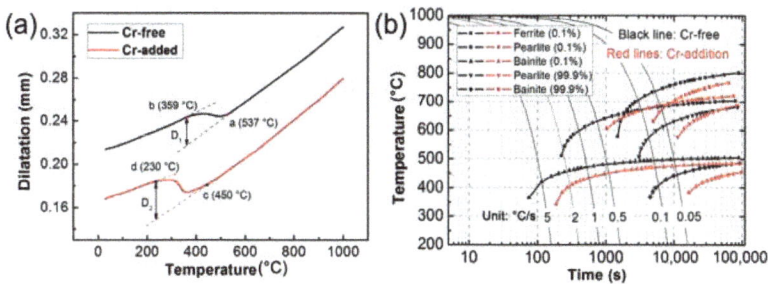

Figure 9. (a) The dilatation versus temperature during cooling process for the two steels treated by continuous cooling; (b) Calculated continuous cooling transformation (CCT) curves for the two steels.

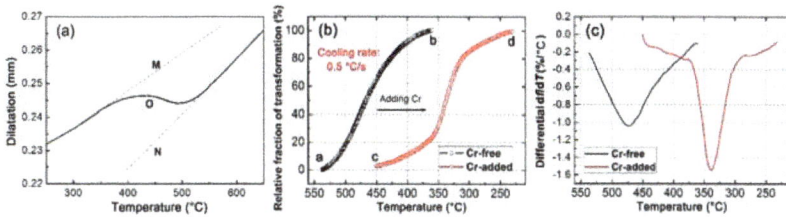

Figure 10. (a) The example of lever rule; (b) The relative change of transformation fraction versus temperature during the cooling process (cooling rate: 0.5 °C/s) from point a to b (Figure 9a) in Cr-free steel, and from point c to d (Figure 9a) in Cr-added steel; (c) The slope of the transformed fraction curves (df/dT).

3.2.2. Microstructure

The typical microstructures of the Cr-free and Cr-added steels treated by continuous cooling are shown in Figure 11. The microstructure is significantly different between the two steels. The microstructure in Cr-free steel contains lath-like bainite (LB), granular bainite (GB), polygon ferrite (PF), and martensite/austenite (M/A) constitution, whereas the microstructure in Cr-added steel contains LB, a large amount of M, and a small amount of PF and RA. No carbides are observed. The volume fractions of different phases are determined using the software Image-Pro Plus 6.0 (Media Cybernetics, Rockville, MD, USA) according to the grayscale and morphology of different phases. There is about 48 vol. % bainite (GB + LB) in Cr-free steel, whereas the volume fraction of bainite (mainly LB) in Cr-added steel is about 24 vol. %. This is consistent with the result obtained by the lever rule (Figure 10b), in which the volume fraction bainite is about 22 vol. % in Cr-added steel. The amount of PF decreases with Cr addition due to the solute-drag effect of Cr [31]. The amount of M significantly increases because the transformation temperature region decreases by Cr addition and large amount of transformation occurs below M_S temperature (Figure 9a). PF presents more irregular morphology. The regions showing bainite morphologies in the two steels are magnified and presented in Figure 11c,d. It is obvious that bainite is coarse in Cr-free steel, whereas it is significantly refined in Cr-added steel due to the lower transformation temperature. In addition, XRD results show that the volume fraction of RA decreases from 8.7 vol. % to 2.3 vol. % with the addition of Cr. This is because there is more ferrite and bainite in Cr-free steel. It is known that carbon diffuses into the surrounding untransformed austenite during ferritic and bainitic transformation, which stabilizes the untransformed austenite [20].

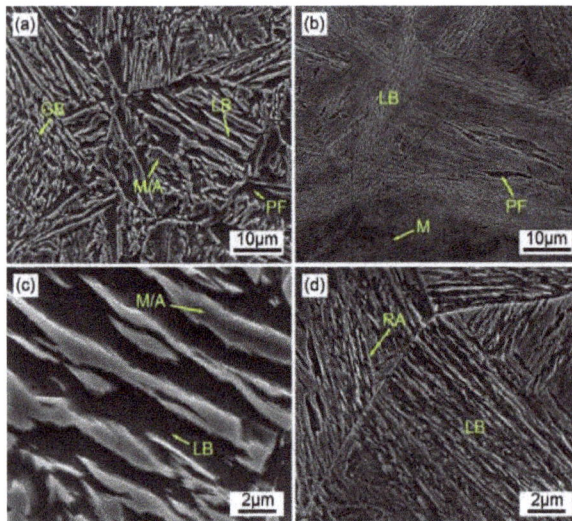

Figure 11. The typical microstructures of the Cr-free and Cr-added steels treated by continuous cooling: (**a**,**c**) Cr-free steel; (**b**,**d**) Cr-added steel. LB, lath-like bainite; GB, granular bainite; PF, polygon ferrite; M/A, martensite/austenite.

3.2.3. Mechanical Properties

The tensile results of Cr-free and Cr-added steels treated by continuous cooling are given in Table 4. The engineering strain-stress curves of the two steels are shown in Figure 12. Compared with Cr-free steel, the yield strength and tensile strength in Cr-added steel is increased because there is more hard martensite in Cr-added steel. However, more amounts of martensite in Cr-added steel significantly decreases the elongation of the steel. As a result, the PSE in Cr-added steel obviously decreases. Moreover, the tensile fractures of the two steels are shown in Figure 13. The fracture in the Cr-free steel mainly consists of dimples, whereas a large amount of quasicleavage morphologies are observed in the Cr-added steel, indicating that the plasticity is better in Cr-free steel. Therefore, Cr addition is harmful for the properties, especially ductility, of low carbon bainitic steel under a continuous cooling treatment.

Table 4. The tensile results and RA characteristics of two tested steels treated by continuous cooling.

Steel	YS (MPa)	TS (MPa)	TE (%)	PSE (GPa%)	V_γ (vol. %)	C_γ (wt. %)
Cr-free	662 ± 13	1054 ± 15	13.2 ± 0.8	13.9 ± 0.56	8.7 ± 0.7	1.05 ± 0.05
Cr-added	812 ± 15	1145 ± 21	6.9 ± 0.2	7.9 ± 0.15	2.3 ± 0.2	0.82 ± 0.04

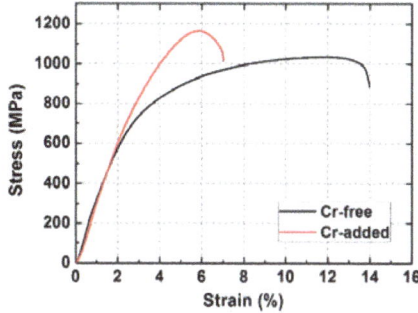

Figure 12. The engineering strain-stress curves of Cr-free and Cr-added steels treated by continuous cooling.

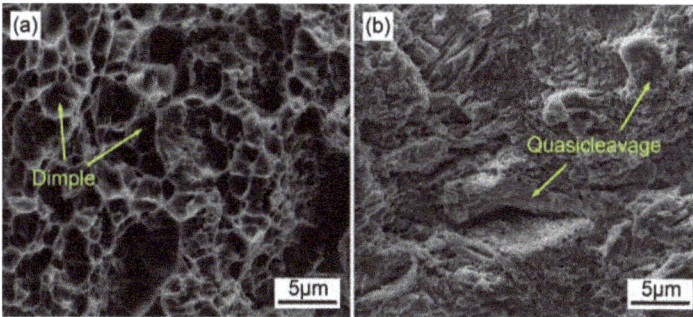

Figure 13. The tensile fractures of two steels treated by continuous cooling: (**a**) Cr-free steel; (**b**) Cr-added steel.

4. Conclusions

Two kinds of heat treatment procedure (austempering and continuous cooling) are designed. The effects of Cr addition on the bainitic transformation, the microstructure, and the properties of low carbon carbide-free bainitic steels are investigated. The following conclusions can be drawn:

(1) Chromium addition hinders the isothermal bainitic transformation kinetics and decreases the amount of isothermal transformation due to the decrease in chemical driving force for nucleation and growth of bainite. The hindrance of Cr on bainitic transformation is more significant at higher transformation temperatures because the decreased chemical driving force accounts for a larger proportion.

(2) Chromium addition increases the strength and elongation simultaneously for austempering treatment at a lower temperature. However, when the austempering temperature is higher, the strength increases and the elongation obviously decreases by Cr addition, resulting in the decrease in the product of tensile strength and elongation. In addition, the austempering temperature should be lower in Cr-added steel than that in Cr-free steel in order to obtain better comprehensive properties.

(3) For continuous cooling treatment in the present study, the amount of RA decreases, and the yield strength and tensile strength increases, but the total elongation obviously decreases in Cr added steel due to more amounts of martensite. The product of tensile strength and elongation significantly decreases.

Acknowledgments: The authors gratefully acknowledge the financial supports from National Natural Science Foundation of China (NSFC) (No. 51274154), the National High Technology Research and Development Program of China (No. 2012AA03A504), and the Special Fund of Wuhan University of Science and Technology for Master Students' Short-Term Studying Abroad.

Author Contributions: Guang Xu and Mingxing Zhou conceived and designed the experiments; Mingxing Zhou and Junyu Tian performed the experiments; Mingxing Zhou, Junyu Tian, and Haijiang Hu analyzed the data; Qing Yuan contributed materials tools; and Mingxing Zhou wrote the paper.

Conflicts of Interest: The authors declare no conflict of interest. The founding sponsors had no role in the design of the study; in the collection, analyses, or interpretation of data; in the writing of the manuscript, and in the decision to publish the results.

Abbreviations

The following abbreviations are used in this manuscript:

SEM	scanning electron microscope
XRD	X-ray diffraction
BF	bainite ferrite
RA	retained austenite
M	martensite
LB	lath-like bainite
GB	granular bainite
PF	polygon ferrite
YS	yield strength
TS	tensile strength
TE	total elongation
PSE	product of strength and elongation
TRIP	transformation induced plasticity

References

1. Jafari, M.; Ziaei-Rad, S.; Saeidi, N.; Jamshidian, M. Micromechanical analysis of martensite distribution on strain localization in dual phase steels by scanning electron microscopy and crystal plasticity simulation. *Mater. Sci. Eng. A* **2016**, *670*, 57–67. [CrossRef]
2. Huang, Y.Y.; Li, Q.G.; Huang, X.F.; Huang, W.G. Effect of bainitic isothermal transformation plus Q&P process on the microstructure and mechanical properties of 0.2 C bainitic steel. *Mater. Sci. Eng. A* **2016**, *678*, 339–346.
3. Zhou, M.X.; Xu, G.; Wang, L.; Hu, H.J. Combined effect of the prior deformation and applied stress on the bainite transformation. *Met. Mater. Int.* **2016**, *22*, 956–961. [CrossRef]
4. Hu, H.J.; Zurob, H.S.; Xu, G.; Embury, D.; Purdy, G.R. New insights to the effects of ausforming on the bainitic transformation. *Mater. Sci. Eng. A* **2015**, *626*, 34–40. [CrossRef]
5. Zhou, M.X.; Xu, G.; Wang, L.; Xue, Z.L.; Hu, H.J. Comprehensive analysis of the dilatation during bainitic transformation under stress. *Met. Mater. Int.* **2015**, *21*, 985–990. [CrossRef]
6. Ou, M.G.; Yang, C.L.; Zhu, J.; Xia, Q.F.; Qiao, H.N. Influence of Cr content and Q&P&T process on the microstructure and properties of cold-coiled spring steel. *J. Alloys Compd.* **2017**, *697*, 43–54.
7. Hu, H.J.; Xu, G.; Zhou, M.X.; Yuan, Q. Effect of Mo content on microstructure and property of low-carbon bainitic steels. *Metals* **2016**, *6*, 173. [CrossRef]
8. Hu, H.J.; Xu, G.; Wang, L.; Xue, Z.L.; Zhang, Y.L.; Liu, G.H. The effects of Nb and Mo addition on transformation and properties in low carbon bainitic steels. *Mater. Des.* **2015**, *84*, 95–99. [CrossRef]
9. Han, Y.; Kuang, S.; Liu, H.S.; Jiang, Y.H.; Liu, G.H. Effect of chromium on microstructure and mechanical properties of cold rolled hot-dip galvanizing DP450 steel. *J. Iron Steel Res. Int.* **2015**, *22*, 1055–1061. [CrossRef]

10. Wan, S.L.; Gao, H.Y.; Li, Z.Y.; Nakashima, H.; Hata, S.; Tian, W.H. Effect of lower bainite/martensite/retained austenite triplex microstructure on the mechanical properties of a low-carbon steel with quenching and partitioning process. *Int. J. Min. Met. Mater.* **2016**, *23*, 303–313.

11. Jirková, H.; Kučerová, L.; Mašek, B. The effect of chromium on microstructure development during Q-P process. *Mater. Today Proc.* **2015**, *2S*, S627–S630. [CrossRef]

12. Caballero, F.G.; Bhadeshia, H.K.D.H.; Mawella, K.J.A.; Jones, D.G.; Brown, P. Design of novel high strength bainitic steels. Part 1. *Mater. Sci. Technol.* **2001**, *17*, 512–516. [CrossRef]

13. Caballero, F.G.; Bhadeshia, H.K.D.H.; Mawella, K.J.A.; Jones, D.G.; Brown, P. Design of novel high strength bainitic steels. Part 2. *Mater. Sci. Technol.* **2001**, *17*, 517–522. [CrossRef]

14. Caballero, F.G.; Bhadeshia, H.K.D.H. Very strong bainite. *Curr. Opin. Solid State Mater.* **2004**, *8*, 251–257. [CrossRef]

15. Caballero, F.G.; Santofimia, M.J.; Garcia-Mateo, C.; Chao, J.; de Garcia Andres, C. Theoretical design and advanced microstructure in super high strength steels. *Mater. Des.* **2009**, *30*, 2077–2083. [CrossRef]

16. Hasan, H.S.; Peet, M.J.; Avettand-Fènoël, M.-N.; Bhadeshia, H.K.D.H. Effect of tempering upon the tensile properties of a nanostructured bainitic steel. *Mater. Sci. Eng. A* **2014**, *615*, 340–347. [CrossRef]

17. Bracke, L.; Xu, W. Effect of the Cr content and coiling temperature on the properties of hot rolled high strength lower bainitic steel. *ISIJ Int.* **2015**, *55*, 2206–2211. [CrossRef]

18. Tian, J.Y.; Xu, G.; Zhou, M.X.; Hu, H.J.; Wan, X.L. The effects of Cr and Al addition on transformation and properties in low-carbon bainitic steels. *Metals* **2017**, *7*, 40. [CrossRef]

19. Bhadeshia, H.K.D.H. A rationalisation of shear transformations in steels. *Acta Metall.* **1981**, *29*, 1117–1130. [CrossRef]

20. Bhadeshia, H.K.D.H. *Bainite in Steels*, 2nd ed.; The Institute of Materials: London, UK, 2001; pp. 117–187.

21. Shipway, P.H.; Bhadeshia, H.K.D.H. The effect of small stresses on the kinetics of the bainite transformation. *Mater. Sci. Eng. A* **1995**, *201*, 143–149. [CrossRef]

22. Zhang, X.X.; Xu, G.; Wang, X.; Embury, D.; Bouaziz, O.; Purdy, G.R.; Zurob, H.S. Mechanical behavior of carbide-free medium carbon bainitic steels. *Metall. Mater. Trans. A* **2014**, *45A*, 1352–1361. [CrossRef]

23. Wang, C.Y.; Shi, J.; Cao, W.Q.; Dong, H. Characterization of microstructure obtained by quenching and partitioning process in low alloy martensitic steel. *Mater. Sci. Eng. A* **2010**, *527*, 3442–3449. [CrossRef]

24. Zhou, M.X.; Xu, G.; Wang, L.; He, B. Effects of austenitization temperature and compressive stress during bainitic transformation on the stability of retained austenite. *Trans. Indian Inst. Met.* **2016**, *69*, 1–7. [CrossRef]

25. Cooman, B.C.D. Structure-properties relationship in TRIP steels containing carbide-free bainite. *Curr. Opin. Solid State Mater.* **2004**, *8*, 285–303. [CrossRef]

26. García-Mateo, C.; Caballero, F.G. The role of retained austenite on tensile properties of steels with bainitic microstructures. *Mater. Trans.* **2005**, *46*, 1839–1846. [CrossRef]

27. Sugimoto, K.; Kobayashi, J.; Ina, D. Toughness of Advanced Ultra High-Strength TRIP-Aided Steels with Good Hardenability. In Proceedings of the International Conference on Advanced Steels 2010 (ICAS 2010), Guilin, China, 9–11 November 2010; Metallurgical Industry Press: Beijing, China, 2010.

28. Kobayashi, J.; Ina, D.; Yoshikawa, N.; Sugimoto, K. Effects of the addition of Cr, Mo and Ni on the retained austenite characteristics of 0.2% C–Si–Mn–Nb ultrahigh-strength TRIP-aided bainitic ferrite steels. *ISIJ Int.* **2012**, *52*, 1894–1901. [CrossRef]

29. Huang, J.; Poole, W.J.; Militzer, M. Austenite formation during intercritical annealing. *Metal. Mater. Trans. A* **2004**, *35*, 3363–3375. [CrossRef]

30. Quidort, D.; Brechet, Y.J.M. A model of isothermal and non-isothermal transformation kinetics of bainite in 0.5% C Steels. *ISIJ Int.* **2002**, *42*, 1010–1017. [CrossRef]

31. Béché, A.; Zurob, H.S.; Hutchinson, C.R. Quantifying the solute drag effect of Cr on ferrite growth using controlled decarburization experiments. *Met. Mater. Trans. A* **2007**, *38*, 2950–2955. [CrossRef]

metals

MDPI

Article

Resistance Upset Welding of ODS Steel Fuel Claddings—Evaluation of a Process Parameter Range Based on Metallurgical Observations

Fabien Corpace [1,2], Arnaud Monnier [1], Jacques Grall [1], Jean-Pierre Manaud [2], Michel Lahaye [3] and Angeline Poulon-Quintin [2,*]

[1] Den-Service d'Etudes Mécaniques et Thermiques (SEMT), CEA, Université Paris-Saclay, F-91191 Gif-sur-Yvette, France; fabien.corpace@cea.fr (F.C.); arnaud.monnier@cea.fr (A.M.); jacques.grall@cea.fr (J.G.)
[2] Centre National de la Recherche Scientifique (CNRS), University Bordeaux, ICMCB, UPR 9048, F-33600 Pessac, France; manaud@icmcb-bordeaux.cnrs.fr
[3] Surface Analyses Departement, University Bordeaux, Placamat, UMS 3626, F-33600 Pessac, France; michel.lahaye@placamat.cnrs.fr
* Correspondence: angeline.poulon@icmcb.cnrs.fr; Tel.: +33-540-006-260

Received: 18 July 2017; Accepted: 8 August 2017; Published: 29 August 2017

Abstract: Resistance upset welding is successfully applied to Oxide Dispersion Strengthened (ODS) steel fuel cladding. Due to the strong correlation between the mechanical properties and the microstructure of the ODS steel, this study focuses on the consequences of the welding process on the metallurgical state of the PM2000 ODS steel. A range of process parameters is identified to achieve operative welding. Characterizations of the microstructure are correlated to measurements recorded during the welding process. The thinness of the clad is responsible for a thermal unbalance, leading to a higher temperature reached. Its deformation is important and may lead to a lack of joining between the faying surfaces located on the outer part of the join which can be avoided by increasing the dissipated energy or by limiting the clad stick-out. The deformation and the temperature reached trigger a recrystallization phenomenon in the welded area, usually combined with a modification of the yttrium dispersion, i.e., oxide dispersion, which can damage the long-life resistance of the fuel cladding. The process parameters are optimized to limit the deformation of the clad, preventing the compactness defect and the modification of the nanoscale oxide dispersion.

Keywords: ODS steel; PM2000; oxide dispersion strengthened; welding; resistance welding; fuel cladding; sodium fast reactor; dynamical recrystallization

---

1. Introduction

Today, energetic needs are increasing and will keep increasing in the coming decades. Nuclear energy is an interesting solution to produce part of this energy. Fourth-generation reactors are being studied and they are foreseen to be operational by the year 2040. The Generation IV International Forum (GIF Symposium, 9–10 September 2009, Paris, France) has identified six concepts of reactors. Among these concepts, Sodium Fast Reactor (SFR) technology is studied in France due to large feedback in Phénix and Superphénix. The generation IV innovative concept of SFR introduces some specifications that are different from those of the past SFR which are not sustained by present materials. Consequently, new materials have to be qualified [1].

In order to increase the efficiency of the new SFR, the fuel cladding material may have to undergo higher dpa (displacement per atom) than the stainless steels previously used and presented in Nuclear systems of the future generation IV. Potential substitutes are Oxide Dispersion Strengthened (ODS)

alloys due to promising corrosion and neutron behavior and good high temperature mechanical properties. These properties are the consequence of a homogeneous dispersion of nanoscale oxide particles inside the metallic matrix [2].

If the ODS material transitions to a molten phase during the fabrication route and especially during the assembling step, the nanoscale particles may be reallocated in the matrix and the initial oxide dispersion may be modified. Consequently, solid-state welding processes are promising [3,4].

Among the solid-state welding processes, resistance upset welding has been successfully applied to various ODS alloys for several fuel pin cladding applications [5–10]. These studies focused on mechanical test results. The welds are characterized using mechanical testing up to 800 °C by traction, burst test, creep or fatigue. Little or no information is related to microstructure evolution or oxide distribution evolution. However, metallurgical analyses have been conducted on several ODS alloys welded by different processes such as inertia friction welding [11], friction stir welding [12–14] and diffusion bounding [15]. It appears that these processes can induce modification of the metallurgical state, including dynamic recrystallisation, modification in grain orientation or modification in the oxide dispersion. Some rupture in these modified areas is reported [4,11].

The modifications of the microstructure, occurring during solid-state bonding, may lead to possible changes in the mechanical properties. A lack of metallurgical studies on resistance upset welding on ODS alloys has been identified. This study focuses on the metallurgical evolution during resistance upset welding applied to cladding made in a ferritic ODS alloy (PM2000). Evidence of modifications of the nanoscale oxide dispersion is shown. Therefore, the process parameters have to be optimized in order to produce welds with good compactness and no modification of the oxide distribution.

2. Materials and Methods

2.1. Material and Geometry

The studied material is PM2000 in the recrystallized form (millimeter-sized grains) produced by PLANSEE (Metallwerk Plansee GmbH, Lechbruck am See, Germany). It is a ferritic ODS alloy strengthened by nano-oxides composed of yttria (size <50 nm) or Aluminum and yttria (size < 100 nm). Its composition is presented in Table 1.

Table 1. PM2000 nominal composition (wt %).

Fe	Cr	Al	Ti	Y_2O_3
Balanced	20	5.5	0.5	0.5

The pieces to be welded are a clad and an end plug. The clad has an outside diameter of 10.5 mm and a thickness equal to 0.5 mm. The end plug has a diameter of 10.5 mm. The contact surfaces are chamfered with an angle of 45°. Figure 1 shows the pieces to be welded: clad and end plug.

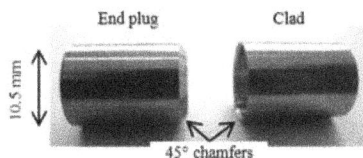

Figure 1. Pieces to be welded: end plug and clad.

2.2. Resistance Welding Process

The resistance welding process is composed of three main steps [16] represented in Figure 2: the squeezing phase, the welding phase and the forging phase. The two pieces to be welded are circled by two electrodes. The length of the clad that is out of the electrode, called "clad stick-out" (C_s), is considered as a process parameter as demonstrated in a previous paper [17] and observed by other authors on other materials [7,9]. During the squeezing phase, the two pieces to be welded are put and held in contact by a contact force F_w. The next step is the welding phase: a current with an intensity I_w is imposed through the two pieces in contact for a duration t_w. The Joule effect generates heat in the materials, especially near the contact surfaces between the two pieces. Then, during the forging phase, the current is stopped and the contact force F_w is upheld until the cooling of the weldment is complete.

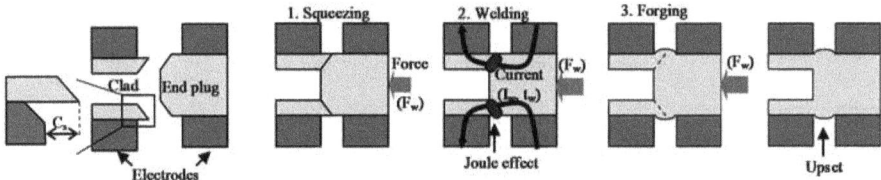

Figure 2. Schematic scenario of the resistance upset welding steps.

2.3. Experimental Device and Measurements

The experimental device has been developed for the considered geometry and is fabricated by the French company "TECHNAX industrie" (Saint-Priest, France). The device is composed of three parts:

- An electrical device: This part is composed of transformers and rectifiers delivering a rectified smoothed current with a frequency of 2 kHz;
- A control device: This part allows the setting of the process parameters (contact force F_w, current intensity I_w, welding time t_w);
- A welding device: A schematic view of this part is shown in Figure 3.

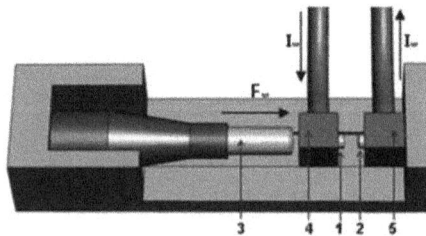

Figure 3. Schematic view of the welding device (F_w = welding force; I_w: welding current intensity; 1: end plug; 2: clad; 3: pneumatic jack; 4 and 5: electrodes).

The end plug (1) and the clad (2) to be welded are circled by two electrodes (4 and 5). The end plug and its electrodes are attached to the moving part of the device whereas the clad and its electrode are attached to the frame of the device and are motionless. The contact force is applied by means of a pneumatic jack (3) that put in contact the two pieces to be welded (1 and 2). Once the mechanical contact is established (the squeezing phase), the electrical device imposes the current through the two pieces by means of the electrodes.

The device is instrumented to measure the current intensity $I(t)$, the axial position of the mobile part $x(t)$, the electrical potential between the electrodes $U_{elec}(t)$ and the force $F(t)$ every 10 μs. From these measurements, it is possible to compute two values:

The dissipated energy at the end of the process E_{end} represents the amount of electrical energy dissipated in the welded pieces and the electrodes at the end of the welding process. This value is computed according to the following equation:

$$E_{end} = \int_{t=0}^{t=t_{end}} U_{elec}(t) \times I(t) \times dt \tag{1}$$

The collapse value is the length change of the pieces during the welding phase.

2.4. Sample Preparation and Observation

The welded pieces are cut along one diameter. Therefore, the weld can be observed in two locations, one face located at 180° of the other. All welds are observed using an optical microscope (Zeiss, München, Germany) before and after a metallographic etch (20 s in 20 mL H_2O–20 mL HCl–15 mL HNO_3).

The yttrium dispersion is characterized (localization and quantification) using a CAMECA SX100 (CAMECA, Gennevilliers, France) microprobe at 20 kV with a step of 1 μm and a dwelling time of 7 ms. Therefore, only the modifications at a microscopic level are observed. Back Scattering Electron (BSE) observations have been conducted on some welds using the same device.

3. Results

In this study, we deliberately made different kinds of weld—from good quality weld to weld with obvious defects. The aim is to have a good understanding of the formation of the welds in order to avoid the defects. The process parameter range is listed in Table 2. In the following, the process parameters will be quoted as a quadruplet (F_w; I_w; t_w; C_s).

Table 2. Process parameter range.

Name	Description	Min.	Max.
F_w	Force (N)	1800	2200
I_w	Current intensity (kA)	14	18
t_w	Welding time (ms)	10	15
C_s	Clad stick-out (mm)	0.2	0.8

It has to be noted that some welds present a compactness defect on one of their faces due to material ejection. These material ejections are under investigation but their location seems to incriminate the electrode—clad electrical contact. The faces presenting these defects are not taken into account and are rejected from the metallurgical analysis.

3.1. Welding Mechanism on Typical Welds

Figure 4 shows a typical weld with good compactness, a continuous interface and no modification of the yttrium distribution except in the internal upset. The modifications in the upsets are supposed to have no importance in the mechanical properties of the weld due to their positions outside the welded joint. The observations in the end plug, "far" from the joint, are representative of the base material due to the low temperature and the negligible deformation reached in this part [17]. Cavities are due to the fabrication route.

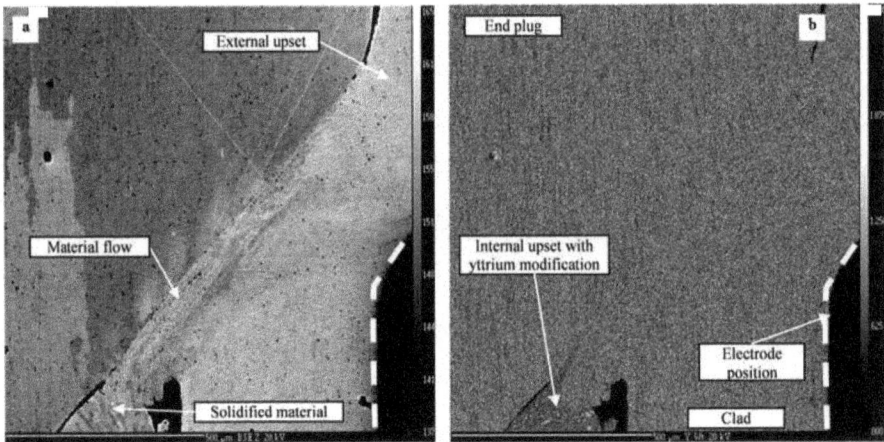

Figure 4. Wavelength-dispersive spectroscopy (WDS) analysis of a welded zone with $(F_w; I_w; t_w; C_s)$ = (1800 N; 14 kA; 15 ms; 0.8 mm): (**a**) Back Scattering Electron (BSE) picture; (**b**) Corresponding yttrium distribution.

The thinness of the clad compared to the bulk end plug creates a thermal unbalance, leading to higher temperatures in the clad than in the plug. Therefore, the clad is always more deformed than the end plug. The electrode achieves both the cooling and the clamping of the clad and therefore the highest deformations are localized in the clad part that is not in contact with the electrode (clad part out of the electrode).

3.2. Typical Deformation

The typical deformation of the clad is highlighted in Figure 5 where an elongated grain can be used to observe the deformation. The shape of the outer upset shows that the clad is sliding along the 45° chamfer of the end plug. This upset will be removed for the SFR application. The sliding creates a high deformation area (with an S-shaped grain) near the contact between the clad and the electrode.

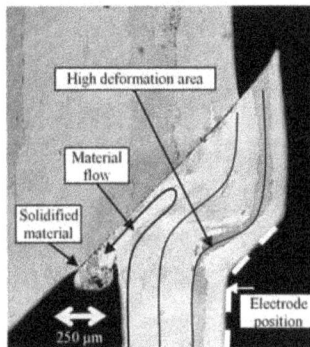

Figure 5. Evidence of the material deformation with the selected parameters $(F_w; I_w; t_w; C_s)$ = (2600 N; 14 kA; 10 ms; 0.8 mm).

In the internal side, the upset is composed of solidified grains due to expulsed molten materials during welding. Heated materials (below melting temperature) seem to have flown from the clad to the internal upset and are then trapped between the pieces. The deformation of the clad can be monitored by the measurement of the collapse value. The collapse value versus the dissipated energy for 32 experiments within the process parameter range is plotted in Figure 6 for two different clad stick-outs. The collapse (i.e., the deformation) increases when the dissipated energy increases for a given C_s value. A high stick-out-value leads to a higher collapse value. It has to be noted that the collapse can be higher than C_s due to an irregular hand-made chamfer on the electrode (cf. Figure 2).

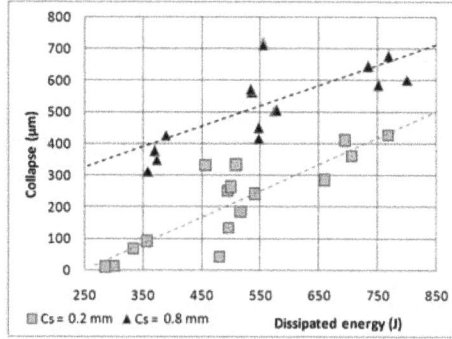

Figure 6. Collapse value as a function of the dissipated energy for 32 experiments.

3.3. Welds with a Low Dissipated Energy

A weld achieved with process parameters leading to a low dissipated energy (370 J) is shown in Figure 7. In the outer part of the weld joint, the faying surfaces are insufficiently joined. This spacing between the faying surfaces will be called a lack of joining in the following.

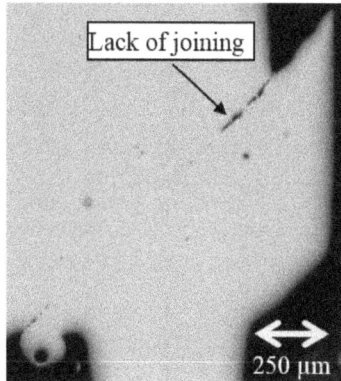

Figure 7. Example of a lack of joining (Mirror polish without etching) for a sample welded with the selected parameters $(F_w; I_w; t_w; C_s)$ = (2600 N; 14 kA; 10 ms; 0.8 mm).

For all the welds presenting a lack of joining, their localization is always in the outer part of the weld, roughly at the same place as shown in Figure 7. For each weld, two faces are observed and the length of this lack of joining can be measured.

The "not welded length" plotted in Figure 8 as a function of the dissipated energy is the average value of these two measures. It appears that an increase in the dissipated energy leads to a decrease of the not welded length. For a shorter clad stick-out, the dissipated energy required to achieve welds with no lack of joining on both sides is lower (450 J) than the dissipated energy required for a high value of clad stick-out (530 J).

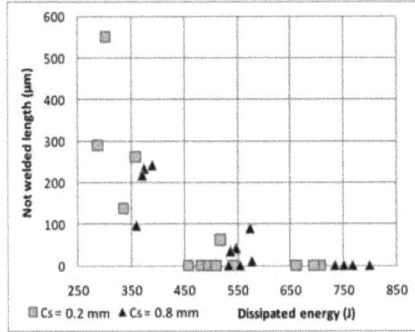

Figure 8. Not welded length as a function of the dissipated energy for 32 experiments.

3.4. Welds with a High Collapse

A weld achieved with process parameters leading to a high dissipated energy (537 J) is shown in Figure 9. Some modifications of the microstructure and of the yttrium dispersion are observed in the inner upset, at the interface and in the clad. The modified part in the clad starts from the electrode–clad contact area and extends to the joint. The black dots on the etched pictures correspond to a fine grain microstructure that is characteristic of a dynamical recrystallisation phenomenon. As shown in the yttrium distribution mapping, the modifications of the oxide dispersion occur specifically in the area where a recrystallisation phenomenon occurred. Clusters with a high density of yttrium surrounded by an area with a deficit in yttrium contents are noticeable.

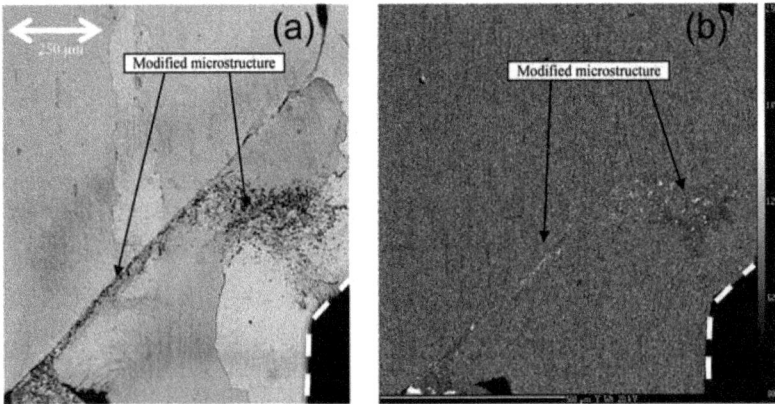

Figure 9. Evidence of modification of the metallurgical state of the sample welded using the parameters $(F_w; I_w; t_w; C_s)$ = (2600 N; 18 kA; 10 ms; 0.8 mm). (**a**) Optical micrograph of an etched sample; (**b**) WDS Yttrium distribution.

It has to be noted that all the welds presenting a modification of the yttrium distribution at the interface also presented a modification near the electrode–clad contact. It shows that recrystallization may first take place at the electrode–clad contact and then spread to the interface and to the inner upset. As a result, in Table 3, the modified area has been ranged from 0 to 4 (arbitrary unit) according to its shape and its extent.

Table 3. Classification used for the yttrium distribution modification.

Arbitrary Unit	Shape and Extent
0	No yttrium modifications
1	Modification localised in the outer part of the joint with possible extension to the outer upset
2	Modification in the clad near the electrode piece contact
3	2 + spreading to the interface
4	3 + spreading through the interface to the inner upset

Two faces of the same weld are observed and two values are assigned. The "yttrium modification" plotted in Figure 10 as a function of the collapse value, is the average of these two values.

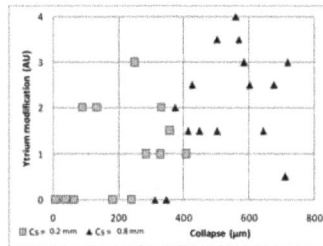

Figure 10. Yttrium modification (arbitrary unit defined in Table 3) as a function of collapse value.

It appears that the yttria modification is more important when the collapse increases. A short clad stick-out reduces the collapse and the welds with no yttrium modification are more numerous in this configuration.

4. Discussion

The objective is to find a range of dissipated energy (i.e., process parameters) to achieve welds with no lack of joining between the pieces and no modification of the oxide dispersion.

4.1. Lack of Joining between the Pieces

A lack of joining can appear in the outer part of the joint due to two phenomena:

- An electro-thermal phenomenon: The current flow is constricted in the inner part of the weld joint because of the bulk end plug in contact with the thin clad. A higher current density in the internal part of the welded joint leads to a higher temperature in this part than in the outer part of the joint. This phenomenon has been simulated by Zirker [8] using electrothermal computation on tube-rod geometry but no experimental evidence was shown. The observations of the material flow on the inner part of the joint give credit to Zirker's simulation.
- A thermo-mechanical phenomenon: The clad heats up faster than the end plug (due to the thermal unbalance) resulting in softening the part of the clad out of the electrode. The applied force and the 45° chamfer create a radial force, resulting in the sliding of the clad on the 45° chamfer. This sliding can open the contact in the outer part, preventing its welding.

The lack of joining can be avoided with an increase of the dissipated energy (Figure 8), leading to a higher collapse. Therefore, the end plug penetrates farther in the clad, closing the gap between the clad and the end plug created during the sliding of the clad. Moreover, increasing the energy increased the overall heat-up, resulting in a higher temperature in the outer part of the joint, preventing the defects.

A decrease in the clad stick-out reduces the not welded length. Decreasing the clad stick-out can increase the stiffness of the clad part out of the electrode and should prevent the opening of the contact by limiting the tube deformation and its sliding.

4.2. Yttrium Modifications

A modification of the yttrium dispersion can lead to a long-term failure of the weldment due to weakness points created by the lack of anchors to dislocation movement. The yttrium modification is observed in areas where recrystallization occurred.

An increase in the modification extent occurs when the collapse increases (Figure 10). The collapse value is highly influenced by the clad stick-out and the dissipated energy (Figure 6). As the collapse increases, the clad deformation increases which combined with a high temperature can trigger the dynamical recrystallization. Modified area localizations observed in Section 3.4 are consistent with the area of high deformation observed in Section 3.2.

Some authors have proposed some mechanisms in order to correlate the recrystallization and the modification of the Yttrium. Zhang et al. [15] proposed an interaction mechanism between oxides and dislocations to explain the modification of the oxide dispersion during diffusion bonding of an ODS alloy. Based on the works of Yazawa et al. [18], Yamamoto et al. [19] proposed that the oxide dispersion can be modified during phase transformation of a 9% Cr ODS-alloy due to a modification in the matrix-oxide coherency. Both dislocation movement and modification of the grain orientations occur during dynamical recrystallization. Therefore, both hypotheses may explain the modification in the yttrium distribution where the dynamical recrystallization occurred.

4.3. Optimizing Resistance Upset Welding for ODS Steel Fuel Cladding

Two types of defects to avoid are identified: a lack of joining between the pieces and a significant modification of the Yttrium dispersion observable in WDS analyses.

The lack of joining between the pieces can be avoided by increasing the dissipated energy. None of the welds below $E = 450$ J are free of this lack of joining (this value reached 530 J for $C_s = 0.8$ mm). Therefore, for $C_s = 0.2$ mm, the dissipated energy must be higher than 450 J.

The Yttrium modification can be avoided by decreasing the collapse. The collapse is mainly influenced by the dissipated energy and the clad stick-out. None of the welds above $E = 550$ J are free of the modification in the yttrium distribution (this value drops to 375 J for $C_s = 0.8$ mm). Therefore, for $C_s = 0.2$ mm, the dissipated energy must be lower than 550 J. Below 550 J, some welds still present a modification of the yttrium distribution and other parameters have to be adjusted to avoid all the defects in this area.

For a clad stick-out of 0.2 mm, a working area exists where welds free of both defects are produced. This working area does not exist for a clad stick-out of 0.8 mm.

The working area is illustrated in Figure 11. The working area is located in a range of dissipated energy between 450 and 550 J. The energy range may be highly dependent on the process route.

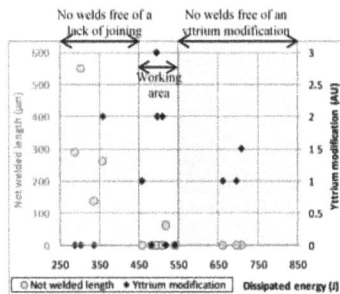

Figure 11. Range of the allowable dissipated energy to achieve welds with no lack of joining and no yttrium modifications ($C_s = 0.2$ mm).

However, some welds in the working area present an yttrium modification and/or a lack of joining due to other process parameters that should be optimized as well. A weld produced in this range and free of both defects is presented in Figure 12. After optimization, the reproducibility of the process for the optimized process routes is observed.

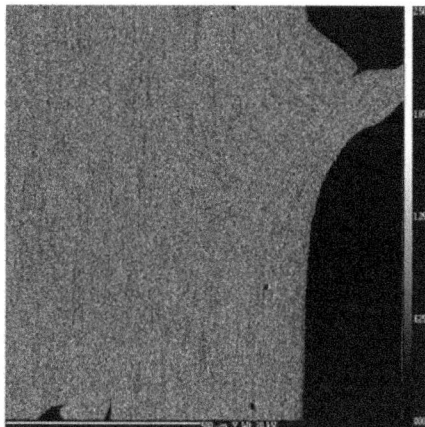

Figure 12. WDS yttrium distribution for a weld within the working area with the selected parameters $(F_w; I_w; t_w; C_s)$ = (2400 N; 14 kA; 15 ms; 0.2 mm).

5. Conclusions

Resistance upset welding has been applied with success to the ODS steel PM2000 (20% Cr) for the SFR fuel cladding application. For the given process parameters range, welds with different qualities have been produced. The main conclusions are:

- The clad part out of the electrode is highly deformed during the process due to thermal unbalance between the clad and the plug.
- For a high dissipated energy and a high collapse, the deformation of the clad can generate recrystallization phenomena associated with a modification of the yttria distribution.
- For a low dissipated energy, the faying surfaces can be insufficiently joined on the outer part of the join.
- For a clad stick-out of 0.2 mm, a range of dissipated energies exists where weldments with no significant yttria distribution modification and no lack of joining between pieces can be produced.

Other process parameters, such as the geometry of the contact surfaces, could be optimized in order to limit the deformation and avoid the lack of joining between pieces. The mechanical properties of the modified area should also be studied because a given amount of yttrium modification may be acceptable, rising the upper limit of the dissipated energy.

Acknowledgments: This work is part of PhD between the Institut de Chimie de la Matière Condensée de Bordeaux (ICMCB) and the Commissariat à l'énergie atomique et aux énergies alternatives (CEA-Saclay). The PhD is part of a research program for materials suitable for Sodium Fast Reactor (SFR) financed by CEA, Areva NP (Areva Nuclear Power) and EDF (Électricité De France).

Author Contributions: Fabien Corpace is the Ph.D. student. He wrote the paper, Arnaud Monnier, Jean-Pierre Manaud and Angeline Poulon-Quintin directed his work. Michel Lahaye performed the WDS experiments; Jacques Grall contributed to sample preparation.

Conflicts of Interest: The authors declare no conflict of interest.

References

1. Dubuisson, P.; de Carlan, Y.; Garat, V.; Blat, M. ODS Ferritic/martensitic alloys for Sodium Fast Reactor fuel pin cladding. *J. Nucl. Mater.* **2012**, *428*, 6–12. [CrossRef]
2. Yvon, P.; Le Flem, M.; Cabet, C.; Seran, J.L. Structural materials for next generation nuclear systems: Challenges and the path forward. *Nucl. Eng. Des.* **2015**, *294*, 161–169. [CrossRef]
3. Hedrich, H.D. Joining of ODS-superalloys. In *High Temperature Materials for Power Engineering Part 1*; Bachelet, E., Ed.; Kluwer Academic Publishers: Liège, Belgium, 1990; pp. 789–799, ISBN 0-7923-0925-1.
4. Wright, I.; Tatlock, G.; Badairy, H.; Chen, C. *Summary of Prior Work on Joining of Oxide Dispersion-Strengthened Alloys, Task 8*; Oak Ridge National Laboratory (ORNL): Oak Ridge, TN, USA, 2009. Available online: https://digital.library.unt.edu/ark:/67531/metadc932605/ (accessed on 8 August 2017).
5. Seki, M.; Hirako, K.; Kono, S.; Kihara, Y.; Kaito, T.; Ukai, S. Pressurized resistance welding technology development in 9Cr-ODS martensitic steels. *J. Nucl. Mater.* **2004**, *329–333*, 1534–1538. [CrossRef]
6. Ukai, S.; Kaito, T.; Seki, M.; Mayorshin, A.A.; Shishalov, O.V. Oxide Dispersion Strengthened (ODS) Fuel Pins Fabrication for BOR-60 Irradiation Test. *J. Nucl. Sci. Technol.* **2005**, *42*, 109–122. [CrossRef]
7. Zirker, L.; Bottcher, J.; Shikakura, S.; Tsai, C.; Hamilton, M. Fabrication of Oxide Dispersion Strengthened Ferritic Clad Fuel Pins. In Proceedings of the International Conference on Fast Reactors and Related Fuel Cycles, Kyoto, Japan, 28–31 October 1991.
8. Zirker, L.; Tyler, C. Pressure Resistance Welding of High Temperature Metallic Materials. In Proceedings of the ANS Decomissioning, Decontamination & Reutilization Conference, Idaho Falls, ID, USA, 29 August–2 September 2010.
9. De Burbure, S. Resistance Butt Welding of Dispersion-Hardened Ferritic Steels. In Proceedings of the Advances in Welding Processes 3rd International Conference, Harrogate, UK, 7–9 May 1974; pp. 216–228.
10. De Burbure, S. Resistance welding of pressurized capsules for in-pile creep experiments. *Weld. J.* **1978**, *57*, 23–30.
11. Shinozaki, K.; Kang, C.Y.; Kim, Y.C.; Aritoshi, M.; North, T.H.; Nakao, Y. The metallurgical and mechanical properties of ODS alloy MA 956 friction welds. *Weld. J.* **1997**, *76*, S289–S299.
12. Chen, C.L.; Wang, P.; Tatlock, G.J. Phase transformations in yttrium–aluminium oxides in friction stir welded and recrystallised PM2000 alloys. *Mater. High Temp.* **2009**, *26*, 299–303. [CrossRef]
13. Mathon, M.H.; Klosek, V.; de Carlan, Y.; Forest, L. Study of PM2000 microstructure evolution following FSW process. *J. Nucl. Mater.* **2009**, *386–388*, 475–478. [CrossRef]
14. Legendre, F.; Poissonnet, S.; Bonnaillie, P.; Boulanger, L.; Forest, L. Microstructural Characterizations in Friction Stir Welded Oxide Dispersion Strengthened Ferritic Steel Alloy. *J. Nucl. Mater.* **2009**, *386–388*, 537–539. [CrossRef]
15. Zhang, G.; Chandel, R.S.; Seow, H.P.; Hng, H.H. Microstructural Features of Solid State Diffusion Bonded Incoloy MA 956. *Mater. Manuf. Process.* **2003**, *18*, 599–608. [CrossRef]
16. Zhang, H.; Senkara, J. *Resistance Welding: Fundamentals and Applications*, 2nd ed.; CRC/Taylor & Francis: Boca Raton, FL, USA, 2006; p. 53, ISBN 978-1-4398-5371-9.
17. Corpace, F.; Monnier, A.; Poulon-Quintin, A.; Manaud, J.-P. Simulation of Resistance Upset Welding for ODS Steel Fuel Cladding. In Proceedings of the Conference Proceeding-JOM16/ICEW-7, Tisvildeleje, Denmark, 10–13 May 2011.
18. Yazawa, Y.; Furuhara, T.; Maki, T. Effect of matrix recrystallization on morphology, crystallography and coarsening behavior of vanadium carbide in austenite. *Acta Mater.* **2004**, *52*, 3727–3736. [CrossRef]
19. Yamamoto, M.; Ukai, S.; Hayashi, S.; Kaito, T.; Ohtsuka, S. Reverse Phase Transformation from α to γ in 9Cr-ODS Ferritic Steels. *J. Nucl. Mater.* **2011**, *417*, 237–240. [CrossRef]

metals

MDPI

Article

Carbides Evolution and Tensile Property of 4Cr5MoSiV1 Die Steel with Rare Earth Addition

Hanghang Liu [1,2], Paixian Fu [1,2,*], Hongwei Liu [1,2], Chen Sun [1,2], Jinzhu Gao [1,2] and Dianzhong Li [1,2]

[1] Institute of Metal Research, Chinese Academy of Sciences, 72 Wenhua Road, Shenyang 110016, China; hhliu15b@imr.ac.cn (H.L.); hwliu@imr.ac.cn (H.L.); csun15s@imr.ac.cn (C.S.); jzgao11b@imr.ac.cn (J.G.); dzli@imr.ac.cn (D.L.)

[2] School of Materials Science and Engineering, University of Science and Technology of China, 72 Wenhua Road, Shenyang 110016, China

* Correspondence: pxfu@imr.ac.cn; Tel.: +86-24-2397-1973

Received: 17 September 2017; Accepted: 13 October 2017; Published: 18 October 2017

Abstract: Studies of 4Cr5MoSiV1 die steel suggest that under appropriate conditions, additions of rare earth (RE) can enhance tensile property. This improvement is apparently due to the more uniform distribution of carbides and the enhancement of precipitation strengthening after RE additions. In this present work, the effect of the RE addition on the carbides evolution and tensile property of 4Cr5MoSiV1 steel with various RE contents (0, 0.018, 0.048 and 0.15 wt %) were systematically investigated. The two-dimensional detection techniques such as optical microscopy (OM), scanning electron microscopy (SEM), transmission electron microscopy (TEM), and X-ray diffraction (XRD) were used to investigate the carbides evolution of as-cast, annealed and tempered with RE addition. The results indicated that the carbides in 4Cr5MoSiV1 steels were modified by adding the suitable amount of RE. The eutectic structure and coarse eutectic carbides were all refining and the morphology of the annealed carbides initiated change from strip shape to ellipsoidal shape compared with the unmodified test steel (0RE). In addition, the amount of the tempered M_8C_7 carbides increased initially and then decreased with the alteration of RE addition from 0.018 to 0.15 wt %. Notably, the tensile test indicated that the average value of ultimate tensile strength (UTS) and elongation rate of 0.048RE steel increased slightly to 1474 MPa and 15%, higher than the 1452 MPa and 12% for the unmodified test steel (0RE), respectively. Such an addition of RE (0.048 wt %) would have a significant effect on the carbides evolution of as-cast, annealed and tempered and resulting in the tensile property of 4Cr5MoSiV1 die steel.

Keywords: rare earth; 4Cr5MoSiV1 die steel; carbides evolution; tensile strength; elongation rate

1. Introduction

4Cr5MoSiV1 is an excellent hot-worked die steel, which is widely applied in fields such as hot forging, hot extrusion and die-casting. Usually, the working surface temperature of dies can get up to 550 °C, which is very close to the tempering temperature of die steel. Continuous evolution of the microstructure will occur and significantly affect the various properties of the die steel [1]. Actually, the cracks as a network are normally observed on the die surface, which results in more than 80% failure of hot-work dies [2]. To date, researchers found that the uniform hardness, impact toughness, tensile strength and high temperature fatigue strength will be beneficial to prolong the service life when the die steel is subjected to intense friction and mechanical shock in service [3–9]. In addition, N. Mebarki [10], S. Kheirandish [11] and X. Hu [12] found that the coarse eutectic carbides in the process of thermal fatigue could decrease the cyclic softening behavior and lead to fatigue failure. Due to the segregation of chemical constituents, coarse eutectic carbides can be formed during solidification, which can promote crack growth and early failure, and should be reduced by appropriate methods.

The effect of rare earth (RE) in steels is well known as a deoxidizer and desulphuriser. The size and morphology of non-metallic inclusions in steels have changed dramatically after adding Ce, La, Y and Ga [13–17]. Fu [18] and Gao [19] also reported that RE elements like Ce, La and Y can form highly stable oxides, oxy-sulphides and sulphides, which precipitate as solid particles in the melt due to their high melting temperatures. It was shown that the grain size of steel could be refined with a suitable amount of RE and resulted in enhanced impact toughness [16]. In addition, adding RE elements in steel also was an effective method to improve the hot ductility, corrosion properties and abrasion resistance [20–23]. However, there were only a few reports on the improvement of mechanical properties and related changes in carbides due to RE addition. It was shown that the morphology of eutectic carbides changed from network-like structures to granular carbides and resulted in improved mechanical properties in cast steel [15,18,24,25]. However, the influence of RE on the evolution of as-cast, annealed and tempered carbides in 4Cr5MoSiV1 steel is still controversially discussed because of the lack of quantitative experimental data. In general, the reason for the enhancement of the tensile property points to morphological changes of the phase constituents and phase fractions [15].

In this study, the influence of RE additions (0, 0.018, 0.048 and 0.15 wt %) on the carbides evolution and tensile property of 4Cr5MoSiV1-RE die steel were investigated. Intensive investigations were done regarding the changes of the microstructure constituents as well as the influence of the changing carbides phase fractions on the tensile property.

2. Experimental Procedures

The four experimental raw materials of 4Cr5MoSiV1 steel were fabricated with 25 kg capacity medium frequency induction furnace, and the chemical compositions were shown in Table 1. The RE (mainly containing 30 wt % La and 70 wt % Ce) were added into the molten steel under the vacuum atmosphere when the oxygen content of the steel was reduced to a low level (lower than 10 ppm). Different amounts of RE were added into the melt, wrapped in pure iron foil, stirred to ensure the homogeneity of compositions. The residual amounts of RE in steel were 0, 0.018, 0.048 and 0.15 wt %, respectively.

Table 1. Chemical composition of test steels (wt %).

Alloy	C	Si	Mn	Cr	Mo	V	P	S	O	N	H	RE
0 rare earth (RE)	0.37	1.18	0.50	5.00	1.42	1.05	0.005	0.005	0.0008	0.0045	1.7	0
0.018RE	0.35	1.20	0.48	4.90	1.40	1.04	0.005	0.003	0.0006	0.0044	1.3	0.018
0.048RE	0.35	1.18	0.52	5.00	1.39	1.04	0.003	0.003	0.0006	0.0046	1.0	0.048
0.15RE	0.36	1.20	0.50	4.92	1.40	1.05	0.003	0.003	0.0005	0.0045	0.8	0.15

Four raw materials were homogenized at 1200 °C for two hours, and hot-forged at the temperature ranging from 950 °C to 1150 °C. All the ingots were air-cooled down to room temperature after forging. The final size of forging ingots was 70 mm × 70 mm × 500 mm. Subsequently, the heat treatment process of the four forged materials was carried out. Firstly, the samples were heated to 870 °C for two hours, then cooled to 740 °C and isothermal annealing for four hours, and eventually furnace-cooled to room temperature. Subsequently, the quenching and the two times high-temperature tempering process were performed at 600 °C and 610 °C, respectively. The quenching needed to remain one hour at 1040 °C, then was oil-cooled. The remaining time of tempering temperature was two hours, and then air-cooling was carried out.

After conventional metallographic preparation, polished surfaces were etched with 4% nital solution (4 mL HNO_3, 96 mL C_2H_5OH) for subsequent optical microscopy (OM) (Leica Co, DM ILM, Wetzlar, Germany) and scanning electron microscopy (SEM) (JSM-6301F, Japan Electronics Corporation, Tokyo, Japan). Metallographic observations were carried out on the specimens' subjected to as-cast and annealed state. The image analyses were performed by using Image-Pro Plus 6.0 software (6.0, Media Cybernetics Inc., Rockville, MD, USA). Image-Pro Plus is designed for processing, enhancing and analyzing pictures; it has exceptionally rich measurement and customization features. In the present trials, 15 large fields of view (5000×) and 15 small fields of view (10,000×) of the SEM micrographs were randomly selected in order to analyze the average diameter, the distribution, and quantity of the annealed carbides of the test steels with different additions of RE. Similarly, 15 large fields of view (100×) and 15 small fields of view (200×) of the optical microscopys (OM) were randomly selected in order to analyze the secondary dendrite arm spacing of the as-casted specimens and these statistics are based on reference [26]. In addition, the types of the annealed carbides were extensively analyzed by SEM equipped with energy dispersive X-ray spectroscopy (EDX).

The tempered carbides substructure of the test steels was observed using an X-ray diffraction (XRD) with Cu-Kα (λ = 1.5406 Å) with a scanning angle from 15° to 85° and a scanning speed of 2°/min and a field-emission transmission electron microscope (TEM) operated at 200 kV. The TEM observation was conducted by using F20 (FEI company, Hillsboro, OR, USA), which is also equipped with an Oxford INCA type spectrometer (Japan Electronics Corporation, Tokyo, Japan) and GATAN 832 CCD image recorder (Japan Electronics Corporation, Tokyo, Japan). TEM samples were mechanically thinned to the thickness of approximately 50 μm by SiC paper, then punched into disks of 3 mm in diameter and further thinned by twin-jet electro polishing, mixing in of 10 vol % perchloric acid ethanol solution at a voltage of 20 V and a temperature between −30 °C and −20 °C.

The X-ray diffraction (XRD; Cu-Kα; λ = 1.5406 Å; with a scanning angle from 15° to 85° and a scanning speed of 2°/min) combined with Reference intensity ratio analysis (RIR) [27] and MDI Jade 6.5 software(6.5, Materials Data Inc., Livermore, CA, USA) were applied for the qualitative and quantitative phase analysis.

Round bar tensile specimens were prepared in the longitudinal direction with the gage length and diameter of 25 mm and 5 mm, respectively. They were then tested at a strain rate of 0.5 mm/min by using an AG-100KNG tensile machine from Shimadzu (Kyoto, Japan).At least three tensile tests for each testing condition are adopted here for the average value. The tensile properties were tested at room temperature.

3. Results

3.1. Carbides Evolution

3.1.1. Influence of RE Addition on As-Cast Carbides

Figure 1a,e,f shows the metallographic and SEM micrographs of the as-cast microstructure of unmodified test steel (0RE). The results show that most of the eutectic carbides distribute in net shape and are generally coarse, Energy dispersive spectrometer (EDS) analysis shows these are $(V, Mo)_xC_y$ (Figure 1g). In addition, the as-cast microstructures of modified test steels are shown in Figure 1b–d,h–j. The structural difference between unmodified and modified test steels are that the dendritic spacings turn into fine structures and eutectic carbides are refining or even disappearing. The secondary dendrite arm spacing of the as-casted specimens are quantitatively analyzed, as shown in Figure 2. The results show that the spacing of secondary dendrite arm was decreased from 72 μm to 25 μm with different additions of RE. It indicates that RE has a role in refining the secondary dendrite arm and the precipitation of eutectic carbides. A similar refining of the precipitation of eutectic carbides with a certain amount of RE addition was also described by J. Hufenbach [15] and Fu [18].

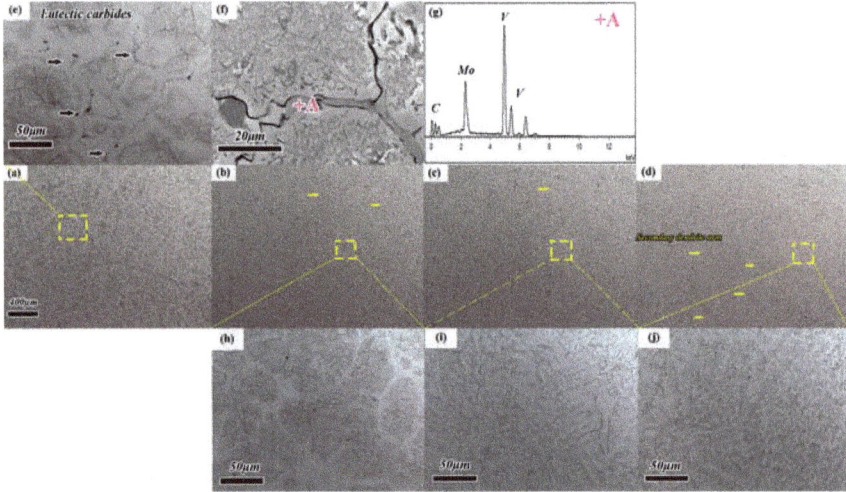

Figure 1. Metallographic and SEM micrographs of the specimens as-casted: (**a**) Metallographic observations in 0RE; (**b**) Metallographic observations in 0.018RE; (**c**) Metallographic observations in 0.048RE; (**d**) Metallographic observations in 0.15RE; (**e**) Metallographic observations in 0RE; (**f**) SEM micrographs in 0RE; (**g**) Energy dispersive spectrometer (EDS) analysis of eutectic carbides at A region in (**f**); (**h**) Metallographic observations in 0.018RE; (**i**) Metallographic observations in 0.048RE and (**j**) Metallographic observations in 0.15RE.

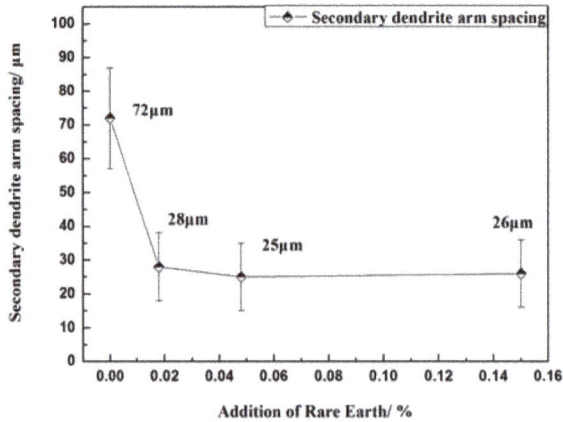

Figure 2. The secondary dendrite arm spacing of the specimens as-casted with different additions of RE.

3.1.2. Influence of RE Addition on the Annealed Carbides

The morphology of the annealed carbides in test steels was systematically investigated by metallographic and SEM micrographs, as shown in Figures 3 and 4. The results show that the morphology of the annealed carbides gradually becomes ellipsoid and the chain carbides become less when adding the RE. Most notably, some large clusters of the annealed carbides are observed, as shown in Figure 4a,b in the unmodified test steel (0RE). However, dispersion distributions of ellipsoid annealed carbides were observed in the observation field when adding RE (Figure 4d–f). In addition, coarse eutectic carbides are commonly found in the unmodified test steel (Figure 4b,c),

EDS analysis shows that these are $(V, Mo)_xC_y$. However, less coarse eutectic carbides are found in steel with RE addition (Figure 4d–f).

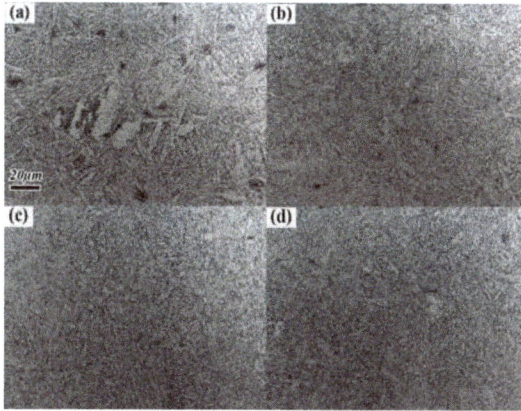

Figure 3. Metallographic observations of the annealed specimens: (**a**) 0RE; (**b**) 0.018RE; (**c**) 0.048RE and (**d**) 0.15RE.

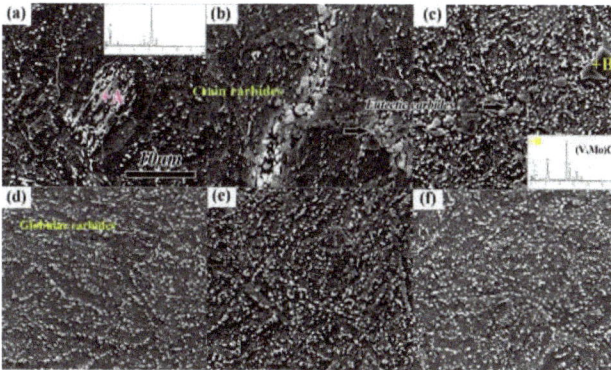

Figure 4. SEM and EDS observations of the annealed specimens: (**a**) Carbides aggregation in 0RE; (**b**) chain carbides in 0RE; (**c**) the eutectic carbides in 0RE; (**d**) ellipsoid carbides in 0.018RE; (**e**) ellipsoid carbides in 0.048RE and (**f**) ellipsoid carbides in 0.15RE. (The insert in (**a**) is the EDS analysis of carbides at A region, the insert in (**c**) is the EDS analysis of eutectic carbides at B region).

The relationship between the content of RE and the average diameter, the distribution, and quantity of the annealed carbides of the test steels are quantitatively analyzed, as shown in Figure 5. The results show that the average diameter of the annealed carbides almost unchanged with the increase of RE content, whereas the percentage is increased from 11.8% to 19.6% (Figure 5a). In addition, the annealed carbides distribution is discrete in the unmodified test steel (0RE), and the aggregate carbides with diameters greater than 1.5 µm still account for more than 3% of the area fraction, as shown in Figure 5b. However, the size distribution of the annealed carbides is more uniform when RE is added, which indicates that RE has the role of spheroidizing and uniformly dispersing of the annealed carbides. In addition, the quantity of the annealed carbides increases from 1682 to 2078 in the same statistical region (773.77 µm^2) with the addition of 0.018 to 0.15 wt % RE (Figure 5c).

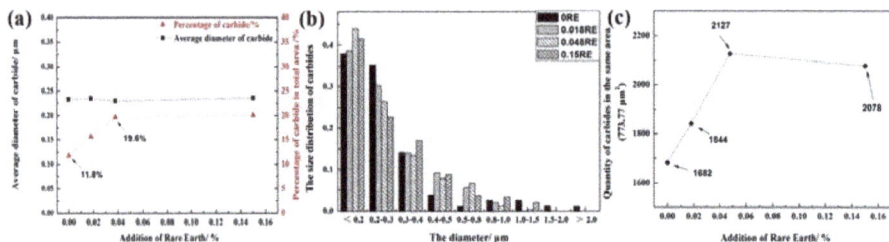

Figure 5. (a) Average diameter and percentage of the annealed carbides with different additions of RE; (b) The distributions of the annealed carbides with different additions of RE; and (c) The quantity of the annealed carbides in the same area (773.77 μm^2) with different additions of RE.

3.1.3. Influence of RE Addition on the Tempered Carbides

The quantitative analysis and crystal structure types of the tempered carbides of the test steels are examined by XRD. The corresponding patterns are shown in Figure 6 and the results are summarized in Table 2. The results show that the carbides of M_7C_3, M_8C_7 and $M_{23}C_6$ are detected in all test steels. In addition, the amount of the tempered M_8C_7 carbides slightly increases initially and then decreases with the alteration of RE addition from 0.018 to 0.15 wt %. The effect of RE on the amount of the M_8C_7 carbides was also described by J. Hufenbach for high-strength Fe85Cr4Mo8V2C1 cast steel [15].

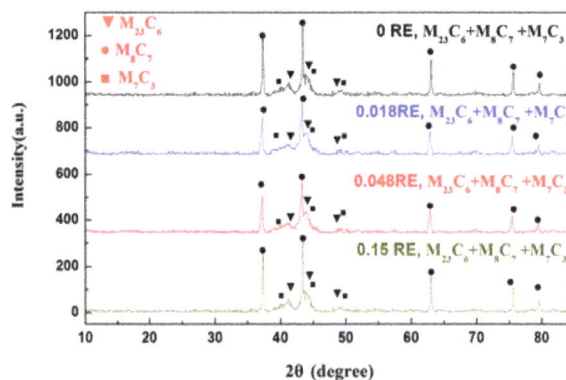

Figure 6. X-ray diffraction patterns of the 4Cr5MoSiV1-RE die steel with different additions of RE.

Table 2. Phase composition, space group and lattice parameters of carbides of the 4Cr5MoSiV1-*x*RE (*x* = 0, 0.018, 0.048, 0.15) steels determined by the analysis of XRD data.

Steel	Structure Type	Space Group	*a* (nm)	*b* (nm)	*c* (nm)	Phase Content (wt %)
0RE	V_8C_7	P4332(212)	8.340	8.340	8.340	40
	$M_{23}C_6$	Fm-3m(225)	10.660	10.660	10.660	29
	M_7C_3	Pmcm(51)	7.015	12.153	4.532	31
0.018RE	V_8C_7	P4332(212)	8.340	8.340	8.340	48
	$M_{23}C_6$	Fm-3m(225)	10.660	10.660	10.660	36
	M_7C_3	Pmcm(51)	7.015	12.153	4.532	16
0.048RE	V_8C_7	P4332(212)	8.340	8.340	8.340	52
	$M_{23}C_6$	Fm-3m(225)	10.660	10.660	10.660	36
	M_7C_3	Pmcm(51)	7.015	12.153	4.532	12
0.15RE	V_8C_7	P4332(212)	8.340	8.340	8.340	44
	$M_{23}C_6$	Fm-3m(225)	10.660	10.660	10.660	28
	M_7C_3	Pmcm(51)	7.015	12.153	4.532	28

Meanwhile, the tempered carbides in unmodified (0RE) and modified test steels (0.15RE) are observed using both TEM bright-field (BF) observation and selected area electron diffraction (SAED) pattern in order to investigate the crystal structure and chemical composition, as shown in Figure 7. The results show that the particle in Figure 7b,c is the $M_{23}C_6$ carbides with the face-centered cubic crystal structure. The larger irregular block $M_{23}C_6$ carbides will be precipitated along the grain and subgrain boundaries when the tempering temperature is above 500 °C [28]. Further, a large spherical particle in Figure 7a,d is identified to be the M_7C_3 carbides with the hexagonal close-packed crystal structure. Finally, a large number of small spherical particles in Figure 7a–c were identified to be the M_8C_7 carbides. In addition, from EDS analysis, it is known that these carbides are complex mixed carbides, whereby the M represents Cr, V, Mn, Mo and Fe in unmodified steel (Figure 7e). However, when RE reaches 0.15 wt %, the chemical compositions are changed to Cr, V, Mn, Mo, Fe, La and Ce (Figure 7f). It shows that RE mainly influences the volumetric fraction and chemical composition of the tempered carbides, which has little influence on the crystal structure.

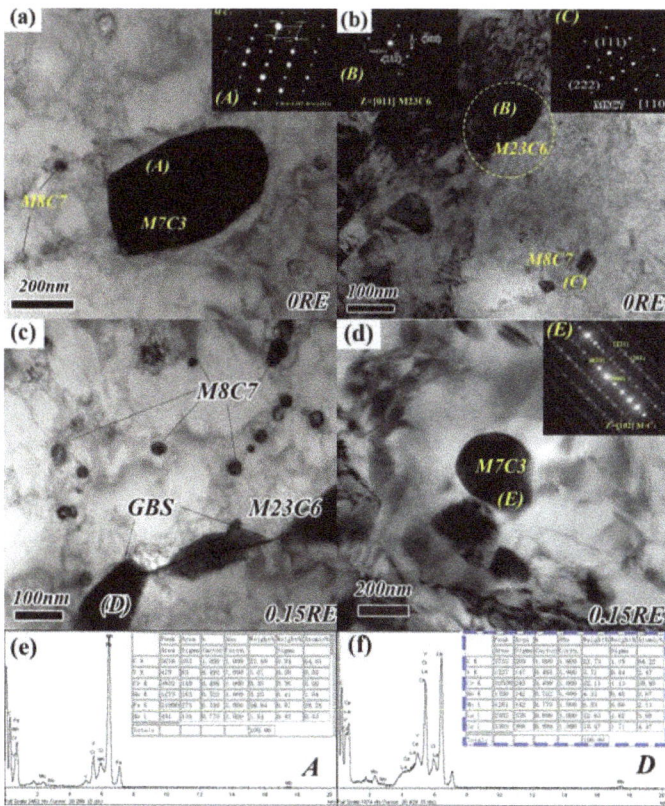

Figure 7. TEM microstructure with EDS analysis and diffraction pattern analysis of the types of carbides in the 4Cr5MoSiV1-RE die steel in 0RE and 0.15RE: (**a**) M_7C_3 and M_8C_7 in 0RE; (**b**) $M_{23}C_6$ and M_8C_7 in 0RE; (**c**) $M_{23}C_6$ and M_8C_7 in 0.15RE; (**d**) M_7C_3 in 0.15RE; (**e**) the EDS analysis of M_7C_3 at A region in (**a**) and (**f**) the EDS analysis of $M_{23}C_6$ at D region in (**c**). (The insert in (**a**) is the selected area electron diffraction (SAED) pattern of M_7C_3 at A region), (The insert in (**b**) is the SAED pattern of $M_{23}C_6$ at B region), (The insert in (**b**) is the SAED pattern of M_8C_7 at C region), and (the insert in (**d**) is the SAED pattern of M_7C_3 at E region).

In summary, adding RE leads to a refining of eutectic structure and coarse eutectic carbides. In addition, the morphology of the annealed carbides initiates changes from strip shape to ellipsoidal shape, and the amount of the tempered M_8C_7 carbides increases initially and then decreases with the alteration of RE addition from 0.018 to 0.15 wt %.

3.2. Tensile Property

Table 3 presents the yield strength (YS), ultimate tensile strength (UTS), elongation and reduction of area of each sample of 4Cr5MoSiV1 steels with different addition of RE. The average value of yield strength (YS) and ultimate tensile strength (UTS) slightly increases from 1227 MPa to 1254 MPa and 1452 MPa to 1474 MPa when the content of RE is 0.048 wt %. However, when the RE content reaches 0.15 wt %, the average value of yield strength (YS) and ultimate tensile strength (UTS) decreases to 1213 MPa and 1430 MPa, respectively. A similar decrease of the tensile strength after reaching a certain amount of RE addition was also described by J. Hufenbach [15] and Wang [25]. In addition, Figure 8 shows the engineering stress-strain curves of the four specimens. The results show that all four test steels exhibit a similar tensile response, the elongation rate of 0.048RE steel increases slightly to 15%, higher than 12% for the 0RE steel.

Table 3. Variation in tensile strength of each sample in response to different additions of RE.

Alloy	No.	Yield Strength (YS) ($\sigma_{0.2}$) (Mpa)		Ultimate Tensile Strength (UTS) (σ_b) (Mpa)		Elongation Rate (δ)		Reduction of Area (ψ)	
		Vaule	Average Vaule	Vaule	Average Vaule	Vaule	Average Vaule	Vaule	Average Vaule
	1#	1228		1453		12.5		48	
0RE	2#	1228	1227	1455	1452	12	12	47.5	48
	3#	1225		1448		13		48.5	
	1#	1238		1465		13		50	
0.018RE	2#	1235	1236	1458	1461	13	13	48	48.5
	3#	1235		1460		13		48	
	1#	1257		1478		16.5		54.5	
0.048RE	2#	1250	1254	1470	1474	14	15	51	53
	3#	1255		1474		14.5		53.5	
	1#	1210		1425		13.5		50	
0.15RE	2#	1212	1213	1430	1430	14	13.5	49	50
	3#	1217		1435		13		51	

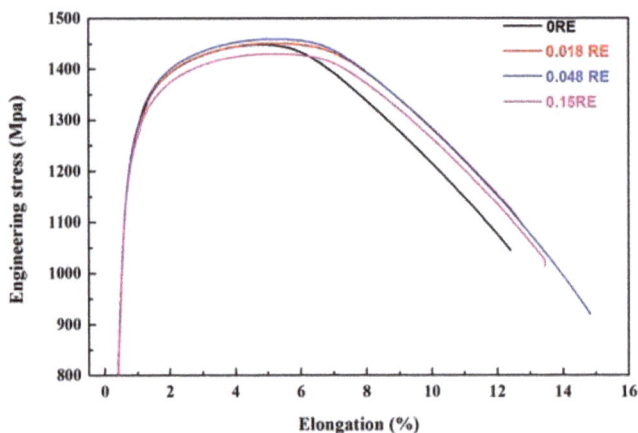

Figure 8. The representative engineering tensile stress-strain curves in response to different additions of RE.

4. Discussion

4.1. Carbides Evolution Induced by RE Addition

The reasons for the refinement of eutectic carbides with different addition of RE can be summarized as follows: on the one hand, the solubility of RE in the matrix is very small, and most of the RE in steel will be obviously segregated and enriched into the front of dendrites during the solidification process, and leading to high composition supercooling [29]. This is advantageous to the decrease of dendrite spacing. Therefore, the dendrite structure and the eutectic carbides formed in the residual melts at the end of the solidification process were refined (Figures 1 and 2). On the other hand, according to the heterogeneous nucleation theory proposed by Turnbull [30], whether the additive contributes to nucleation of molten metal depends on the following two conditions. First, the additive must remain at the melting point above the solid phase when the substrate is nucleated. Second, the index of the lattice surface of the substrate phase and the matrix should be less mismatched. The low mismatching, smaller surface energy, leads to a decrease of super-cooling required for nucleation, and results in refining of the grains. RE-inclusion possesses both high melting point and low mismatching between the matrix through the preliminary research work on the influence of RE on inclusions in 4Cr5MoSiV1 die steel by our group [19]. Moreover, the most common form of inclusions are found believed to be RE oxy-sulfides (Ce_2O_2S) with various RE contents (0, 0.015, 0.025 and 0.10 wt %) in 4Cr5MoSiV1 die steel [19]. In addition, the lattice disregistry between (0001) Ce_2O_2S and (111) δ-Fe was only 3.5% [19]. Simultaneously, Bramfitt [31] has shown that the additive is effective heterogeneous nuclei if the lattice misfit between the inclusions and the matrix is less than 6%. Therefore, RE-inclusion like Ce_2O_2S can be used as inoculants to greatly enhance nucleation, which is advantageous to the development of dendritic crystals to polycrystals and reduces dendrite spacing [18]. During the process of austenite growth, the dendritic crystals contact with each other to form a framework, which hinders the continuous growth of eutectic carbides [32]. In addition, the reason for the coarse ellipsoid eutectic carbides was commonly found after annealing in 0RE samples as the carbides dissolved progressively into matrix when the heat treatment temperature increased. The sharp corners of the carbides became more round and some vulnerable areas began to crack.

In addition, the evolution of the annealed carbides from chain forms to ellipsoid forms are explained in that not only the solubility of RE in steel is very limited but also the RE atoms tend to segregate on carbides/matrix interfaces during the heat treatment process. In addition, the much larger atomic radius of RE, 0.3745 nm and 0.3637 nm for La and Ce, respectively, than that of Fe (0.254 nm) will cause lattice distortion and thus decreasing carbides stability. As a result, the preferred growth rate of the annealed carbides is hindered, thereby preventing the formation of the chain carbides [33].

In order to accurately investigate the effect of RE on the amount of V_8C_7-type carbides precipitation, it is necessary to quantify the relationship between the amount of RE in the steel and the solution temperature of V_8C_7-type carbides. To date, a large number of experimental results showed that V_xC_y-type carbides had a certain degree of carbon atom vacancy, so the chemical formulas of V_xC_y-type carbides were generally $VC_{0.75}$, $VC_{0.875}$ and VC and so on. In this part, we assume that the possibility of V_xC_y-type carbides in steel is the absence of interstitial atoms. The formula of the equilibrium solid solubility of VC-type carbides in steel is shown in Formula (1) [34]:

$$\lg\{[V] \cdot [C]\} = 6.72 - \frac{9500}{T}, \tag{1}$$

In addition, the activity coefficients of the RE and other solute elements in steel on V and C elements were derived into the Formula (1), and the solubility product of VC-type carbides with RE and without RE experimental steels was calculated. Table 4 shows the interaction coefficient e_i^j of various elements in liquid steel at 1873 K from Wagner's relation [35,36]. In addition, the formula of solid solubility of binary second phases MC-type carbide in steel is Formula (2) [37]:

$$\lg\{[M]\cdot[C]\} = A - \frac{B}{T} - \sum_{j=1}^{n}\frac{A_{Fe}}{100A_j\ln 10}e_M^j w_j - \sum_{j=1}^{n}\frac{A_{Fe}}{100A_j\ln 10}e_C^j w_j, \tag{2}$$

where [M], [C] is the mass fractions of M and C elements, which are solid soluble in steel, respectively. A and B are constants, where A is 6.72, B is 9500, A_j for atomic weight, and w_j for mass fraction in steel. According to the chemical composition of experimental steels in Table 1, the solid solubility product of VC-type carbides in the unmodified test steel (0RE) is calculated by using the Formula (3):

$$
\begin{aligned}
\lg\{[V]\cdot[C]\} = {}& 6.72 - \frac{9500}{T} - \frac{56}{100\times 12\times\ln 10}e_C^C w_C - \frac{56}{100\times 28\times\ln 10}e_C^{Si}w_{Si} - \frac{56}{100\times 55\times\ln 10}e_C^{Mn}w_{Mn} \\
& - \frac{56}{100\times 52\times\ln 10}e_C^{Cr}w_{Cr} - \frac{56}{100\times 51\times\ln 10}e_C^V w_V - \frac{56}{100\times 96\times\ln 10}e_C^{Mo}w_{Mo} \\
& - \frac{56}{100\times 140\times\ln 10}e_C^{Ce}w_{Ce} - \frac{56}{100\times 12\times\ln 10}e_V^C w_C - \frac{56}{100\times 28\times\ln 10}e_V^{Si}w_{Si} \\
& - \frac{56}{100\times 55\times\ln 10}e_V^{Mn}w_{Mn} - \frac{56}{100\times 52\times\ln 10}e_V^{Cr}w_{Cr} - \frac{56}{100\times 51\times\ln 10}e_V^V w_V \\
& - \frac{56}{100\times 96\times\ln 10}e_V^{Mo}w_{Mo} - \frac{56}{100\times 140\times\ln 10}e_V^{Ce}w_{Ce} \\
= {}& 6.582951 - \frac{9187.7954}{T}
\end{aligned}
\tag{3}
$$

Similarly, the solid solubility product of VC-type carbides in steel in 0.018RE, 0.048RE and 0.15RE experimental steels are as follows (4)–(6):

$$\lg\{[V]\cdot[C]\} = 6.582731 - \frac{9186.3783}{T}, \tag{4}$$

$$\lg\{[V]\cdot[C]\} = 6.58071 - \frac{9172.5043}{T}, \tag{5}$$

$$\lg\{[V]\cdot[C]\} = 6.57911 - \frac{9174.6018}{T}, \tag{6}$$

Further, the total solution temperature of VC-type carbides in steel can be calculated by the following Formula (7) [36].

$$T_{AS} = \frac{B}{A - \lg(M\cdot X^x)}, \tag{7}$$

In Formula (7), the constants A and B are the same as in Formula (3), M and X are the mass fractions of the second phase (VC-type carbide) in steel, respectively (%), and the total solution temperature of VC-type carbide in 0RE steel can be calculated by Formula (7), and the results are as shown below:

$$T_{AS} = \frac{9187.7954}{6.582951 - \lg(0.37\times 1.05)} = 1313.750855129\ \text{K}, \tag{8}$$

Similarly, the total solution temperature of VC-type carbide in 0.018RE, 0.048RE and 0.15RE steel is 1308 K, 1306 K and 1310 K, respectively. The calculation results show that the addition of RE elements in steel can promote the dissolution of V and reduce the total solution temperature of VC-type carbides significantly and the maximum temperature difference can be 7 K (1313 K to 1306 K). After tempering at the same temperature, the saturated V element in the matrix will be dispersed in the form of VC-type carbide precipitation, and the strength of the steel can be improved to a certain extent.

Table 4. Interaction coefficient e_i^j of various elements in liquid steel at 1873 K.

Element (i, j)	C	Si	Mn	Cr	V	Mo	Ce
V	−3.4	4.2	0	0	1.3	0	−2836/T + 1.40
C	8890/T	4.84 + 7370/T	−5070/T	7.02 − 21,800/T	23,900/T − 22.9	3.86 − 17,870/T	−150/T + 0.05

4.2. Tensile Property Induced by RE Addition

The dimensions, morphology and volumetric fraction of carbides play a critical role in improving the strength properties of 4Cr5MoSiV1 die steel. In this study, the main reasons for the increase in strength properties from the aspect of carbides evolution can be summarized as follows: first, the proeutectoid carbides not only influence the uniformity of microstructure, but also have a very significant influence on the strength properties [38]. The cracks tend to grow and spread around chain proeutectoid carbides [37]. Second, the effect of eutectic carbides modification by adding RE results in the finer morphologies observed in Figure 1. Hence, more carbon and other alloy elements dissolve into matrix at the high-temperature austenitizing treatment, and enhance the precipitation of alloy carbides during tempering. However, a further increase of the RE content to 0.15 wt % results in a slightly decrease of yield and ultimate strength. The trend partly can be explained by the changing phase fractions which are described above (Table 2). The fraction of the tempered M_8C_7 carbides increased from 40 to 52 wt % by adding 0.048 wt % RE in the test steels. Further additions of RE (0.15 wt %) led to a significant decrease of the tempered M_8C_7 carbides fraction from 52 to 44 wt %. As a kind of nano-carbides, M_8C_7 carbides have the function of improving precipitation strengthening according to research proposed by Hojun Gwon [39] and J. Hufenbach [15]. In addition, the ductility was better with various RE contents (0, 0.018, 0.048 and 0.15 wt %) addition (Table 3). One possible explanation for the observed enhanced ductility of the RE addition may be found in the arrangement of the carbides [15]. The complex chain carbides became increasingly interrupted with the RE addition; therefore, crack propagation along this path was hampered [40,41].

5. Conclusions

In this study, the influence of RE additions (0, 0.018, 0.048 and 0.15 wt %) on the carbides evolution and tensile property of 4Cr5MoSiV1-RE die steel were investigated. The main results are summarized as follows:

1. The microstructure observation of as-cast shows that after adding RE, it will lead to a refining of eutectic structure and coarse eutectic carbides.
2. The morphology of the annealed carbides initiates changes from strip shape to ellipsoidal shape, and the quantity of the annealed carbides increases from 1682 to 2078 in the same statistical region (773.77 μm^2) with the alteration of RE addition from 0.018 to 0.15 wt %.
3. The amount of the tempered M_8C_7 carbides increases initially and then decreases with the alteration of RE addition from 0.018 to 0.15 wt %. The addition of RE influences the chemical composition and morphology of tempered carbides, which has little influence on the lattice structure.
4. Tensile test shows that ultimate tensile strength (UTS) and elongation rate of 0.048RE steel increases slightly to 1474 MPa and 15%, higher than the 1452 MPa and 12% for the unmodified test steel (0RE), respectively.
5. Adding 0.048 wt % RE content to the 4Cr5MoSiV1 die steel can obtain tool steels with better tensile strength and elongation rate, which is promising for advanced tool design.

Acknowledgments: The authors acknowledge the financial support by the National Key Research and Development Program of China (2016YFB0300401).This work also was supported by the Cooperation Program of Hubei province and Chinese Academy of Sciences (The Research and Development of Key Technologies for Special Steel of Homogeneous High Performance).

Author Contributions: Hanghang Liu carried out the experiments and analysis of the results. Paixian Fu, Hongwei Liu, Chen Sun and Jinzhu Gao prepared and revised the manuscript. Dianzhong Li contributed to the interpretation and discussion of results.

Conflicts of Interest: The authors declare no conflict of interest.

References

1. Zhou, Q.C.; Wu, X.C.; Shi, N.N.; Li, J.; Min, N. Microstructure evolution and kinetic analysis of DM hot-work die steels during tempering. *Mater. Sci. Eng. A* **2011**, *528*, 5696–5700. [CrossRef]

2. Tsujii, N.; Abe, G.; Fukaura, K.; Sunada, H. High temperature low cycle fatigue behaviour of a 4.2 Cr-2.5 Mo-V-Nb hot work tool steel. *J. Mater. Sci. Lett.* **1996**, *15*, 1251–1254. [CrossRef]

3. Li, G.; Li, X.; Wu, J. Study of the thermal fatigue crack initial life of H13 and H21 steels. *J. Mater. Process. Technol.* **1998**, *74*, 23–26. [CrossRef]

4. Arif, A.; Sheikh, A.; Qamar, S. A study of die failure mechanisms in aluminum extrusion. *J. Mater. Process. Technol.* **2003**, *134*, 318–328. [CrossRef]

5. Telasang, G.; Majumdar, J.D.; Padmanabham, G.; Manna, I. Wear and corrosion behavior of laser surface engineered AISI H13 hot working tool steel. *Surf. Coat. Technol.* **2015**, *261*, 69–78. [CrossRef]

6. Wei, M.X.; Wang, S.Q.; Wang, L.; Cui, X.H.; Chen, K.M. Effect of tempering conditions on wear resistance in various wear mechanisms of H13 steel. *Tribol. Int.* **2011**, *44*, 898–905. [CrossRef]

7. Sola, R.; Giovanardi, R.; Parigi, G.; Veronesi, P. A novel method for fracture toughness evaluation of tool steel with post-tempering cryogenic treatment. *Metals* **2017**, *7*, 75. [CrossRef]

8. Perez, M.; Rodrigez, R.; Belzunce, F.J. The use of cryogenic treatment to increase the fracture toughness of the hot work tool steel used to make forging dies. *Procedia Meteralia Sci.* **2014**, *3*, 604–609. [CrossRef]

9. Poli, R.S.G.; Defanti, S.; Veronesi, P.; Parigi, G. Effect of deep cryogenic treatment on the properties of AISI M2 steel. In Proceedings of the European Conference on Heat Treatment and 22nd IFHTSE Congress, Venice, Italy, 20 May 2015.

10. Mebarki, N.; Delagnes, D.; Lamesle, P.; Delmas, F.; Levaillant, C. Relationship between microstructure and mechanical properties of a 5% Cr tempered martensitic tool steel. *Mater. Sci. Eng. A* **2004**, *387–389*, 171–175. [CrossRef]

11. Kheirandish, S.; Noorian, A. Effect of niobium on microstructure of cast AISI H13 hot work tool steel. *J. Iron Steel Res. Int.* **2008**, *15*, 61–66. [CrossRef]

12. Hu, X.; Li, L.; Wu, X.; Zhang, M. Coarsening behavior of $M_{23}C_6$ carbides after ageing or thermal fatigue in AISI 4Cr5MoSiV1 steel with niobium. *Int. J. Fatigue* **2006**, *28*, 175–182. [CrossRef]

13. Ha, H.; Park, C.; Kwon, H. Effects of misch metal on the formation of non-metallic inclusions and the associated resistance to pitting corrosion in 25% Cr duplex stainless steels. *Scr. Mater.* **2006**, *55*, 991–994. [CrossRef]

14. Cai, Y.C.; Liu, R.P.; Wei, Y.H.; Cheng, Z.G. Influence of Y on microstructures and mechanical properties of high strength steel weld metal. *Mater. Des.* **2014**, *62*, 83–90. [CrossRef]

15. Hufenbach, J.; Helth, A.; Lee, M.H.; Wendrock, H.; Giebeler, L.; Choe, C.Y.; Kim, K.H.; Kühn, U.; Kim, T.-S.; Eckert, J. Effect of cerium addition on microstructure and mechanical properties of high-strength Fe85Cr4Mo8V2C1 cast steel. *Mater. Sci. Eng. A* **2016**, *674*, 366–374. [CrossRef]

16. Kim, S.T.; Jeon, S.H.; Lee, I.S. Effects of rare earth metals addition on the resistance to pitting corrosion of super duplex stainless steel-Part 1. *Corros. Sci.* **2010**, *52*, 1897–1904. [CrossRef]

17. Ahn, J.H.; Jung, H.D.; Im, J.H.; Jung, K.H.; Moon, B.-M. Influence of the addition of gadolinium on the microstructure and mechanical properties of duplex stainless steel. *Mater. Sci. Eng. A* **2016**, *658*, 255–262. [CrossRef]

18. Fu, H.; Xiao, Q.; Li, Y. A study of the microstructures and properties of Fe-V-W-Mo alloy modified by rare earth. *Mater. Sci. Eng. A* **2005**, *395*, 281–287. [CrossRef]

19. Gao, J.; Fu, P.; Liu, H.; Fu, P.; Li, D. Effects of rare earth on the microstructure and impact toughness of H13 steel. *Metals* **2015**, *5*, 383–394. [CrossRef]

20. Liu, H.L.; Liu, C.J.; Jiang, M.F. Effect of rare earths on impact toughness of a low-carbon steel. *Mater. Des.* **2012**, *33*, 306–312. [CrossRef]

21. Sun, Z.; Zhang, C.S.; Yan, M.F. Microstructure and mechanical properties of M50NiL steel plasma nitrocarburized with and without rare earths addition. *Mater. Des.* **2014**, *55*, 128–136. [CrossRef]

22. Chen, L.; Ma, X.; Wang, L.; Ye, X. Effect of rare earth element yttrium addition on microstructures and properties of a 21Cr-11Ni austenitic heat-resistant stainless steel. *Mater. Des.* **2011**, *32*, 2206–2212. [CrossRef]

23. Jiang, X.; Song, S.H. Enhanced hot ductility of a Cr-Mo low alloy steel by rare earth cerium. *Mater. Sci. Eng. A* **2014**, *613*, 171–177. [CrossRef]

24. Fu, H.; Xiao, Q. Effect of rare earth and titanium additions on the microstructures and properties of low carbon Fe-B cast steel. *Mater. Sci. Eng. A* **2007**, *466*, 160–165. [CrossRef]
25. Wang, M.; Mu, S.; Sun, F.; Wang, Y. Influence of rare earth elements on microstructure and mechanical properties of cast high-speed steel rolls. *J. Rare Earth* **2007**, *25*, 490–494. [CrossRef]
26. Ning, A.; Mao, W.; Chen, X.; Guo, H.; Guo, J. Precipitation behavior of carbides in H13 hot work die steel and its strengthening during tempering. *Metals* **2017**, *7*, 70. [CrossRef]
27. Hubbard, C.R.; Evans, E.H.; Smith, D.K. The reference intensity ratio, I/I_c, for computer simulated powder patterns. *J. Appl. Crystallogr.* **1976**, *9*, 169–174. [CrossRef]
28. Dong, J.; Zhou, X.; Liu, Y.; Li, C.; Liu, C.; Guo, Q. Carbide precipitation in Nb-V-Ti microalloyed ultra-high strength steel during tempering. *Mater. Sci. Eng. A* **2017**, *683*, 215–226. [CrossRef]
29. Xu, Z.M. Influence of Ce and Al on nodularization of eutectic in austenite-bainite steel. *Mater. Res. Bull.* **2000**, *35*, 1261–1268. [CrossRef]
30. Turnbull, D.; Vonnegut, B. Nucleation catalysis. *Ind. Eng. Chem.* **2002**, *44*, 1292–1298. [CrossRef]
31. Bramfitt, B.L. Planar lattice disregistry theory and its application on heterogistry nuclei of metal. *Metall. Trans.* **1970**, *1*, 1987–1995. [CrossRef]
32. Yang, Q.X.; Liao, B.; Liu, J.H.; Yao, M. Effect of rare earth elements on carbide morphology and phase transformation dynamics of high Ni-Cr alloy cast iron. *J. Rare Earth* **1998**, *1*, 37–41.
33. Xu, Y.W.; Song, S.H.; Wang, J.W. Effect of rare earth cerium on the creep properties of modified 9Cr-1Mo heat-resistant steel. *Mater. Lett.* **2015**, *161*, 616–619. [CrossRef]
34. Narita, K. Physical chemistry of the groups IVa (Ti, Zr), Va (V, Nb, Ta) and the rare earth elements in steel. *Trans. ISIJ* **1975**, *15*, 145–152.
35. Sharma, R.C.; Lakshmanan, V.K.; Kirkaldy, J.S. Solubility of niobium carbide and niobium carbonitride in alloyed austenite and ferrite. *Metall. Mater. Trans. A* **1984**, *15*, 545–553. [CrossRef]
36. Wang, L.M.; Du, T.; Lu, X.L.; Li, Z.B.; Gai, Y.C. Thermodynamics and application of rare earth elements in steel. *J. Chin. RE Soc.* **2013**, *21*, 251–254.
37. Long, Q.L. *Second Phases in Structural Steels*; Metallurgical Industry Press: Beijing, China, 2006; p. 119. (In Chinese)
38. Mudali, U.K.; Raj, B. *High Nitrogen Steels and Stainless Steels-Manufacturing, Properties and Applications*; Narosa Publishing House: New Delhi, India, 2004.
39. Gwon, H.; Kim, J.K.; Shin, S.; Cho, L.; de Cooman, B.C. The effect of vanadium micro-alloying on the microstructure and the tensile behavior of TWIP steel. *Mater. Sci. Eng. A* **2017**, *696*, 416–428. [CrossRef]
40. Wang, K.H.; Lee, S.; Lee, H. Effects of alloying elements on microstructure and fracture properties of cast high speed steel rolls: Part II. Fracture behavior. *Mater. Sci. Eng. A* **1998**, *254*, 296–304. [CrossRef]
41. Picasa, I.; Cuadrado, N.; Casellas, D.; Goez, A.; Llanes, L. Microstructural effects on the fatigue crack nucleation in cold work tool steels. *Procedia Eng.* **2010**, *2*, 1777–1785. [CrossRef]

metals

MDPI

Article

Effect of Weld Current on the Microstructure and Mechanical Properties of a Resistance Spot-Welded TWIP Steel Sheet

Mumin Tutar , Hakan Aydin * and Ali Bayram

Engineering Faculty, Mechanical Engineering Department, Uludag University, 16059 Gorukle-Bursa, Turkey; mumintutar@uludag.edu.tr (M.T.); bayram@uludag.edu.tr (A.B.)
* Correspondence: hakanay@uludag.edu.tr; Tel.: +90-224-294-0652

Received: 30 September 2017; Accepted: 14 November 2017; Published: 23 November 2017

Abstract: In this study the effect of the weld current on the microstructure and mechanical properties of a resistance spot-welded twinning-induced plasticity (TWIP) steel sheet was investigated using optical microscopy, scanning electron microscopy–electron back-scattered diffraction (SEM–EBSD), microhardness measurements, a tensile shear test and fractography. Higher weld currents promoted the formation of a macro expulsion cavity in the fusion zone. Additionally, higher weld currents led to a higher indentation depth, a wider heat-affected zone (HAZ), coarser grain structure and thicker annealing twins in the HAZ, and a relatively equiaxed dendritic structure in the centre of the fusion zone. The hardness values in the weld zone were lower than that of the base metal. The lowest hardness values were observed in the HAZ. No strong relationship was observed between the hardness values in the weld zone and the weld current. A higher joint strength, tensile deformation and failure energy absorption capacity were obtained with a weld current of 12 kA, a welding time of 300 ms and an electrode force of 3 kN. A complex fracture surface with both brittle and limited ductile manner was observed in the joints, while the base metal exhibited a ductile fracture. Joints with a higher tensile shear load (TSL) commonly exhibited more brittle fracture characteristics.

Keywords: TWIP steel; resistance spot welding; weld current; microstructure; mechanical properties; fractography

---

1. Introduction

Due to strict energy-efficiency regulations aimed at reducing exhaust emissions, researchers are making an effort to reduce vehicle weights to enhance vehicle fuel efficiency. Innovative high-strength steels are frequently used to both reduce the vehicle weight and to improve passenger safety. In recent years, considerable efforts have focused on high manganese twinning-induced plasticity (TWIP) steels for car body manufacturing, which are composed of a fully austenitic microstructure with a high amount of manganese and a significant percentage of carbon. The characteristics of TWIP steels, such as their outstanding mechanical properties, including high tensile strength, large ductility, high work-hardening rates, non-magnetism and high impact resistivity, have made these steels attractive to the automobile sector since they fulfil the requirements for safety, energy economy and environmental protection. The predominant deformation mechanism of TWIP steels is twinning, which is determined by the stacking fault energy (SFE) value, depending on the Mn, Al, Si and C content as well as on temperature [1–4]. Medium SFE values (between 20 and 35 mJ/m^2) provide mechanical twinning inside the grains [4–6].

Numerous studies have been carried out on the microstructure, mechanical properties and plastic deformation mechanisms of TWIP steels [1,4–10]. The weldability of TWIP steel sheets is another matter of concern that needs to be studied in more detail since the use of metal sheets in the automobile sector inevitably involves welding. Resistance spot welding (RSW), in which two or three superposed sheets are welded by means of local heating caused by the Joule effect, is the most commonly used method of joining metal sheets in car-body manufacturing because of its low cost, speed, cleanliness and ease of automation [11–14]. However, the designed microstructure of steel sheets is destroyed during the RSW process due to the welding thermal cycle. As a result, the microstructural evolution during RSW is highly dependent on the welding parameters. The microstructural changes during the RSW process strongly influence the mechanical properties. In previous studies of resistance spot-welded TWIP steels, Saha et al. [15] investigated the heat-affected zone (HAZ) liquation crack and segregation behaviour of resistance spot-welded TWIP steel and reported that the crack length and crack opening widths increased with heat input. Furthermore, Saha et al. [11] reported the microstructure, mechanical properties and fracture morphology of resistance spot-welded TWIP steels. Yu et al. [16] identified the RSW characteristics of 1 GPa grade TWIP steel. Spena et al. [17] investigated the effects of the process parameters on the mechanical and microstructural properties of resistance spot-welded TWIP sheets and reported that an improper clamping force and weld current promoted excessive metal expulsions and the formation of welding defects in the weld spots. Yu et al. [18] considered improvements to the weldability of 1-GPa grade TWIP steel, and found that a larger nugget size (NS) and higher tensile shear strength were obtained in constant power-control welding compared to constant current-control welding. Razmpoosh et al. [19] also evaluated the microstructure and mechanical properties of resistance spot-welded TWIP steel and reported that the HAZ exhibited a significant grain growth in the narrow band, and an increase in the weld current and time led to a significant drop in the peak load during tensile shear tests.

In recent years, TWIP sheet steels have gained popularity in the modern automotive industry. Therefore, the application of these steels requires a more complete understanding of the issues associated with RSW. It is thus important to study the welding behaviour of the resistance spot-welded joints of TWIP sheet steels. In this experimental study, the microstructural changes, microhardness, tensile shear properties and fracture morphologies of TWIP steel sheets welded at different weld currents were investigated in detail.

2. Experimental Procedure

In this study, 1.3 mm thick commercial TWIP steel sheets were used to study the microstructure and mechanical behaviour of resistance spot-welded TWIP steel joints. The chemical composition and tensile properties of the TWIP steel used in this investigation are shown in Tables 1 and 2, respectively. The tensile properties of the TWIP steel sheets used in this study were determined using a standard tensile test according to the ASTM E8 [20] (Figure 1). The tensile-test curves of the investigated TWIP steel can be seen in Figure 2.

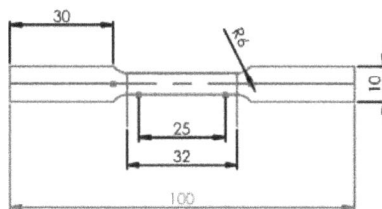

Figure 1. Tensile-test specimen for the base metal (unit: mm).

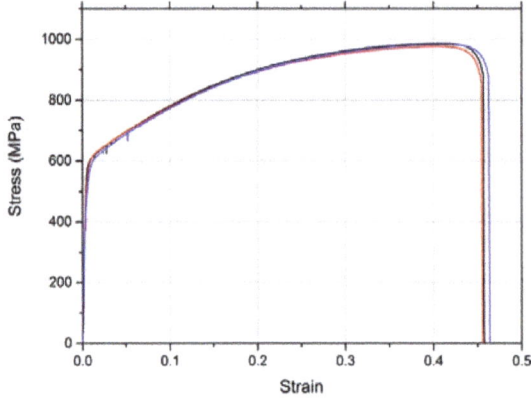

Figure 2. Tensile-test curves for the twinning-induced plasticity (TWIP) steel used in this investigation (three TWIP samples with the same characteristics were tested).

Table 1. The chemical composition (wt. %) of the experimental sheet steel used in this investigation.

Steel	Fe	C	Mn	Si	Al	Cr	Ti
TWIP	Balance	0.28	15.6	1.06	1.89	0.564	0.1

Table 2. The tensile properties of the experimental steel sheet used in this investigation.

Base Metal	0.2% Proof Strength (MPa)	Ultimate Tensile Strength (MPa)	Total Elongation (%)
TWIP Steel	640	982	46

The samples were cut into pieces with dimensions of 50 mm × 20 mm using a laser-cutting machine and RSW was performed by overlapping the sheets (Figure 3). The joining processes were carried out using a medium-frequency direct current (MFDC) RSW machine connected to an ABB robot arm (Figure 4). Copper alloy electrodes with a face diameter of 6 mm were employed. Before welding, the sheets were cleaned with acetone in order to remove oil, oxide and surface scale on both sides of the sheets. The welds were performed with welding currents of 8, 9, 10, 11 and 12 kA, while keeping the other parameters constant (Table 3). A parameter chart of the welding process used in this study can be seen in Figure 5. To avoid effects due to the electrode tip diameter, the electrode tip was changed or subjected to precision machining after each set of 20 welds.

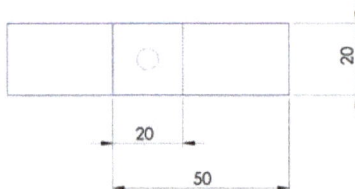

Figure 3. The welding sample installed with overlapping (unit: mm).

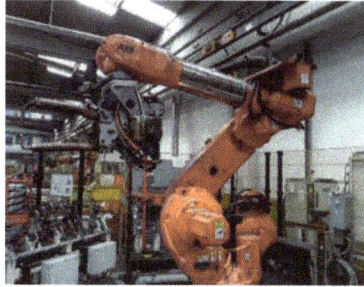

Figure 4. The medium-frequency direct current (MFDC) resistance spot-welding (RSW) machine connected to the ABB robot arm used in this investigation.

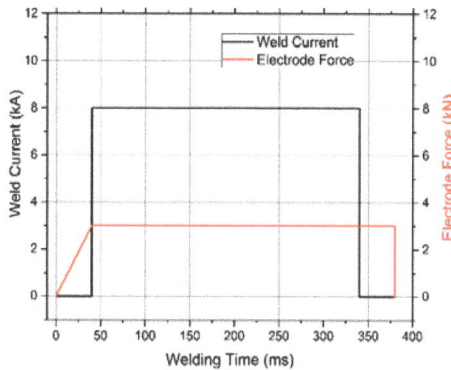

Figure 5. A parameter chart of the welding process used in this study.

Table 3. The fixed welding parameters used in this study.

Welding Time (ms)	Electrode Force (kN)	Squeeze Time (ms)	Holding Time (ms)	Weld Atmoshere
300	3	40	40	Ambient

The characterization of the microstructural evolution of the welded specimens was performed using optical microscopy and scanning electron microscopy–electron back-scattered diffraction (SEM–EBSD). The welded joints were cross-sectioned through the weld nugget centre using an electrical-discharge cutting machine. The transverse weld sections were mounted, ground and mechanically polished successively with a 0.25 μm diamond paste. The microstructure was revealed successively with a two-step tint-etching method using Nital (3%) and a $Na_2S_2O_5$ solution (10 g $Na_2S_2O_5$ in 100 mL H_2O), respectively. The optical microscopy studies of the welded sections were carried out using a Nikon DIC microscope under polarized light with a Clemex image analysis system. For EBSD analyses, the welded sections were polished mechanically with a 1-μm diamond paste. Then, the specimens were polished with an active oxide polishing suspension through 0.05 μm colloidal silica particles for 10 min. Scans with step sizes of 0.1 μm for the base metal (BM) and 0.2 μm for the HAZs were carried out on a Zeiss Merlin FEG-SEM microscope equipped with an EDAX/TSL EBSD system, using a Hikari EBSD camera. An electron beam with a 15 kV accelerating voltage and a beam current of 5 nA was used with a sample working distance of 13 mm. The EBSD data was processed with TSL Orientation Imaging Microscopy (OIM) analysis version 7.2. In the EBSD analyses, initially, the grain confidence index standardization (GCIS) clean-up method was used. Grains with an average GCIS < 0.1 were excluded. Crystallographic data were expressed as inverse pole figure (IPF) maps with image quality (IQ) maps, misorientation angle profiles and the distribution of coincidence site lattice (CSL) type boundaries. The grains in the IPF and IQ maps were defined as clusters of neighbouring points with a misorientation of less than 5° from each other. The CSL-type boundaries and misorientation angles were calculated using only the identified grains.

To describe the mechanical behaviour of the joints, microhardness measurements and tensile–shear tests were conducted. Vickers microhardness tests were performed on the metallographic specimens using a DUROLINE-M microhardness tester with a dwell time of 10 s and an indenter load of 200 g for the HAZ and the BM, and 500 g for the fusion zone (FZ) to avoid inaccurate measurements of the coarse dendritic structure in the FZ-containing micro-pores. The tensile shear tests were performed with a fully computerized UTEST-7014 tensile testing machine in laboratory conditions using a crosshead displacement speed of 5 mm/min. The maximum tensile shear load (TSL), the tensile deformation and the failure energy absorption capacity of the joints were extracted from the TSL-deformation curves of the joints. The fracture surfaces of the joints were also examined using a Zeiss EVO 40 XVP type SEM operated at 20 kV. The NS of the welds was measured through the fractured specimens in all welding conditions using a Mitutoyo digital calliper.

3. Results and Discussion

3.1. Microstructure

The weld cross-sections of the resistance spot-welded TWIP joints revealed three main distinct microstructural zones, including the BM, the HAZ and the FZ, as shown in Figure 6. The BM microstructure is mainly composed of a fully fine-grained austenite phase with an average grain size of 3–5 μm. It includes mechanical twins as well as a few transgranular annealing twins, which are generally located within relatively large-sized grains. The weld zone consists of a fully austenitic microstructure, which has different morphologies at different zones of the joint depending on the peak temperature of the relevant regions during the RSW process. The FZ microstructure shows a cast microstructure, which was melted during the RSW process and then rapidly resolidified, and has mainly a coarse columnar dendritic structure owing to the directional and columnar solidification from the fusion boundary towards the centre. On the other hand, the welding thermal cycle during the RSW process produced significant grain growth in the HAZ, including annealing twins (Figure 6). Razmpoosh et al. [19] have reported that this extreme grain growth in the HAZ resulted from the low thermal conductivity of TWIP steels. The HAZ can be divided into a coarse-grained heat-affected zone (CGHAZ), which occurs in areas immediately adjacent to the FZ, and a fine-grained heat affected zone (FGHAZ), which is still relatively coarser than the BM (Figure 6). Similarly, the FZ can also be divided into a fine dendritic fusion zone (FDFZ) and a coarse dendritic fusion zone (CDFZ) caused by different cooling rates in different regions of the FZ. Weld imperfections, such as macro-pores, micro-pores and expulsion cavities, may occur in the FZ of the resistance spot-welded TWIP joints [5,21,22]. All the spot-welded joints had interdendritic macro- and micro-pores in the FZ owing to solidification shrinkage in this special material (Figures 7–9). Additionally, the highest weld current (12 kA) led to the formation of a macro expulsion cavity in the FZ during the RSW process as a result of an extensive expulsion phenomenon (molten material loss) due to the highest heat input.

Higher weld currents led to a coarser grain structure and thicker annealing twins in the HAZ, a wider HAZ, and a relatively equiaxed dendritic structure in the centre (interface of the sheets) of the FZ due to higher heat input leading to the slower cooling rate (Figures 9–11). The HAZ width increased with increasing weld currents (Figure 12). It should be noted that there is a strong linear correlation between the HAZ width and the weld current. The coefficient of determination (R^2) from fitting a linear model using regression analysis to the obtained results for the linear correlation between the HAZ width and the weld current is 0.86.

Figure 6. The typical macrostructure and microstructures of a resistance spot-welded TWIP steel joint (12 kA). (CDFZ: coarse dendritic fusion zone, FDFZ: fine dendritic fusion zone, FGHAZ: fine-grained heat-affected zone, CGHAZ: coarse-grained heat affected zone).

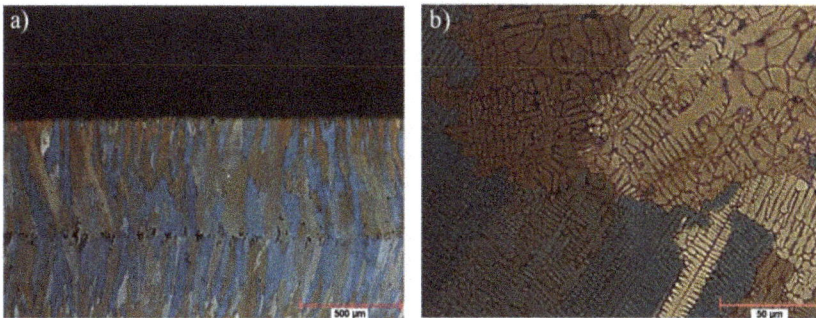

Figure 7. Interdendritic macro-pores (**a**) and micro-pores (**b**) in the fusion zone (FZ) of a resistance spot-welded TWIP steel joint ((**a**) 9 kA, (**b**) 12 kA).

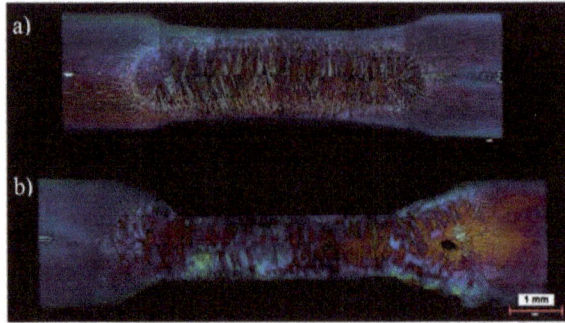

Figure 8. The effect of the weld current on the formation of a macro expulsion cavity [23]: (**a**) 8 kA, (**b**) 12 kA.

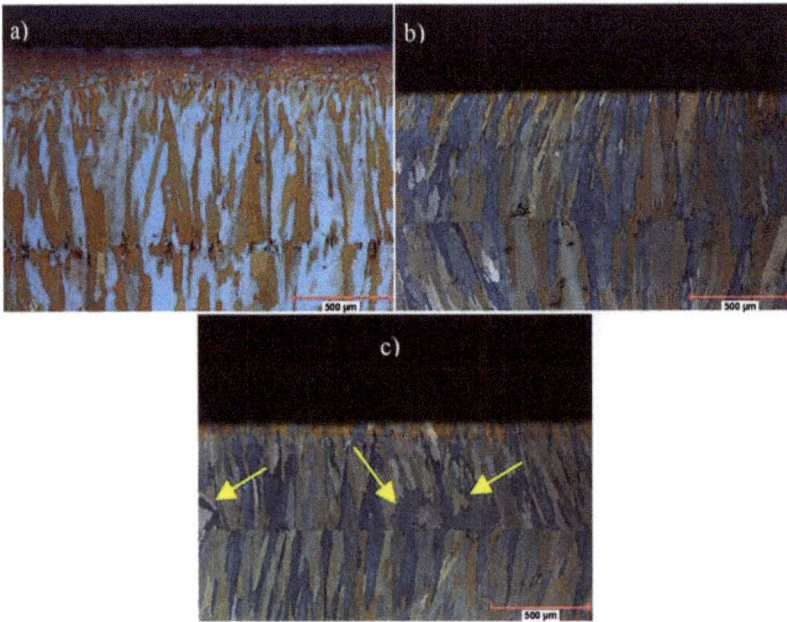

Figure 9. The FZs of the resistance spot-welded TWIP joints: (**a**) 8 kA; (**b**) 10 kA; (**c**) 12 kA.

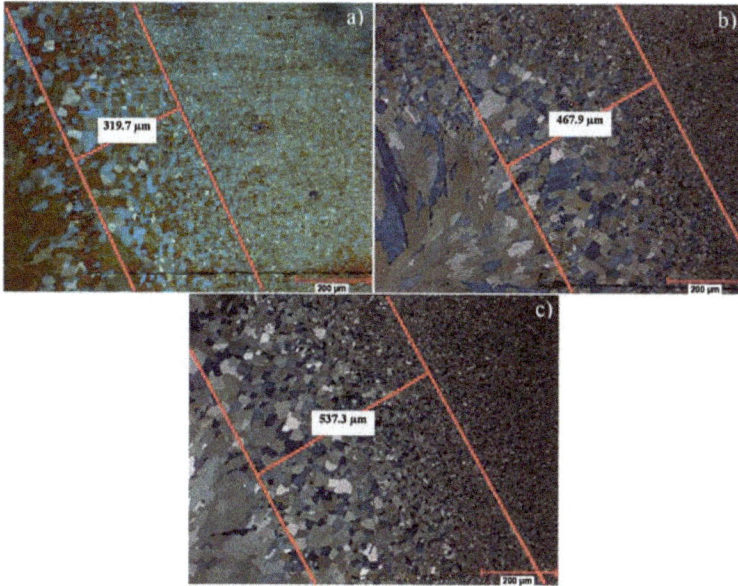

Figure 10. The heat-affected zones (HAZs) of the resistance spot-welded TWIP joints: (**a**) 8 kA; (**b**) 10 kA; (**c**) 12 kA.

Figure 11. The HAZ microstructures of the resistance spot-welded TWIP joints: (**a**) 8 kA; (**b**) 10 kA; (**c**) 12 kA.

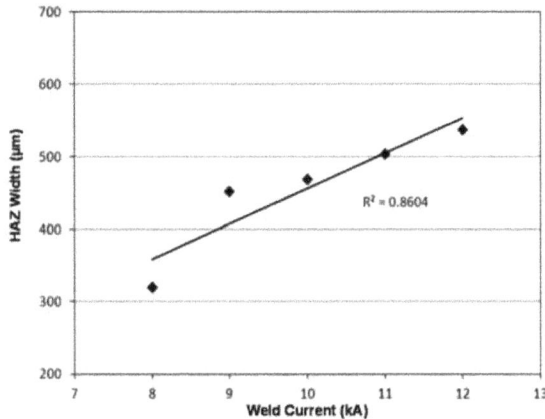

Figure 12. The HAZ width of the resistance spot-welded TWIP joints versus the weld current.

The evaluation and characterization of the BM and the HAZ microstructures was performed with SEM–EBSD microstructural crystallography (the OIM technique). Due to the limitations of the EBSD technique for resolving nanoscale twins, nano-sized deformation twins in the BM microstructure could not be mapped with this technique. However, the EBSD crystallographic orientation map (IPF + IQ map) given in Figure 13a could be used to analyse deformation twins in the BM microstructure, since the deformed substructures, such as dislocations and deformation twins, in the grains are strongly dependent on the crystallographic orientation [24,25]. In Figure 13a, the grains oriented close to the <001>//rolling directions (RD) in the crystallographic orientation map can be evaluated as grains exhibiting low deformation-twinning activity and a well-developed dislocation substructure, while the grains oriented close to the <111>//RD indicate recovered or recrystallized grains. The EBSD maps of the HAZ of the spot-welded specimens are presented in Figure 13b,c. The HAZ of the specimens showed a fairly coarse-grained structure (secondary recrystallization) near the FZ and a recrystallized grain structure near the BM. From these figures, annealing twins can also be seen in the HAZ. The increase in the HAZ width and grain coarsening in the HAZ with the increase of weld current, due to the higher heat input, can also be seen.

The misorientation angle profiles and the distribution of CSL-type boundaries obtained from the SEM–EBSD studies could also be used to analyse the microstructure and twin-structure. The misorientation angle of 60° (high-angle grain boundaries) in the misorientation profile can be attributed to a twin structure (twin boundaries) or recovered and recrystallized grain structure having high-angle grain boundaries [26–31]. The misorientation profile of the BM revealed the presence of a misorientation angle of 60° at a fraction of 23.83%, resulting from recovered or recrystallized grain structures having high-angle grain boundaries and some twin boundaries (annealing and deformation twins) (Figure 14a). The HAZ microstructures of the spot-welded joints revealed a misorientation angle of 60° at fractions of 15.77% in the 8 kA joint and 17.88% in the 12 kA joint, arising from recrystallized grains with high-angle grain boundaries and annealing twins (Figure 14b,c). These values are lower than that of the BM (Figure 14). This may be due to a fairly fine recovered grain structure (more grain boundaries) and the deformation twins in the EBSD map of the BM shown in Figure 13a. On the other hand, although the HAZ EBSD map of the 12 kA joint contains fewer grain boundaries than that of the 8 kA joint, as expected, the higher weld current (12 kA joint) resulted in more annealing twin boundaries and in recrystallized grains having high-angle grain boundaries due to the higher heat input. In addition, from the EBSD maps in Figure 13b,c, no significant difference in the density of annealing twins was observed with heat input variation during the RSW process. Therefore, this may be related to an increase in the length of the annealing twins through grain coarsening in the HAZ rather than an increase of the density of the annealing twins with increasing heat input owing to the higher weld current. Similar data can be seen in the distribution of CSL-type boundaries in the specimens (Figure 15). The fractions of $\Sigma 3$ type CSL boundaries corresponding to 60°<111> misorientation angles in the BM and the HAZs of the 8 kA and 12 kA joints are 26.97%, 19.4% and 21.07%, respectively. On the other hand, low-angle grain boundaries ($2° < \theta < 10°$) in the misorientation profiles may indicate a deformed grain structure or recovery and recrystallized grains having low-angle grain boundaries [26,31]. Low-angle grain boundaries in the BM microstructure occur at a fraction of 4.38%, resulting from low deformation-twinning activity and some recovery and recrystallized grains having low-angle grain boundaries (Figures 13a and 14a). The elongated shape grains, which may possibly be recovered grains having low deformation twinning activity, and recrystallized grains can be observed in Figure 13a. The HAZ microstructures of the spot-welded joints revealed the presence of low-angle grain boundaries at fractions of 11.24% for the 8 kA joint and 11.93% for the 12 kA joint, arising from recovery and recrystallized grains with low-angle grain boundaries, presumably in the HAZs near the BM (Figure 14b,c). As expected, these values are higher than that of the BM due to heat exposure during the RSW process. Higher weld currents lead to more recovery and recrystallized grains having low-angle grain boundaries.

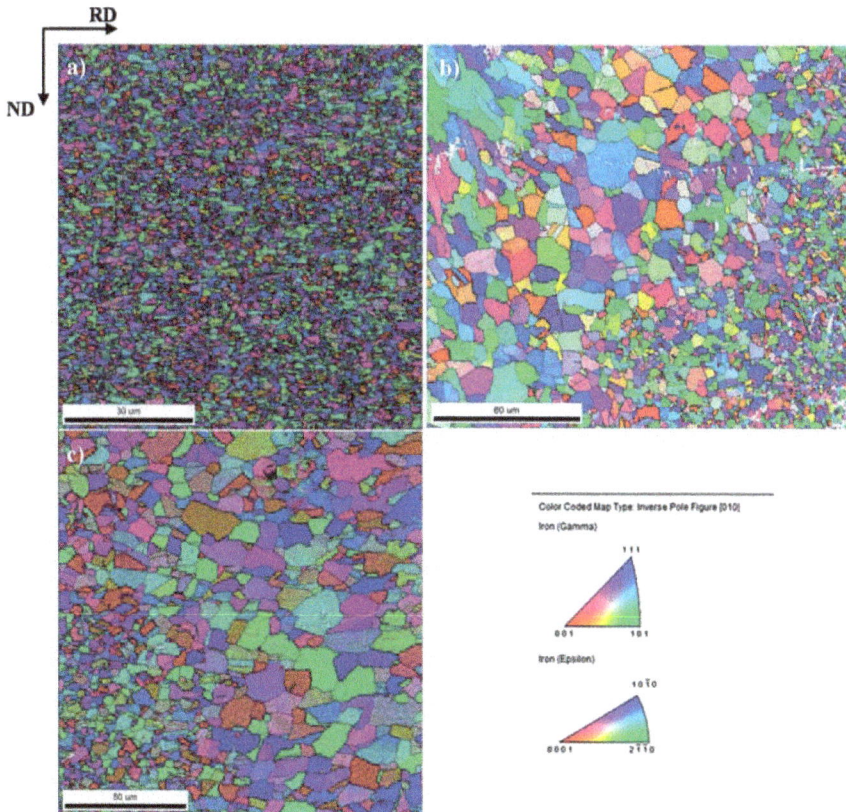

Figure 13. The IPF + IQ maps obtained from the scanning electron microscopy–electron back-scattered diffraction (SEM–EBSD) technique: (**a**) the base metal; (**b**) the HAZ of the joint welded at 8 kA; (**c**) the HAZ of the joint welded at 12 kA. (RD: Rolling Direction; ND: Normal Direction)

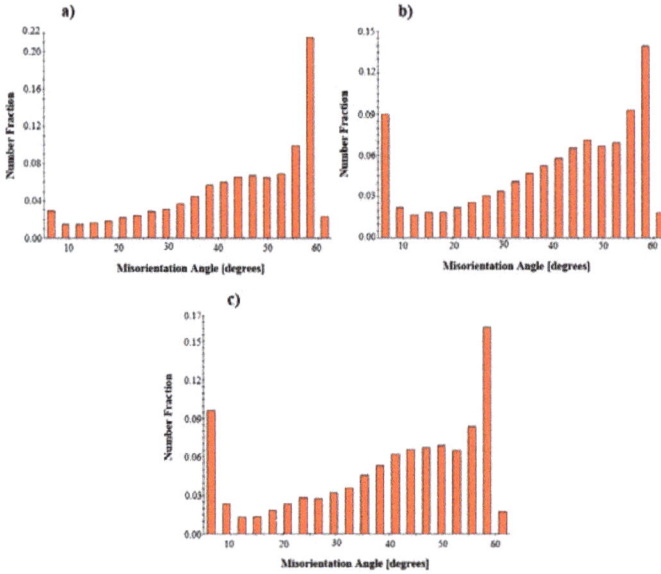

Figure 14. The misorientation angle profiles obtained from the SEM–EBSD analysis: (**a**) the base metal; (**b**) the HAZ of the joint welded at 8 kA; (**c**) the HAZ of the joint welded at 12 kA.

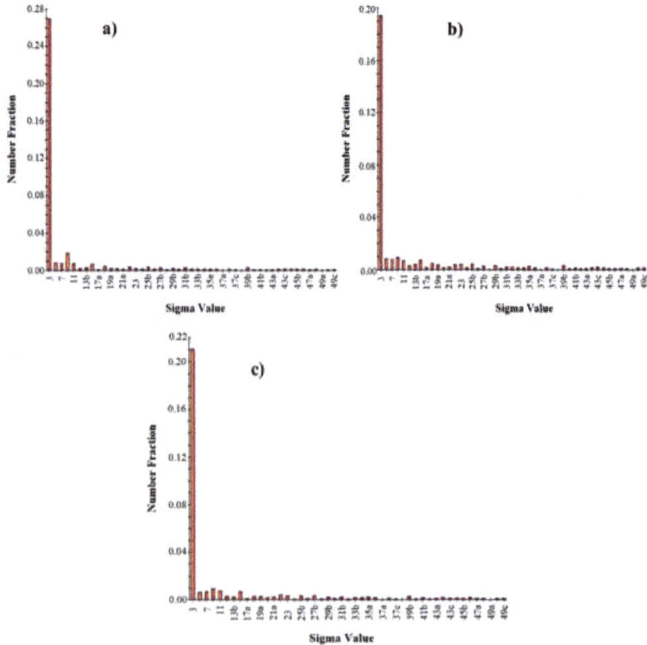

Figure 15. The distribution of coincidence site lattice (CSL) type boundaries obtained from the SEM–EBSD analysis: (**a**) the base metal; (**b**) the HAZ of the joint welded at 8 kA; (**c**) the HAZ of the joint welded at 12 kA.

To ensure safe design standards for the durability and crashworthiness of vehicles, there are various industrial standards that recommend a minimum NS for a given sheet thickness. The following equation is suggested by the American Welding Society (AWS) standard [32]:

$$D = 4t^{1/2}$$

(1)

where D and t are the NS and the sheet thickness, respectively.

In the present study, the minimum NS obtained was approximately 5.5 mm and all the joints exhibited a pullout failure (PO) mode, ensuring high load-bearing capacity. For the 1.3-mm thick TWIP sheets used in this study, this minimum value is compatible with the AWS. The NS is closely related to the generated heat energy at the sheet interface during the RSW process. The relationship between the NS and the weld current can be seen in Figure 16a. For a welding current of up to 11 kA, the NS varied within a relatively narrow range (between 5.5 and 6.0 mm). The maximum NS was obtained with the highest weld current (12 kA) owing to the increased heat input [23]. The relationship between the indentation depth and the weld current is shown in Figure 16b. The indentation depth of the joints increased with increasing weld current due to the increased heat input during the RSW process.

3.2. Microshardness

Vickers microhardness tests were performed on the FZ, the HAZ and the BM of the welded specimens. The BM hardness of the TWIP sheets used in this study was approximately 260 $HV_{0.2}$. In general, the hardness of the weld zone, including the FZ and the HAZ, is lower than that of the BM (Figure 17). This can be attributed to the coarser grains in the weld zone (HAZ and FZ), segregation of the alloying elements in these zones, and a lower carbon percentage in the FZ due to decarbonisation during the RSW process [11,22,33]. The lowest hardness values for all the specimens were obtained at the HAZ. This is considered to be caused by the significant coarsening of the austenitic grains and the decrease of dislocations and deformation twins in the HAZ owing to the heat input during the welding process. The HAZ hardness increased slightly with increasing weld current. This may be attributed to the effect of annealing twins owing to the increased heat input [34–36]. On the other hand, as expected, the hardness of the FDFZ was always higher than that of the CDFZ. However, no relationship was obtained between the hardness values in the FZ and weld current.

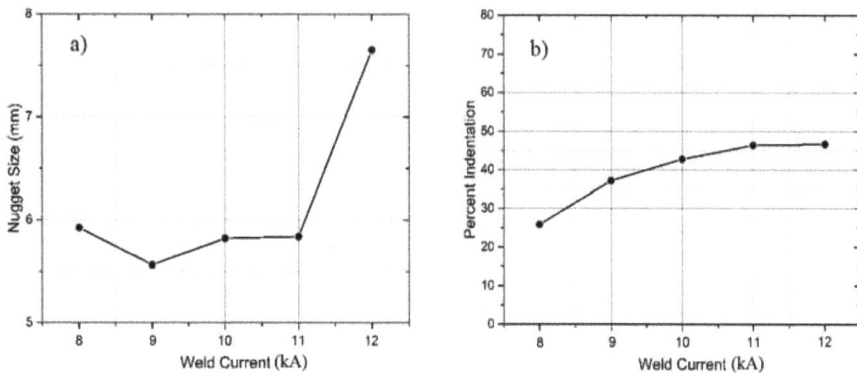

Figure 16. Nugget size (**a**) and indentation depth (**b**) of the resistance spot-welded TWIP joints versus the weld current [23].

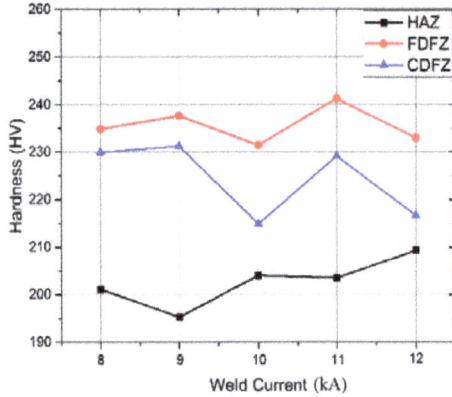

Figure 17. The effect of the weld current on the hardness of the distinct microstructural zones in the weld regions of the spot-welded TWIP joints.

3.3. Tensile Shear Properties and Fractography

In the quality evaluation of resistance spot-welded joints, the static load-carrying capability of the welds, which is determined by tensile shear testing, is an important parameter. The tensile shear curves of the spot welds fabricated with different weld currents are shown in Figure 18. The average values of the tensile properties and the failure energy absorption capacities of the joints can be seen in Table 4. The acceptable tensile shear load (ATSL) of the welded specimens, according to AWS D8.1 M, can be calculated using following function [37]:

$$\text{ATSL} = [(-6.36 \times 10^{-7} \times \text{UTS}^2 + 6.58 \times 10^{-4} \times \text{UTS} + 1.674) \times \text{UTS} \times 4 \times t^{1.5}]/1000 \text{ [kN]} \quad (2)$$

where UTS is the ultimate tensile strength, and t is the thickness of the sheets. An ATSL of 9.94 kN can be calculated by considering the UTS and the thickness of the sheets, 982 MPa and 1.3 mm, respectively. The TSLs of all the fabricated joints were roughly 12–38% higher than the ATSL due to the optimization studies performed before this study.

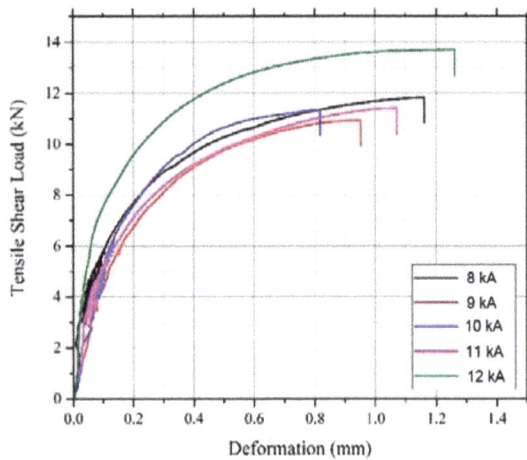

Figure 18. The effect of the weld current on the tensile curves of the resistance spot-welded TWIP joints.

Table 4. The tensile properties and failure energy absorption capacities of the resistance spot welds fabricated with different weld currents (average values). (SD: Standard Deviation).

Welding Current (ms)	Tensile Shear Load (kN)	SD	Tensile Shear Deformation (mm)	SD	Failure Energy Absorption (J)	SD
8	11.83	0.17	1.16	0.10	4.86	0.54
9	11.14	0.40	0.99	0.04	3.68	0.56
10	11.45	0.59	0.93	0.16	3.42	0.69
11	11.13	0.50	0.93	0.19	3.51	1.02
12	13.72	0.17	1.38	0.17	7.22	1.43

The weld current significantly affected the tensile properties of the resistance spot-welded TWIP joints (Figure 19). For a weld current of up to 11 kA, the TSLs of the joints varied within a relatively narrow range of 11.13–11.83 kN. The maximum TSL of 13.72 kN was obtained when the weld current was 12 kA. This significant increase can be explained by the significant increase in the NS due to the higher heat input [23]. The TSLs obtained with different weld currents are in good agreement with the NS values (Figures 16a and 19). Interestingly, the joint that had a macro expulsion cavity in the FZ had the highest TSL (Figures 6, 8 and 18). This demonstrates that the presence of this macro defect did not profoundly affect the shear–tension strength of the spot welds. The effect of the weld current on the tensile shear deformation of the resistance spot-welded TWIP joints is shown in Figure 19 and Table 4. For a weld current of up to 10–11 kA, the tensile shear deformations of the joints decreased with increasing weld current. This may be attributed to the NS, the increased brittleness in the HAZ (relatively lower plastic deformation) due to grain coarsening and the formation of annealing twins with the higher weld current owing to the higher heat input. The peak tensile shear deformation was obtained with 12 kA. The joint having the highest TSL had the highest tensile deformation. The effect of the weld current on the failure energy absorption capacity of the spot-welded TWIP joints can be seen in Figure 19 and Table 4. The failure energy decreased with increasing weld currents up to 10 kA, then it increased with increasing weld current. This behaviour is almost completely consistent with the tensile deformation values.

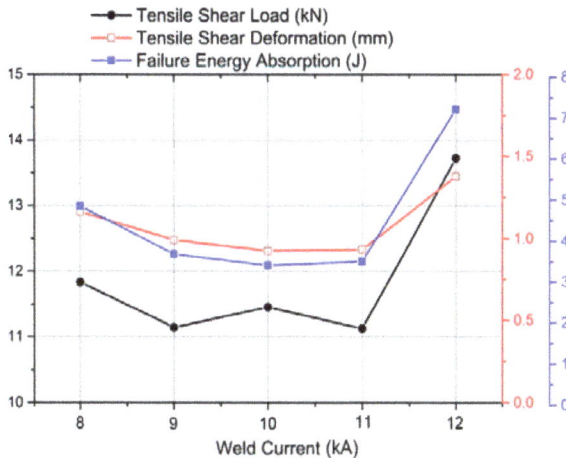

Figure 19. The tensile shear properties of the resistance spot-welded TWIP joints versus the weld current.

The tensile shear fracture modes of the joints are presented in Figure 20. As is well known, the failure mode is a good quantitative indicator of the quality of the spot welds. Resistance spot welds fail with two major failure modes, the interfacial failure (IF) mode and the pullout failure (PO) mode.

To ensure the durability and crashworthiness of vehicles, the welding parameters in the RSW should be adjusted so that the pullout failure mode, in which the fracture occurs in the HAZ at the edge of the spot weld, is guaranteed. In this study, all joints exhibited a full button pullout failure mode due to the higher deformation energy. Additionally, expulsion phenomena were observed in all the joints and increased with increasing weld currents (Figure 20).

Figure 20. The fracture modes of the tensile shear test samples: (**a**) 9 kA; (**b**) 11 kA; (**c**) 12 kA.

The fracture surface of the BM shows a ductile manner with the formation of fairly equaxial and deep dimples with conical shapes (without any cleavage facets) elongated along the direction of the tensile load (Figure 21). The dimple structure is relatively finer than the BM grain structure, predominantly suggesting that microcracks and micro-void nucleation during tensile testing mostly initiates at the small discontinuities formed at the twin boundaries and inclusions [6]. The SEM images of the fracture surfaces of the spot-welded joints are shown in Figures 22 and 23. A brittle fracture surface with areas of fairly limited ductility was observed in the flank side of the 11 kA joint which had the low fracture load (Figure 22). It should also be emphasized that the intra-layered fractures in this zone are quite brittle. The fracture surface near the weld nugget consists of some shallow voids, exhibiting some characteristics of a ductile failure, and facets among voids in a larger area, indicating a relatively brittle fracture (Figure 22). In the 12 kA joint with the highest fracture load, the fracture surface near the weld nugget indicates a brittle cleavage fracture with a few shallower small voids (Figure 23). On the flank side, brittle intra-layered fractures with some void areas of fairly limited ductility were observed. The boundaries, shown as the dashed lines in Figure 23, with dense deformation streaks may be associated with the extreme grain growth in the HAZ. The joints that had higher strength commonly showed more brittle fracture characteristics, especially on the fracture surface near the weld nugget.

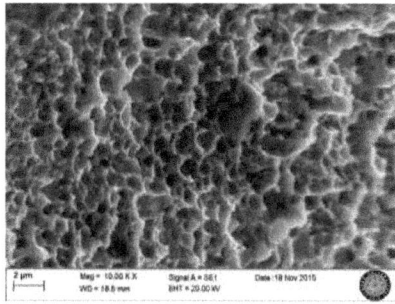

Figure 21. SEM image of the fracture surface of the base metal.

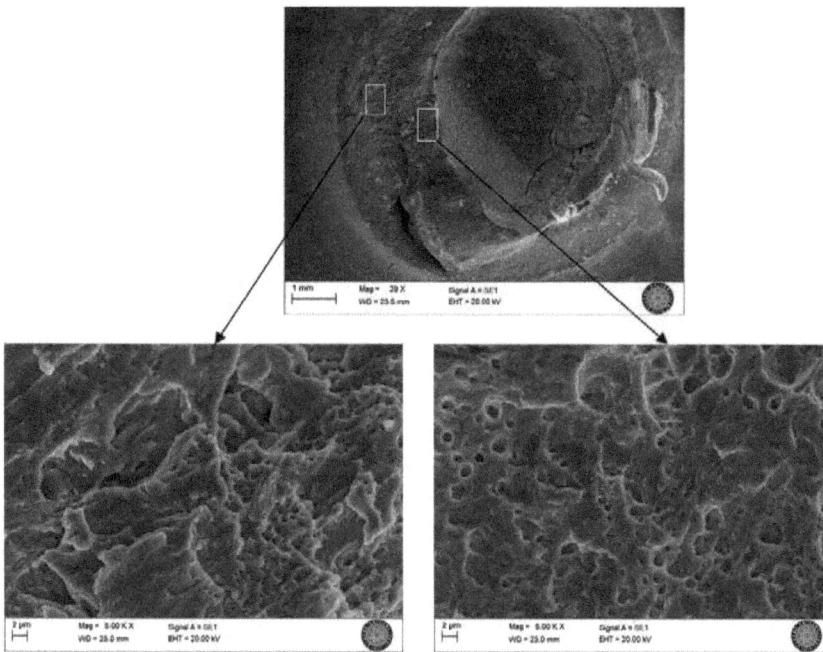

Figure 22. SEM images of the fracture surface of the joint welded at 11 kA.

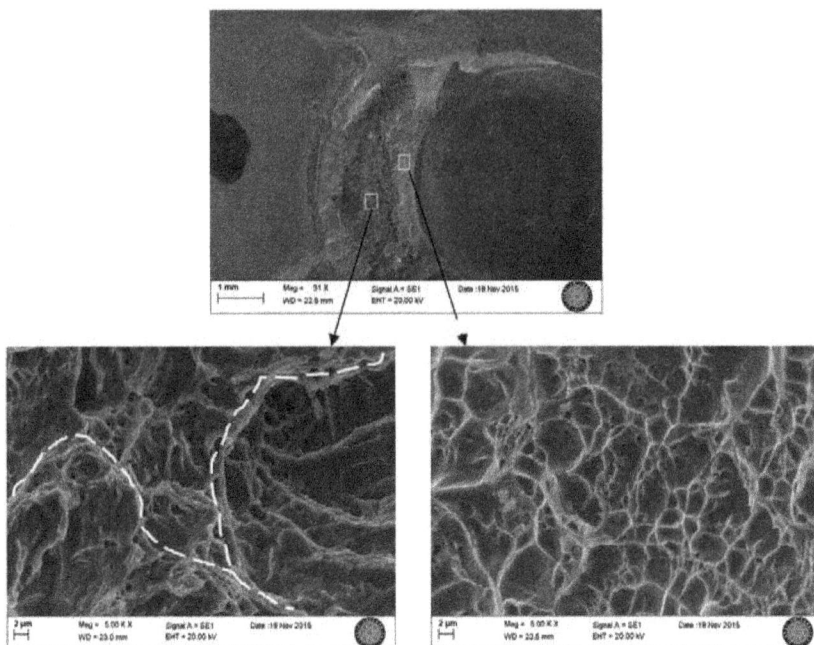

Figure 23. SEM images of the fracture surface of the joint welded at 12 kA.

4. Conclusions

The present study focused on the microstructure and the mechanical properties of resistance spot-welded TWIP sheet steels fabricated at different weld currents. From this investigation, the following conclusions can be derived:

1. Higher weld currents lead to the formation of a macro expulsion cavity in the fusion zone, a coarser grain structure and thicker annealing twins in the HAZ, a wider HAZ, a higher indentation depth, and a relatively equiaxed dendritic structure in the centre of the fusion zone.
2. The HAZ width increases almost linearly with increasing weld current.
3. The hardness values in the weld zone are lower than that of the base metal. The HAZ has the lowest hardness values. However, there is no significant relationship between the hardness values in the weld zone and the weld current.
4. A higher joint strength, tensile deformation and failure energy absorption capacity can be obtained using the following welding parameters: a weld current of 12 kA, a welding time of 300 ms, and an electrode force of 3 kN.
5. The joints exhibit a complex fracture surface with both brittle and limited ductile manners, while the base metal shows a ductile fracture. Joints with the highest strength commonly show more brittle fracture characteristics.

Acknowledgments: The authors are grateful to the Scientific and Technological Research Council of Turkey (TUBITAK) for its financial support to this research (Project number: MAG 213M597). The authors are also grateful to Ermetal Inc. for providing facilities for the resistance spot-welding processes.

Author Contributions: Mumin Tutar, Hakan Aydin and Ali Bayram conceived and designed the experiments; Mumin Tutar and Hakan Aydin performed the experiments; Mumin Tutar, Hakan Aydin and Ali Bayram analyzed the data; and Hakan Aydin and Mumin Tutar wrote the paper.

Conflicts of Interest: The authors declare no conflict of interest.

References

1. Reyes-Calderón, F.; Mejía, I.; Boulaajaj, A.; Cabrera, J.M. Effect of microalloying elements (Nb, V and Ti) on the hot flow behavior of high-Mn austenitic twinning induced plasticity (TWIP) steel. *Mater. Sci. Eng. A* **2013**, *560*, 552–560. [CrossRef]
2. Saeed-Akbari, A.; Imlau, J.; Prahl, U.; Bleck, W. Derivation and Variation in Composition-Dependent Stacking Fault Energy Maps Based on Subregular Solution Model in High-Manganese Steels. *Metall. Mater. Trans. A* **2009**, *40*, 3076–3090. [CrossRef]
3. Allain, S.; Chateau, J.P.; Bouaziz, O.; Migot, S.; Guelton, N. Correlations between the calculated stacking fault energy and the plasticity mechanisms in Fe-Mn-C alloys. *Mater. Sci. Eng. A* **2004**, *387–389*, 158–162. [CrossRef]
4. Hamada, A.S. *Manufacturing, Mechanical Properties and Corrosion Behaviour of High Mn TWIP Steels*; University of Oulu: Oulu, Finland, 2007; Volume 281, ISBN 9789514285837.
5. Saha, D.C.; Park, Y. Do Weldability and liquation cracking characteristics on resistance-spot-welded high-Mn austenitic steel. In Proceedings of the ASM International Conference: Trends in Welding Research, Chicago, IL, USA, 4–8 June 2012; pp. 330–335.
6. Ma, L.; Wei, Y.; Hou, L.; Yan, B. Microstructure and Mechanical Properties of TWIP Steel Joints. *J. Iron Steel Res. Int.* **2014**, *21*, 749–756. [CrossRef]
7. Busch, C.; Hatscher, A.; Otto, M.; Huinink, S.; Vucetic, M.; Bonk, C.; Bouguecha, A.; Behrens, B.A. Properties and Application of High-manganese TWIP-steels in Sheet Metal Forming. *Procedia Eng.* **2014**, *81*, 939–944. [CrossRef]
8. Dan, W.J.; Liu, F.; Zhang, W.G. Mechanical behavior prediction of TWIP steel in plastic deformation. *Comput. Mater. Sci.* **2014**, *94*, 114–121. [CrossRef]
9. Grässel, O.; Krüger, L.; Frommeyer, G.; Meyer, L.W. High strength Fe-Mn-(Al, Si) TRIP/TWIP steels development-properties-application. *Int. J. Plast.* **2000**, *16*, 1391–1409. [CrossRef]
10. Ghasri-Khouzani, M.; McDermid, J.R. Effect of carbon content on the mechanical properties and microstructural evolution of Fe-22Mn-C steels. *Mater. Sci. Eng. A* **2015**, *621*, 118–127. [CrossRef]
11. Saha, D.C.; Cho, Y.; Park, Y. Metallographic and fracture characteristics of resistance spot welded TWIP steels. *Sci. Technol. Weld. Join.* **2013**, *18*, 711–720. [CrossRef]
12. Grässel, O.; Rommeyer, G.; Derder, C.; Hofmann, H. Phase Transformations and Mechanical Properties of To cite this version. *J. Phys. IV Fr.* **1997**, *7*, 383–388. [CrossRef]
13. Aydin, H. The mechanical properties of dissimilar resistance spot-welded DP600-DP1000 steel joints for automotive applications. *Proc. Inst. Mech. Eng. Part D J. Automob. Eng.* **2014**, *229*, 599–610. [CrossRef]
14. Aydin, H.; Durgun, İ.; Tutar, M.; Bayram, A. Correlations between welding time and mechanical properties of spot welded dissimilar joints for high strength steels. In Proceedings of the 7th Congress of Automotive Technologies, OTEKON, Bursa, Turkey, 26–27 May 2014.
15. Saha, D.C.; Chang, I.; Park, Y.D. Heat-affected zone liquation crack on resistance spot welded TWIP steels. *Mater. Charact.* **2014**, *93*, 40–51. [CrossRef]
16. Yu, J.; Shim, J.; Rhee, S. Characteristics of Resistance Spot Welding for 1 GPa Grade Twin Induced Plasticity Steel. *Mater. Trans.* **2012**, *53*, 2011–2018. [CrossRef]
17. Russo Spena, P.; De Maddis, M.; Lombardi, F.; Rossini, M. Investigation on Resistance Spot Welding of TWIP Steel Sheets. *Steel Res. Int.* **2015**, *86*, 1480–1489. [CrossRef]
18. Yu, J.; Choi, D.; Rhee, S. Improvement of weldability of 1 GPa grade twin-induced plasticity steel. *Weld. J.* **2014**, *93*, 78s–84s.
19. Razmpoosh, M.; Shamanian, M.; Esmailzadeh, M. The microstructural evolution and mechanical properties of resistance spot welded Fe-31Mn-3Al-3Si TWIP steel. *Mater. Des.* **2015**, *67*, 571–576. [CrossRef]
20. American Society for Testing and Materials. ASTM E8/E8M standard test methods for tension testing of metallic materials. *Annu. Book ASTM Stand.* **2010**, *4*, 1–27.
21. Zhang, H.; Senkara, J. *Resistance Welding: Fundamentals and Applications*; CRC Press Taylor & Francis Group: Boca Raton, FL, USA, 2006; ISBN 0-8493-2346-0.
22. Saha, D.C.; Han, S.; Chin, K.G.; Choi, I.; Park, Y. Do Weldability evaluation and microstructure analysis of resistance-spot-welded high-Mn steel in automotive application. *Steel Res. Int.* **2012**, *83*, 352–357. [CrossRef]

23. Tutar, M.; Aydin, H.; Bayram, A. Weld Current Effect on the Mechanical Properties of Resistance Spot Welded TWIP980 Steel Sheets. In Proceedings of the International Research Symposium on Recent Innovations in Engineering Science and Technology, Paris, France, 9–10 March 2016.

24. Gutierrez-Urrutia, I.; Zaefferer, S.; Raabe, D. The effect of grain size and grain orientation on deformation twinning in a Fe-22 wt. % Mn-0.6 wt. % C TWIP steel. *Mater. Sci. Eng. A* **2010**, *527*, 3552–3560. [CrossRef]

25. Gutierrez-Urrutia, I.; Raabe, D. Study of Deformation Twinning and Planar Slip in a TWIP Steel by Electron Channeling Contrast Imaging in a SEM. *Mater. Sci. Forum* **2011**, *702–703*, 523–529. [CrossRef]

26. Haase, C.; Barrales-Mora, L.A.; Molodov, D.A.; Gottstein, G. Tailoring the mechanical properties of a twinning-induced plasticity steel by retention of deformation twins during heat treatment. *Metall. Mater. Trans. A Phys. Metall. Mater. Sci.* **2013**, *44*, 4445–4449. [CrossRef]

27. Kumar, B.R.; Das, S.K.; Mahato, B.; Das, A.; Ghosh Chowdhury, S. Effect of large strains on grain boundary character distribution in AISI 304L austenitic stainless steel. *Mater. Sci. Eng. A* **2007**, *454–455*, 239–244. [CrossRef]

28. Saleh, A.A.; Gazder, A.A.; Pereloma, E.V. EBSD observations of recrystallisation and tensile deformation in twinning induced plasticity steel. *Trans. Indian Inst. Met.* **2013**, *66*, 621–629. [CrossRef]

29. Hielscher, R.; Bachmann, F.; Schaeben, H.; Mainprice, D. *Features of the Free and Open Source Toolbox MTEX for Texture Analysis*; Géosciences Montpellier: Montpellier, France, 2014.

30. Lee, S.Y.; Chun, Y.B.; Han, J.W.; Hwang, S.K. Effect of thermomechanical processing on grain boundary characteristics in two-phase brass. *Mater. Sci. Eng. A* **2003**, *363*, 307–315. [CrossRef]

31. Yuan, X.; Chen, L.; Zhao, Y.; Di, H.; Zhu, F. Influence of annealing temperature on mechanical properties and microstructures of a high manganese austenitic steel. *J. Mater. Process. Technol.* **2015**, *217*, 278–285. [CrossRef]

32. American Welding Society (AWS) D8 Committee on Automotive Welding. *SAE D8.9M: Recommended Practice for Test Methods for Evaluating the Resistance Spot Welding Behavior of Automotive Sheet Steel Materaials*; American Welding Society: Miami, FL, USA, 2002.

33. Pouranvari, M.; Marashi, S.P.H. Critical review of automotive steels spot welding: Process, structure and properties. *Sci. Technol. Weld. Join.* **2013**, *18*, 361–403. [CrossRef]

34. Razmpoosh, M.H.; Zarei-Hanzaki, A.; Heshmati-Manesh, S.; Fatemi-Varzaneh, S.M.; Marandi, A. The Grain Structure and Phase Transformations of TWIP Steel during Friction Stir Processing. *J. Mater. Eng. Perform.* **2015**, *24*, 2826–2835. [CrossRef]

35. Tang, Z.Y.; Misra, R.D.K.; Ma, M.; Zan, N.; Wu, Z.Q.; Ding, H. Deformation twinning and martensitic transformation and dynamic mechanical properties in Fe-0.07C-23Mn-3.1Si-2.8Al TRIP/TWIP steel. *Mater. Sci. Eng. A* **2015**, *624*, 186–192. [CrossRef]

36. Abadi, M.M.K.; Najafizadeh, A.; Kermanpur, A.; Mazaheri, Y. Effect of annealing process on microstructure and mechanical properties of high manganese austenitic TWIP steel. *Int. J. ISSI* **2011**, *8*, 1–4.

37. AWS D8.1M. *Specification for Automotive Weld Quality–Resistance Spot Welding of Steel*; American Welding Society (AWS): Miami, FL, USA, 2013.

MDPI AG

St. Alban-Anlage 66

4052 Basel, Switzerland

Tel. +41 61 683 77 34

Fax +41 61 302 89 18

http://www.mdpi.com

Metals Editorial Office

E-mail: metals@mdpi.com

http://www.mdpi.com/journal/metals